MATERIALS SCIENCE RESEARCH
Volume 3
The Role of Grain Boundaries and Surfaces in Ceramics

University Conferences on Ceramic Science

*This conference, held at North Carolina State University at Raleigh, was
conducted by the School of Engineering and the
Division of Continuing Education of North Carolina State University
in cooperation with the U.S. Army Research Office (Durham)
and the U.S. Office of Naval Research*

MATERIALS SCIENCE RESEARCH
Volume 3

The Role of Grain Boundaries and Surfaces in Ceramics

*Proceedings of the Conference held November 16-18, 1964
at North Carolina State University at Raleigh*

Edited by
W. Wurth Kriegel
Department of Mineral Industries, North Carolina State University

and

Hayne Palmour III
Department of Engineering Research, North Carolina State University

SPRINGER SCIENCE+BUSINESS MEDIA, LLC 1966

ISBN 978-1-4899-6162-4   ISBN 978-1-4899-6311-6 (eBook)
DOI 10.1007/978-1-4899-6311-6

Library of Congress Catalog Card Number 63-17645

Foreword

This book is broadly concerned with physical ceramics, an increasingly well-defined research and academic discipline. In our concept, physical ceramics includes all of the interrelationships existing between constitution, microstructure, and responses of the material to stresses and energetic environments. Grain boundaries and surfaces attract much study in physical ceramics, for it is at these discontinuities that atomistic processes tend to be localized and stress concentrations and environmental interactions tend to be most severe. The useful macroscopic properties of polycrystalline materials are often established at these interfaces. Increasing recognition of the importance in ceramics of events occurring at free and bound surfaces, together with significant research advancements in their systematic study, provided the background and the unifying theme for the November 16–18, 1964, Conference and for these proceedings.

The Conference was held on the campus of North Carolina State University at Raleigh under the auspices of its School of Engineering and its Division of Continuing Education and was sponsored jointly by the University, the U. S. Army Research Office – Durham, and the U. S. Office of Naval Research. The Conference was attended by more than 150 scientists and engineers. It was international in scope, featuring eight foreign scientists as speakers or session chairmen and including several other participants from the United Kingdom and Australia. It also was the first in a new series of University Conferences on Ceramic Science, arranged by North Carolina State University, the University of Notre Dame, the University of California at Berkeley, and the College of Ceramics of the State University of New York at Alfred University.

The substantive content of the papers presented fell logically into the following five parts: (I) Interfaces — Sites for Kinetic Processes; (II) Electromagnetic Wave Behavior Near Interfaces; (III) Grain Boundary Contributions to Deformation; (IV) Grain Boundary Contributions to Strength and Thermomechanical Behavior; and (V) Surface and Environmental Contributions to Mechanical Behavior.

This organization of topics has been preserved in preparing the proceedings for publication. There are 31 chapters, four of which are concerned principally with kinetics associated with interfaces (Part I) and seven of which treat the interactions of a broad spectrum of electromagnetic waves at and near boundaries (Part II). Extensive interest in the mechanical behavior of ceramic solids accounted for the majority of papers. Six are concerned principally with the influence of boundaries upon deformation and plasticity (Part III); seven papers emphasize the effects of grain boundaries upon strength, fracture initiation, and other structural considerations (Part IV), and the last six (Part V) consider structural flaws occurring at free surfaces in terms of their contributions to chemical stability, fracture processes, adherence, and mechanical strength.

The keynote speaker (Chapter 1) identified several themes which pervaded much of the Conference: We were admonished to remember that ceramics are inherently complex, involving binary systems even in the simplest cases, that reactions with atmospheric environments may determine which chemical equilibria really prevail and which transport mechanisms actually operate, and that it is principally at surfaces and interfaces where these decisive chemical events are initiated. We were encouraged, above all, to capitalize upon the novel aspects in ceramics which find little or no parallel in metallic systems.

Progress toward adequate understanding and effective utilization of modern ceramic materials is being effected by the skills and talents of many scientists and engineers. Interdisciplinary interplay between basic research and engineering synthesis in the creation of new ceramic processes and products has been particularly significant within the last half-decade. Much of the driving force in the rapidly advancing field of physical ceramics stems from the excitement of discovery and the exhilaration of solving difficult problems in basic research; however, the ultimate "payoff" inevitably involves the development of reliable, effective, and economical ceramic components and devices.

Advancement in this challenging field can be likened to the experience of the mountain climber who, upon reaching each "false" crest, sees a yet higher ridge rising above the intervening valley. Certainly for ceramics, a real resting point is not yet in sight. Four and one half years earlier, this University organized a ceramic conference with a similar research-oriented theme ("Mechanical

Properties of Engineering Ceramics," Interscience Publishers, New York—London, 1961). It seems clear to us, after editing the proceedings of both conferences, that several major ridges have been successfully scaled in the intervening years. Even for those of us who seem to spend most of our time laboring to get out of the valleys, the present level of attainment and understanding is, in fact, much higher than it was in 1960. We sense a steep upward gradient and significant forward progress, which, hopefully, this volume can assist in documenting.

We wish to express our gratitude to the authors and conference participants; their knowledge, experience, and enthusiasm for communication of their scientific interests made both the conference and this volume possible. We and the others who heard him also are indebted to the dinner speaker, Dr. Ralph G. H. Siu, of the Army Research Council, for the opportunity to learn about the relationship between research, economics, and politics and "Chinese Baseball," a game in which the bases remain fixed only as long as the ball is not in motion!

The cooperative efforts of the conference staff were essential ingredients in the success of the conference, and these are gratefully acknowledged. We especially wish to thank the representatives of the sponsoring agencies—Dr. H. M. Davis and Col. Nils M. Bengtson of AROD, Dr. Cyrus Klingsberg and Dr. I. Salkovitz of ONR, and Dean R. E. Fadum and Dr. R. G. Carson of NCSU—for their interest, encouragement, and assistance.

In the preparation of this volume, we have relied heavily upon the secretarial assistance of Mrs. Marion S. Rand, help in indexing by Mrs. J. M. Waller, and the sound advice of Miss Evelyn Grossberg and her associates at Plenum Press.

W. W. Kriegel

Hayne Palmour III

Raleigh, North Carolina
February 1966

Conference Staff

Conference Co-Chairmen

W. Wurth Kriegel, Professor-in-Charge, Ceramic Engineering
Dept. of Mineral Industries, N.C.S.U.

Hayne Palmour III, Research Professor of Ceramic Engineering
Dept. of Engineering Research, N.C.S.U.

Session Chairmen

H.M. Davis, Director, Metallurgy – Ceramic Division
U. S. Army Research Office – Durham

G.C. Kuczynski, Professor of Metallurgical Engineering and
    Materials Science
University of Notre Dame

S. Amelinckx, Visiting Foreign Scientist
Carnegie Institute of Technology
(Studiecentrum voor Kernenergie, S. C. K., Mol, Belgium)

R. J. Stokes, Staff Scientist
Honeywell Research Center

J. A. Pask, Professor of Ceramic Engineering
University of California at Berkeley

N. M. Parikh, Scientific Advisor
Illinois Institute of Technology Research Institute

Advisors on Program and Publication

W. W. Austin, Professor of Metallurgy and Dept. Head
Dept. of Mineral Industries, N.C.S.U.

C. B. Alcock, N.S.F. Visiting Foreign Scientist
Dept. of Mineral Industries, N.C.S.U.
(Professor of Chemical Metallurgy, University of London)

L. Slifkin, Professor of Physics
University of North Carolina at Chapel Hill

H. H. Stadelmaier, Research Professor of Metallurgy
Dept. of Engineering Research, N.C.S.U.

Arrangements

David B. Stansel, Assistant Director
Division of Continuing Education, N.C.S.U.

N. W. Conner, Director of Engineering Research
N.C.S.U.

Hospitality and Tours

W. C. Hackler, Professor of Ceramic Engineering
Dept. of Mineral Industries, N.C.S.U.

John V. Hamme, Associate Professor of Ceramic Engineering
Dept. of Mineral Industries, N.C.S.U.

R. F. Stoops, Research Professor of Ceramic Engineering
Dept. of Engineering Research, N.C.S.U.

Transportation

M. Paul Davis, Research Associate
Dept. of Engineering Research, N.C.S.U.

J.M. Waller, Instructor
Dept. of Mineral Industries, N.C.S.U.

Publicity

Motte V. Griffith, Jr., Public Information Officer
Division of Continuing Education, N.C.S.U.

Mary N. Yionoulis, Publications and Information
School of Engineering, N.C.S.U.

Contents

PART III. Grain Boundary Contributions to Deformation

PART IV. Grain Boundary Contributions to Strength and Thermomechanical Behavior

PART V. Surface and Environmental Contributions
to Mechanical Behavior

Chapter 1

Keynote Address: Equilibria and Transport in Ceramic Oxide Interfaces

C. B. Alcock[*]

Imperial College of Science and Technology
London, England

The oxide ceramics are important technologically and conform reasonably well to the ideal model of an ionic solid. Even in the simplest case, they should be regarded as binary chemical systems in which the chemical potentials of the components are determined by the oxygen potential of the environment, as well as by the temperature. The properties of typical oxide systems at interfaces, such as vaporization, segregation, electrical and material transport, are discussed in relation to this binary nature. The differences and similarities between ceramic systems and elementary metallic systems in this respect are outlined also.

INTRODUCTION

In recent years, the scientific study of relatively well-characterized ceramics has grown to a scale comparable with that of metallic systems. The student of ceramics has benefited considerably from the fact that the corresponding development in the study of metals preceded that in ceramics by about ten years, and the devising of theoretical models and experimental techniques had, to a large extent, already been done. However, the time is now ripe for the ceramist to begin to emerge from the "metallurgical phase," since there are novel aspects to the study of ceramics which find little or no parallel in metallic systems. This paper will be largely concerned with the effects of the chemical nature of the environment on the properties of ceramics and the way in which this might be expected to influence the physico-chemical properties of interfaces. At the present stage of development, oxide systems are the ones that have been most exhaustively studied. For this reason, the discussion in this paper will be exclusively centered on these technologically important materials.

[*]The author was N. S. F. Visiting Scientist, N. C. State University at Raleigh, at the time of the conference.

SURFACE ENERGY AT HIGH TEMPERATURES

In making a simplified model of oxides which is mathematically tractable, the most promising approximation is to regard the oxides as ionic crystalline materials. One immediate advantage in making this approximation is that surface energies may then be calculated for each crystal face by means of well-established techniques along the lines first evolved by Born and Stern [1]. The energies calculated in this way for the (100) and (110) faces of the sodium chloride lattice, which is the crystal structure of some ceramic oxides, can be expressed by simple equations, such as

$$\gamma_{(100)} = 0.0145 \, (Ze)^2/r^3 \tag{1}$$

$$\gamma_{(110)} = 0.0394 \, (Ze)^2/r^3 \tag{2}$$

where Z is the charge on each ion and r is the internuclear distance. Equations (1) and (2) show that the surface energy of the (100) face is about one-third of that of the (110) face.

This approach suggests then a qualitative difference between metallic systems and the anticipated behavior of oxide ceramic systems. In metallic systems, where a satisfactory procedure for the a priori calculation of surface energies has not yet been devised, the results of experiment suggest a smaller difference between low-index planes; thus, Sundquist [2] has shown that the difference in surface energy between the (111) and (100) planes of gold, silver, and copper is only about 30%.

Shuttleworth's calculations [3] for solid argon, where the binding is principally by van der Waals forces, suggest an even smaller ratio, $\gamma_{(100)}$ being within 5% of $\gamma_{(111)}$.

It is problematical whether we should expect the results of the ionic model to be accurate in relating the surface energies of different crystal faces in the high-temperature region, which is of interest to ceramists, because it is known that most oxides become semiconductors at high temperatures. These systems are then somewhere between the ionic and metallic extremes, and clearly it is of importance to establish whether the oxides at high temperature can be described significantly by the ionic model in such a way as to make a clear distinction, with respect to relative surface energies, between ceramics and metallic systems.

The complicating factor of semiconduction in oxides at high

temperature is the oxygen chemical potential of the system. Thus, it is well-known that the magnitudes of the electrical conductivity of oxides depend, in the case of simple oxide systems, on the oxygen partial pressure in the atmosphere surrounding the material. It is even possible by changing the chemistry of the gaseous environment to alter the quality of semiconduction in a given oxide from p-type to n-type with a continuous transition. There are also oxide solid solutions which do behave as ideal electrolytes to a very good approximation over a wide range of oxygen pressures, such as calcia–zirconia which becomes an n-type semiconductor only below 10^{-18} atm at 1000°C and calcia–thoria which becomes a p-type semiconductor above 10^{-6} atm at 1000°C and is electrolytic down to at least 10^{-25} atm at 1000°C.

It would seem fruitful to devise methods whereby the direct measurement of relative surface energies could be made at high temperature as a function of oxygen potential, in order to obtain the effects of the deviation from purely ionic behavior. Of the experimental procedures which have been developed for metallic systems, Mykura's method [4], in which the morphology of twin boundary grooves is studied, would seem to be most valuable. Techniques depending on room-temperature observation, such as cleavage [5] and microcalorimetry [6], would seem to be too indirect for use in this important area, unless surface entropy data are also available. These might then be applied to the room-temperature behavior to yield extrapolations to the more important high-temperature region.

VAPORIZATION RATES AND MODES

The surface properties of oxides under conditions of high-temperature service are directly related to the way and rate at which vaporization occurs. It is now well-established that the equilibrium vaporization mechanisms include decomposition to atomic species and also, in some cases, to suboxides. Thus, alumina, magnesia, and ytterbia vaporize principally to atoms, whereas yttria, hafnia, and thoria vaporize as gaseous oxides under high-vacuum conditions [7]. The stoichiometry of these high-vacuum vaporization modes may be used to predict what the principal species will be under an atmosphere containing oxygen. Thus, the higher the oxygen pressure, the more will AlO predominate in the vaporization of alumina and the less important the atomic vaporization will be, whereas thoria will evaporate mainly as ThO_2 from conditions of high vacuum to those of the normal atmosphere. UO_2, however, by virtue of the variable valency of uranium, will evaporate as UO_2 in high vacuum, but mainly as UO_3 in atmospheres with a relatively high oxygen pressure.

It is clear then that, in order to characterize the rate of surface evaporation in oxide systems, it is necessary to specify the temperature, the total pressure, and the partial oxygen pressure of the gaseous environment.

For a typical polycrystalline material, it does not appear necessary to specify the crystallographic nature of the surface in a discussion of vaporization rates, since very little faceting is normally observed. In single-crystal studies, however, this matter is quite important, since the rate of vaporization and the nature of the morphology of the surfaces resulting from vaporization do vary from face to face.

Recent measurements of rates of vaporization in vacuo of alumina and magnesia single-crystal plates have shown how the rates vary among some simple crystallographic directions, and the morphology of the surfaces which were obtained after a considerable amount of vaporization varied markedly from face to face.

The pyramidal (1012) face of alumina has a quite liquidlike appearance when compared with the well-developed faceting of the basal and prism planes. These findings are in line with the calculations of Frank, Burton, and Cabrera [8], who suggest that a considerable degree of surface roughening should occur on high-index planes at temperatures near the melting point as a result of surface vaporization and migration.

In connection with these observations, Andrade and Randall [9] have found a similar structure on metallic cadmium which had been evaporated to about the same extent (30 mg/cm^2). They made the further observation that, on plastically deforming the cadmium samples, basal plane slip occurred along the plates which had been developed by vacuum-etching. This suggested that the process of vaporization in vacuo showed the inherent distribution of the (regularly spaced) planes which would glide under deformation, the thickness of the "packets" being about 0.75μ. Application of this concept to the present results for alumina indicates these planes would appear to be much closer together, since some twenty individual steps could be counted in the $2-\mu$-deep morphology on the basal plane samples.

SEGREGATION AT GRAIN BOUNDARIES

Since this is such a broad subject from the chemical point of view, it seems appropriate to restrict the discussion to one principal solvent, magnesia.

In principle, the calculation of the limiting amounts of impurities which can be sustained by magnesia without phase separation is a relatively simple exercise in applied thermodynamics when the impurity cation is one of constant valency, e.g., CaO, SiO_2, or Al_2O_3. However, the actual appearance of a segregate when the impurity is present at a low level of concentration probably depends largely on other factors which are very difficult to calculate at present.

A degree of disregistry between the segregate and host lattice oxygen-ion sublattices will add to the difficulty of nucleation already present due to the fact that a new phase is growing. At present, the energetics of disregistry is not accurately calculable, and, furthermore, the "catalytic" effect of dislocation arrays at grain boundaries and sub-boundaries probably will be of significance in deciding the site of precipitation. Thus, Hornbogen [10] has shown that, for solutions in solid iron, internal boundaries are important sites for impurity segregation.

It is not surprising that there exists a good deal of confusion in the literature concerning solubility limits in ceramic systems, even those where only constant-valency cations are present. The difficulties of segregate nucleation referred to above and the relatively low diffusion coefficients which are frequently found for cations in oxide systems suggest that the conventional methods may be unreliable, unless long annealing periods are allowed for equilibrium to be achieved. It seems we should either bend our efforts toward more direct and sensitive observation of segregation at high temperatures, such as a possible development of high-temperature microprobe analysis, or we should rely more heavily on calculations using thermodynamic data. There is at least one case in the metals literature where this procedure was shown to be more reliable than the existing differential thermal-analysis method [11].

When the cation has a variable valency, e.g., iron, manganese, and chromium, the prediction of segregation is made more difficult by the fact that near interfaces the oxygen potential of the environment will probably determine the ratio of higher to lower valencies of impurity ions. The region of the solid over which this surface effect extends will increase with time at constant temperature, depending on, for example, the oxygen diffusion coefficient in the host oxide. The concentration of a multivalent impurity at or near a boundary thus may be markedly different from that of the impurity in the bulk phase, depending on the atmosphere. The increase in the number of higher-valency ions near the interfaces will lower the activity coefficient of the impurity species in these regions, and will cause diffusive flow from the bulk to the interface regions. It follows that unless sufficient time is given for equilibration of the whole solid with the gaseous phase, the distribution of impurities will change during an experiment.

At present, we have very little bulk thermodynamic information to aid us in calculating such ratios as Fe^{+2}/Fe^{+3} in dilute solution in MgO as a function of iron content, temperature, and oxygen potential, nor do we know the rates at which oxygen potentials are transmitted along grain boundaries in typical systems.

Hahn and Muan [12] have demonstrated that, in solid solutions in MgO, the dilute solutions of NiO, MnO, and FeO are close to ideal in the Raoultian sense. We might assume, therefore, that the process

$$\tfrac{1}{2}O_2 + 2M^{+2} \rightarrow 2M^{+3} + M_{vac}^{+2} + O^{-2} \tag{3}$$

has the same contributory factors to the energetics in solution in magnesia as in the pure oxide phases. Thus we would require knowledge of the energy of ionization,

$$M^{+2} \rightarrow M^{+3} + e^- \tag{4}$$

the energy of vacancy formation and relaxation around the vacancy,

$$M^{+2} \rightarrow M_{vac}^{+2} + M^{+2}(g) \tag{5}$$

and the energy of dissociation of the positive hole–vacancy complex.

Moore [13] has calculated these for a number of transition-metal oxides, and the only alteration which must be made in his calculations is the substitution of the dielectric constant of magnesia for that of the transition-metal oxide in the dissociation energy of the vacancy–positive hole complex. The energy for vacancy formation may be left constant, since the lattice energy of magnesia is quite close to those of the transition-metal oxides.

GRAIN BOUNDARY DIFFUSION MEASUREMENTS

Since, in most applications, ceramic materials will be poly-crystalline in microstructure, increased attention is being given to the relative rates of diffusive transport through the volume and along grain boundaries. Methods, therefore, must be devised which allow the analysis of mass transfer in terms of the two contributions.

It is tempting to compare the decrease in surface activity of an isotopically labelled element in diffusion, studied on polycrystalline and single-crystal material, and to assume from these measurements that the difference will reveal the grain boundary contribution. This procedure is not unequivocal, since the coherency of the isotopic material with the substrate can vary from one surface to another, and this might vitiate any comparison. The most satisfactory experimental solution is to devise a technique in which the diffusing isotope is supplied from a vapor phase, as, for example, in the measurement of oxygen self-diffusion by exchange between gas and solid, or by using a sectioning technique.

This latter method has its difficulties when low diffusion coefficients of the order of 10^{-15}–10^{-17} cm^2/sec must be measured. New methods of particle spectrometry may find application, as was shown in some recent work on uranium diffusion in UO$_2$ in the author's laboratory. The isotope that was used was U^{233}, which is an a-particle emitter, the energy being 4.5 MeV. Since there is a linear range–energy curve for a particles in matter, energy spectrometry of the particles, those leaving the surface of a diffusion specimen normally, revealed the diffusion profile. It was found that the plot of log concentration versus distance squared was linear only for the first micron, the first power of the penetration against the logarithm of concentration yielding a linear plot at lower particle energies. The separation of the two regions into volume and grain boundary contributions was confirmed by comparison of diffusion profiles in single and polycrystalline samples.

The difficulties to be overcome in the analysis of grain boundary diffusion at present are to a large extent mathematical. Techniques, such as those of Fisher [14], are only useful when the grain boundary diffusion coefficient is much greater than the volume diffusion coefficient. The experimental difficulties will be considerably reduced when techniques of joining single crystals to form artificial grain boundaries have been satisfactorily developed.

TRANSPORT OF MATTER UNDER COMPRESSION

Several studies have been made recently of creep in oxides at high temperatures, in which the rate has been studied as a function of temperature. The expectation, in the light of the Nabarro—Herring model of diffusive creep [15], was that the rate of the process would be dependent on the rate of diffusion of the slower-moving ionic species. Experiment has shown that the process often occurs at a rate comparable to that of the faster-moving species [16].

In considering these results, the following analysis was evolved which couches the problem in simple physico-chemical terms and which may shed some light on a number of similar problems.

The flux of a species in a multicomponent system may be expressed by the general equation

$$i = - B_i C_i \left(\frac{1}{N} \frac{\partial \mu_i}{\partial x} + z_i e \frac{\partial E}{\partial x} \right) \tag{6}$$

where B_i is the mobility coefficient of species i of concentration C_i, N is Avogadro's number, $\partial \mu_i / \partial x$ is the gradient of chemical potential in the direction opposite to that of the flux, z_i is the charge of the particle, and $\partial E / \partial x$ is the electric-field gradient. This equation may be used for a nonisothermal system as a basis for calculating the thermal electromotive forces, as was shown by Lidiard [17] in the form

$$i = - B_i C_i \left(\frac{1}{N} \frac{\partial \mu}{\partial T} \frac{\partial T}{\partial x} + z_i e \frac{\partial E}{\partial x} \right) \tag{7}$$

It can also serve as a basis for the calculation of transport rates under a pressure gradient by writing

$$i = - B_i C_i \left(\frac{1}{N} \frac{\partial \mu_i}{\partial p} \frac{\partial p}{\partial x} + z_i e \frac{\partial E}{\partial x} \right) \tag{8}$$

and applying the overall electroneutrality requirement.

The simple case of transport in a binary alloy when no electric field is present can readily be solved by noting that a steady state of transport is achieved when the fluxes of the two components are in the ratio of their respective mole fractions (X_i). This can occur only when a concentration gradient has been built up along the pressure gradient, which slows down the more mobile component and speeds up the more slowly moving one. In the steady state, it can be shown that in an A + B alloy

$$i_A = \frac{B_A B_B}{B_A X_B + B_B X_A} V_m \frac{\partial p}{\partial x} X_A \tag{9}$$

where V_m is the molar volume of the alloy. The flux equation for a simple ionic substance when $Z_A = Z_B$ is identical, with Z replacing X_A.

Now an oxide at high temperatures must be regarded as a three-component system—cations, anions, and electrons or electron defects — but, if we regard one component, say, the anions, as having negligible mobility, then the fluxes of cations and electrons or defects must be balanced. In this case, transport equivalent to the migration of one atomic species occurs under the pressure gradient, the stoichiometry of the oxide changing correspondingly. There will now exist an oxygen potential difference between the two interfaces from which and to which transport is occurring, and this must be compensated by the migration of oxygen along the surfaces of the grain. Providing this process occurs more rapidly than the volume diffusion of the cations, the latter process will be rate-determining.

One final deduction which can be made from this analysis concerns the edge dislocation model of grain boundaries. According to this model, the difference in inclination of corresponding planes of two crystals joined at a low-angle tilt boundary is accommodated by the introduction of edge dislocations spaced equally apart. Since each side of the dislocation core is a region of compression or tension with respect to the bulk phase, the effects of the resulting pressure gradient from core to core should result in an electric potential difference between them and in a migration of the cores along the boundary at a rate determined by the migration of the slower ionic species in the case of a purely ionic solid. The charges experimentally observed to be associated with edge dislocations in ionic crystals may well originate from such a source as that suggested here.

CONCLUSION

This discussion may be summarized by stating that ceramic systems are, by their very nature, multicomponent chemical systems. It has long been known that the properties of solids are sensitive to stoichiometry, and the ceramist must take account of this fact if he wishes to work with well-defined systems. The requirement of chemical equilibrium means that the possible interactions with the environment of the solid, be that environment liquid or gas, must be evaluated and allowed to go to completion. This is particularly relevant to a symposium in which the properties of the surfaces of grains, both internal and external, are to be discussed, since, even though the environmental effect may be of small significance to the consideration of bulk properties, it is at boundaries that the effect is most marked.

REFERENCES

1. M. Born and E. Stern, Sitzber. preuss. Akad. Wiss. 901 (1919).
2. B. E. Sundquist, Acta Met. 12:67 (1964).
3. R. Shuttleworth, Proc. Phys. Soc. 62A:167 (1949).
4. H. Mykura, Acta Met. 5:346 (1957).
5. J. J. Gilman, J. Appl. Phys. 31:2208 (1960).
6. G. C. Benson, H. P. Schreiber, and F. van Zeggeren, Can. J. Chem. 34:1553 (1956).
7. R. J. Ackermann and R. J. Thorn, Progress in Ceramic Science, Vol. I, Pergamon Press (New York), 1961, p. 39.
8. W. K. Burton, N. Cabrera, and F. C. Frank, Phil. Trans. Roy. Soc. 243A:299 (1951).
9. E. N. da C. Andrade and R. F. Y. Randall, Proc. Phys. Soc. 63B:198 (1950).
10. E. Hornbogen, Trans. A. S. M. 55:719 (1962).
11. C. Wagner, Acta Met. 2:242 (1954).
12. W. C. Hahn and A. Muan, Trans. AIME, 224:416 (1962); and J. Phys. Chem. Solids 19:338 (1961).
13. W. J. Moore, J. Electrochem. Soc. 100:302 (1953).
14. J. C. Fisher, J. Appl. Phys. 22:74 (1951).
15. F. R. N. Nabarro, Rept. Conf. Strength of Solids, (University of Bristol), 1948, p. 75.
16. A. E. Paladino and R. L. Coble, J. Am. Ceram. Soc. 46:133 (1963).
17. A. B. Lidiard, Handbuch der Physik, Vol. 20, Springer-Verlag (Berlin), 1957, p. 339.

PART I. Interfaces — Sites for Kinetic Processes

G.C. KUCZYNSKI, Presiding

University of Notre Dame
Notre Dame, Indiana

Chapter 2

Oxygen-18 Diffusion in Surface Defects on MgO as Revealed by Proton Activation

J. Birch Holt and Ralph H. Condit

Lawrence Radiation Laboratory
University of California
Livermore, California

A study of the effect of surface defects on the self-diffusion of oxygen in MgO is reported. The technique involves exchange with oxygen-18 enriched gas followed by proton bombardment to yield the radioisotope, fluorine-18. This may be treated as a normal radioactive tracer. The penetration distribution of the oxygen isotope is indicated by autoradiography. Preferential diffusion of oxygen down bicrystal grain boundaries is definitely proven. Dislocations introduced by cold-working are found to contribute to the overall rate of oxygen diffusion. Surface fractures are shown to be important channels for the rapid movement of oxygen. The implications of these findings are discussed.

INTRODUCTION

Nearly all of the studies dealing with the self-diffusion of oxygen in oxides have been accomplished by use of the gaseous-exchange technique. In this type of experiment, oxide samples are annealed at high temperatures in an oxygen atmosphere enriched with oxygen-18. The decrease in the concentration of oxygen-18 in the ambient atmosphere is followed as a function of time using a mass spectrometer. This experimental technique, which is required by the lack of any suitable radioactive isotopes of oxygen, measures only the total mass flow into or out of a crystal. The concentration distribution of the oxygen-18 must then be inferred from the kinetic data. Without a concentration profile, an unambiguous interpretation of the data becomes untenable, particularly in the refractory oxides where the volume diffusion coefficients and the penetration distances are apparently very small, even up to 1800°C. This limit is imposed by the melting point of the platinum—40% rhodium container. The low penetration (less than 10μ) magnifies the distorting influence of surface defects, such as grain boundaries, fractures, and dislocations.   The purpose of this paper is to present our observations of the effect of these defects on oxygen self-diffusion in magnesium oxide.   This is made possible by a new experimental technique [1] in which the presence of oxygen-18 is revealed by proton activation combined with autoradiographic procedures.

EXPERIMENTAL METHODS

A typical spectrographic analysis of large-grained, fused magnesium oxide crystals supplied by Semi-Elements Company revealed the presence of the following elements (in ppm): calcium, 300; silicon, 50; aluminum, 30; iron, 10; chromium, 5; and copper, < 5.   Cylinders 1 in. in length by $\frac{1}{2}$ in. in diameter were bored out of these crystals and cut into bicrystal specimens $\frac{1}{8}$–$\frac{1}{16}$ in. thick with a grain boundary running roughly parallel to the c-axis of the disk.   These disks were used to determine the effect of grain boundaries on the diffusion of oxygen into magnesium oxide. For assessing the importance of dislocations as channels for rapid diffusion, other specimens were cleaved from the large single-crystal pieces and were of various sizes.

The prepared specimens were annealed in an inductively heated platinum–40% rhodium crucible in a water-cooled Vycor chamber. An oxygen atmosphere enriched with 40% oxygen-18 was used to fill the chamber to approximately 120 mm Hg. The times and temperatures of these diffusion anneals will be indicated with the results from the individual specimens. After the oxygen-18 high-temperature exchange, the bicrystal disks were irradiated with a 2.8-MeV proton beam in the 90-in. cyclotron at the Lawrence Radiation Laboratory, Livermore, California.

The location of the oxygen-18 in the crystalline magnesium oxide is detected by making use of the $O^{18}(p, n)F^{18}$ nuclear reaction. The F^{18} is a positron emitter and has a half-life of 1.87 hr, which allows conventional radioactive tracer techniques to be employed. The recoil distance of the F^{18} is less than 0.1 μ, thus ensuring that the F^{18} is essentially in the same place as was the oxygen-18.

Impurities, such as lithium, boron, calcium, and nickel, may be activated and thus produce some background activity. However, they were not present in sufficient amounts in these specimens to cause any interference.   A general evaluation of the proton activation of a wide range of impurities is contained in a previous paper [1]. Immediately after the proton bombardment, the crystals were counted with a beta counter to determine the half-life of the radioactive species, which, in every instance, agreed with that of fluorine-18. The activation of oxygen-18 was confined to about a 5-μ surface region, thus permitting the oxygen distribution to be observed with autoradiography. Kodak Type A autoradiographic plates and AR 10 stripping film extended the range of resolution from 30 down to 5 μ, respectively. Results of these autoradiographs will be shown in later sections.

DIFFUSION IN GRAIN BOUNDARIES

Laurent and Bénard [2], in their classic diffusion experiments with both single-crystal and polycrystalline alkali halides, noted that the apparent diffusion coefficient of the anions varied with the size of grains, i.e., increasing with decreasing grain size. To verify this conclusion, they prepared autoradiographs of polycrystalline samples in which the iodine ion had been diffused, and observed that indeed preferential anion diffusion did occur at the grain boundaries. Oishi and Kingery [3] measured oxygen diffusion in single-crystal and polycrystalline aluminum oxide using the gaseous-exchange technique. The apparent diffusion coefficient for the single crystal was lower than the polycrystalline material, suggesting enhanced oxygen diffusion at the grain boundaries. Paladino et al. [4], using the same procedures as Oishi and Kingery but substituting yttrium iron garnet, found the diffusion rate was not dependent on the presence of grain boundaries. Tien [5], in studying the electrical conductivity of $Zr_{0.84}Ca_{0.16}O_{1.84}$ solid solution, suggests that the higher conductivity of polycrystals is due to the higher diffusion rate of the oxygen ion along the grain boundaries. Several papers [6,7] have been written explaining such solid-state reactions as creep and sintering of oxides on the assumption that there is rapid diffusion of oxygen along grain boundaries which requires the cation diffusion to become rate-controlling. To our knowledge, no direct observation of this preferential diffusion of oxygen in oxide systems has been demonstrated. The use of oxygen-18 coupled with proton activation offers a powerful method for examining this type of behavior.

The magnesium oxide bicrystals were model systems for the investigation of the existence of oxygen diffusion in grain boundaries.

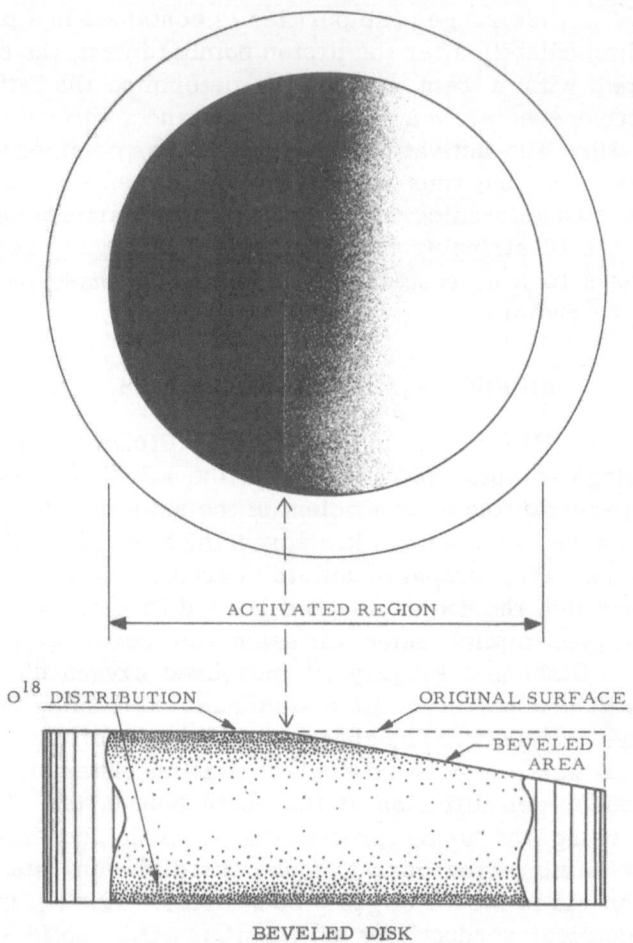

ACTIVATED REGION

O^{18} DISTRIBUTION   ORIGINAL SURFACE

BEVELED
AREA

BEVELED DISK

Fig. 1. Top—Figure representing an autograph of the proton-irradiated disk. Bottom—
Schematic drawing showing a beveled MgO disk after annealing in oxygen-18.

Subsequent to the oxygen-18 anneal of the bicrystals, they are
beveled, as shown in the cross section in the bottom part of Fig. 1.
This shallow bevel runs parallel to the grain boundary and represents a drop of 30 to 40 μ from the line of the bevel to the edge of
the disk. After the proton irradiation, an autoradiograph is taken
of the beveled section as described previously. Since the optical
density of the autoradiograph, as measured by a microdensitom-
eter, is directly proportional to the concentration of oxygen-18,

a concentration profile may be constructed not only of the grain boundary portion, but also of that part diffusing within the crystals. Significant volume penetration would give an autoradiograph as schematically illustrated in the top of Fig. 1.

An actual autoradiograph of a MgO bicrystal annealed at 1650°C for 16.5 hr is shown in Fig. 2. The most obvious feature of the autoradiograph is the distinct appearance of the grain boundary. It was necessary to eliminate any possibility of misinterpretation of the autoradiograph which might be caused by proton activation of impurities segregated at the boundary. This was accomplished by grinding off $500\,\mu$ of the top of the bicrystal disk used for the auto-radiograph in Fig. 2. The disk was bombarded again with protons, and a second autoradiograph showed no darkening at the boundary. The autoradiograph in Fig. 2 is then direct proof that there is en-hanced diffusion of oxygen down the grain boundary to a depth of at least 30 μ. The large-angle boundary in this particular bicrystal has both tilt and twist components. Other important features of the autoradiograph include the overexposed area, extending the width of the disk, which contains any contribution to the volume diffusion as well as part of the initial interface. During the oxygen-18 anneal, some platinum plated out on the surface and was flattened during the beveling operation and, thus, appears as the light section to the top. This platinum deposit apparently does not hinder the oxygen-18 diffusion, but it shields the MgO from the proton beam. The rate of diffusion of oxygen into the bulk of the two crystals was very small, giving a depth of penetration of less than $2\,\mu$. The penetration was so small that any precise measure-ments were impossible. To obtain volume diffusion coefficients, oxygen anneals must be made at much higher temperatures at longer periods of time and will require a different experimental approach, which is being developed at this time. Figure 3 is an autoradiograph of another bicrystal again distinctly showing the grain boundary diffusion. Both of these autoradiographs were taken with Type A film.

It has been proven with some metals that grain boundary dif-fusion is a function of the orientation of the boundary [8,9]. The boundary possesses an interfacial energy that increases up to a maximum of approximately 45° in cubic systems as the degree of misorientation increases. Variation in the grain boundary dif-fusion coefficient should take place in the same manner, i.e., in-creasing with increasing misorientation. Leymonie and Lacombe [10]

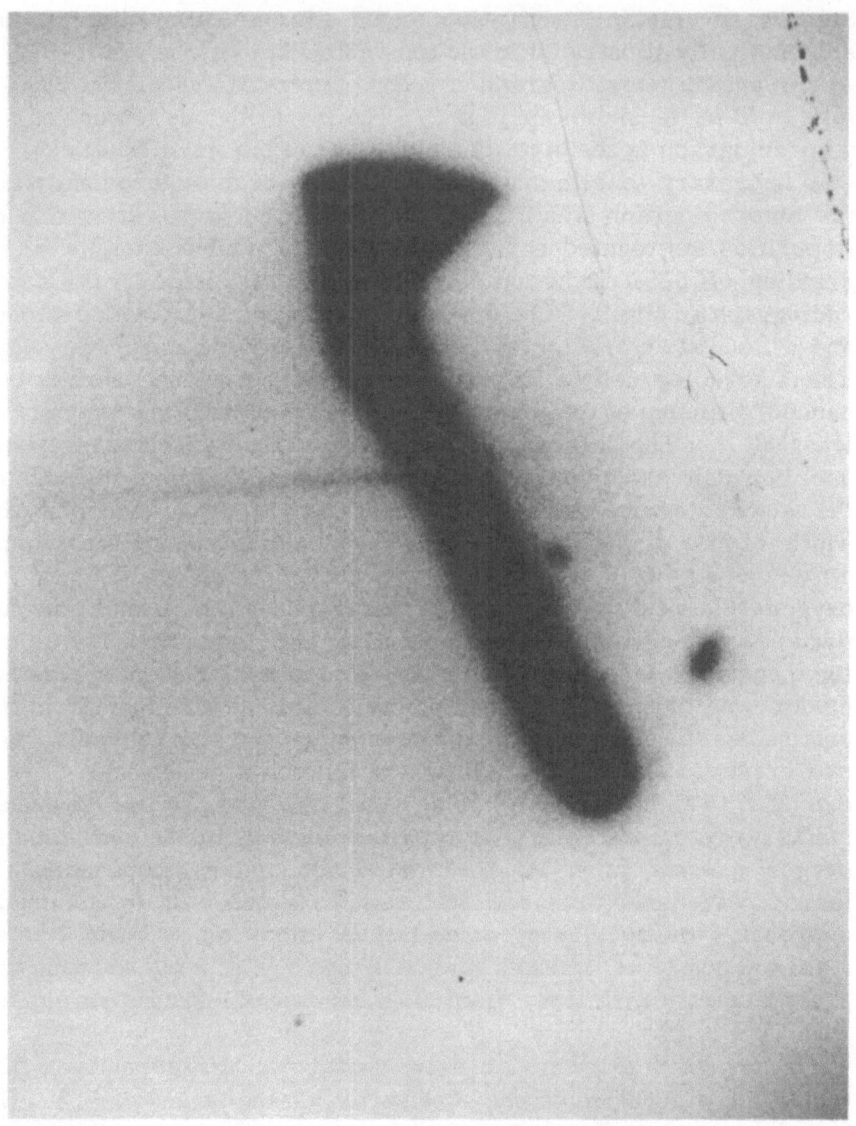

Fig. 2. Autoradiograph exposed by a MgO bicrystal disk which was annealed in 40% enriched oxygen-18 at 1650°C for 16.5 hr and then beveled prior to irradiation with a
2.8-MeV proton beam.

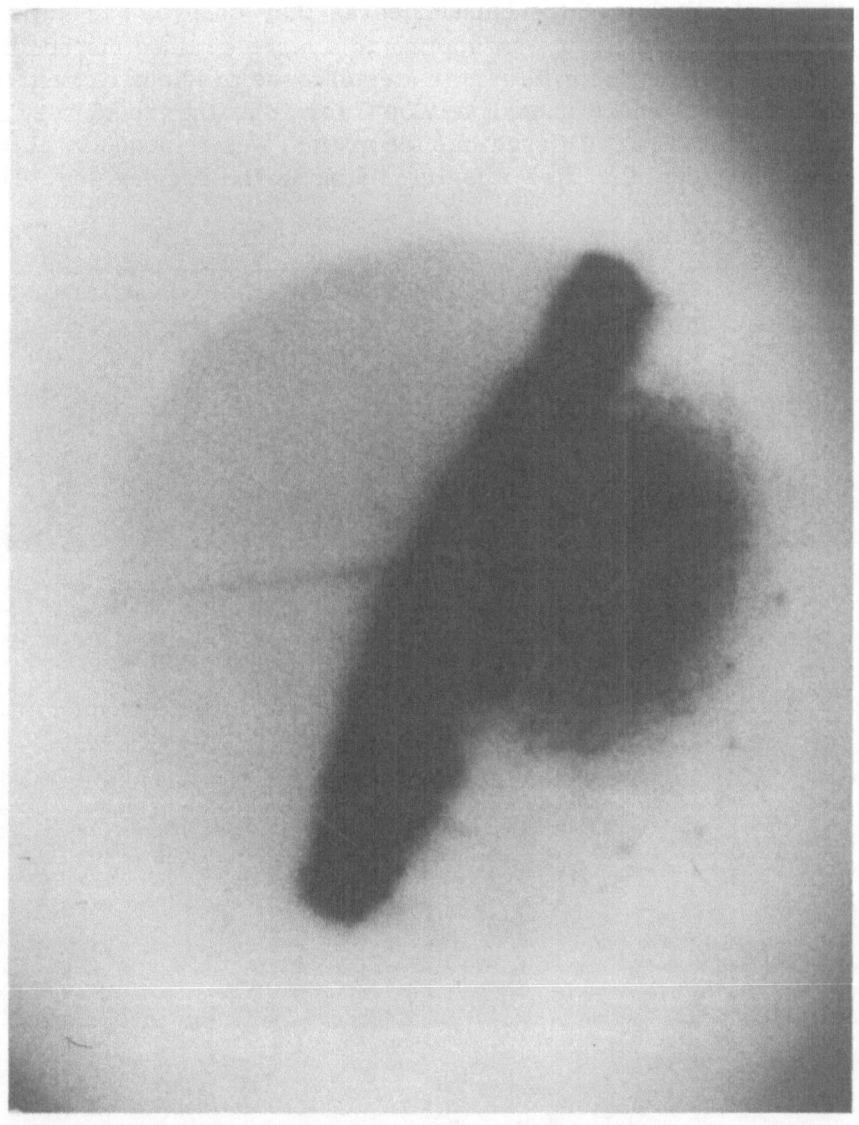

Fig. 3. An autoradiograph taken of a bicrystal disk different from that of Fig. 2, but one which again demonstrates preferential oxygen diffusion at the grain boundary.

found this to be the case for the diffusion of radioactive iron into polycrystalline iron and observed on the autoradiograph the difference in darkening of the boundaries as they changed orientation.

This same method was applied to a large-grained magnesium oxide sample cut from the same material as the previous bicrystals. After the oxygen-18 anneal at 1650°C for 16 hr the top 15 to 20 μ, containing that part diffused into the grains, was removed by abrasion, and then the disk was irradiated in the regular way. The

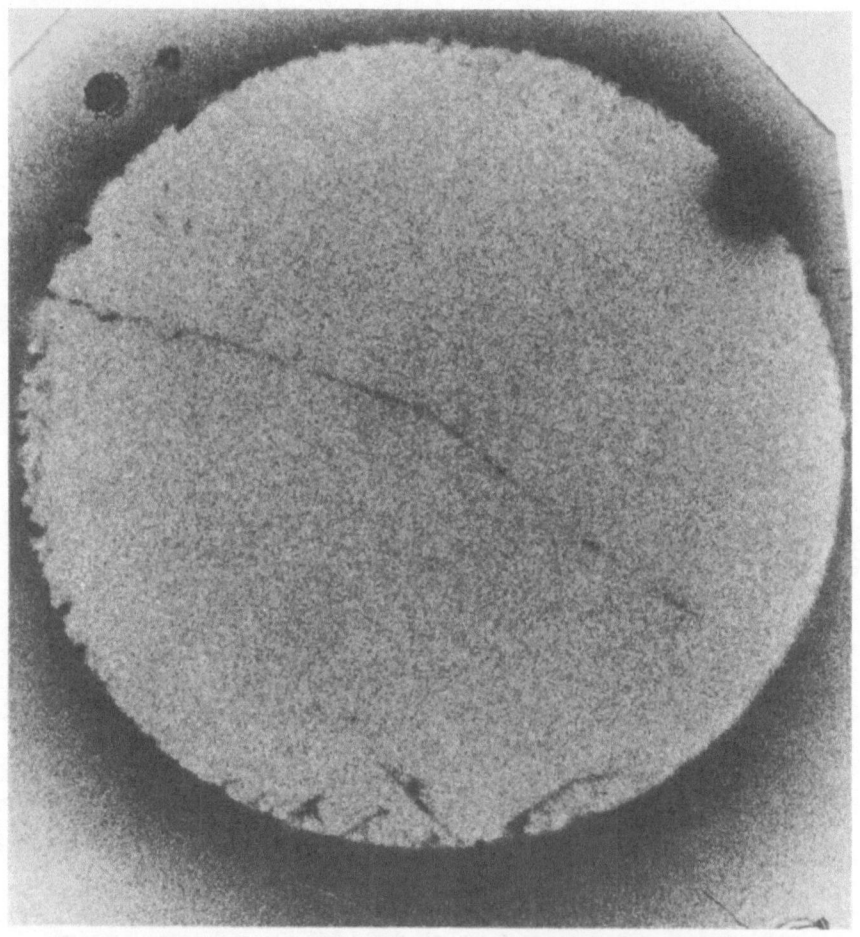

Fig. 4. Impurity segregation at the grain boundary is revealed by the autoradiograph of an as-received MgO bicrystal that was irradiated with $3 \cdot 10^{18}$ nvt neutrons.

grain boundaries were clearly outlined, but any effect of orientation per se was inconclusive, especially in view of the importance
of impurities. This was brought to our attention when a disk without
any prior treatment was irradiated with neutrons at the low-power
test reactor at Livermore, with a neutron irradiation of $3 \cdot 10^{18}$ nvt.
The neutrons activated the impurities throughout the bicrystal, and
the autoradiograph in Fig. 4 vividly shows this segregation at the
boundary. This appears to be a very promising technique for
investigating this general phenomenon of segregation. A rather
cursory experiment in which iron was diffused into a bicrystal,
then annealed in oxygen-18, and irradiated in the usual manner
qualitatively showed that the extent of oxygen diffusion in the boundary was greater in this case. A parallel experiment in which cobalt
was substituted for the iron gave no evidence of an increased rate
of oxygen diffusion. Apparently, the presence of a second phase,
such as $MgFe_2O_4$, at the boundary contributes to the greater mobility of the oxygen.

Cabañe [11], in following up the work of Laurent and Bénard,
discovered that when water vapor was excluded from the halide
crystals, no enhancement of the diffusion of the anion at the boundary was evident. The control exerted by impurities at the boundaries on the diffusion of ions seems very general in nature and
must be strongly emphasized, particularly in the oxide systems
where oxygen diffusion is being studied. If relatively clean
boundaries could be produced, it is likely that little if any preferential diffusion would be observed. Further research is now being
conducted on the effect of various impurities on oxygen diffusion
in grain boundaries of magnesium oxide bicrystals and will be
reported later.

A closer examination might now be made of the observed grain
boundary diffusion. The analysis of grain boundary diffusion measurements was aided greatly by the solution derived by Fisher [12].
His approach was relatively simple, but inexact, while later treatments by Whipple [13], Levine and MacCallum [14], and others [15,16]
were the opposite — more rigorous, but also difficult to apply
experimentally. LeClaire [17] made a comparison of the Fisher and
Whipple analyses and found certain regions in which their solutions
agree with each other. For our present considerations, we may
turn to the approximate solution of Fisher, which is

$$C/C_0 = \exp\left(-\pi^{-\frac{1}{4}} \eta \beta^{-\frac{1}{2}}\right) \operatorname{erfc}\left(\xi/2\right) \qquad (1)$$

where

$$\eta = \frac{y}{(D_v t)^{1/2}} \qquad \xi = \frac{x - \delta/2}{(D_v t)^{1/2}} \qquad \beta = \frac{D_b}{2D_v (D_v t)^{1/2}}$$

and C is the concentration of the diffusing specie in the boundary at time t and penetration y, C_0 is the initial concentration of diffusing specie, y is the penetration distance down the boundary, D_v is the volume diffusion coefficient, D_b is the grain boundary diffusion coefficient, δ is the grain boundary width, and t is the time.

It would be interesting to estimate, if only roughly, the grain boundary width δ, which has been the subject of much speculation in grain boundary diffusion studies. In metals, δ is usually regarded as being of the order of 10–100 A, while, in the diffusion in the alkali halides to which we previously alluded, the width was reported to be several microns. For our estimation, we may use the parameter

$$\beta = \frac{D_b \delta}{2D_v (D_v t)^{1/2}} \tag{2}$$

Both Whipple and Fisher point out that the constant concentration contours at the boundary are very shallow when β is about 1. If β is 10 or larger, sharp profiles are evident, such as the one shown in Fig. 2. In this case, β most certainly must be greater than 10. The data of Oishi and Kingery [18] on the self-diffusion of oxygen in magnesium oxide by the gaseous-exchange technique show that $D_v \approx 3 \cdot 10^{-13}$ cm^2/sec at 1650°C. However, from our experimental observations, D_v should be 10^{-14} cm^2/sec, if not smaller. The reason for this order of magnitude difference will be discussed in the next section. It is known that D_b is important in relation to D_v only at lower temperatures, but D_b is not observed until about 1600°C, so at the operating temperature D_v is at least several orders of magnitude smaller than D_b. With the values $D_v = 10^{-14}$ cm^2/sec and $t = 5.8 \cdot 10^4$ sec, equation (2) becomes

$$D_b/D_v > 4.8 \cdot 10^{-4}/\delta \tag{3}$$

If $\delta = 10$ A, the ratio $D_b/D_v > 4.8 \cdot 10^3$, which is a reasonable value in comparison with metal systems. If a width of 1 μ is assumed, then $D_b/D_v \approx 5.0$, which means that $D_v \approx D_b$, which is obviously not the case from Fig. 2. This rather crude estimate would seem to indicate that the experimental findings are more compatible with a width in the angstrom range than the micron range, which is rather surprising when compared to other diffusion studies in ionic solids. This also seems to be at variance with the importance of impurities segregated along the boundaries. It should be possible to determine, from the autoradiographs, the values of δD_b at various temperatures, from which grain boundary activation energy may be computed. Once this is known, then better estimates of δ will be possible.

DIFFUSION IN DISLOCATIONS

Since the importance of grain boundaries as avenues for oxygen diffusion in MgO has been firmly established, we might naturally inquire into the role of dislocations as channels for the rapid diffusion of oxygen. For purposes of this paper, the dislocations in magnesium oxide will be divided into two general categories: (1) grown-in dislocations, and (2) dislocations introduced by some form of mechanical working. We considered only the latter group since the possibility of elucidating their contribution seemed more likely. Some of the methods which have been used to produce dislocations in magnesium oxide are electrical discharge, pulling, compressing or bending the crystals, and a ball in contact with the surface.

Studies [19,20] of mechanical properties have shown the existence of six { 110 } slip planes with one <110> slip direction on each plane, forming a rather simple slip system. If the surface of a crystal is considered a (001) plane, then four { 110 } planes will be inclined 45° to this surface, while the remaining two intersect the surface at 90°. Dislocation loops which extend to the surface on the 45° inclined planes are screw dislocations, while conversely those in the 90° plane will be edge dislocations. Etching of an indented surface [21] has demonstrated the presence of these two types with asymmetrical etch pits characteristic of screw dislocations and symmetrical pits belonging to edge dislocations.

Lines were impressed on the surface of cleaved crystals with a ball point pen using only moderate pressure. This crude method produced gross numbers of dislocations of both types, as revealed by etching with an $AlCl_3$ solution at 50°C for 5 min. After this treatment, the cleaved sample was annealed in oxygen-18, beveled, and irradiated in the same manner as described for the bicrystals.

For all investigations of oxygen diffusion in the refractory oxides (Al_2O_3 [3], MgO [18], and BeO [22]) using the gaseous-exchange method reported in the literature, crushed single crystals were used to increase the area-to-volume ratio. Groves and Kelly [23]

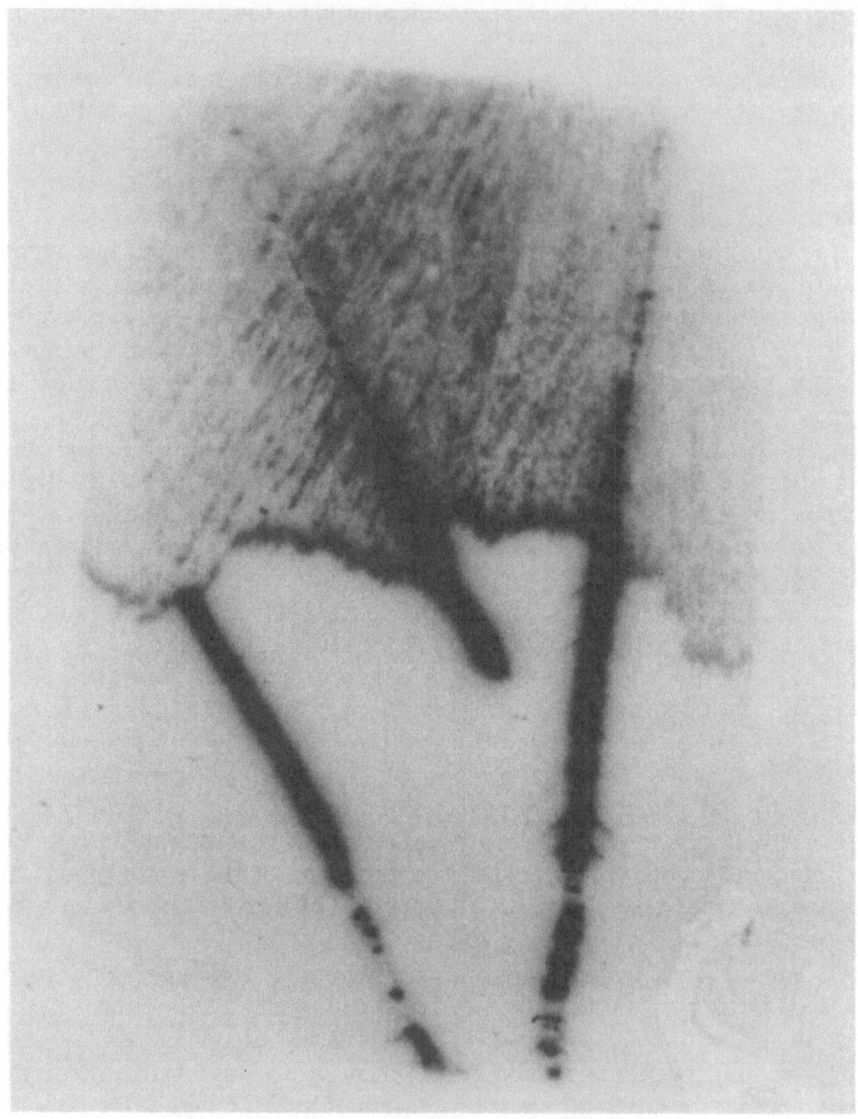

Fig. 5. MgO crystal, with three lines impressed on the surface by a ball point pen, which was annealed in oxygen-18, beveled, and irradiated with protons. The autoradiograph of this crystal is shown above.

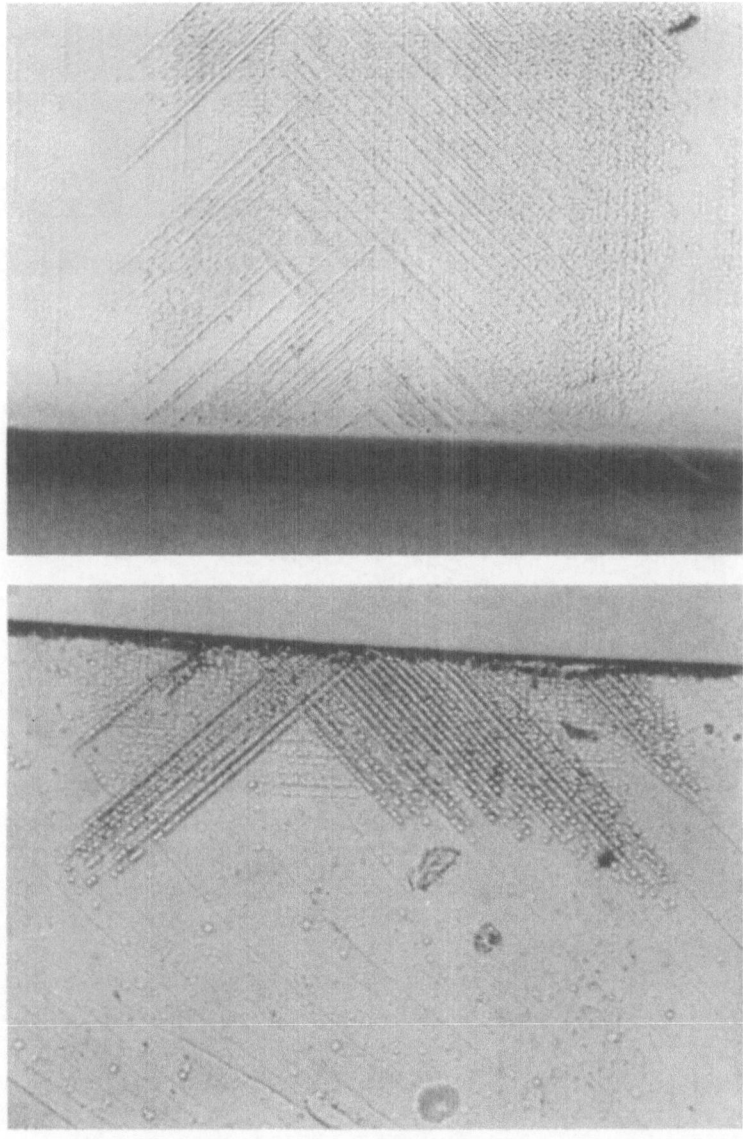

Fig. 6. The micrographs show the etched surface (top) and cross section (bottom) of a MgO single crystal in which dislocations were generated by pressure from a ball point pen.

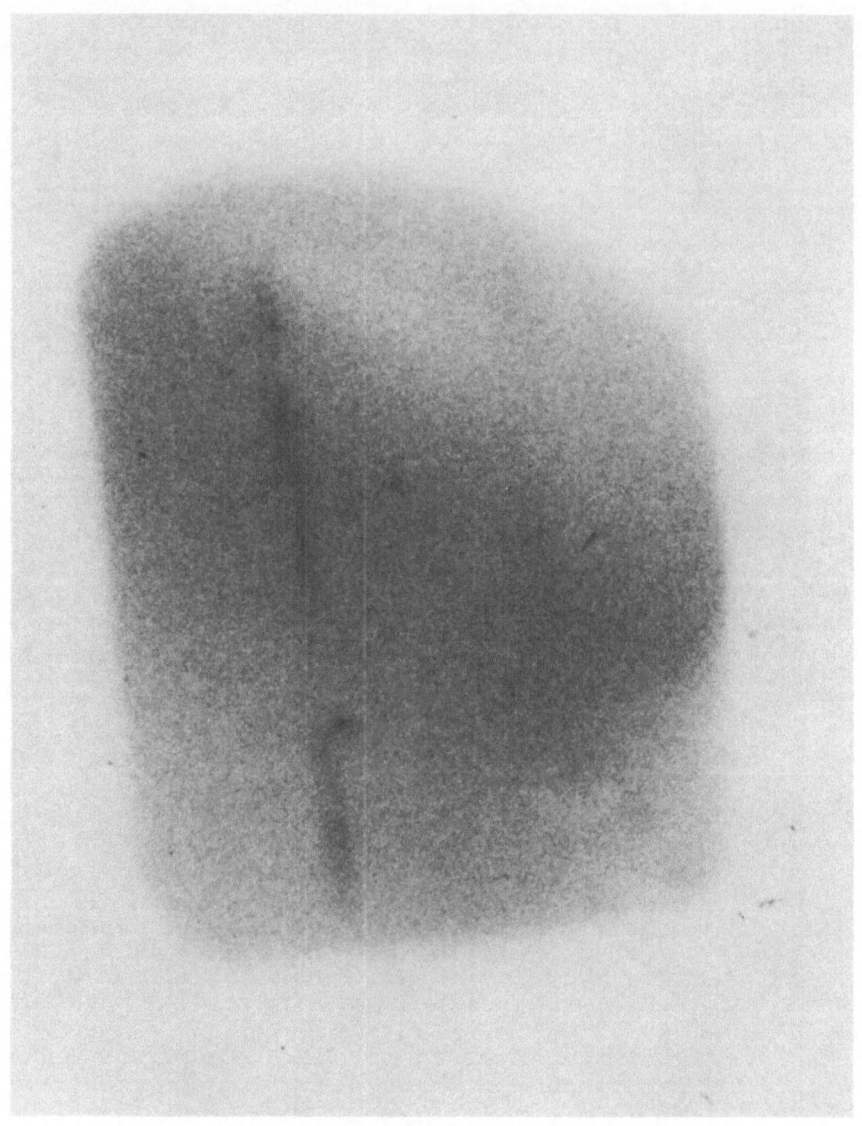

Fig. 7. This autoradiograph displays the effect of dislocations, introduced at the surface, on the diffusion of oxygen-18 in a MgO single crystal.

suggest that oxygen diffusion may have been affected by diffusion along dislocations in the case of Oishi and Kingery's [18] work with magnesium oxide where their specimens were undoubtedly in a cold-worked condition. It was to examine this contingency that the preceding experimental approach was taken. Dislocations of this type are known to anneal out at high temperatures in a fairly short time, but the question remained as to whether in this time the oxygen could have moved down these line defects.

Figure 5 is the autoradiograph taken on AR 10 stripping film of one of the crystals on which three lines had been impressed and then annealed in oxygen-18 at 1630°C for 6 hr. The beveled area lies to the bottom, and there is a $50-\mu$ drop from the line of bevel to the crystal edge. Close examination of the autoradiograph shows that the lines are composed of numerous small fractures, and the darkening on the autoradiograph is due to oxygen diffusion down these fractures. This obvious conclusion, however, emphasizes a fact which heretofore has not been mentioned in diffusion studies dealing with crushed oxide crystals. Undoubtedly, such material contains a significant number of these fractures, and, in the study of magnesium oxide [18], this would give rise to a surface area larger than that used for computing the diffusion coefficients. Although this factor may not change the activation energy, it will affect the intercept term D_0, which means a lower value of D_v. For this reason, along with the extremely small penetration in the bicrystals observed in the autoradiographs, we concluded that D_v at 1650°C was at least 10^{-14} cm^2/sec.

Since the fractures were obscuring any effect of the dislocations, a crystal was prepared in the usual manner, except that the surface on which the lines had been placed was gently ground down at least 50μ with 2/0 paper and another 20μ was removed by chemically polishing with hot orthophosphoric acid. This new surface and a cross section were etched to verify the complete removal of the fractures and to reveal the dislocations, as shown in Fig. 6. The dislocations extended down 70 to 100μ as measured from the cross section. The etched surface was chemically polished again for a short time, and the crystal was annealed in oxygen-18 and irradiated with the resulting autoradiograph appearing in Fig. 7. The vestiges of the dislocation array are still visible but only superficially, as the oxygen diffused down the dislocations barely 2 to 3μ. The dislocations make only a minor contribution, as compared to grain boundaries, to the overall rate of oxygen diffusion; however, they may be of greater significance at short times and lower temperatures.

In a separate test, cleaved pieces of magnesium oxide 1 in. long by $\frac{1}{4}$ in. wide were manually bent in an oxygen-gas flame and then annealed at 1700°C in dry air for 72 hr to polygonize the dislocations generated by the mechanical working. The crystal sections were prepared so the orientations of the long sides were either in the <100> or <110> directions. Since the density of dislocations was increased almost threefold by this method, the optical density of autoradiographs of these crystals should be increased by a similar factor if such dislocations serve as pipes for the rapid diffusion of oxygen. The results were inconclusive, however, and show the need for further refinement of the technique for this type of study.

SUMMARY

Tracing the transport of oxygen in oxides by the proton activation of O^{18} to give the radioactive isotope F^{18} is seen to be entirely feasible. The importance of surface defects on the overall rate of oxygen diffusion in MgO was firmly established by this new technique. At operating temperatures between 1600–1670°C, the volume diffusion coefficient of oxygen in MgO was very small, giving penetration distances less than 2μ for times up to 17 hr. Preferential diffusion of oxygen down grain boundaries occurred quite readily, particularly when a second phase impurity was segregated along the boundary. The roles of impurities and misorientation of the boundaries in grain boundary diffusion phenomena are not clearly understood, mainly due to insufficient experimental data. Any attempt to analyze high-temperature, solid-state reactions in polycrystalline magnesium oxide must take into account the rapid diffusion of oxygen along these boundaries. By the same token, further investigation is necessary to evaluate volume diffusion coefficients of oxygen in magnesium oxide, since the presence of fractures and dislocations influences any diffusional study dealing with crushed single-crystal material.

ACKNOWLEDGMENT

The authors wish to thank the 90-in.-cyclotron staff at the Lawrence Radiation Laboratory for their help in carrying out these experiments. The work was performed under the auspices of the U. S. Atomic Energy Commission, Contract No. W-7405-eng-48.

REFERENCES

1. R. H. Condit and J. B. Holt, J. Electrochem. Soc. 111:1192 (1964).
2. J. F. Laurent and J. Bénard, Compt. Rend. 241:1204 (1955); and J. F. Laurent and J. Bénard, Phys. Chem. Solids 7:218 (1958).
3. Y. Oishi and W. D. Kingery, J. Chem. Phys. 33:905 (1960).
4. A. E. Paladino, E. A. Maguire, and L. G. Rubin, J. Am. Ceram. Soc. 47:280 (1964).
5. T. Y. Tien, J. Appl. Phys. 35:122 (1964).
6. A. E. Paladino and R. L. Coble, J. Am. Ceram. Soc. 46:133 (1963).
7. R. L. Coble, J. Appl. Phys. 34:1679 (1963).
8. S. Yukawa and M. J. Sinnott, Trans. AIME 203:996 (1955).
9. D. Turnbull and R. E. Hoffman, Acta Met. 2:419 (1954).
10. C. Leymonie and P. Lacombe, Rev. Met. 54:653 (1957).
11. J. Cabañe, J. Chem. Phys. 59:1123 (1962).
12. J. C. Fisher, J. Appl. Phys. 22:74 (1951).
13. R. T. P. Whipple, Phil. Mag. 45:1225 (1954).
14. H. S. Levine and C. J. MacCallum, J. Appl. Phys. 31:595 (1960).
15. T. Suzuoka, Trans. Japan Inst. Metals 2:25 (1961).
16. S. M. Klotsman and A. N. Orlov, Issled. po Zharoproch. Splavam, Akad. Nauk SSSR, Inst. Met. 4:90 (1959).
17. A. D. LeClaire, Brit. J. Appl. Phys. 14:351 (1963).
18. Y. Oishi and W. D. Kingery, J. Appl. Phys. 33:905 (1960).
19. R. J. Stokes, T. L. Johnston, and C. H. Li, Phil. Mag. 3:718 (1958).
20. R. J. Stokes, T. L. Johnston, and C. H. Li, Trans. AIME 215:437 (1959).
21. A. S. Keh, J. Appl. Phys. 31:1538 (1960).
22. J. B. Holt, J. Nucl. Mater. 11:107 (1964).
23. G. W. Groves and A. Kelly, J. Appl. Phys. Suppl. 33:456 (1962).

Chapter 3

Properties of Grain Boundaries in Spinel Ferrites

Max Paulus

Laboratoire de Magnétisme et de Physique du Solide
Centre National de la Recherche Scientifique
Bellevue (S. & O.), France

It is shown that all ferrites with the spinel structure should have approximately the same value of the activation energy for grain boundary migration, irrespective of their composition, provided migration of grain boundaries is not hindered by voids or inclusions. Furthermore, the mean free energy λ of the boundaries (as calculated from the kinetics of crystal growth) doubles when a reducing atmosphere is replaced by an oxidizing one. The transition temperature between the two values of λ varies with oxygen partial pressure. Special attention is paid to the equilibrium segregation of some ions at the grain boundaries of spinel ferrite during cooling. This leads to a displacement of the reduction–oxidation equilibrium at the grain boundary and changes the resistivity and eddy-current losses.

INTRODUCTION

In our first works on the magnetic and electric properties of ferrites with a spinel structure of the general formula $2Fe^{+3}$, Fe^{+2}, $(1 - a)\,Me^{+2}$, $4\,O^{-2}$ where Me is a bivalent ion, such as Ni^{+2}, Zn^{+2}, or Mn^{+2}, we have shown [1—3] that the microstructure has a great influence on magnetic and electric properties. The permeability is strongly dependent on the grain size of the aggregate, and the eddy-current losses can be reduced by a factor greater than ten by adding a small amount of calcium to the ferrite.

Some of these ferrites have a great importance for telecommunication devices because of their high magnetic permeability in weak fields joined with low eddy-current losses. Therefore, we studied the grain growth, the grain boundary energy, and the segregation of Ca^{+2} and Y^{+3} at the grain boundary of various ferrites, as functions of temperature, time, atmosphere of sintering, and rate of cooling and we studied their consequences on the reduction–oxidation equilibrium at grain boundaries and on their electric properties.

DEFINITION OF A POLYCRYSTALLINE AGGREGATE OF FERRITE [4]

Given that the diameter of ferrite crystals in the polycrystalline aggregates of ferrites prepared without prefiring and without abnormal recrystallization are lognormally distributed, we define the aggregate by the median $\bar{d}_3$ expressed in microns or in decibels $\bar{\bar{z}}_3 = 10 \log (\bar{d}_3/d_0)$ where $d_0 = 1 \mu$, and by the standard deviation of the crystal diameters

$$\sigma_3 = \sqrt{\frac{\Sigma (z_3 - \bar{\bar{z}}_3)^2}{N}} \tag{1}$$

where σ_3 is the standard deviation expressed in decibels, $\bar{\bar{z}}_3$ is the mean of z_3, and N is the number of measured crystals.

GROWTH OF FERRITE CRYSTALS WITHIN POLYCRYSTALLINE AGGREGATES

The specimens of ferrite are prepared by wet-mixing of the starting oxides (Fe_2O_3, ZnO, Mn_3O_4, or NiO) in a ball mill. The mixture is dried and toroids are pressed at 5000 kg/cm². The pressed products are sintered at a temperature between 1125 and 1425°C for various times. The time range covered was 1–30 hr.

In Fig. 1—relative to a nickel–zinc ferrite (50 Fe_2O_3, 15 NiO, 35 ZnO)—the square of the median crystal diameter is plotted versus time of sintering for different temperatures. Figure 1 shows that for the first stages of sintering, where the square of the median crystal diameter is a linear function of sintering time, the growth of ferrite crystals within the polycrystalline aggregates of ferrites is represented by the following relation [5]:

$$\bar{\bar{d}}^2_{3(t_1)} - \bar{\bar{d}}^2_{3(t_0)} = \frac{22 \, \nu \sigma_3 \, \phi^4 \, \lambda}{KT \cdot 10^{0.6\sigma_3}} (t_1 - t_0) \exp - \frac{Q}{RT} \tag{2}$$

$$\bar{\bar{d}}^2_{3(t_1)} - \bar{\bar{d}}^2_{3(t_0)} = \frac{K_0}{T} (t_1 - t_0) \exp - \frac{Q}{RT} \tag{3}$$

where Q is the activation energy for the ionic jumps, T is the temperature of sintering, t is the time of sintering, $\bar{d}_3$ is the median of the crystal diameters, σ_3 is the standard deviation of the crystal

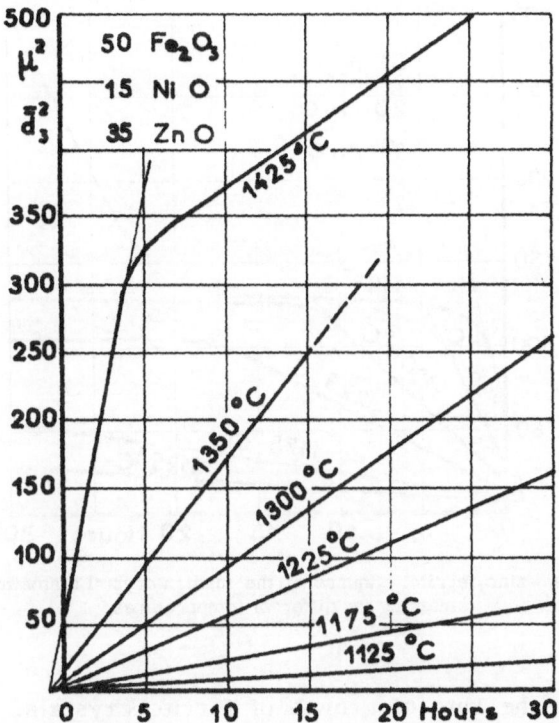

Fig. 1. Nickel–zinc ferrite.   Square of the median crystal diameter versus time of
sintering for different temperatures.

diameters, ϕ is the ionic diameter (oxygen), ν is the frequency of
ionic vibration, λ is the specific grain boundary energy, $K_0/T = 6.5$
cm/sec when $\theta > \theta_{tr}$, $K_0/T = 15.0$ cm/sec when $\theta < \theta_{tr}$, and $Q = 83,000$
cal/mole.

This relation is valid as long as the migration of grain boundaries is not hindered by inclusions or voids. In all cases where
the rate of crystal growth is not in agreement with this equation,
we were able to observe the presence of inclusions or voids which
hindered the displacement of grain boundaries [6]. This is the case
for the manganese–zinc ferrite (Fig. 2), where the limiting grain
size is about 10μ at which value the crystal growth stops due to
the voids.

Since diffusion of interstitial cations is more rapid than that
of oxygen ions [7], diffusion of the latter should have a decisive

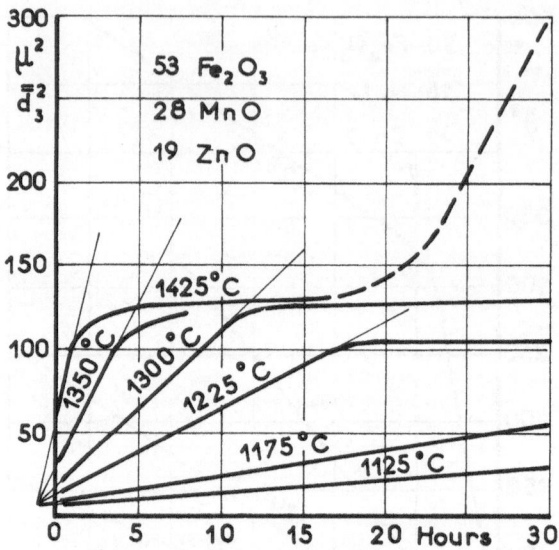

Fig. 2. Manganese—zinc ferrite. Square of the median crystal diameter versus time of
sintering for different temperatures.

influence in the rate of growth of ferrite crystals. As a result,
the rate of growth must be the same for all ferrites with spinel
structure provided migration of grain boundaries is not hindered
by voids or inclusions. The identity of experimental values of the
pre-exponential coefficient K_0/T and of the activation energy Q for
various ferrites is a clue to this assertion.

For example, Fig. 3 shows the square of the median of the
crystal diameter as a function of the inverse of the sintering temperature, for a sintering time of 4 hr. It appears that the experimental points for manganese—zinc ferrite sintered in a nitrogen
atmosphere with 1% oxygen and for nickel—zinc ferrite sintered in a
pure oxygen atmosphere·are represented by the same curve. The
value of the activation energy $Q \approx 83,000$ cal/mole should be considered as an activation energy of the self-diffusion of oxygen ions
along the grain boundaries between 1125 and 1425°C.

The transition temperature θ_{tr} between the two values of the
pre-exponential coefficient, $K_0/T \approx 6.5$ cm/sec for the higher temperatures and 15.0 cm/sec for the lower temperatures, corresponds
to the temperature at which the partial pressure of oxygen is just
equal to the equilibrium oxygen pressure of the stoichiometric

ferrite. When the partial pressure of oxygen varies, the transition temperature between the two values of K_0/T varies with the equilibrium temperature of the ferrite. Figure 3 shows three curves for a manganese–zinc ferrite sintered in atmospheres of nitrogen containing 0.5, 1, and 5% oxygen. Therefore, the rate of crystal growth doubles when a reducing atmosphere is replaced by an oxidizing one.

FREE ENERGY OF THE GRAIN BOUNDARY

All terms implicitly involved in the pre-exponential coefficient K_0/T [equation (2)], with the exception of the free energy of the boundary λ, can be taken as constant at the transition temperature. Therefore, the value of the boundary free energy as calculated from the kinetics of crystal growth doubles when a reducing atmosphere is replaced by an oxidizing one.

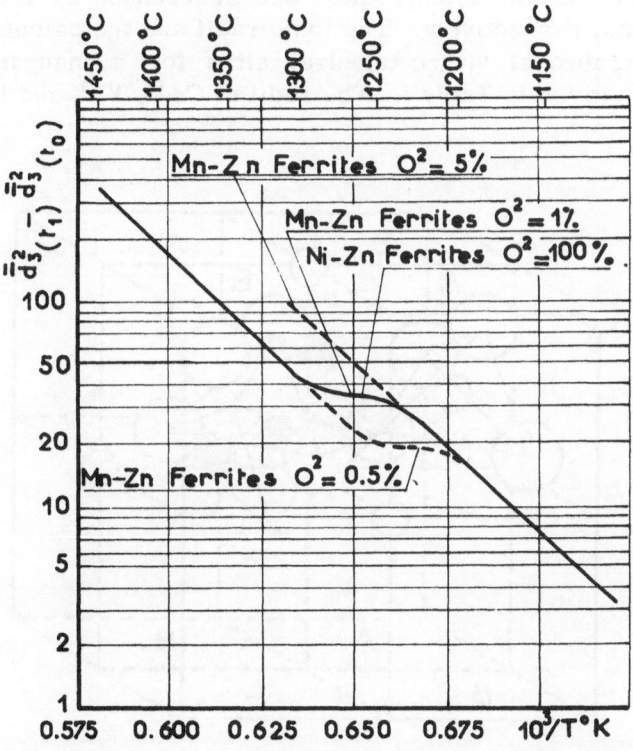

Fig. 3. $\bar{d}^2_{3(t_1)} - \bar{d}^2_{3(t_0)}$ as a function of $1/T$ °K. Time of sintering, 4 hr.

The apparent increase of the grain boundary energy may be attributed either to a real increase of the distortion at the grain boundaries or to a decrease of the lattice energy in the crystal. In order to make a choice between these two assumptions, we are studying the intensity of the X-ray scattering background for the same ferrite, sintered at the same grain size, either in an oxidizing or reducing atmosphere. The same specimens also are being observed by transmission electron microscopy. This problem has not yet been solved completely.

EQUILIBRIUM SEGREGATION OF Ca^{+2} AT THE CRYSTAL BOUNDARIES

In the spinel lattice, the relatively large oxygen ions form a face-centered cubic lattice (Fig. 4). In this cubic close-packed structure, two kinds of interstitial sites occur, the tetrahedral and the octahedral sites, which are surrounded by four and six oxygen ions, respectively. The ionic radii and the calculated radii of the tetrahedral and octahedral sites for a manganese–zinc ferrite are given in Table I. The radii of Ca^{+2}, Y^{+3}, and Mg^{+3} have

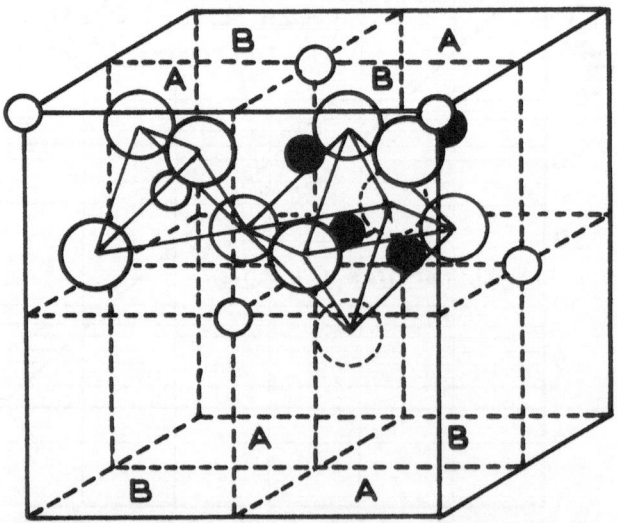

Fig. 4. Arrangement of cations in the spinel lattice. Only two octants of the spinel structure are represented. The large spheres represent the oxygen ions. The small black and white spheres represent the cations on tetrahedral and octahedral sites, respectively.

TABLE I

Ionic Radii, Calculated Site Radii, and Distortion Energy for a Manganese–Zinc Ferrite (53 Fe_2O_3, 28 MnO, 19 ZnO mol.%)

Ions	R_G (A)	Distortion Energy (cal/g-ion)	
		tetragonal	octagonal
O^{-2}	1.32		
Fe^{+3}	0.67		
Fe^{+2}	0.83		
Zn^{+2}	0.83		
Mn^{+2}	0.91		
Ca^{+2}	1.06	12,700	2,800
Y^{+3}	1.06	25,200	5,800
Mg^{+2}	0.78	3,700	0

Values for R for the tetragonal and octagonal sites are 0.52 and 0.80 A, respectively.

been added for comparison. It is seen that the ionic radii of the Ca^{+2} and Y^{+3} are the same and far bigger than those of either the tetrahedral or octahedral sites.

It is obvious that the addition of a small amount of these large cations will introduce a distortion energy in the grain interior which can be estimated from elastic theory. The equation for the distortion energy (in erg/g-ion) caused by a small sphere placed in a spherical hole in a block of material large compared with the sphere is given by McLean [8]:

$$Q = \frac{24 \pi KGr^3 \left[(r_1 - r_0)/r_1 \right]^2 N_0}{3K + 4G} \tag{4}$$

where K is the bulk modulus of the added ion, r_0 is the radius of the unoccupied site, r_1 is the radius of the isolated ion, r is the radius of the ion in situ, G is the shear modulus of the matrix, and N_0 is Avogadro's number.

The distortion energies caused by Ca^{+2} and Y^{+3} in the spinel lattice of the manganese–zinc ferrite are very high, even in the large octahedral site.   On the other hand, addition of Mg^{+2}, which has the same valence as Ca^{+2}, introduces practically negligible distortion energy (Table I).

The effect of this energy is shown by the distortion of the spinel (311) X-ray diffraction peak.   Curve I of Fig. 5 illustrates the normal peak given by: pure spinel when quenched or slowly cooled,

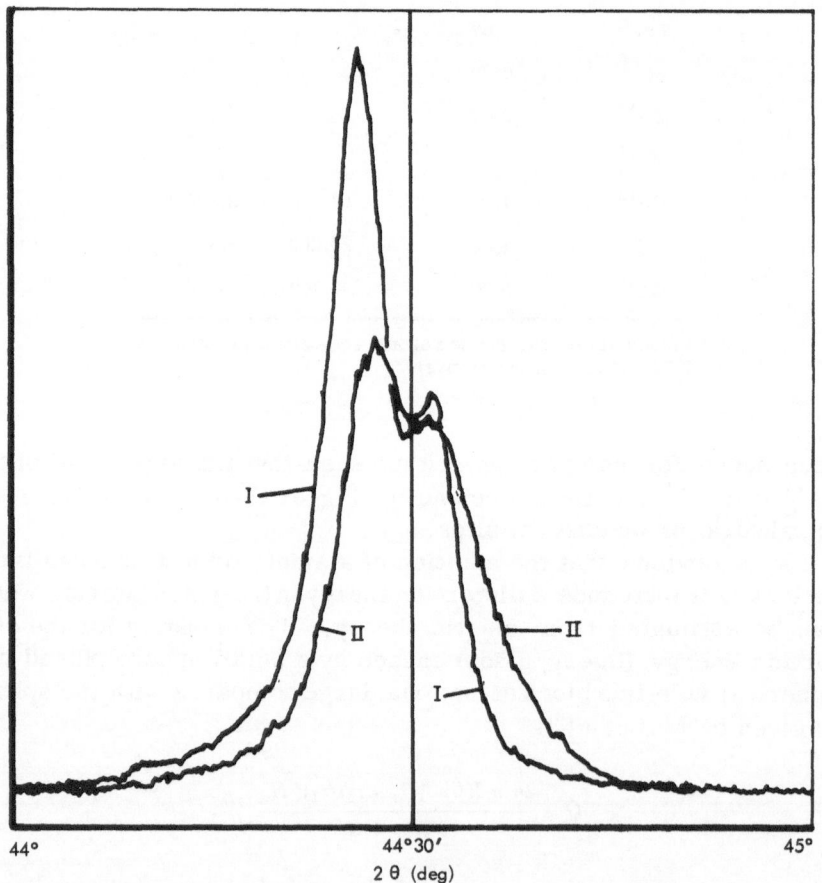

Fig. 5. X-ray diffraction line (311) for a manganese–zinc ferrite (53.0 Fe_2O_3, 28 MnO, 19 ZnO).   Curve I — Pure ferrite or with 0.12 mol.% MgO quenched or slowly cooled and ferrite with 0.12 mol.% CaO or 0.06 mol.% Y_2O_3 slowly cooled; Curve II—Ferrite with 0.12 mol.% CaO or 0.06 mol.% Y_2O_3 quenched.

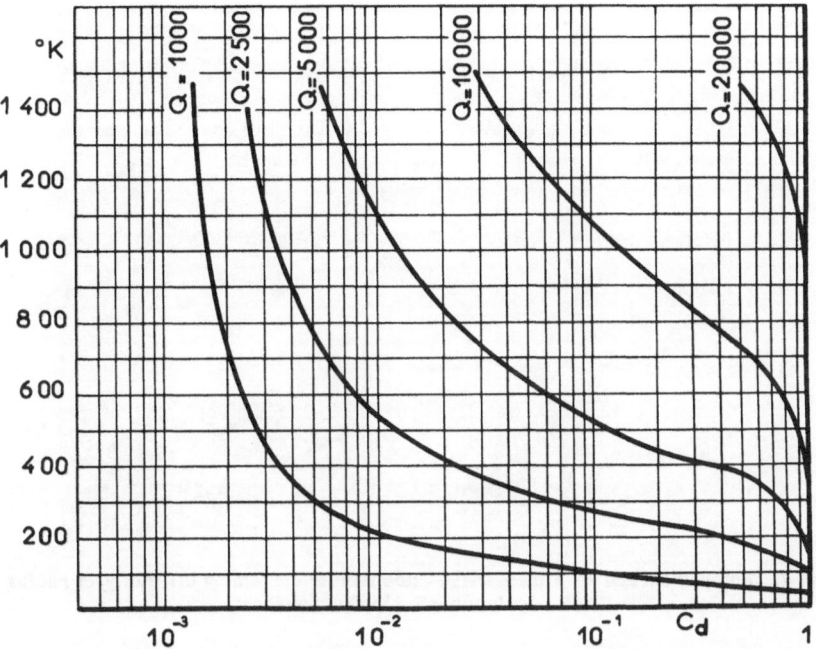

Fig. 6. Relation between grain boundary concentration and temperature for different values of distortion energy Q in cal/mole in each case for a lattice concentration of 0.1% of deforming ions.

spinel containing Mg^{+2} both quenched or slowly cooled, and spinel containing Ca^{+2} when slowly cooled. Curve II shows the distortion and shifting of the peak by Ca^{+2} or Y^{+3} when the spinel is quenched.

Since the introduction of Ca^{+2} and Y^{+3} in the spinel lattice entails an important increase in the lattice distortion energy, these ions have to segregate toward the boundaries during the cooling process. The grain boundary concentration is sensitive to the magnitude of the distortion energy. At equilibrium, the relative concentration of deforming ions at the boundaries increases with the distortion energy and decreases with increase in temperature (Fig. 6). A high resolution autoradiograph of a ferrite containing 0.1 mol.% of radioactive calcium-45 slowly cooled (Fig. 7a) supports this view. Figure 7b is a micrograph of another part of the same sample after etching. Figures 8a and 8b are two micrographs of etched ferrites with and without calcium, slowly cooled. We observe grooves at the boundaries of ferrites containing 0.1% of

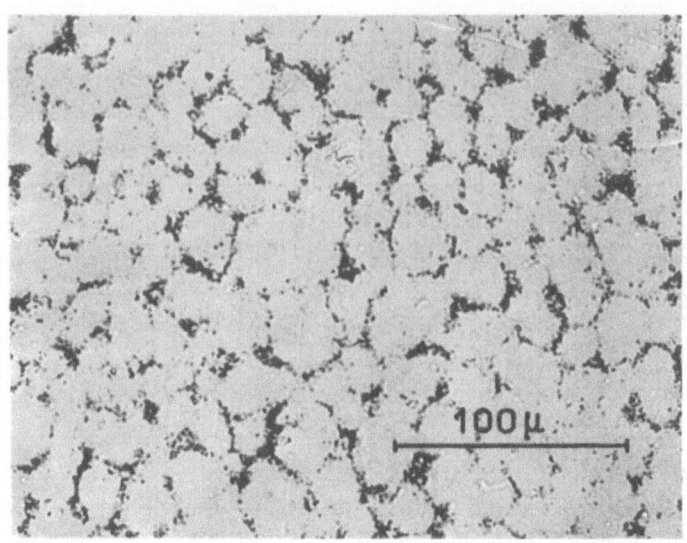

Fig. 7a. Autoradiograph of a manganese–zinc ferrite containing 0.1 mol.% of radioactive calcium-45, slowly cooled.

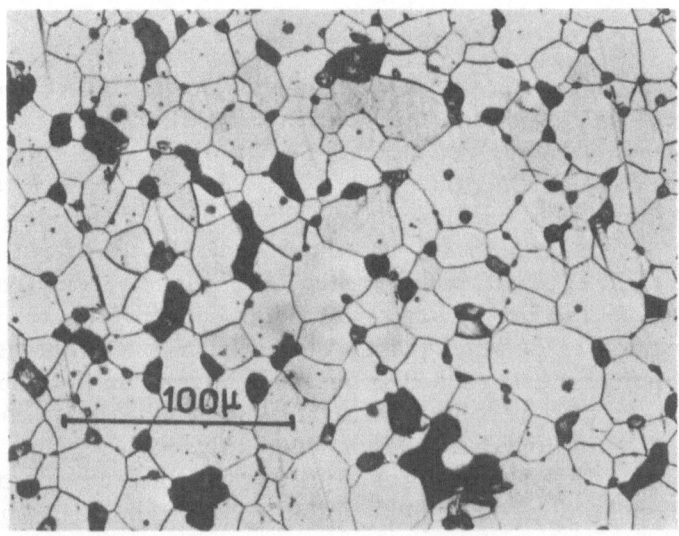

Fig. 7b. Micrograph after etching of another part of the sample in Fig. 7a.

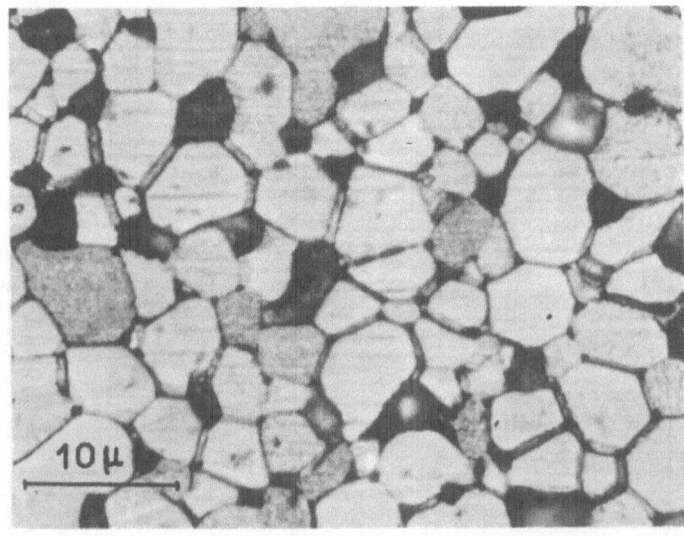

Fig. 8a. Micrograph of a sample of ferrite (53 Fe_2O_3, 28 MnO, 19 ZnO) with 0.3 mol.% $CaCO_3$. Etching time, 3 min; reagent, 75% HCl–25% C_2H_5OH; temperature, 20°C.

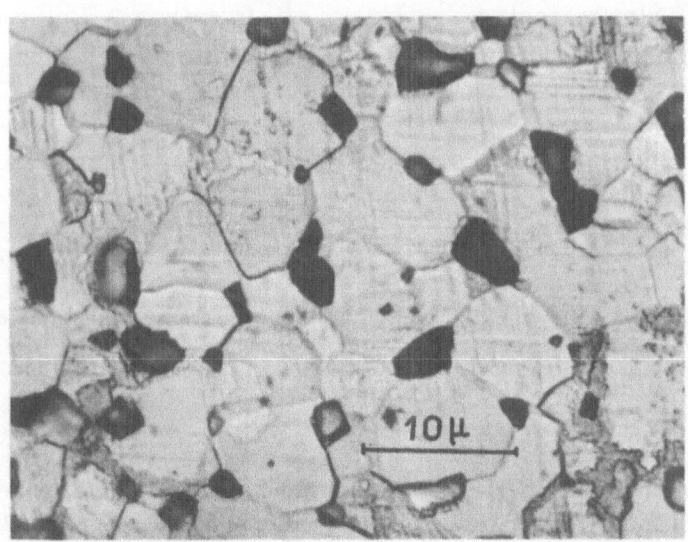

Fig. 8b. Micrograph of a sample of ferrite as in Fig. 8a, but without any addition of calcium.

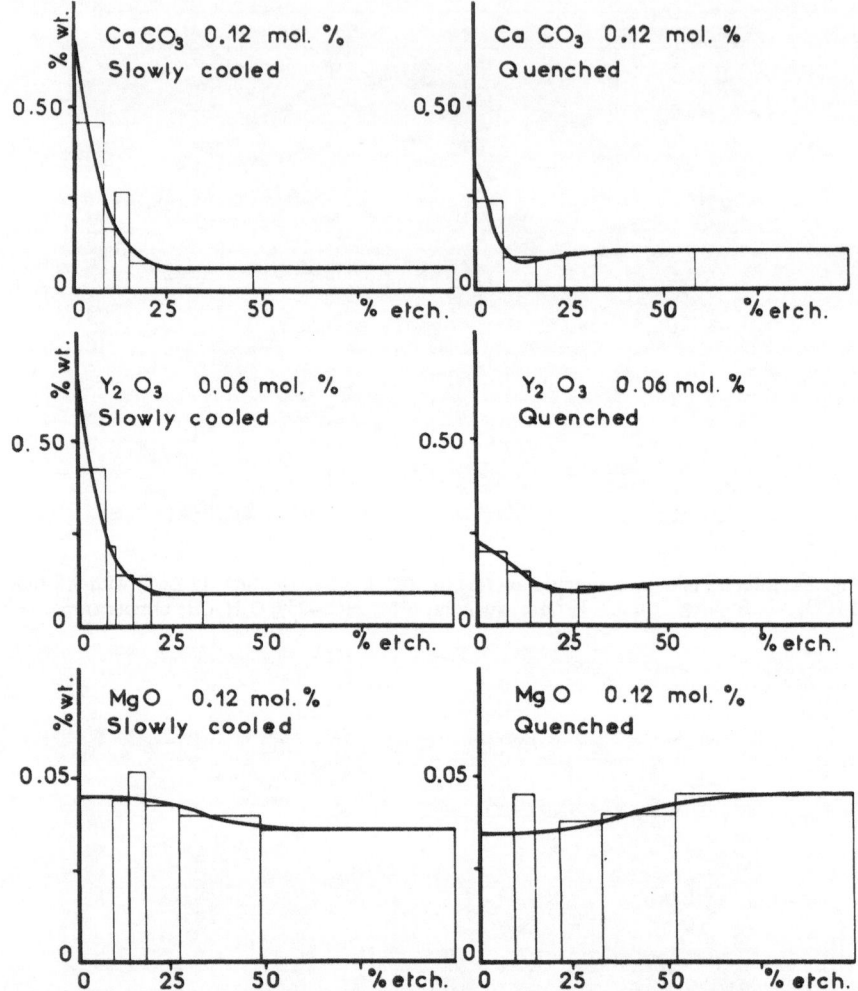

Fig. 9. Chemical analysis of calcium, yttrium, and magnesium in a solution obtained by successive chemical etchings on ferrites ground down to a fineness comparable to the average grain size. The ion content is plotted as a function of the etching percentage of the ferrite powder.

calcium, but not in the ferrite without calcium. The same differences in metallographic effect at grain boundaries appear also after ionic bombardment.

These results are confirmed by chemical analysis whereby the calcium, yttrium, and magnesium content was estimated in solution obtained by successive chemical etchings on ferrites ground down to a fineness comparable to the average crystal size. Figure 9 represents the calcium, yttrium, and magnesium content of quenched ferrites and that of those slowly cooled as a function of the etching percentage of the ferrite powder. These curves show well the segregation of calcium and yttrium during cooling. The valence of yttrium differs from that of calcium, but the deformation energy associated with its presence in the spinel lattice is even more important than that of calcium. The estimation of yttrium content in solutions obtained by successive chemical etching shows that the behavior of yttrium is the same as that of calcium, as far as segregation is concerned. On the other hand, addition of magnesium, which has the same valence as calcium, but is practically without deformation energy, has not given rise to any equilibrium segregation at crystal boundaries.

REDUCTION – OXIDATION EQUILIBRIUM AT THE BOUNDARIES AND IN THE CRYSTALS

The segregation at the grain boundaries of ferrites modifies the reduction–oxidation equilibrium. Modifications due to calcium segregation at the boundaries can be considered to take place by three different processes.

Process A

Calcium ions segregate toward boundaries simultaneously with the back diffusion of another ion with low deformation energy, namely, the iron ion, moving into the crystals. As the number of divalent ions increases, the equilibrium requires a decrease in the number of Fe^{+2} ions. At the grain boundary, one may write

$$2Fe^{+3}, \alpha Fe^{+2}, \delta Ca^{+2}, (1 - \alpha - \delta) Me^{+2}, 4O^{-2} + \beta Ca^{+2} - \beta Fe^{+2} \longrightarrow$$

$$2Fe^{+3}, (\alpha - \beta) Fe^{+2}, (\delta + \beta) Ca^{+2}, (1 - \alpha - \delta) Me^{+2}, 4O^{-2} \quad (5)$$

Process B

Calcium ions segregate toward boundaries without back diffusion, but with oxidation of the boundaries. The decrease in the number of Fe^{+2} ions at the boundaries may be written as

$$2Fe^{+3}, \alpha Fe^{+2}, \delta Ca^{+2}, (1 - \alpha - \delta) Me^{+2}, 4O^{-2} + \beta Ca^{+2} + \tfrac{2}{3}\beta O^{-2} \longrightarrow$$

$$(2 + \tfrac{2}{3}\beta) Fe^{+3}, (\alpha - \tfrac{2}{3}\beta) Fe^{+2}, (\delta + \beta) Ca^{+2}, (1 - \alpha - \delta) Me^{+2}, (4 + \tfrac{3}{4}\beta) O^{-2} \quad (6)$$

It can be noticed that in the inner parts of the crystals, there is also a lowering of the Fe^{+2} content, giving rise to the formation of divalent cation vacancies $\square^{+2}$:

$$2Fe^{+3}, \, aFe^{+2}, \, \delta Ca^{+2}, \, (1 - a - \delta) \, Me^{+2}, \, 4O^{-2} - \epsilon Ca^{+2} \longrightarrow$$

$$(2 + 2\epsilon) \, Fe^{+3}, \, (a - 2\epsilon) \, Fe^{+2}, \, (1 - a - \delta) \, Me^{+2}, \, (\delta - \epsilon) \, Ca^{+2}, \, \epsilon \, \square^{+2}, \, 4O^{-2} \quad (7)$$

Process C

The segregation occurs with neither back diffusion nor oxidation of the grain boundary. The situation in the crystals is the same as in process B; however, at the boundaries, the Fe^{+2} content increases as β calcium ions go into interstitial positions.

$$2Fe^{+3}, \, aFe^{+2}, \, \delta Ca^{+2}, \, (1 - a - \delta) \, Me^{+2}, \, 4O^{-2} + \beta Ca^{+2} \longrightarrow$$

$$(2 - 2\beta)Fe^{+3}, \, (a + 2\beta) \, Fe^{+2}, \, (1 - a - \delta) \, Me^{+2}, \, \delta Ca^{+2}, \, \beta Ca^{+2} \bullet, \, 4O^{-2} \quad (8)$$

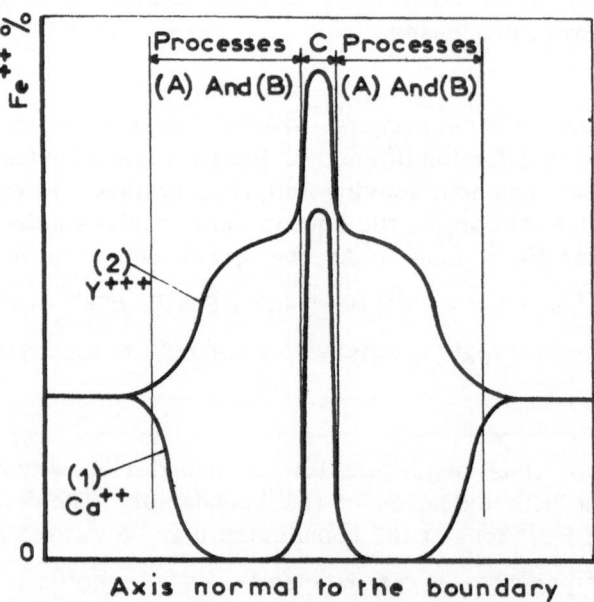

Fig. 10. Schematic representation of the number of Fe^{+2} ions across the boundary.

While processes A and B lead to a decrease in the number of Fe^{+2} ions at the grain boundary, process C occurs at the very center of the boundary,. where the calcium content can reach such high values (about 100 times the mean value) that oxidation and back diffusion are negligible.

In the close neighborhood of the boundary, the Ca^{+2} content is less high and processes A and B reduce the number of Fe^{+2} ions. Figure 10 is a schematic representation of the variation of the number of Fe^{+2} ions through a boundary. In the case of Y^{+3} addition, the three processes lead to an increase of the number of Fe^{+2} ions at the boundary.

CHANGES IN RESISTIVITY AT THE GRAIN BOUNDARIES BY SEGREGATION OF Ca^{+2}, Mg^{+2}, or Y^{+3}

It is well known that the conductivity of ferrites is caused, in particular, by the simultaneous presence of ferrous and ferric ions on equivalent lattice sites (octahedral sites). The double layer, practically free of ferrous ions in the presence of calcium segregation, should possess a resistivity of $10^4-10^5 \Omega$/cm corresponding to the resistivity of a ferrite with some deficiency in iron.

Resistivity micromeasurements performed on the very core of the ferrite crystals of polycrystalline aggregates have made it possible to compute the ratio (ρ_z/ρ_g) of the resistivity of the polycrystalline aggregate ρ_z to that of the crystal ρ_g. Figure 11 diagrammatically summarizes numerous measurements of the ratio (ρ_z/ρ_g) which confirms our assertions concerning the proposed processes.

For pure ferrites (without any additions), values of the ratio (ρ_z/ρ_g) are the same whether they are quenched or slowly cooled in the presence of pure nitrogen. Slow cooling of the ferrite in the presence of an atmosphere containing 1% of oxygen increases the values of the ratio (ρ_z/ρ_g) by a factor of four. Resistivity increases by oxidation of the boundary. Results for Mg^{+2} additions are similar. The magnesium ions do not segregate at the boundaries.

The behavior of the ferrite to which calcium was added is completely different. Although after quenching the value of (ρ_z/ρ_g) is almost equal to that of a pure ferrite, cooling in pure nitrogen increases (ρ_z/ρ_g) by a factor of about ten. Segregation of calcium at the boundaries modifies the reduction–oxidation equilibrium on the lines of processes A and C. When cooling is carried

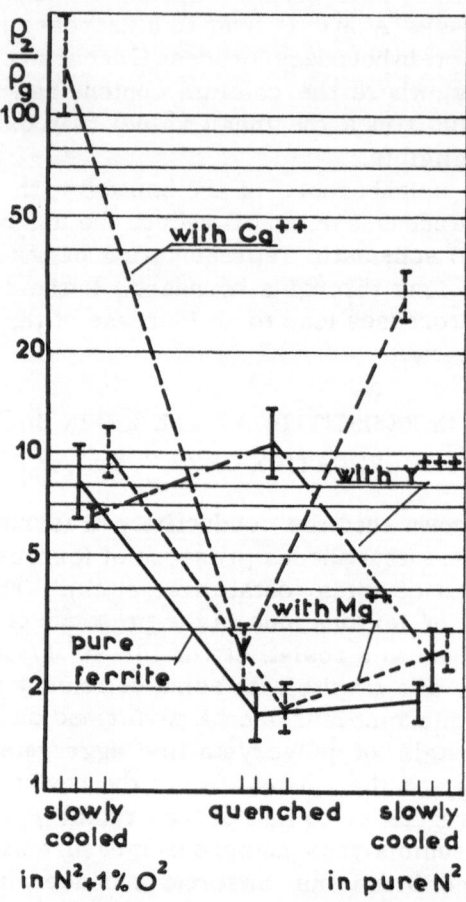

Fig. 11. Ratio (ρ_z/ρ_g) of the resistivity of the polycrystalline aggregate ρ_z to that of the crystal ρ_g versus thermochemical treatments for a pure ferrite, and for ferrites with additions of 0.12 mol.% CaO, 0.06 mol.% Y_2O_3, or 0.12 mol.% MgO.

out in presence of nitrogen containing 1% of oxygen, the ratio (ρ_z/ρ_g) is increased by a factor of about fifty. It is thus seen that the effect of oxygen was to initiate process B and also a simple oxidation with the formation of vacancies of divalent ions, as is the case for a pure ferrite.

Yttrium additions bring about opposite results. By slow cooling, the number of Fe^{+2} ions at the boundaries increases according to processes A and C which for yttrium proceed in the same sense and the ratio (ρ_z/ρ_g) decreases.

All these results show there is a good agreement between segregation, oxidation–reduction equilibrium, and the electric properties at the grain boundaries.

REFERENCES

1. C. Guillaud and M. Paulus, C. R. Acad. Sci. Fr. 21(242):2525–2528 (1956).
2. C. Guillaud, M. Paulus, and R. Vautier, C. R. Acad. Sci. Fr. 23(242):2712–2715 (1956).
3. C. Guillaud, G. Villers, A. Marais, and M. Paulus, Solid State Physics in Electronics and Telecommunication, Vol. 3, Academic Press (London), 1960, Part 1, pp. 71–90.
4. M. Paulus, Metaux Corrosion Ind. 448:447–468 (1962); and 449:14–34 (1963).
5. M. Paulus, Phys. Stat. Solidi 2(9):1181–1194 (1962).
6. M. Paulus, Phys. Stat. Solidi 2(10):1325–1341 (1962).
7. C. Wagner, Z. Physik. Chem. 34B:309–316 (1936).
8. D. McLean, Grain Boundaries in Metals, Clarendon Press (Oxford), 1957.

Chapter 4

The Kinetics of Grain-Boundary Groove Growth on Alumina Surfaces

Wayne M. Robertson and Roger Chang

North American Aviation Science Center
Thousand Oaks, California

The growth of grain-boundary grooves on polished surfaces of polycrystalline aluminum oxide has been studied over the temperature range 1100–1700°C. Groove widths were measured interferometrically after annealing in air for varying times. From the kinetics of groove growth, it is shown that diffusion on the crystal surfaces controls the process. The surface diffusion coefficient (in cm^2/sec) is given by the relation

$$D_s = (7 \pm 5) \cdot 10^2 \exp\left[-(75,000 \pm 5,000)/RT\right]$$

The surface diffusion coefficient at 1600°C is about 10^{-6} cm^2/sec, which is nearly six orders of magnitude larger than the lattice diffusion coefficient of aluminum ions and from seven to nine orders of magnitude larger than that of oxygen ions at this temperature. Lattice diffusion does not contribute appreciably to the formation of grain-boundary grooves. Possible mechanisms of surface diffusion on aluminum oxide are discussed.

INTRODUCTION

Thermal etching has proved to be a useful tool for studying the surface-related properties of crystalline materials. It has been used on aluminum oxide to obtain information about the dislocation structure of crystals [1], to reveal sub-boundaries [2], and to measure the relative grain boundary-to-surface energy [3]. A study of the kinetics of a thermal etching process can also yield useful information, although such studies have not been performed using alumina. The present paper reports a study of the kinetics of grain-boundary groove growth on polycrystalline aluminum oxide samples over a range of temperatures. This study yields information on the mechanism and kinetics (diffusion constants) of material transport in the grooving process, which may lead to an understanding and control of the surface structure and the surface-related properties of this material.

49

THEORY

The theory of grain-boundary groove formation driven by capillary forces has been developed by Mullins in a series of papers [4-7]. Exact solutions to the kinetic problems for groove formation were obtained by several transport mechanisms. In the analysis of these problems, the shape of the surface profile is used as a boundary condition for the kinetic problem. The transport processes that have been considered are: (1) surface diffusion over the crystal surface; (2) volume diffusion, either of defects in the crystal or of the crystal material in a surrounding medium; (3) evaporation and condensation of the crystal. A fourth possible mechanism is viscous flow of the material. The latter two mechanisms appear to be unimportant in the present case, because of the low vapor pressure and the crystalline nature of alumina.

Mullins' theory predicts that for the surface diffusion and volume diffusion processes a grain boundary–surface intersection, which was initially plane, will develop a shape similar to that shown in Fig. 1. The equilibrium between the grain boundary and surface tensions establishes an equilibrium angle at the root of the groove. Material is transported from the groove area and forms small humps beside the groove above the level of the original surface. The distance w between these two humps is taken as the width of the groove.

Mullins' theory predicts that the groove profile will have a constant shape which expands uniformly with time t in all of its linear dimensions. For a surface diffusion mechanism [4] the width w and depth d are given by

$$w = 4.6 \ (Bt)^{1/4} \tag{1}$$

and

$$d = 0.973 \ m \ (Bt)^{1/4}$$

where

$$B = D_s n \ \gamma \Omega^2 / kT \tag{1a}$$

and where γ is the surface free energy of the solid–gas interface, Ω is the atomic volume of the diffusing species, D_s is the surface diffusion coefficient of the diffusing species, n is the number of atoms per square centimeter of surface which take part in the diffusion process, m is the tangent of the root angle, and kT has the

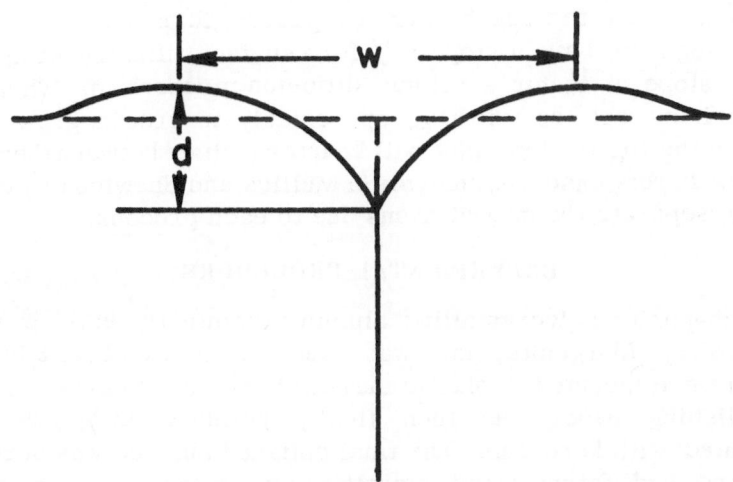

Fig. 1. Schematic profile of grain-boundary groove.

usual meaning. For a volume diffusion mechanism [5], the width and depth are given by

$$w = 5.0 \ (Ct)^{\frac{1}{3}} \tag{2}$$

and

$$d = 1.01 \ m \ (Ct)^{\frac{1}{3}}$$

where

$$C = D_v \ \gamma \Omega / kT \tag{2a}$$

and where D_v is the volume diffusion coefficient of the diffusing species and the other symbols denote the same quantities as above. For the evaporation–condensation and viscous flow mechanisms, the groove width will increase in proportion to the square root of time and the first power of the time, respectively.

There are several assumptions in the theory. The most important of these are: (1) The surface diffusion coefficient and surface free energy are independent of the crystallographic orientation of the surface; and (2) the slope of the surface with respect to the original flat surfaces is small. The applicability of these assumptions to the present case will be considered further in the discussion.

It can be seen that measuring the time dependence of the development of the linear dimensions of a groove can furnish information on the mechanism of groove formation. A plot of $\log w - \log t$ will have a slope of $\frac{1}{4}$ for a surface diffusion mechanism and a slope of $\frac{1}{3}$ for a volume diffusion mechanism. When both mechanisms are contributing appreciably to groove growth, the slope of the $\log w - \log t$ plot will be intermediate between these two values. In this case, the analysis of Mullins and Shewmon [7] can be used to separate the contributions due to each process.

EXPERIMENTAL PROCEDURE

High-purity polycrystalline alumina (nominally 99.7 + % Al_2O_3) supplied by Morganite, Inc. was used. Samples about 3 by 5 by 10 mm were mounted in plastic and polished through various grades of polishing paper and then finally polished on $\frac{1}{4}$-μ diamond lubricated with kerosene. The final polished surface was scratch-free and had fairly large smooth areas where grain-boundary grooves could be observed after annealing, though it contained quite a few pits and holes where pieces of the alumina had broken out during the polishing. After polishing, the sample was broken out of the mount and washed carefully in acetone to remove all traces of the mounting material and the kerosene from polishing.

The polished sample was placed in a covered high-purity alumina crucible and placed on a hearth in a Globar electric resistance furnace. The temperature was manually maintained constant to within ± 20°C by controlling the power input to the furnace. Temperatures were measured with a platinum—platinum-10% rhodium thermocouple. After annealing for a length of time, the crucible was removed from the furnace and cooled in air to room temperature. The sample was examined microscopically and then annealed for further periods at the same temperature.

The microscopic examinations were made with a Zeiss inter-ference microscope, which shows the contours of the groove on the surface. The grain-boundary grooves were either photographed on 35-mm film and enlarged in printing to about 750 × magnification, or photographed directly on Polaroid film at a magnification of about 350 ×. Measurements of groove widths were made on the enlarged micrographs.

RESULTS

Typical micrographs of grain-boundary grooves are shown in Fig. 2. The humps beside the grooves can be seen, as well as the sharp groove root.

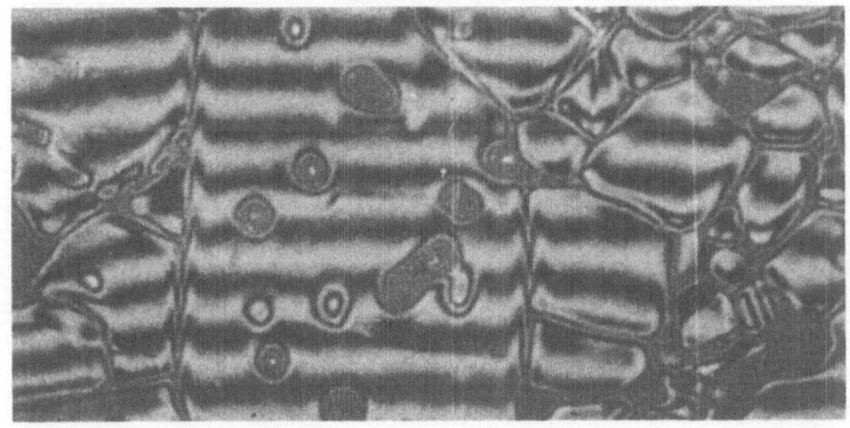

$\vdash\!\!\!-\!\!\!\dashv$ 10 μ

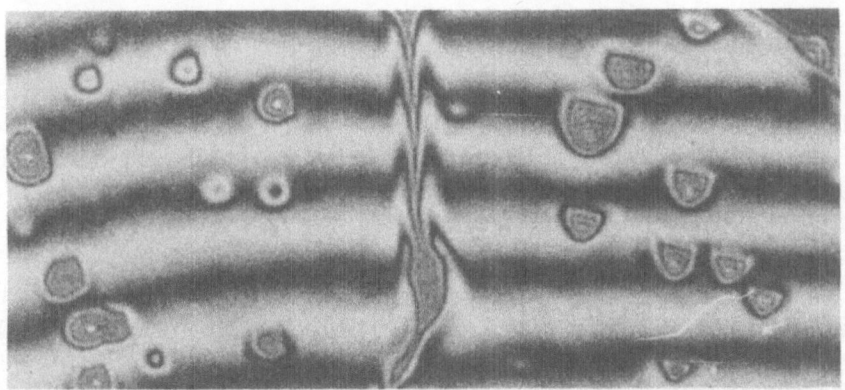

Fig. 2. Interference micrographs of grain–boundary grooves in aluminum oxide at 1500°C. Fringe spacing is 0.27 μ. Top—2.2 hr. Bottom—4.2 hr.

The grooves grew with time as predicted by Mullins' theories. Plots of log w versus log t are shown in Fig. 3 for a range of temperatures. Each point is the arithmetic average of the groove widths measured on four or more micrographs; the scatter in each point is about ±0.2 μ. The number near each line represents the value of its slope; the slope of the line equals the exponent of the time

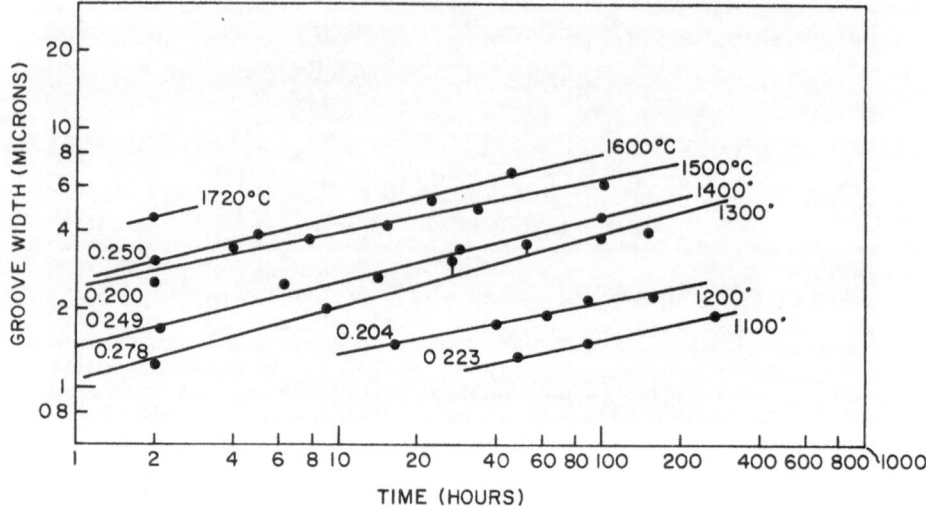

Fig. 3. Growth of grain-boundary grooves on aluminum oxide in air. Numbers give slopes of least-squares lines.

dependence of the groove width. The values of the slopes are scattered around a value of 0.25. From this value of the slopes, it is concluded that the mechanism of groove formation is surface diffusion over the crystal surfaces. Much of the scatter in the slopes is attributed to temperature variations during the annealing runs. Appreciable grooving was observed at temperatures as low as 1100°C, more than nine hundred degrees below the melting point of aluminum oxide.

From the measured values of width and time, equation (1) yields values for the constant B at each temperature. From equation (1a), it is possible to calculate values for the surface diffusion coefficient. The surface free energy of alumina has been measured at 1850°C by Kingery [3]; he obtained a value of 905 ergs/cm². With the assumption this is independent of temperature, and by use of the following values:

$$\Omega = 2.11 \cdot 10^{-23} \text{ cm}^3/\text{atom}$$

and

$$n = \Omega^{-2/3} = 1.31 \cdot 10^{15} \text{ atoms/cm}^2$$

the surface diffusion coefficient was calculated for each temperature. Figure 4 shows a plot of log D_s versus $1/T$. The solid line is given by the following relation for D_s (in cm^2/sec):

$$D_s = (7 \pm 5) \cdot 10^2 \exp\left[-(75{,}000 \pm 5{,}000)/RT\right] \qquad (3)$$

calculated by least-squares analysis. The errors given in the pre-exponential term and in the activation energy are those estimated by taking the extremes of the values of D_s. The probable errors calculated from the points of Fig. 4 are much less than the estimated errors given. It should be noted that the absolute value of the pre-exponential factor of D_s depends on the accuracy of the measurement of γ and on the value taken for Ω. It is a little unclear

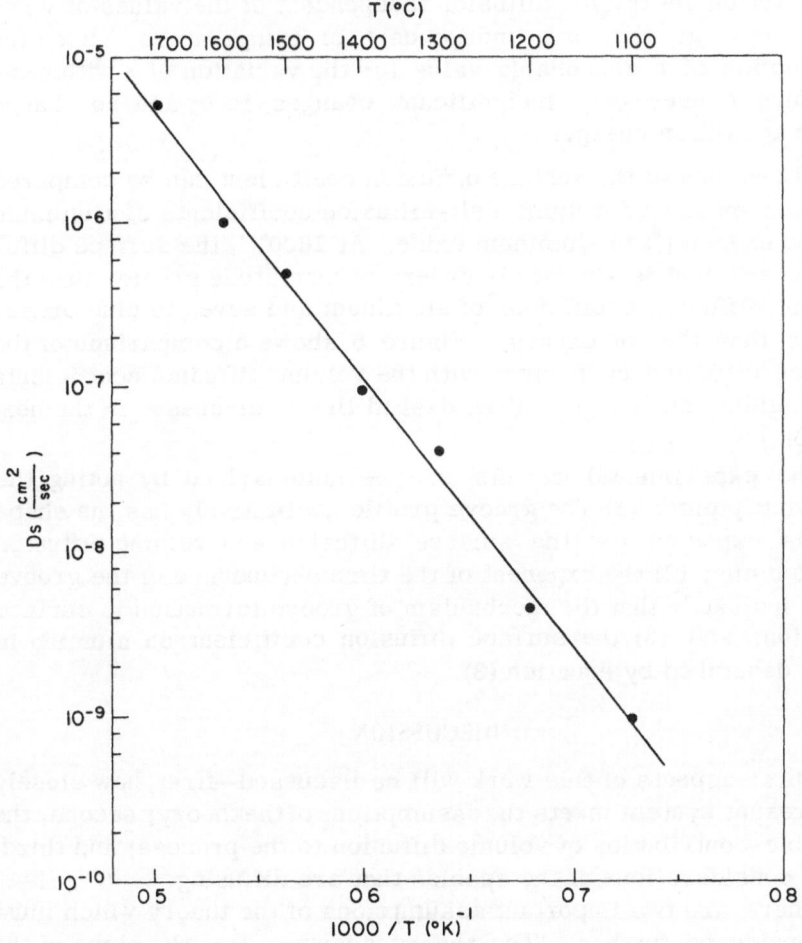

Fig. 4. Surface diffusion coefficient of aluminum oxide in air.

just what values should be taken for the atomic volume of the diffusing species, since it is not known just what species is diffusing. The volume taken here is that of one-half of the Al_2O_3 molecule in an alumina crystal. The value for γ obtained by Kingery is quite reasonable, but it was obtained in a rather roundabout fashion and could be inaccurate by as much as 50%. However, the activation energy obtained for the process is independent of the values taken for the atomic parameters of the system. It depends only on the measured values for B; from equation (1a), it can be seen that a plot of log BT versus $1/T$ would have a slope directly proportional to activation energy for diffusion independent of the values of γ and Ω, as long as they are independent of temperature. Even the assumption of a reasonable value for the variation of γ with temperature causes only insignificant changes in D_s and no change in the activation energy.

The values of the surface diffusion coefficient can be compared with the measured volume self-diffusion coefficients of aluminum [8] and oxygen [9] in aluminum oxide. At 1600°C, the surface diffusion coefficient is almost six orders of magnitude greater than the volume diffusion coefficient of aluminum and seven to nine orders larger than that of oxygen. Figure 5 shows a comparison of the surface diffusion coefficient with the volume diffusion coefficients of aluminum and oxygen. (The dashed line is discussed in the next section.)

The experimental results can be summarized by noting the following points: (1) The groove profile qualitatively has the shape that is expected for the surface diffusion and volume diffusion mechanisms; (2) the exponent of the time dependence of the groove width indicates that the mechanism of groove formation is surface diffusion; and (3) the surface diffusion coefficient on alumina in air is described by equation (3).

DISCUSSION

Three aspects of this work will be discussed—first, how closely the present system meets the assumptions of the theory; second, the possible contribution of volume diffusion to the process; and third, some considerations of the species that are diffusing.

There are two important assumptions of the theory which must be considered further. The theory assumes that the slope of the surface is everywhere very small compared to unity. This assump-

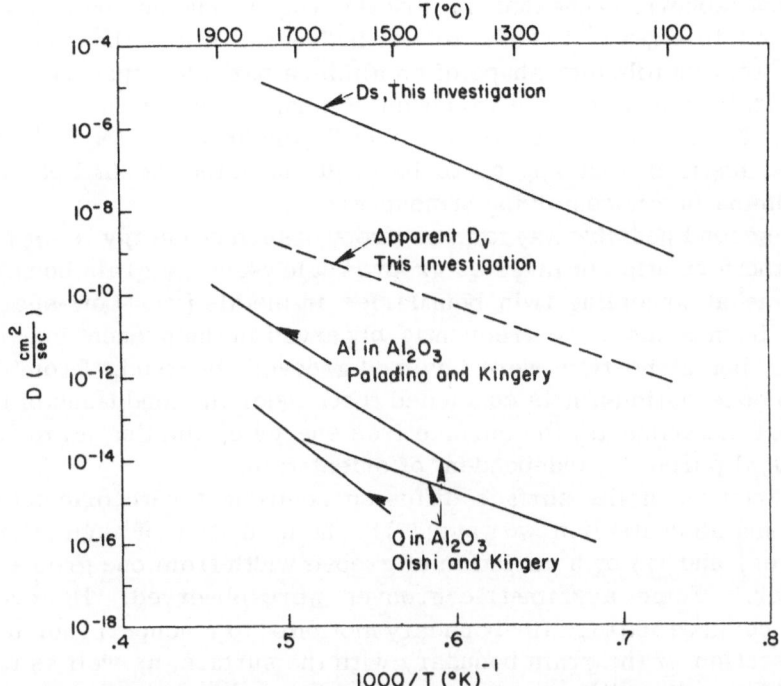

Fig. 5. Comparison of surface diffusion and volume diffusion coefficients of aluminum oxide.

tion is rigorously met everywhere except possibly at the root of the grain-boundary groove. However, the measurements of Kingery [3], which appear to be confirmed by a close examination of the micrographs in the present work, show that the maximum slope at a grain-boundary groove is about $\tan 14° \cong 0.25$, which, of course, is considerably less than unity.

A second assumption of the theory is that the surface free energy and the surface diffusion coefficient are independent of the crystallographic orientation of the surface. A variation of the surface free energy with orientation could cause the formation of facets on low-surface-energy planes [10]. Most of the alumina surfaces in the present experiments did not facet, although a few of the surfaces did have some small facets, and very rarely a grain would completely break up into a stepped surface. Measurements were not made on grain boundaries adjacent to faceted surfaces. In general, the grain-boundary groove profiles were smoothly rounded. Facets were observed by Kuczynski et al. [11] on small alumina spheres annealed in dry hydrogen, but not on spheres annealed in oxygen, helium, or wet hydrogen. Under these conditions, the equilibrium shape of an alumina particle appeared to be spherical, indicating little variation of surface energy with orientation. It is interesting to note that Palmour and Kriegel [1] observed features that appear to be facets upon the thermal etching of alumina in a nonoxidizing atmosphere.

A second possible way for anisotropy of surface energy to appear is in the formation of inverted grooves at low-energy grain boundaries, as at annealing twin boundaries in metals [12]. Low-energy grain boundaries were frequently observed in the present experiments, but at no time were inverted grooves observed. From the above observations, it is concluded that, under the conditions of the present experiments, the surface free energy of alumina is, for all practical purposes, independent of orientation.

Variation of the surface diffusion coefficient with orientation could manifest itself in two ways: (1) by the formation of asymmetric grooves, and (2) by a variation of groove width from one groove to another. Some asymmetric grooves were observed. However, these could be due to grain-boundary motion or to a nonperpendicular intersection of the grain boundary with the surface, as well as to a variable surface diffusion coefficient. Measurements were not made on asymmetric grooves. The widths of symmetric grooves were found to be practically constant from one groove to another. From the constancy of the groove width, it is concluded that the surface diffusion coefficient does not vary appreciably with orientation. (However, the appearance of a small fraction of asymmetric grooves léaves the possibility that the surface diffusion coefficient could be appreciably different from the average over some limited range of orientations.) From the above considerations, it appears that alumina in air meets the assumptions of Mullins' theory quite well.

It is necessary to consider more closely how much volume diffusion could contribute to the formation of the grooves. In the results, it was noted that the time dependence of groove width followed the kinetics that would be expected for surface diffusion, and the surface diffusion coefficients were calculated from these results. Equations (2) and (2a) could be used to calculate the groove widths that would be expected due to volume self-diffusion in the crystal using data for the diffusion coefficients of aluminum ions [8] and oxygen ions [9] in alumina. On the other hand, equations (2) and (2a) could be applied to the data, and, from these equations, values of the diffusion coefficient necessary to explain the results could be derived. On using this latter procedure, it is found that the apparent volume diffusion coefficients obtained are at least two orders of magnitude greater than the measured aluminum diffusion coefficients and four or five orders of magnitude greater than the measured oxygen diffusion coefficients. Also, the activation energy for the apparent volume diffusion coefficient derived in this fashion is about one-half the activation energy found from tracer diffusion studies. The apparent volume diffusion coefficients obtained in this way are given by the dashed curve in Fig. 5. Forcing a fit of the data to volume diffusion theory does not explain the data very well. The curve of the form of equation (2), which best fits the data, gives widths that are too narrow by about 20% at short times and too wide by about the same amount at long times. Thus, we are led to the conclusion that volume diffusion in the crystal does not contribute appreciably to grain-boundary groove growth under these conditions.

It is interesting to speculate about what species controls the rate of surface diffusion on alumina in air. There are several possibilities: aluminum ions, oxygen ions, or some species AlO_x. It should be noted first of all that the present experiments give no direct evidence tying any particular species to the diffusion process; all that is measured is an effective diffusion coefficient. However, it is the opinion of the present authors that the diffusion of aluminum or an aluminum–oxygen species along the surface limits the process in the present case. Since the annealing was done in the presence of air, the exchange of oxygen from one part of the surface to another could take place through the gas phase. The rate of movement of oxygen through the gas is a very rapid process compared to a diffusion process on a crystal. However, in the presence of air, aluminum will not move through the gas phase, but must move over the crystal surface to get a net transfer of material from one place to another. Paladino and Coble [13] found that oxygen diffused rapidly in the vicinity of a grain boundary, so that processes near a grain boundary were controlled by aluminum-ion diffusion, rather than by oxygen-ion diffusion. It would appear that a similar effect occurred near a surface, where, however, the controlling aluminum species has a lower activation energy for diffusion than has aluminum in the bulk of the crystal.

Further thermal grooving work on alumina could indicate more specifically the mechanism of surface diffusion. A useful experiment would be to anneal in atmospheres with variable oxygen pressures, from dry hydrogen to pure oxygen. Another interesting extension of this work would be to anneal the alumina immersed in a liquid metal. These various annealing conditions could change the mechanism of groove formation from surface diffusion to volume diffusion or evaporation–condensation. Even if surface diffusion remained the mechanism of groove formation, the different annealing conditions could change the magnitudes of the diffusion coefficient and the surface energy to greatly alter the rate of groove formation.

ACKNOWLEDGMENTS

The help of F. Ekstrom and R. Spurling in carrying out the experimental work is gratefully acknowledged.

REFERENCES

1. H. Palmour III and W. W. Kriegel, "Brittleness in Ceramics I: Dislocations in Single-Crystal Sapphire As Revealed by Thermal Etching," Engineering Study Report, U.S. Army Contract No. DA-36-034-ORD-2645, North Carolina State College, Raleigh (January 1961).
2. J. E. May, "Polygonization of Sapphire," in: W. D. Kingery (ed.), Kinetics of High-Temperature Processes, John Wiley and Sons (New York and MIT), 1959, pp. 30-37.
3. W. D. Kingery, "Metal—Ceramic Interactions: IV. Absolute Measurement of Metal—Ceramic Interfacial Energy and the Interfacial Adsorption of Silicon from Iron—Silicon Alloys," J. Am. Ceram. Soc. 37 (2):42-45 (1954).
4. W. W. Mullins, "Theory of Thermal Grooving," J. Appl. Phys. 28: 333-339 (1957).
5. W. W. Mullins, "Grain Boundary Grooving by Volume Diffusion," Trans. AIME 218: 354-361 (1960).
6. W. W. Mullins, "Solid Surface Morphologies Governed by Capillarity," Metal Surfaces, ASM (Metals Park), 1963, pp. 17-66.
7. W. W. Mullins and P. G. Shewmon, "The Kinetics of Grain-Boundary Grooving in Copper," Acta Met. 7:163-170 (1959).
8. A. E. Paladino and W. D. Kingery, "Aluminum-Ion Diffusion in Aluminum Oxide," J. Chem. Phys. 37: 957-962 (1962).
9. Y. Oishi and W. D. Kingery, "Self-Diffusion of Oxygen in Single-Crystal and Poly-crystalline Aluminum Oxide," J. Chem. Phys. 33: 480-486 (1960).
10. W. M. Robertson, "The Faceting of Copper Surfaces at 1000°C," Acta Met. 12: 241-253 (1964).
11. G. C. Kuczynski, L. Abernethy, and J. Allen, "Sintering Mechanisms of Aluminum Oxide," in: W. D. Kingery (ed.), Kinetics of High-Temperature Processes, John Wiley and Sons (New York and MIT), 1959, pp. 163-172.
12. W. M. Robertson and P. G. Shewmon, "Variation of Surface Tension with Surface Orientation in Copper," Trans. AIME 224: 804-811 (1962).
13. A. E. Paladino and R. L. Coble, "Effect of Grain Boundaries on Diffusion-Controlled Processes in Aluminum Oxide," J. Am. Ceram. Soc. 46: 133-136 (1963).

Chapter 5

The Role of Phase Boundaries During Reaction Between Single Crystals of Various Oxides

H. G. Sockel and H. Schmalzried

Max Planck Institut für Physikalische Chemie
Göttingen, Germany

Phase-boundary reactions between single crystals of different binary oxides (e.g., $NiO + Al_2O_3 \rightarrow NiAl_2O_4$) are described. The reaction product can be monocrystalline or polycrystalline depending upon crystal structure of reactants and reaction product and on experimental conditions. The reaction product may be a single crystal even if the crystal structures of the reactants differ from that of the reaction product. Examples are given. Epitaxy, structural relationships, and the atomic mechanism of the phase-boundary reaction are discussed. It is concluded that under certain conditions dislocations may play a decisive role for the phase-boundary reaction. The kinetics involving phase-boundary reactions is discussed.

INTRODUCTION

The general scope of the reactions between solids is extremely wide. This discussion will be restricted to reactions between two different solid phases forming one or several product phases. Neither decomposition and recrystallization reactions of single solid substances, nor equilibration processes of solid solutions with different concentrations of the components will be considered. The former aspect of solid-state reactions has been discussed in some detail recently by Brindley [1], and the latter aspect is the topic of many papers in the field of metal diffusion. Special reference is given to the Kirkendall effect [2,3] and recent work on diffusion in ternary systems by Kirkaldy [4,5].

The following discussion deals with heterogeneous reactions in which the reactants and the reaction product are separated from each other by phase boundaries. The theoretical discussion and the experimental examples are restricted to reactions between binary ionic crystals forming ternary compounds. In addition to the fact that this group of reactions is important from a practical point of view, there are thermodynamic reasons for the limitation. In order to fix unambiguously the thermodynamic state of the ternary compound, it is necessary to predetermine, in addition to temperature and overall pressure, two more independent thermodynamic variables. If, for example, the reaction product is of the form AB_2O_4, where A, B, and O are symbols for cations and anions, respectively, it is experimentally convenient to predetermine the partial pressure of the electronegative component p_{O_2}. Assuming local thermodynamic equilibrium during the reaction process, the chemical potential of the reactants at the phase boundaries is also given. Thus, an unambiguous treatment of the kinetics of the reaction process is possible. It is obvious, however, that for ionic crystals with more than three components the predetermination of all independent thermodynamic variables at the phase boundaries may become most difficult [6].

In the following, we shall first study the kinetics of solid-state reactions of the type $AO + B_2O_3 \rightarrow AB_2O_4$, giving special attention to phase-boundary reactions. Thereafter, a number of experimental examples are discussed, and it is concluded that dislocations may play a decisive role for phase-boundary reactions if certain structural relationships between reactants and reaction product are fulfilled.

DISCUSSION OF THE KINETICS

In the idealized one-dimensional reaction between AO and B_2O_3, forming AB_2O_4 as a single-phase reaction product, one has to regard the flux of ions in the ternary compound and across the phase boundaries AO/AB_2O_4 and AB_2O_4/B_2O_3. The phase boundary in the simplest case is thought to be a distorted crystallographic region between the perfect lattices of the reactants and the reaction product. However, there is evidence that, because of insufficient plastic deformation, in a number of cases reactants and reaction product are separated by small gas gaps. This is shown schematically in Fig. 1. Thus, if we regard generally the transport of ions from the left-hand side to the right-hand side (and vice versa) of the reaction pattern, we have to discuss five different reaction steps, as indicated in Fig. 2. Here, the partial pressure p_{AO} of

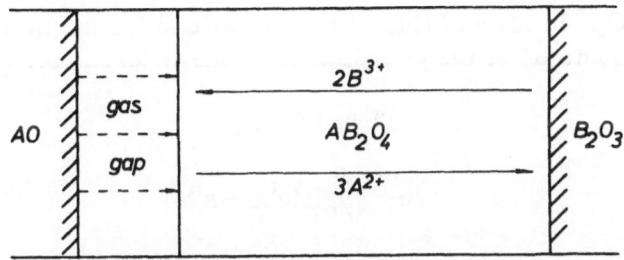

Fig. 1. $AO + B_2O_3 \rightarrow AB_2O_4$: Schematic reaction pattern, involving transport of AO across
a gas gap at the phase boundary $AO \,|\, AB_2O_4$.

one reactant is plotted as a function of the distance from the phase
boundary AO/gas gap. It is obvious how to modify the p_{AO} versus
distance plot if gas gaps are on both sides of the reaction product,
or if ideal contact is established at all interfaces. The five reaction steps are: I—evaporation of solid AO into the gas gap;
II—diffusion of gaseous species across the gap; III—phase-boundary
reaction AO (gas)/AB_2O_4; IV—diffusion of ions across the reaction
layer AB_2O_4; and V—transport of ions across the phase boundary

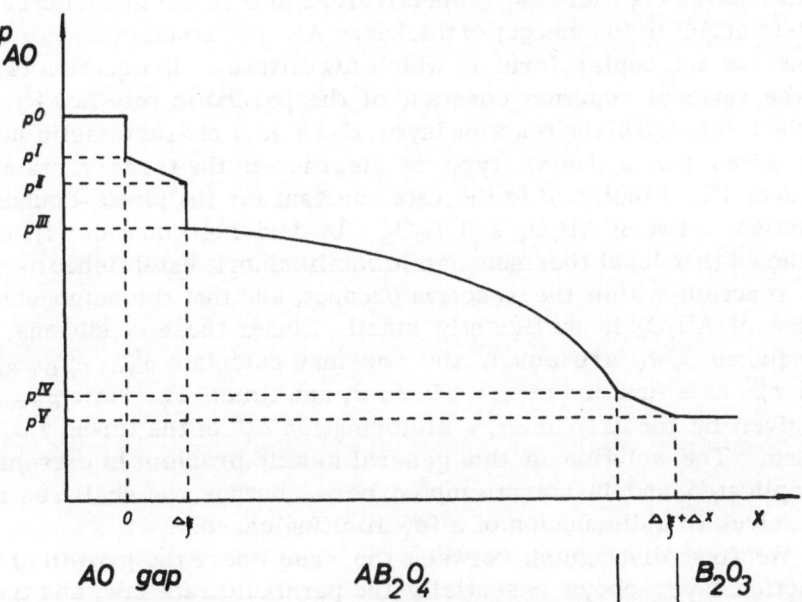

Fig. 2. Activity of AO in terms of p_{AO} as a function of the distance from the interface
$AO \,|\, $ gas gap.

AB_2O_4/B_2O_3. In view of this situation, we can formulate the following flux equations at the coordinates of the different reaction steps:

$$j_I = a \, (2\pi MRT)^{-\frac{1}{2}} \, (p_{AO}^0 - p_{AO}^I) \tag{1}$$

$$j_{II} = \frac{D}{\Delta\xi RT} \, (p_{AO}^I - p_{AO}^{II}) \tag{2}$$

$$j_{III} = a' \, (2\pi MRT)^{-\frac{1}{2}} \, (p_{AO}^{II} - p_{AO}^{III}) \tag{3}$$

$$j_{IV} = \frac{k}{\Delta x} \cdot \frac{(p_{AO}^{III})^{n_i} - (p_{AO}^{IV})^{n_i}}{(p_{AO}^0)^{n_i} - (p_{AO}^V)^{n_i}} \tag{4}$$

$$j_V = \beta \, RT \ln \frac{p_{AO}^{IV}}{p_{AO}^V} \tag{5}$$

Equations (1) and (3) are the Hertz formulas describing evaporation and condensation of AO; a and a' are the evaporation and condensation coefficients, respectively, and D is the diffusion coefficient of AO in the gas gap of thickness $\Delta\xi$. No assumption is made about the molecular form in which AO diffuses. In equation (4), k is the rational reaction constant of the parabolic rate law [7], Δx is the thickness of the reaction layer, and n_i is a characteristic number given for a known type of disorder in the ternary reaction product [8]. Finally, β is the rate constant for the phase-boundary reaction between AB_2O_4 and B_2O_3. In deriving equation (4), it is assumed that local thermodynamic equilibrium is established during the reaction within the reaction product, and that the homogeneity range of AB_2O_4 is sufficiently small. Under these conditions, all the fluxes j_I-j_V are equal, and one may calculate p_{AO}^I, p_{AO}^{II}, p_{AO}^{III}, and p_{AO}^{IV} as a function of a, a', D, k, β, and time. The ratio p_{AO}^V/p_{AO}^0 is given by the free energy of formation ΔG^0 of the ternary compound. The solution of this general kinetic problem is extremely complicated and is not attempted here. Rather, we shall restrict ourselves to a discussion of a few limiting cases.

We first distinguish between the case where the growth of the reaction layer obeys essentially the parabolic rate law, and those cases where phase-boundary reaction steps are rate-determining. By eliminating p_{AO}^I, p_{AO}^{II}, p_{AO}^{III}, and p_{AO}^{IV} from equations (1) – (5), and by assuming $p_{AO}^0 - p_{AO}^{III}/p_{AO}^0 - p_{AO}^V$ and $p_{AO}^{IV} - p_{AO}^V/p_{AO}^0 - p_{AO}^V \ll 5\%$, one obtains for the minimum thickness Δx_m of the reaction product for parabolic growth:

$$\Delta x_m = \frac{20\,k}{p_{AO}^0 - p_{AO}^V} \left[(2\pi MRT)^{\frac{1}{2}} \left(\frac{1}{\alpha} + \frac{1}{\alpha'} \right) + \frac{\Delta \xi RT}{D} + \frac{p^V}{\beta RT} \right] \qquad (6)$$

If the reaction layer thickness can be described by $\Delta x \gg \Delta x_m$, the parabolic rate law holds. In order to calculate the rate constant k, one formulates the flux of ions of kind i in the reaction product at coordinate x as

$$j_i = - \frac{z_i c_i D_i}{RT} \, \mathrm{grad}\ \eta_h \qquad (7)$$

where j_i is given in equivalents/cm^2-sec, if z_i is the absolute valence of ion i, c_i the equivalent concentration; D_i the self-diffusion coefficient (neglecting correlation factors); and η_1 the electrochemical potential per equivalent. If the self-diffusion co-efficient D_i is intermediate between those of the two other sorts of ions and sufficiently different (since one has local thermodynamic equilibrium at the phase boundaries and in the interior of the reaction layer, and the homogeneity range of AB_2O_4 is assumed to be small), one obtains, by integration of equation (7), the parabolic rate law

$$\frac{d\Delta x}{dt} = \frac{kv}{\Delta x} \qquad (8)$$

where v is the increase in volume of the reaction product if one equivalent is transported through AB_2O_4. This is the most important case for practical work. The reaction constant is then given as

$$k = \gamma\, z_i c_i D_i^0\, n_i^{-1} \left(1 - \exp n_i \frac{\Delta G^0}{RT} \right) \qquad (9)$$

where ΔG^0 is the free energy of formation of AB_2O_4 from AO and B_2O_3, γ is a stoichiometric factor of the order of one, and $D_i^0 = D_i$ $(p_{AO} = p_{AO}^0)$.

If not only ions of sort i are rate-determining, the problem can be treated in a similar way. Details of these calculations may be found in previous papers [7, 9, 10]. As an example, Fig. 3 shows experimental and calculated k-values for the formation of $CoCr_2O_4$ from CoO and Cr_2O_3, assuming transport of Co^{+2} and electrons

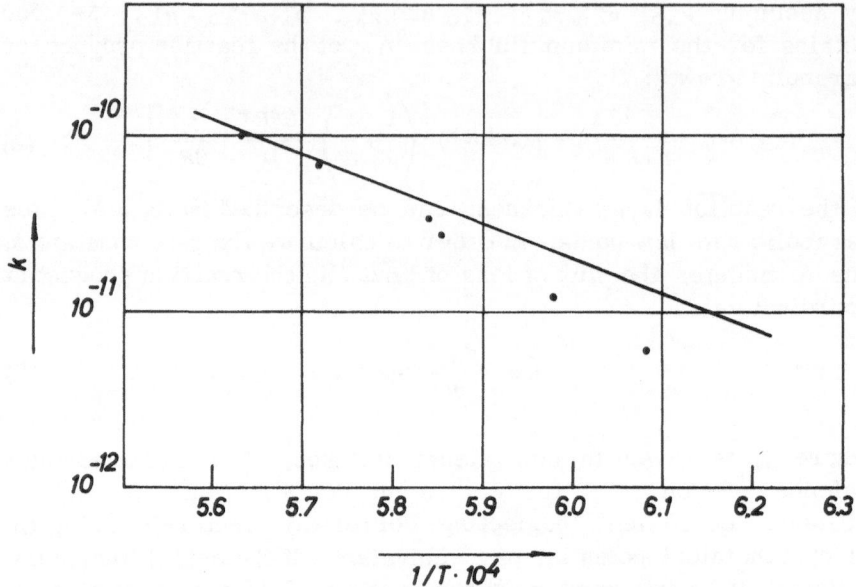

Fig. 3. Experimental (●) and calculated values of the reaction constant k for the formation of $CoCr_2O_4$ from CoO and Cr_2O_3.

(electron holes) across the reaction product, and oxygen via the gas phase.

In case that $\Delta x \ll \Delta x_m$, phase-boundary reactions become rate-determining. The problem is, in many respects, similar to a gas–solid reaction (oxidation of metals), if reaction steps I–III are essentially rate-determining. The transition from linear to parabolic reaction rate can easily be treated in view of equations (1)–(4), if one assumes that $n_i = 1$. (For $n_i \neq 1$, the mathematical treatment becomes very complicated.) The solution for $n_i = 1$ is

$$a\Delta x + \Delta x^2 = 2kvt \tag{10}$$

where $a = 0.1\Delta x_m$. The term $1/\beta RT$ in equation (6) is neglected according to our assumption; the reaction steps I–III control the reaction rate. For a rational treatment of experimental data, it is proposed to plot $t/\Delta x$ as a function of Δx [11]. This then yields a straight line with a slope $1/2kv$ and, for $\Delta x = 0$, $t/\Delta x$ becomes $a/2kv$. Thus, it is possible to determine the rate constants k and a. From preliminary measurements for the two systems, $NiO-Al_2O_3$ and

$NiO-TiO_2$, it is concluded that a is of the order of one micron. The same order of magnitude is obtained if a is calculated from equation (6) by inserting reasonable values of α, α', D, $\Delta\xi$, and p_{AO}. This means that reaction steps I–III are responsible for the linear reaction rate and not the phase-boundary reaction at the interface $AB_2O_4|B_2O_3$ where ideal contact prevails.

Finally, we may drop the assumption of gas transport between reactant and reaction product, assuming instead ideal contact between $AO|AB_2O_4$ and $AB_2O_4|B_2O_3$. If now one of these phase-boundary reactions becomes rate-determining, we can start the mathematical treatment with equations (4) and (5), letting $p_{AO}^{III} \cong p_{AO}^0$. If the chemical potential of AO is defined as $\mu_{AO} = \mu_{AO}^0 + RT \ln p_{AO}$, and if $\mu_{AO}^V - \mu_{AO}^0 = \Delta G^0$, one can rewrite equations (4) and (5) as

$$j_{IV} = \frac{k}{\Delta x} \frac{1 - \exp(n_i\Delta G^0/RT)\, y}{1 - \exp(n_i\Delta G^0/RT)} \tag{11}$$

$$j_V = \beta\Delta G^0 (y-1) \tag{12}$$

where $y = \mu_{AO}^{IV} - \mu_{AO}^0/\Delta G^0$. Since $j_{IV} = j_V$, one obtains

$$\Delta x \frac{\beta\Delta G^0\,[\exp(n_i\Delta G^0/RT) - 1]}{k} = \frac{1 - \exp(n_i\Delta G^0/RT)\, y}{1 - y} \tag{13}$$

From this last equation, one can calculate the relative change of the chemical potential of AO across the phase boundary $1-y$ as a function of the thickness Δx, of the rate constant k, and the disorder type of the reaction product characterized by the number n_i. The example shown in Fig. 4 stresses again the importance of knowing the disorder type of the reaction product in order to evaluate the kinetics of this kind of solid-state reaction. Without treating equations (11) and (12) further, we turn now to the discussion of several examples of phase-boundary reaction mechanisms, if the reactants are single-crystal oxides.

SOME EXPERIMENTAL RESULTS ON PHASE-BOUNDARY REACTIONS

In order to obtain significant results, reaction experiments should be performed using single-crystal reactants. Figure 5 gives two examples of what one may expect. (1) If there is ideal contact

1-y
1
0.8
0.6
0.4
0.2
1
2
5
10
15
Δx
δ

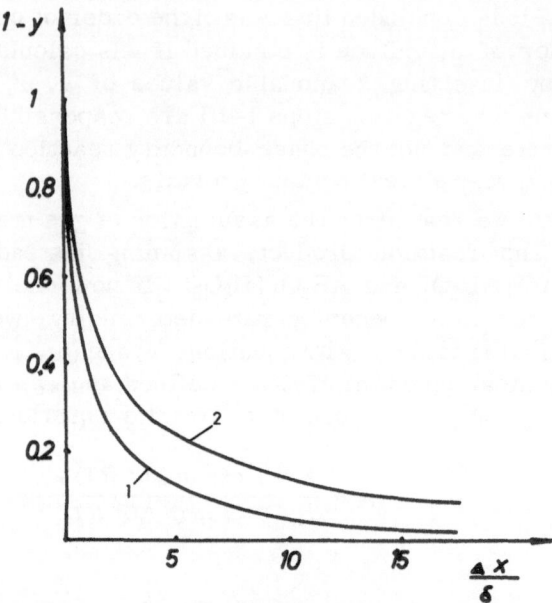

Fig. 4. Relative potential drop of AO at the phase boundary $AB_2O_4 | B_2O_3$ as a function of the reaction layer thickness for (1) vacancy diffusion and (2) interstitial diffusion of the rate-determining ion. $|n_i \Delta G^0/RT| = 1$ and $\delta = k/\beta\Delta G^0 \, [\exp(n_i \Delta G^0/RT) - 1]$.

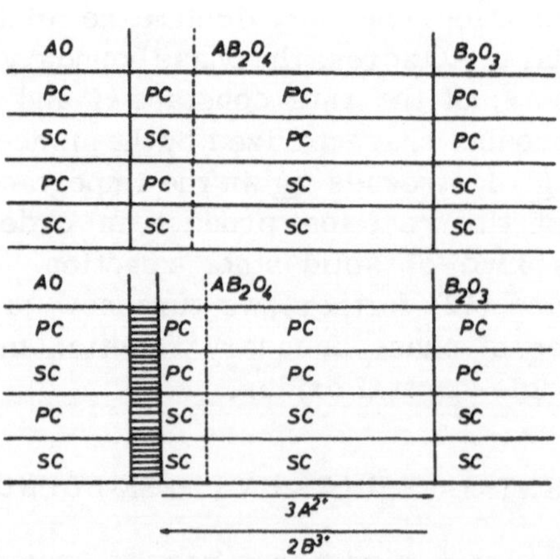

AO
AB2O4
B2O3
PC PC PC PC
SC SC PC PC
PC PC SC SC
SC SC SC SC
AO
AB2O4
B2O3
PC PC PC PC
SC PC PC PC
PC SC SC SC
SC SC SC SC
3A2+
2B3+

Fig. 5. Morphology of the reaction product AB_2O_4 for the reaction between single-crystal (s.c.) and polycrystalline (p.c.) oxides, assuming counter-diffusion of cations.

between reactants and reaction product, the phase-boundary reaction can consist of a number of uncorrelated steps of the different diffusing ions such that no orientational relationship is found between the reactant and the reaction product. (2) If reaction steps at the interface reactant/reaction product are correlated such that one or several sublattices are preserved during the phase-boundary reaction, the reaction product is a single crystal with a certain orientational relationship to the reactant. (3) If gas transport across small gaps prevails at one phase boundary, it depends on epitaxial nucleation and growth, whether the product phase is polycrystalline or single-crystal. All of these cases can be found in solid-state reaction experiments.

After reaction of a single crystal of TiO_2 with a single crystal of CoO or NiO, the reaction product is polycrystalline and has ideal contact with TiO_2 (s. c.). While $NiTiO_3$ is the only reaction phase between TiO_2 and NiO, there are three phases between TiO_2 and CoO at 1300°C — $CoTi_2O_5$, $CoTiO_3$, and Co_2TiO_4. However, since crystal structures between TiO_2 and the reaction products are sufficiently different, no single-crystal reaction product results.

In contrast to these examples, one obtains a single-crystal reaction product from the following reaction at 1400°C:

$$MgO(s.\,c.) + Cr_2O_3(p.\,c.) \rightarrow MgCr_2O_4(s.\,c.) \tag{14}$$

Assuming counter-diffusion of cations in the reaction layer (as can be anticipated regarding the higher diffusivity of cations as compared with anions in crystals with almost closest-packed oxygen sublattices), the phase boundary reaction at the interface $MgO/MgCr_2O_4$ can be described as follows: Mg^{+2} ions in MgO occupy octahedral sites, whereas after reaction the Mg^{+2} ions are found on tetrahedral sites in the normal spinel $MgCr_2O_4$. Thus, one has a simple rearrangement of cations; two Cr^{+3} ions replace three Mg^{+2} ions, while the remaining Mg^{+2} ions are shifted from octahedral to tetrahedral sites. Oxygen ions do not need to move if one disregards a small lattice contraction of about 1% when the higher-charged Cr^{+3} ions enter into the MgO crystal. According to the proposed reaction mechanism (see Fig. 6), one would expect, therefore, one-fourth of the reaction layer to be single-crystal and three-fourths polycrystalline (neglecting small solubilities of MgO and Cr_2O_3 in $MgCr_2O_4$). Experiments show, however, that all the reaction layer $MgCr_2O_4$ is in the form of a single crystal. This fact can be explained by a gas transport of Cr_2O_3 onto the MgO-saturated $MgCr_2O_4$ surface and epitaxial nucleation and growth according to the equation

$$4Cr_2O_3(g) + 3Mg^{+2}(\text{spinel s.c.}) \to 3MgCr_2O_4(\text{spinel s.c.}) + 2Cr^{+3}(\text{spinel s.c.}) \quad (15)$$

as indicated in Fig. 6. In Fig. 7, this explanation is demonstrated by showing the epitaxial growth of $MgCr_2O_4$ pyramids on a cleavage plane (100) of MgO. In agreement with these considerations, one finds that the $\langle 100 \rangle$ and $\langle 010 \rangle$ in MgO are parallel to the $\langle 100 \rangle$ and $\langle 010 \rangle$ in $MgCr_2O_4$.

Finally, we shall discuss the reaction

$$NiO(\text{s.c.}) + a\text{-}Al_2O_3(\text{s.c.}) \to NiAl_2O_4 \quad (16)$$

at 1450°C. There is the following experimental evidence:

(1) As shown in Fig. 6, one-fourth of the reaction product is poly-crystalline, whereas the part of $NiAl_2O_4$ grown from a-Al_2O_3 is monocrystalline.

(2) Independent of the crystallographic orientation of a-Al_2O_3 with respect to the interface $NiAl_2O_4/a$-Al_2O_3, the spinel phase grown from a-Al_2O_3 is a single crystal.

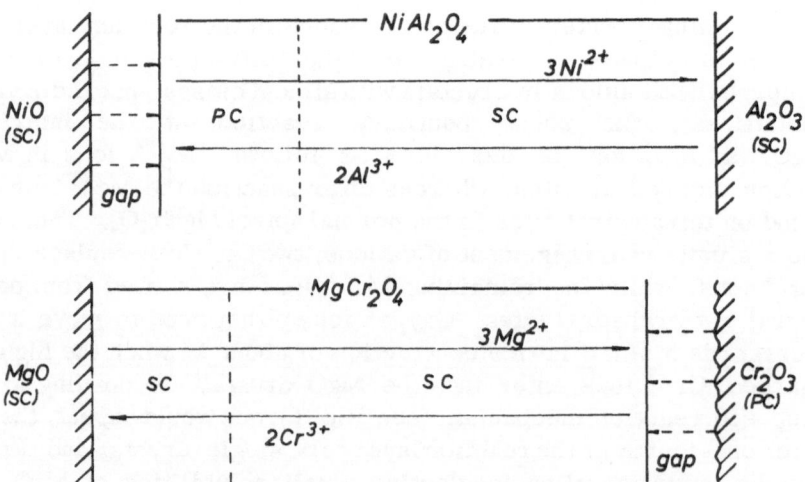

Fig. 6. Reaction mechanism for the solid-state reactions: $NiO(\text{s.c.}) + Al_2O_3(\text{s.c.}) \to NiAl_2O_4$ and $MgO(\text{s.c.}) + Cr_2O_3(\text{p.c.}) \to MgCr_2O_4$.

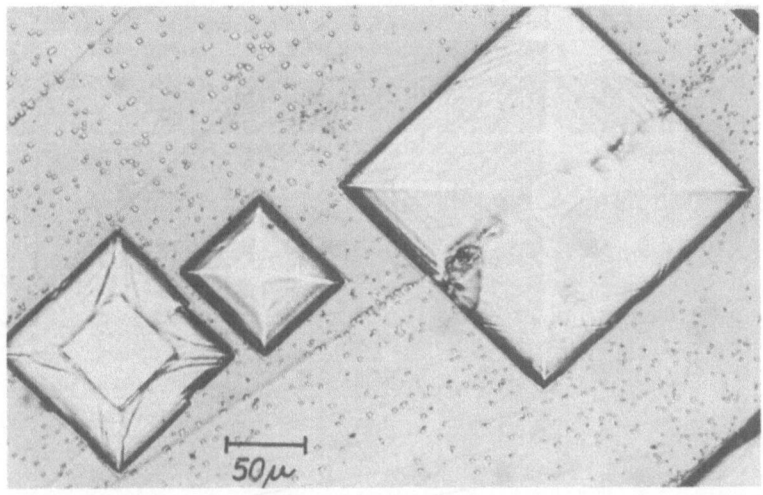

Fig. 7. Beginning of epitaxial growth of $MgCr_2O_4$ on a (100) plane of MgO.

(3) The orientation between α-Al_2O_3 and monocrystalline spinel is $(0001)_{\alpha\text{-}Al_2O_3} \rightarrow (111)_{spinel}$. Thus, the close-packed basal planes of oxygen remain parallel during the phase-boundary reaction.

(4) The interface $NiAl_2O_4/\alpha$-Al_2O_3 is plane and on a microscopical scale parallel to the original interface between α-Al_2O_3 and MgO.

These statements hold in the same way for the system CoO – Al_2O_3. For MgO – Al_2O_3, one may consult the excellent study of Rossi and Fulrath [12]. Interpretation is as follows: At the phase boundary $NiO(s.c.)/NiAl_2O_4(p.c.)$, there is transport of NiO via the gas phase as indicated in Fig. 6. In contrast to the system MgO – Cr_2O_3, no epitaxial nucleation and growth occur.

The phase-boundary reaction at the interface $NiAl_2O_4/\alpha$-Al_2O_3 has to achieve two steps: (1) rearrangement of cations and (2) change of packing of the hexagonal oxygen planes in α-Al_2O_3 (. . .ABABAB. . .) into the cubic face-centered sequence in $NiAl_2O_4$ (. . .ABCABCABC. . .). Since in the systems here discussed the oxygen diffusion coefficient is several orders of magnitude smaller than the diffusion coefficients of cations, it is suggested that a change in the packing sequence of the oxygen sublattice is performed by a correlated effect, rather than by diffusional steps of oxygen ions. It is also known [13,14] that (111) planes in spinel crystals and (0001) planes in α-Al_2O_3 are glide planes for plastic deformation (synchronized shear). In view of this, we may antici-

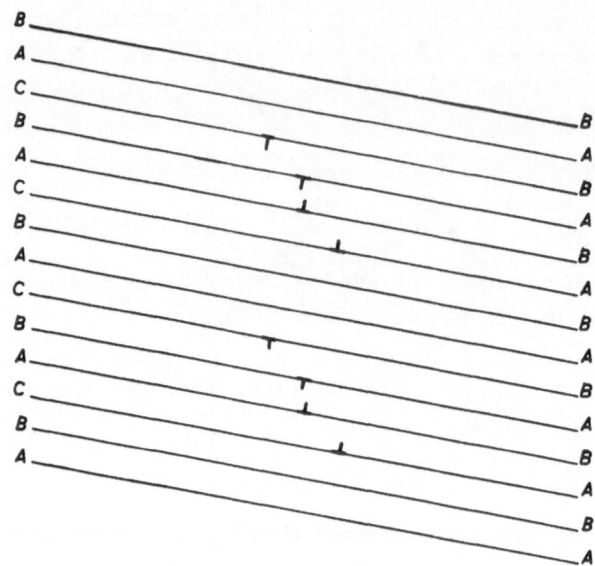

Fig. 8. Schematic pattern of the phase-boundary reaction at the interface $NiAl_2O_4|Al_2O_3$, involving bundles of Shockley partial dislocations.

pate that the simplest correlated phase-boundary reaction steps consist of a movement of Shockley partial dislocations along the glide planes. Figure 8 gives schematically some idea of the proposed mechanism. X-ray topographs taken from the reaction product show that $(111)_{spinel}$ planes, which are parallel to $(0001)_{\alpha-Al_2O_3}$ planes, are rather distorted. This is to be expected since:

(1) The distance between subsequent oxygen planes in $\alpha-Al_2O_3$ and $NiAl_2O_4$ is somewhat different.

(2) Since the packing sequence ...ABABAB... in $\alpha-Al_2O_3$ may be converted into the sequence ...ABCABCABC..., as well as into ...ACBACBACB... by the phase-boundary reaction, the stacking fault probability is high.

(3) The thermal expansion coefficient is different for $\alpha-Al_2O_3$ and $NiAl_2O_4$. This causes strains between $\alpha-Al_2O_3$ and $NiAl_2O_4$ during cooling and makes an investigation of the substructure of $NiAl_2O_4$ at room temperature ambiguous.

It is difficult to predict whether it is the gradient of the electrochemical potential of ions diffusing ahead or the stress field around the phase boundary that keeps the bundles of dislocations moving.   A somewhat similar effect was observed recently [15] during interdiffusion of silicon and phosphorus forming solid solutions.   Changes in lattice parameter as a function of the silicon/ phosphorus ratio create a stress field along the diffusion front, which, in turn, is the source of a dislocation network moving ahead of the diffusion front.

REFERENCES

1. G. W. Brindley, Progr. Ceram. Science 3:1 (1963).
2. C. da Silva and R. F. Mehl, J. Metals 3:155 (1951).
3. A. D. Smigelskas and E. O. Kirkendall, Trans. AIME 171:130 (1947).
4. J. S. Kirkaldy, Can. J. Phys. 36:899 (1958).
5. J. S. Kirkaldy, Can. J. Phys. 36:917 (1958).
6. H. Schmalzried, Ber. Deut. Keram. Ges. 42:11 (1965).
7. C. Wagner, Z. Physik. Chem. B34:309 (1936).
8. H. Schmalzried and C. Wagner, Z. Physik. Chem. NF31:198 (1962).
9. K. Hauffe, Reaktionen in und an festen Stoffen, Springer (Berlin), 1955.
10. H. Schmalzried, Symposium of the Reactivity of Solids, Munich, 1964.
11. G. Wagner, Thermodynamik metallischer Mehrstoffsysteme, Handbuch der Metall-physik, Vol. 1, 1940.
12. R. C. Rossi and R. M. Fulrath, J. Am. Ceram. Soc. 46:145 (1963).
13. M. L. Kronberg, Acta Met. 5:507 (1957).
14. J. Hornstra, Phys. Chem. Solids 15:311 (1960).
15. H. J. Queisser, Discussions Faraday Soc., in press.

PART II. Electromagnetic Wave Behavior Near Interfaces

S. AMELINCKX, Presiding

Studiecentrum voor Kernenergie, S.C.K.
Mol, Belgium

Chapter 6

Surface Reoxidation Phenomena in Certain Ceramics With a Nonstoichiometric Perovskite Structure

J. Dubuisson and P. Basseville

LCC-Stéafix Company
Montreuil-Seine, France

The apparent dielectric constant of a disk of modified barium titanate reaches a permittivity value of 500,000 after successive reduction–reoxidation steps. Reduction and the ensuing reoxidation phenomena are enhanced when ions, such as Ce^{+4}, are present. Starting from a certain degree of reduction, a reorganization of the structure occurs on reoxidation and allotropic transformation points vanish. Evidence of surface reoxidation is shown by the appearance of a barrier layer having a rectifying effect. Structure blocking can be obtained also by action of an oxide, such as Bi_2O_3.

INTRODUCTION

Perovskite-type structures of the general formula ABO_3, where A and B are metal ions, are essentially characterized by the tetrahedral distribution of oxygen atoms around a center ion. Barium titanate, which is the most common compound of this family, may take different allotropic forms according to temperature: a cubic structure, which is the true form of the perovskite structure, stable at temperatures above 120°C and three pseudo-cubic structures— tetragonal in the range -5 to +120°C, orthorhombic in the range -6 to -80°C, and rhombohedral at lower temperatures.

The well-known dielectric properties of barium titanate are generally attributed [1-3] to the interactions, within the crystal lattice, between the titanium ions and the oxygen ions. Any alterations in the titanium–oxygen bonds then produce not only the physical reactions normally occurring in solids (e.g., variations in crystal parameters and changes in allotropic transformation points), but also significant variations in the electrical properties.

Oxygen atoms are only slightly fixed in the lattice. Oxygen deficits leading to a nonstoichiometric structure can be easily

77

produced.   Wyss [4] was first to systematically investigate the
reduction of titanium oxide and titanates. Saburi [5] demonstrated
that reduced barium titanate is an n-type semiconductor in which
conductivity is produced by electron exchange between Ti^{+3} and Ti^{+4}.
Finally, the compositions and techniques for the production of min-
iature capacitors, which use titanates for the dielectric and which
are semiconducting to a varying degree, are described in a number
of patents [6−11].

After a description of the experimental procedure and a study
of the barium titanate reduction kinetics, we will show that:

1. In pure barium titanate, as from a given degree of reduction,
   a structural redistribution occurs; the stable form remains
   the tetragonal form without parameter variation, but the
   allotropic transformation points vanish.

2. The reduced forms of barium titanate, therefore, must be
   ferroelectric semiconductive compounds with a carrier
   concentration approximating $10^{19}/cm^3$ and a mobility of about
   $20 \cdot 10^{-2}$ cm^2/V-sec.

3. The reduction process is enhanced by small additions (from
   0.2 to 0.5 mol.%) of rare earth oxide, but the allotropic
   transformation characteristics of the barium titanate are
   maintained in the tetragonal structure compounds.

4. The reduction and reoxidation of ceramic compounds containing 2.5 mol.% of CeO_2 yield dielectrics with an apparent
   permittivity approximating 200,000, an angle of loss $\leq$
   $400 \cdot 10^{-4}$, and an insulation resistance $\geq 10^8$ Ω at 12 V DC.

5. Higher apparent permittivity values of about 500,000 can be
   achieved by eliminating the reoxidation stage through super-
   ficial blocking of the structure with substances such as
   bismuth oxide or lead borosilicate.

6. Finally, the equivalent diagrams found in the literature explain
   the behavior as a function of frequency or voltage, but do not
   account for phenomena such as the capacitance increase of
   a disk capacitor with the dielectric thickness.

Therefore, we shall conclude by proposing a new equivalent
diagram for barrier-layer capacitors.

EXPERIMENTAL PROCEDURE

Raw Materials

The materials selected were the purest commercial products now available in Europe. Titanium oxide in the anatase form is manufactured by the Société Degussa. Its purity is better than 99.8% and its principal impurities as determined by spectral analysis are silicon and magnesium. The barium carbonate and rare earth oxides produced by the Pechiney Company have a purity better than 99.9%.

Preparation of Barium Titanate

Barium titanate is produced by synthesis in the solid state from titanium oxide and barium carbonate, according to the following general formula:

$$BaCO_3 + TiO_2 \rightarrow BaTiO_3 + CO_2(g) \qquad (1)$$

The conditions required to obtain pure $BaTiO_3$ with a density approximating theoretical were determined by a dilatometric method in a previous study [12]. Briefly, the procedure is as follows: (1) Weigh the dry constituents in the proportions shown by the reaction. (2) Mill the compound under wet conditions for 48 hr in rubber-lined mills with Corindon balls. (3) Oven-dry at 120°C for 2 hr. (4) Mill under dry conditions to a grain size less than $15\,\mu$. (5) Fire in fixed oven in oxidizing atmosphere at 1230°C for 1 hr in silimanite crucibles protected by stabilized zirconia. (6) Mill under wet conditions for 72 hr in rubber-lined Corindon ball mills. The resulting grain size is less than $40\,\mu$ and an X-ray diffraction analysis shows that the product obtained is only barium metatitanate.

For our tests on modified barium titanate, the rare earth oxides are added at this stage. The compound is then mixed and ground under wet conditions for 48 hr. To improve homogeneity of the final product for the low contents, the rare earth oxide was pre-reacted with a certain quantity of titanium oxide. In all cases, the concentrations of titanium and barium ions are such that they correspond to the formula $Ba_{1-x}M_xTiO_3$, where M is the rare earth cation.

Specimen Preparation

The specimens were formed by pressing in a double-acting plug and die system. To facilitate molding, 1.5% polyvinyl alcohol is added. The compacting pressure was $3.5 \, \text{tons/cm}^2$. The disks thus obtained were 20 mm in diameter and about 1.5 mm thick with an apparent density equal to 60% of theoretical.

The compacts were sintered in an oxidizing atmosphere, according to the following thermal cycle: (1) temperature rise rate, 100°C/hr; (2) soaking for 1 or 2 hr at maximum temperature, mainly 1365 or 1380°C, and (3) natural cooling for about 24 hr.

With pure barium titanate, disks with a density higher than 95% of theoretical ($6.02 \, \text{g/cm}^3$) are thus obtained. Spectral analysis showed that the concentrations of impurities do not vary as compared with the initial materials, i.e., the product has not been contaminated by treatment. Finally, an X-ray crystallographic analysis showed that the product is pure, well-crystallized barium titanate.

Reduction and reoxidation operations were carried out in a controlled-atmosphere sealed oven. The specimens to be reduced or reoxidized were placed vertically in grooves of an Al_2O_3 support allowing free reaction of gases. The maximum rate of temperature rise allowed by the oven was 600°C/hr, and this rate was used for all isothermal experiments in obtaining the stabilized temperature. The temperature was kept constant to within ± 2°C by means of a controller.   The pre-dried hydrogen or oxygen was circulated through the oven at the rate of 250 cm^3/min.

Resistance Measurements

Electrodes provided by a ceramic flux or a plastic material (conductive cement) lead to very large deviations either by producing a barrier layer through diffusion, as will be seen later, or by inserting a nonnegligible additional resistance in series. Therefore, the surfaces of the disks were metallized by chemical nickel-plating. In this procedure, the specimens were dipped successively in the following baths: (1) sodium hypophosphite—concentration, 100 g/liter at boiling point; (2) palladium chloride—concentration, 1 g/liter at boiling point; and (3) nickel-plating bath containing 30 g/liter nickel chloride, 10 g/liter sodium hypophosphite, 65 g/liter ammonium citrate, and 50 g/liter ammonium chloride.

As stated by Saburi [13], it was found that the minimum values of the resistance of a specimen can be obtained by firing at 400°C, which is probably due to improved metal-to-ceramic bonding and to the removal of mechanical stresses.

The resistance values were determined by the ammeter-voltmeter method. In view of the resistance variation with voltage, it has been necessary to adopt a fixed mean voltage; otherwise, measurements on reduced parts would have been impossible at high current values (due to temperature rise attributable to the Joule effect). The voltage selected was 12 V, which permits correct measurement on nonreduced and half-reduced barium titanate. The measurement is no longer accurate when the disk resistance is 100 Ω or less, for the resistance varies with temperature due to the Joule effect.

After the test, it was found that the results obtained at 12 V by the above method were in agreement with those obtained at 1.5 V with the Wheatstone bridge, although the resistance measured then was slightly higher than that at 12 V ($\Delta R/R = 3\%$).

It has not been possible to use commercial-type meggers, for they generally operate at variable voltages from 10 to 500 V, according to the order of magnitude of the resistance.

Capacitance Measurements

For capacitance measurements, the electrodes were produced by silver-paint deposit followed by firing to ensure the necessary bonding between the sintered silver film and the ceramic by fluxing. The capacitances are measured at 1 kcps with an AC voltage always below 1 mV. The comparison network was especially designed for this investigation.

EXPERIMENTAL RESULTS

Reduction of Barium Titanate

To determine the barium titanate reduction kinetics, tests were carried out for different periods of time at fixed temperatures suitably selected within the range indicated by various authors as being that during which reduction took place [4, 5, 14-16]. The process was carried out at 900, 975, 1050, and 1150°C for 4, 16, 36, and 121 min.

The tests were made on twenty specimens. For each test, the distribution of the resistance values was plotted on a Gaussian-scale diagram (to obtain a straight-line Henry curve). Figure 1 shows, for example, the statistical distribution of the results obtained after reduction at 825°C. In Table I, the mean and the tenth and ninetieth percentile values of the electrical resistance at 25°C have been noted to give a picture of the dispersion in the results for each test.

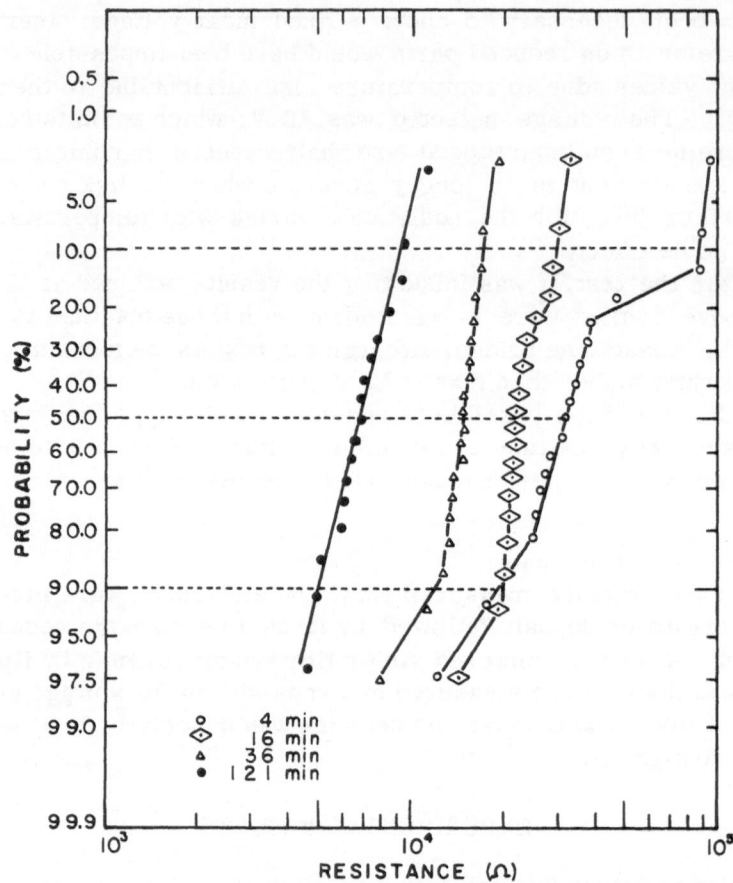

Fig. 1. Typical statistical distribution of resistances obtained by reduction of BaTiO$_3$ disks with hydrogen at 825°C. Diameter, 17.1 mm; thickness, 1.21 mm; tested at 12 V DC.

Variation of Mean Value and Dispersion of Results

Figure 2 shows the experimental results for the variation in resistance as a function of reduction time at various temperatures. As a first approximation, Fig. 3 shows a linear decrease of log R as a function of log t, and allowing for errors in the measurements, the mean value can be calculated, assuming the law log $R = K$ log t is verified. The mean values actually measured, as compared with those which can be calculated from this figure, are given in Table II.

TABLE I

Mean and Extreme Resistance Values for Hydrogen-Reduced $BaTiO_3$ Disks

Time of reduction, min	Resistance, Ω		
	R_{mean}	$R_{10\%}$	$R_{90\%}$
Normally fired	$5.5 \cdot 10^9$	$3.6 \cdot 10^{10}$	$2.75 \cdot 10^9$
825°C			
4	$3.4 \cdot 10^4$	$8.9 \cdot 10^4$	$1.9 \cdot 10^4$
16	$2.2 \cdot 10^4$	$2.95 \cdot 10^4$	$2.0 \cdot 10^4$
36	$1.5 \cdot 10^4$	$1.75 \cdot 10^4$	$1.2 \cdot 10^4$
121	$6.9 \cdot 10^3$	$9.5 \cdot 10^3$	$5.0 \cdot 10^3$
900°C			
4	$4.5 \cdot 10^3$	$5.4 \cdot 10^3$	$3.6 \cdot 10^3$
16	$3.5 \cdot 10^3$	$4.9 \cdot 10^3$	$2.8 \cdot 10^3$
36	$1.1 \cdot 10^3$	$1.3 \cdot 10^3$	$7.8 \cdot 10^3$
121	$1.0 \cdot 10^3$	$1.45 \cdot 10^3$	$4.8 \cdot 10^3$
975°C			
4	$3.3 \cdot 10^3$	$3.6 \cdot 10^3$	$2.5 \cdot 10^3$
16	$3.5 \cdot 10^2$	$5.70 \cdot 10^2$	$6.4 \cdot 10^1$
36	$6.5 \cdot 10^1$	$8.7 \cdot 10^1$	$3.9 \cdot 10^1$
121	$4.9 \cdot 10^1$	$7.2 \cdot 10^1$	$2.2 \cdot 10^1$
1050°C			
16	$3.3 \cdot 10^1$	$7.2 \cdot 10^1$	$1.7 \cdot 10^1$
36	$1.9 \cdot 10^1$	$4.4 \cdot 10^1$	$1.0 \cdot 10^1$
121	9.3	$3.6 \cdot 10^1$	5.4
1150°C			
4	$1.43 \cdot 10^1$	$3.3 \cdot 10^1$	6.2
16	9.4	$2.0 \cdot 10^1$	6.4
36	3.2	$1.85 \cdot 10^1$	1.15
121	2.7	7.5	1.2

Disks: diameter, 17.1 mm; thickness, 1.21 mm; tested at 12 V DC; temperature, 25°C.

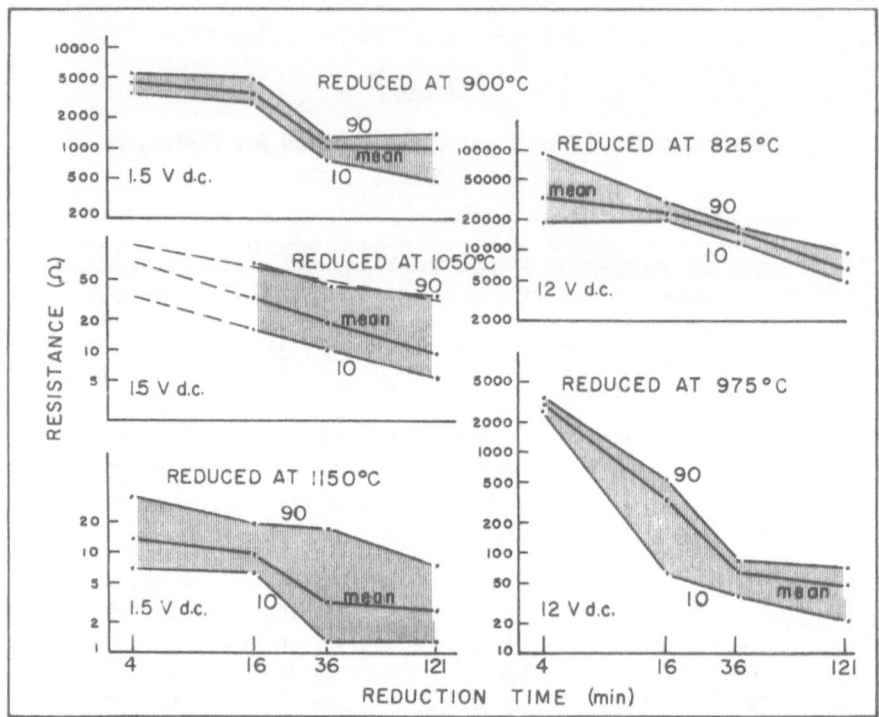

Fig. 2. Effects of time and temperature on electrical resistance of hydrogen-reduced BaTiO$_3$ disks. Mean and extreme values of statistical distributions. Diameter, 17.1 mm; thickness, 1.21 mm.

Calculation of the Activation Energy in Reduction Reactions

If it is assumed (by reasonable inference from the foregoing results) that the law of the variation of the resistance as a function of time at constant temperature is expressed by

$$R = at^n \tag{2}$$

where a and n are constants, then

$$\log R = \log a + n \log t \tag{2a}$$

if $n \neq 0$.

The values thus calculated for each temperature are given in Table III. The value of n is practically constant over the temperature region considered and is approximately equal to $-\frac{1}{2}$.

The rate constant should vary according to the law of Arrhenius—

$$a = k \exp(-W/RT) \qquad (3)$$

where T is the absolute temperature and R is the Boltzmann constant.

Figure 4 shows $\log a = f(1/T)$. The curve is a straight line; on first approximation, its slope provides an approximate value of the activation energy of the reaction. The calculation yields $W \approx -3.30$ eV, which may give an idea of the value of the energy required to extract an atom of oxygen from the structure of $BaTiO_3$.

Variation of the Electrical Resistance of Barium Titanate
with Temperature

The electrical resistance of normally fired barium titanate, measured on a disk 17.1 mm in diameter and 1.21 mm thick, varies with temperature as shown by the curve in Fig. 5. A large variation in the curve slope is noted at each allotropic transformation point.

After reduction, regular curves are obtained (Fig. 6). Discon-

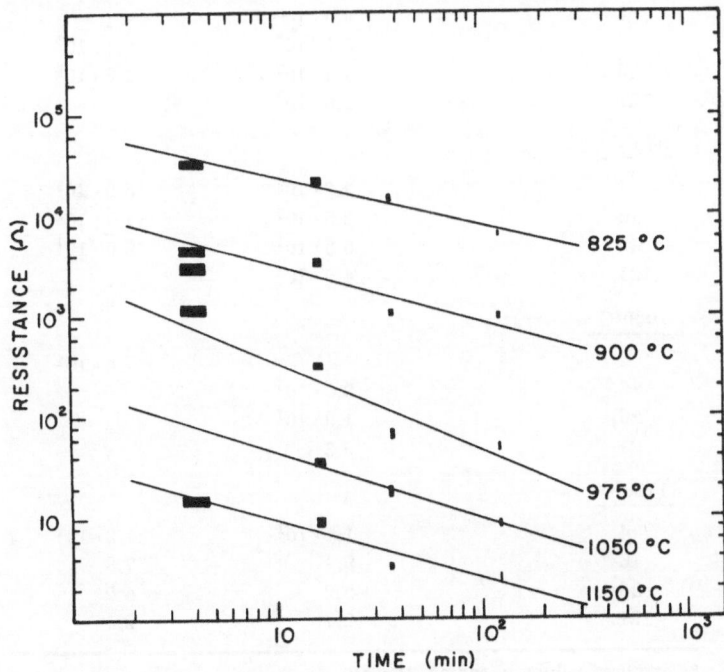

Fig. 3. Variation of electrical resistance with time at different temperatures for hydrogen-reduced $BaTiO_3$ disks. Diameter, 17.1 mm; thickness, 1.21 mm; tested at 12 V DC.

TABLE II

Comparison of Experimental and Computed Mean Resistance Values for Hydrogen-Reduced BaTiO$_3$ Disks

Time of reduction, min	Resistance, Ω	
	Measured	Computed*
Normally fired	$5.5 \cdot 10^9$	—
825°C		
4	$3.4 \cdot 10^4$	$4.8 \cdot 10^4$
16	$2.2 \cdot 10^4$	$2.1 \cdot 10^4$
36	$1.5 \cdot 10^4$	$1.4 \cdot 10^4$
121	$6.9 \cdot 10^3$	$7.1 \cdot 10^3$
900°C		
4	$4.4 \cdot 10^3$	
16	$3.5 \cdot 10^3$	$2.4 \cdot 10^3$
36	$1.1 \cdot 10^3$	$1.6 \cdot 10^3$
121	$1.0 \cdot 10^3$	
975°C		
4	$3.3 \cdot 10^3$	$5.0 \cdot 10^3$
16	$3.5 \cdot 10^2$	$1.7 \cdot 10^2$
36	$6.5 \cdot 10^1$	$9.0 \cdot 10^1$
121	$4.9 \cdot 10^1$	
1050°C		
4		$7.5 \cdot 10^1$
16	$3.3 \cdot 10^1$	
36	$1.9 \cdot 10^1$	
121	9.3	
1150°C		
4	$1.4 \cdot 10^1$	$1.9 \cdot 10^1$
16	9.4	7.8
36	3.2	4.2
121	2.7	2.1

*Computed values based on $R = at^n$.

TABLE III

Computation of the Reaction Rate Constant [Variation of Resistance with Time, equation (2)] at Various Temperatures for Hydrogen-Reduced $BaTiO_3$

Temperature (°C)	t_1 (min)	t_2 (min)	R_1 (Ω)	R_2 (Ω)	n	$\log a$	a
825	4	$2 \cdot 10^2$	$3.6 \cdot 10^4$	$5.9 \cdot 10^3$	−0.463	11.133	$6.84 \cdot 10^4$
900	4	$2 \cdot 10^2$	$5.5 \cdot 10^3$	$5.9 \cdot 10^2$	−0.571	9.394	$1.20 \cdot 10^4$
975	4	$2 \cdot 10^2$	$7.5 \cdot 10^2$	$2.4 \cdot 10^1$	−0.878	7.837	$2.53 \cdot 10^3$
1050	4	$2 \cdot 10^2$	$8.0 \cdot 10^1$	6.2	−0.656	5.292	$1.99 \cdot 10^2$
1150	4	$2 \cdot 10^2$	$1.6 \cdot 10^1$	1.7	−0.572	3.153	$2.34 \cdot 10^1$

tinuities due to allotropic transformations seem to have completely vanished. An X-ray crystallographic analysis of the reduced disks shows that the structure is tetragonal at room temperature and that there is no change in the value of the parameters as compared with a nonreduced ceramic. The results found are $c = 4.030$ A, $a = 3.991$ A, and $c/a = 1.010$. It, therefore, seems that under these conditions, the tetragonal phase has been stabilized to the detriment of the cubic phase, contrary to the opinion of some authors [14]. This is an interesting finding, for it appears that these compounds may be ferroelectric semiconductors.

The determination at room temperature [16] of the Hall constant on a specimen reduced for 121 min at 1150°C yields a carrier concentration of approximately $10^{19}/cm^3$ and a mobility of the order of $20 \cdot 10^{-2} \ cm^2/V$-sec.

Effects of Rare Earth Oxides

Reduction processes in barium titanate are accelerated by the presence of ions, such as Ce^{+4}, Nd^{+3}, Sm^{+3}, and La^{+3}. On firing in a noncontrolled atmosphere, i.e., in open air, the barium titanate, to which small quantities of these oxides have been added, exhibits a bluish-grey color which evinces the presence of Ti^{+3} ions. Such phenomena are confirmed by many authors [4, 5]. Moreover, large variations in electrical resistance are then found as compared with that measured on equivalent pure barium titanate disks. This is actually a special application of the general reduction method

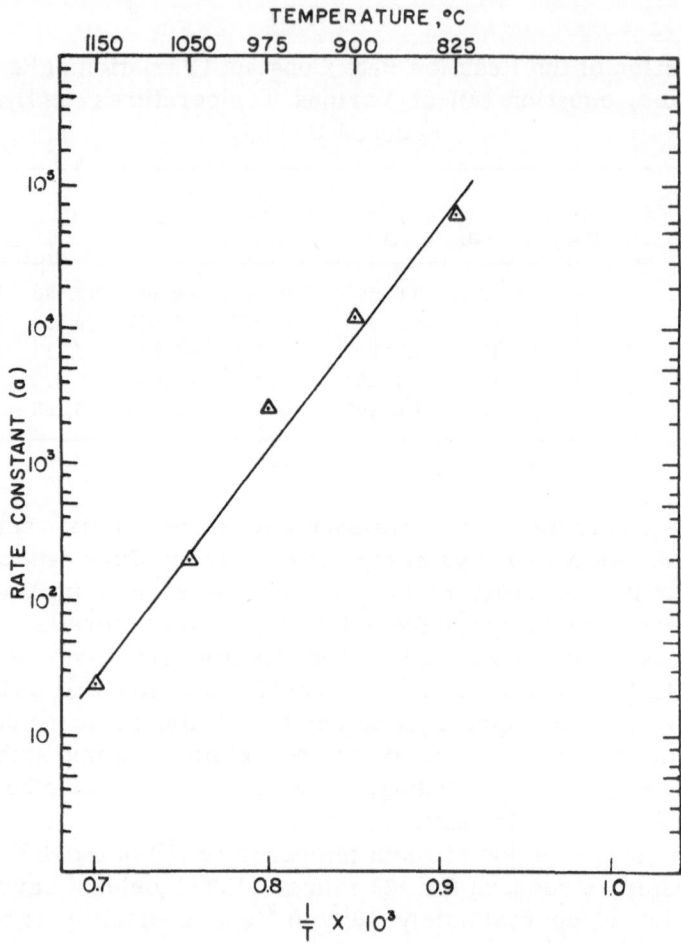

Fig. 4. Variation of reaction rate constant with temperature for hydrogen-reduced BaTiO₃.

devised in 1950 by Verwey [18] known as the "induced valence method," studied by many authors [5, 18, 19]. Figure 7 shows the effects of various oxides, according to their concentration, on the electrical resistance of disks 17.1 mm in diameter and 0.8 mm thick. In all cases, firing had been effected at 1380°C for 1 hr. The values were measured with sintered silver electrodes. The silver paint used contained 70% of silver grains with a diameter less than $1\,\mu$ and 4% of lead borosilicate the purpose of which was to provide the necessary bond between the sintered silver layer and the ceramic base. The latter factor accounts for the comparatively high values of the electrical resistances measured. Nevertheless, the variations with concentration remain applicable.

It has not been possible to produce specimens containing 0.5 and 1.5 mol.% of rare earth oxide free from open pores. From the crystallographic standpoint, a change in structure is noted at room temperature when the concentration exceeds 1.5%. For example, with a 2.5% concentration of cerium oxide, a cubic structure is formed with a lattice parameter of 4.020 ± 0.002 A.

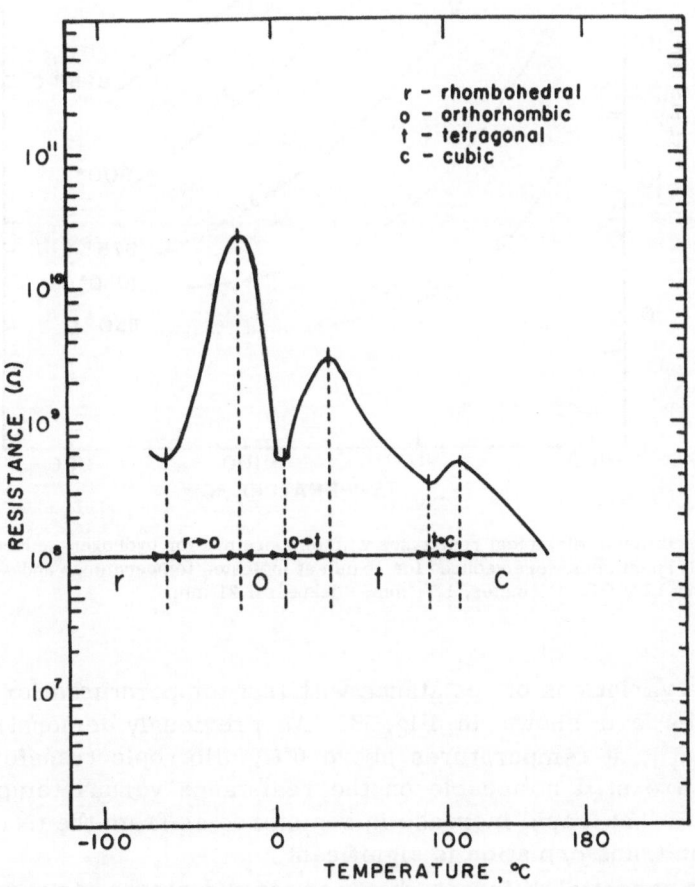

Fig. 5. Variation of electrical resistance with temperature for normally fired BaTiO₃ disks. Diameter, 17.1 mm; thickness, 1.21 mm; tested at 12 V DC. Allotropic phases identified by X-ray crystallography.

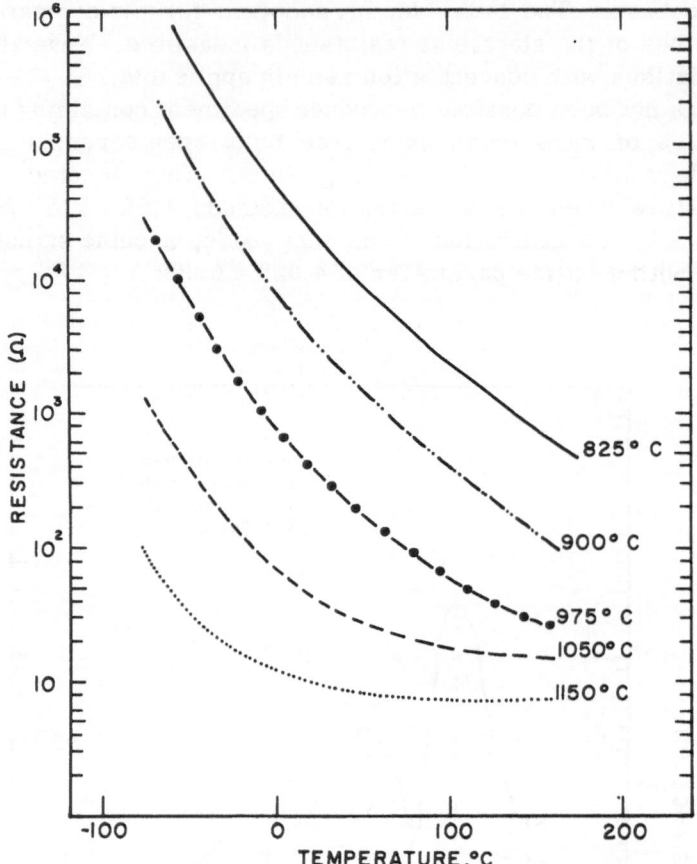

Fig. 6. Variation of electrical resistance with temperature for hydrogen-reduced $BaTiO_3$ disks. All specimens were reduced for 16 min at indicated temperatures and were tested at or below 12 V DC. Diameter, 17.1 mm; thickness 1.21 mm.

The variations of resistance with test temperatures for various additives are shown in Fig. 8. As previously demonstrated by Saburi [5], at temperatures above 0°C, allotropic transformation points are still noticeable on the resistance versus temperature curves. The rapid increase in resistance as from the tetragonal-to-cubic transformation is significant.

The essential difference between the phenomena occurring in the reduction of pure barium titanate by hydrogen and by rare earth additives is that, in the former case, the allotropic transformation points of the various phases of barium titanate vanish (at least they vanish within the temperature region investigated, i.e., -70 to +170°C). In both instances, at normal temperature, the structure remains tetragonal. When the rare earth oxide concentration is higher than 1.5 mol.%, the lattices are cubic at room temperature.

Surface Reoxidation

The action of oxygen on reduced semiconductive barium titanate produces an insulating reoxidized surface layer, the oxygen con-

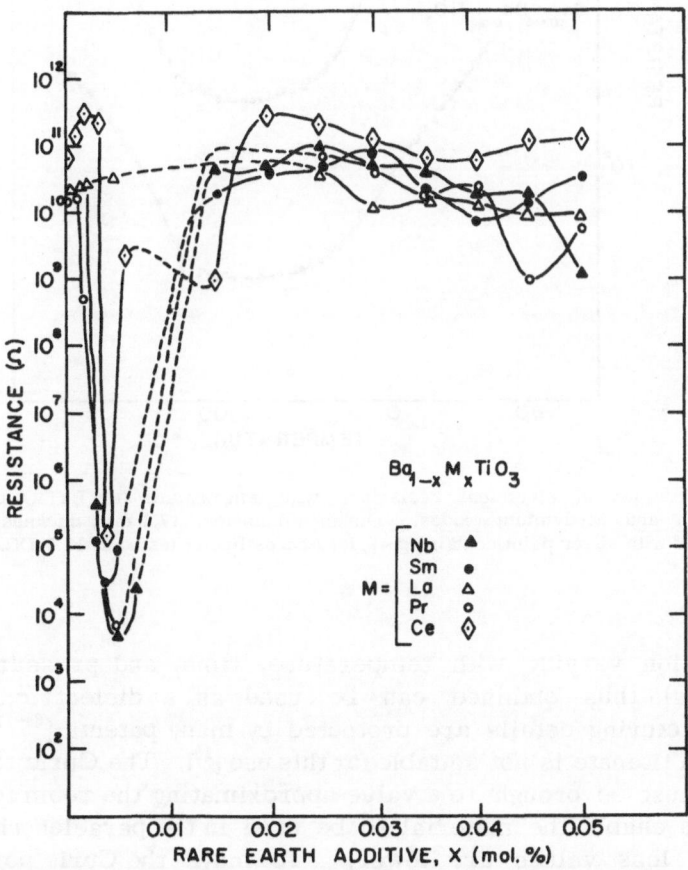

Fig. 7. Variation of electrical resistance of BaTiO₃ doped with increasing contents of rare earth oxides. Disks: diameter, 17.1 mm; thickness, 0.8 mm; air-fired 1 hr at 1380°C.

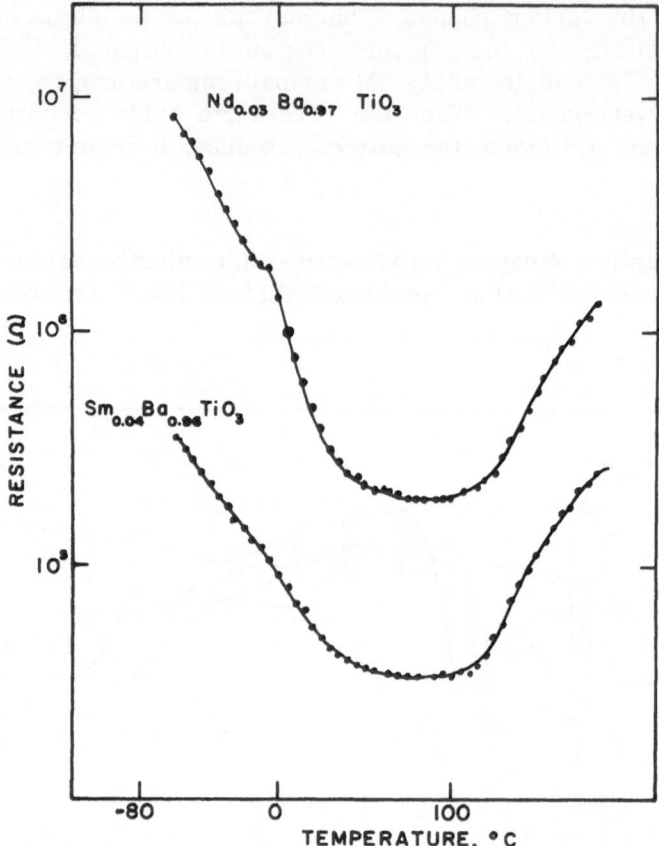

Fig. 8. Variation of electrical resistance with temperature for $BaTiO_3$ doped with samarium and neodymium oxides. Disks: diameter, 17.1 mm; thickness, 0.8 mm; electroded with silver paint containing 4% lead borosilicate; tested at 12 V DC.

centration varying with temperature, time, and pressure. The material thus obtained can be used as a dielectric; specific manufacturing details are protected by many patents [6-11]. Pure barium titanate is not suitable for this use [15]. The Curie temperature must be brought to a value approximating the room temperature to enable the material to be used in the paraelectric region where loss values are lower. To move the Curie point while flattening it, cerium oxide has been used. Figure 9 shows the variation of the curves obtained by plotting $\Delta C/C_{25°C}$ versus temperature for various cerium oxide concentrations, as well as showing the relative dielectric constants ϵ/ϵ_0 measured at 25°C.

The compound selected for this investigation contained 2.5 mol.% of CeO_2. Its Curie point was at about 40°C; its dielectric constant approximated 3500 with $\tan\delta < 300 \cdot 10^{-4}$. The material reduced for 1 hr at 1075°C has a resistivity of about 20 Ω-cm. The curves shown in Fig. 10 were plotted using the statistical method, and only the mean values are shown. In this figure, the simultaneous varia-

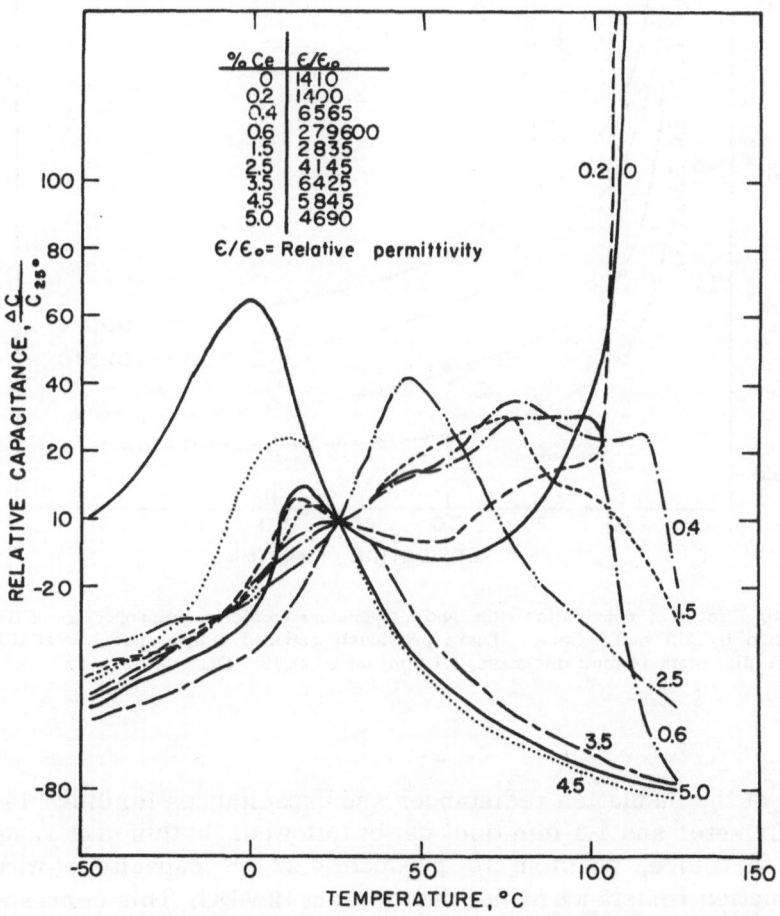

Fig. 9. Influence of CeO_2 doping on the temperature dependence of capacitance and dielectric constant (relative to 25°C) of $BaTiO_3$. Air-fired for 1 hr at 1380°C; CeO_2 concentrations (mol.%) as indicated.

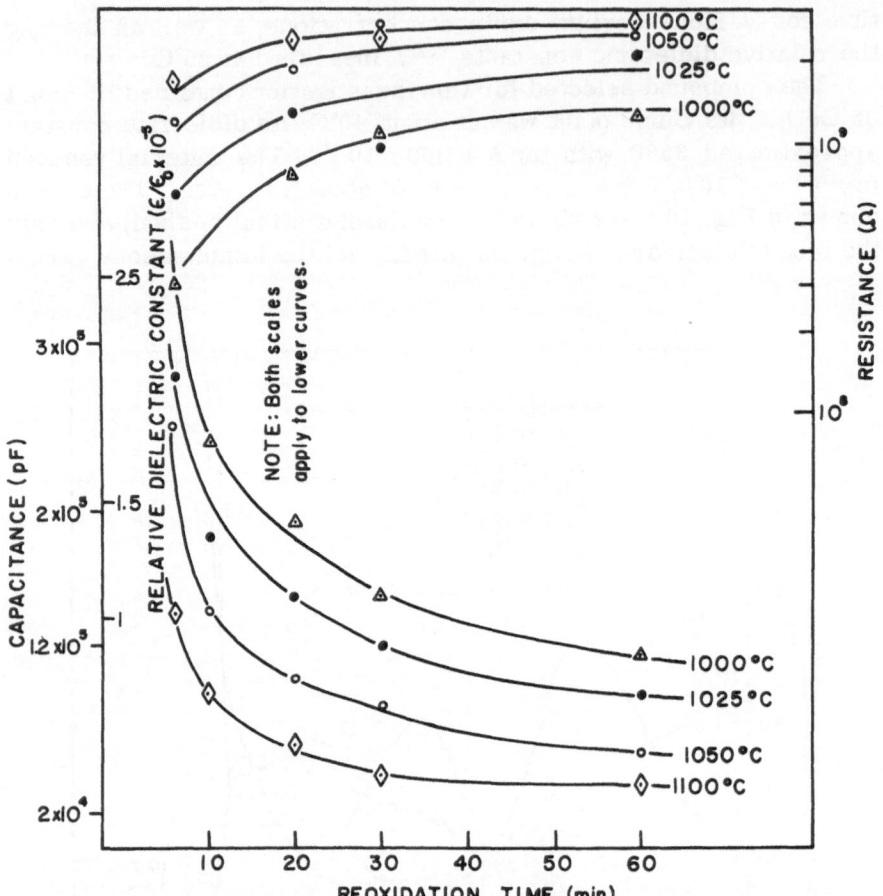

Fig. 10. Effects of reoxidation time and temperature on dielectric properties of $BaTiO_3$ modified by 2.5 mol.% CeO_2. Disks previously reduced in hydrogen (1 hr at 1075°C). Disks: diameter, 14 mm; thickness, 1.1 mm; tested at 12 V DC.

tion of the insulation resistances and capacitances for disks 14 mm in diameter and 1.1 mm thick can be followed. In this size range, it is, therefore, possible to produce 0.22 μF capacitors with an insulation resistance higher than 10^8 Ω at 12 V DC. This corresponds to an apparent dielectric constant of about 200,000.

Higher apparent dielectric constant values can be achieved by eliminating the reoxidation stage. The barrier layer can be obtained through the action of a flux contained in the silver paint used for electroding.   Values exceeding 500,000 can be achieved by adding about 0.5 wt.% Bi_2O_3 or lead borosilicate to the paint.   The operating voltage of such a dielectric is then much lower (in general, 3–4 V).

Figure 11 gives the curves showing the variations in the characteristics of a dielectric obtained by reduction followed by reoxidation, as a function of voltage and frequency.   The equivalent

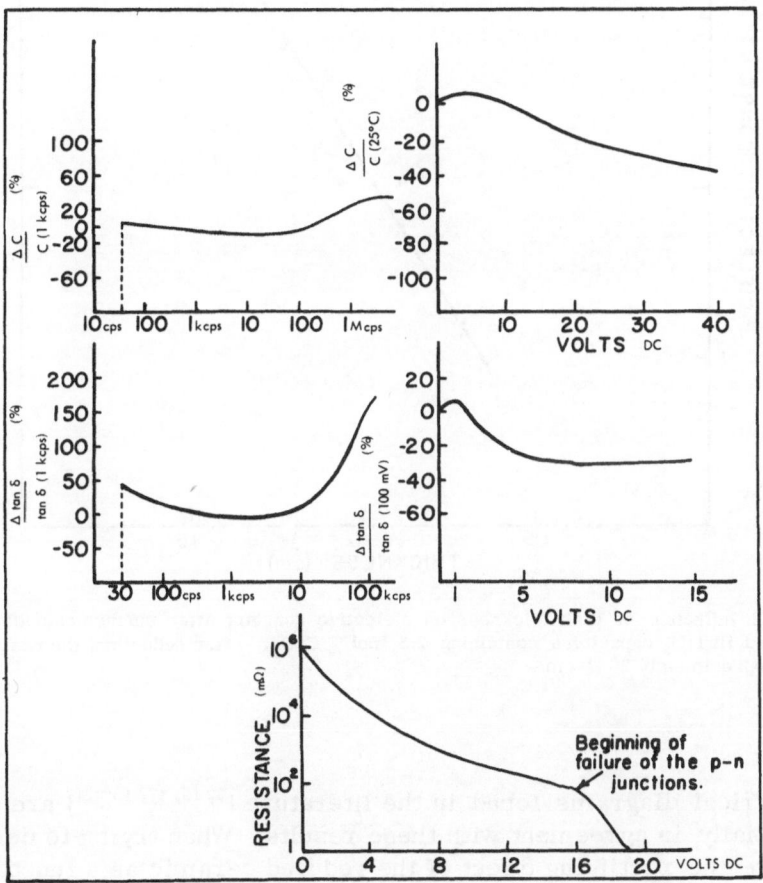

Fig. 11. Variations in relative capacitance, relative dielectric loss angle, and resistance with frequency and voltage after surface reoxidation of reduced $BaTiO_3$ capacitors containing 2.5 mol.% CeO_2.   Disks: diameter, 14 mm; thickness, 0.8 mm.   Curves correspond to 0.22 μF capacitors operating at 12 V DC.

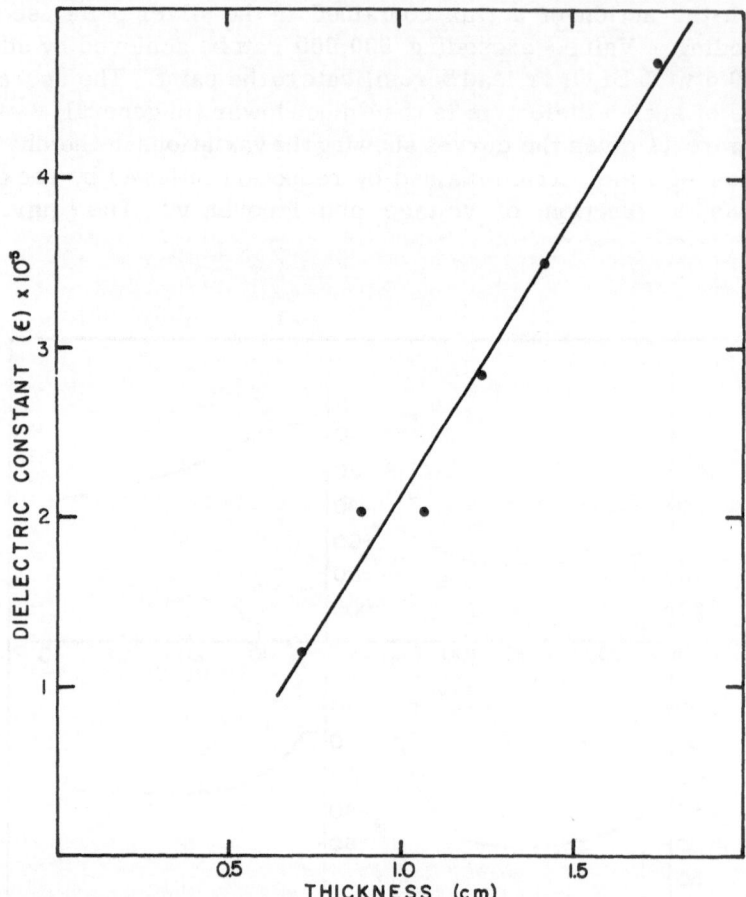

Fig. 12. Influence of disk thickness on dielectric constant after surface reoxidation of reduced BaTiO₃ capacitors containing 2.5 mol.% CeO₂. After reduction, the resistivity was approximately 20 Ω-cm.

electrical diagrams found in the literature [5, 15, 17, 19—21] are substantially in agreement with these results. When trying to demonstrate the rectifying effect of the reduced ceramic as a function of reoxidation [22], it was found that the capacitive term in parallel had a value slightly higher than the capacitance of the capacitor.

Lastly, it was found that the capacitance of reduced—reoxidized ceramic disks increased, all other conditions being equal, with the thickness of the dielectric. As can be seen from Fig. 12, the increase is practically linear for comparatively large variations in thickness.

Therefore, we must admit that, in addition to the surface functions, a phenomenon related to the space charge occurs and sets up a third capacitive component in series, resulting in an effective circuit equivalent to the one shown in Fig. 13. Additional tests are being carried out.

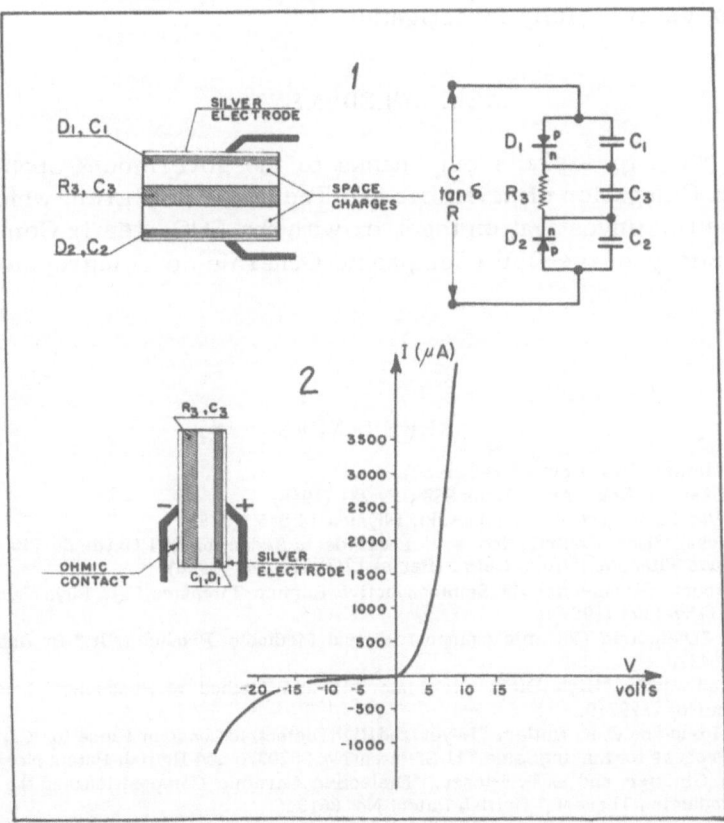

Fig. 13. Capacitors produced by surface reoxidation of reduced titanates. (1) Schematic and equivalent electrical circuit diagrams for disks reoxidized on both surfaces. (2) Schematic diagram and current–voltage curve for a capacitor after removal of one reoxidized surface layer.

CONCLUSIONS

The application of reduction–reoxidation techniques to perovskite-structured ceramics may be of great interest in practice.

When reduced, these ceramics can be considered as resistances that may or may not vary linearly with temperature, according to the preparation technique. Moreover, these semiconductors must have marked ferroelectric properties, which may be of interest in solving certain detection problems.

After reoxidation, dielectrics are obtained with an apparent permittivity approximating 500,000, a value not necessarily restrictive. In our opinion, the basic processes leading to such results have not yet been fully investigated.

ACKNOWLEDGMENTS

We wish to express our thanks to the government-sponsored General Delegation of Scientific and Technical Research, which has financed this investigation together with the LCC-Stéafix Company, a subsidiary of the CSF (Compagnie Générale de Télégraphie Sans Fil).

REFERENCES

1. J. C. Slater, Phys. Rev. 78:748 (1950).
2. G. J. Skanavi, Dokl. Akad. Nauk SSSR 59:231 (1948).
3. J. H. Van Santen and W. Opechowski, Physica 14:545 (1948).
4. R. Wyss, "Une Contribution à l' Étude de la Réduction de l'Oxyde de Titane et de Quelques Titanates," Ann. Chim. (Paris) 12(3):215-242 (1948).
5. O. Saburi, "Properties of Semiconductive Barium Titanates," J. Phys. Soc. Japan 14(9):1159-1174 (1959).
6. RCA, "Dielectric Ceramic Composition and Method of Producing It," British Patent No. 664370.
7. C. Wentworth, "High Dielectric Materials and Method of Producing Them," U.S. Patent No. 2529719.
8. R. R. Roup and C. E. Butler, "Layerized High Dielectric Constant Piece for Capacitors and Process for Making Same," U. S. Patent No. 2520376 and British Patent No. 682794.
9. R. M. Glaister and G. V. Planer, "Dielectric Ceramic Composition and the Method of Production Thereof," British Patent No. 861346.
10. Siemens, German Patent No. 964,020 (1959) and French Patent No. 1262998 (1961).
11. P. Frappart, French Patent No. 853535 (1961).
12. P. Basseville, Marche d'Études DGRST, No. 64 FR·017:7 (1964).
13. O. Saburi and K. Wakind, "Processing Techniques and Applications of Positive Temperature Coefficient Thermistors," IEEE Trans. Compt. Parts CP-10(2):53-67 (1963).

14. R. C. de Vries, "Lowering of Curie Temperature of $BaTiO_3$ by Chemical Reduction," J. Am. Ceram. Soc. 43(4):226 (1960).
15. R. M. Glaister, "Barrier-Layer Dielectrics," Proc. Inst. Elec. Engrs. 109(22B):423-431 (1961).
16. J. Grosvalet and M. Laviron, CSF, personal memorandum (1964).
17. G. G. Blowers, G. T. Hurry, and F. Welsby, "Ceramics Capacitors," Proc. Inst. Elec. Engrs. 109(22B):466-471 (1961).
18. E. J. W. Verwey, P. W. Haayman, F. C. Romeyn, and G. W. Van Costerhout, Philips Res. Rept. 5(3):173 (1950).
19. J. Suchet, Bull Soc. Franc. Elec. 5(53):274-294 (1955); see also J. Suchet, Chimie Physique des Semiconducteurs, Dunod (Paris), 1961, pp. 127-128.
20. C. G. Koops, "On the Dispersion of Resistivity and Dielectric Constant of Some Semiconductors at Audio Frequencies," Phys. Rev. 83(1):121-124 (1951).
21. L. Nicolini, Italian Patent No. 492,404.
22. J. Guyonnet, LCC-Stéafix Internal Report No. 2122 (Feb. 1962) and No. 2181 (March 1962).

Chapter 7

Variation of Dielectric Constant and Spontaneous Strain With Density in Polycrystalline Barium Titanate

H. H. Stadelmaier and S. W. Derbyshire

North Carolina State University
Raleigh, North Carolina

A simple model of a porous ferroelectric explains the change of the dielectric constant with density. This model agrees well with the experimental results. Certain barium titanate powders show a low spontaneous strain so that their structure at room temperature appears to be cubic rather than tetragonal. Compacts prepared from such a powder exhibit an increase in spontaneous strain with increasing density, and this is explained with a model that is based on Devonshire's thermodynamic theory.

INTRODUCTION

The complexities of the field distribution at phase interfaces defy accurate analysis. Yet it is often feasible to obtain a reasonable model with the help of averaging procedures. Such a method was explored for the ordinary grain boundary in an earlier paper [1]. Here a similar method is applied to the porous ferroelectric.

POLARIZATION CURVE AND DENSITY

To describe the heterogeneous mixture of ferroelectric particles and voids, we introduce a simple model borrowed from electrical conductivity. There, heterogeneous bodies show a behavior that lies between additivity of resistances and additivity of conductances [2]. The electrostatic equivalent is a series of disks parallel to the plates of a capacitor and a number of cylindrical rods parallel to the field. In these limiting cases, the field in the ferroelectric is nearly homogeneous, and the analysis is greatly simplified. That the rods come closer to a correct description of the porous ferroelectric will be demonstrated below.

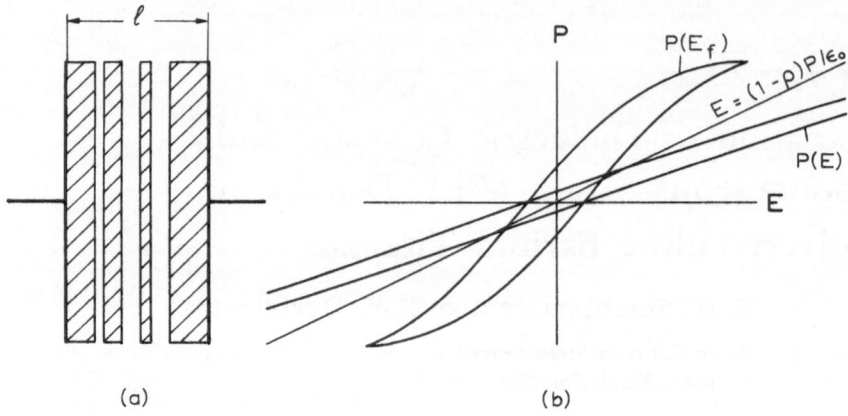

(a)   (b)

Fig. 1. Parallel-plate model of ferroelectric resulting in shearing of hysteresis loop.

For the model shown in Fig. 1(a), the line sum of the field strength is

$$E_a l_a + E_f l_f = El \tag{1}$$

where l_a is the sum of all air gap widths, l_f is the sum of the disk widths, and l is the plate-to-plate distance in the capacitor. The subscripts a and f refer to air and ferroelectric, respectively, and E is the applied field. The flux density must be continuous; hence, it follows that

$$D_a = \epsilon_0 E_a = D_f \tag{2}$$

and, observing that $D = \epsilon_0 E + P$,

$$\epsilon_0 E_a = \epsilon_0 E_f + P \tag{2a}$$

By substituting equation (2a) into equation (1), E_a is eliminated, and one finally obtains

$$\epsilon_0 E = \epsilon_0 E_f + P l_a/l \tag{3}$$

For an ellipsoidal body in the field $E, l_a/l$ is equal to the depolarization factor N. For our model, l_a/l is also equal to $1-\rho$, where ρ is the fraction of theoretical density. The effect of the air gaps is to shear the polarization curve $P = f(E_f)$ along the abscissa because the field $\epsilon_0 E_f$ is increased by $(1 - \rho)P$, as shown in Fig. 1(b). Because of the large value of P in ferroelectrics, the polarization curve is completely turned into the abscissa for densities as high as $\rho = 0.99$. This does not agree with the experiment, and, therefore, the disks do not represent a good approximation to the behavior of a porous ferroelectric. We have merely succeeded in demonstrating the well-known fact that even a small air gap between the ferroelectric and the electrodes will have disastrous effects on the electrical properties.

The model of parallel rods is shown in Fig. 2(a). Because the potential on either electrode is constant, the charge distribution cannot be uniform, and, hence, the flux DA (where A is the electrode area) is an average. It is given by

$$DA = D_f A_f + D_a (A - A_f) \tag{4}$$

where the subscripts a and f refer to air and ferroelectric, respectively, and A_f is the total cross-sectional area of the rods. The average flux density is then

$$D = D_f A_f/A + D_a(1 - A_f/A)$$

or, with $D = \epsilon_0 E + P$,

$$\epsilon_0 E + P = (\epsilon_0 E_f + P_f)A_f/A + (\epsilon_0 E_a + P_a)(1 - A_f/A) \tag{5}$$

Because of the constant potential on the electrodes, it follows that $E_a = E_f = E$. Since the polarization outside the ferroelectric P_a

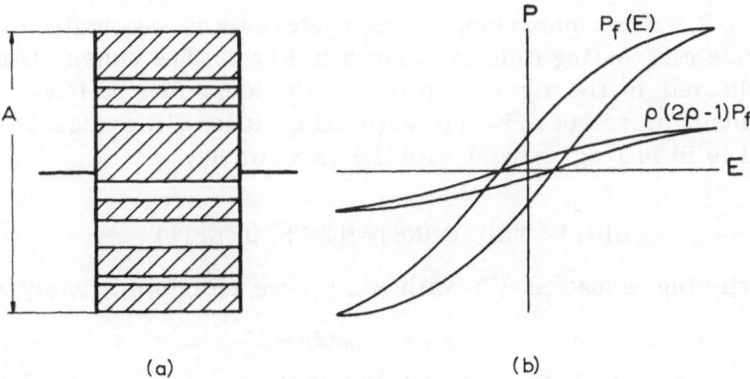

(a)   (b)

Fig. 2. Parallel-rod model of ferroelectric resulting in vertical contraction of hysteresis loop.

must be zero, equation (5) can be written as

$$P = P_t A_t / A = P_t \rho \tag{6}$$

where P is the average polarization and ρ is the density fraction introduced previously. To account for the fact that we do not actually have rod-shaped particles extending from plate to plate, we add a shape factor $s(\rho)$ so that equation (6) now becomes

$$P = P_t s(\rho) \rho$$

The simplest estimate of $s(\rho)$ is obtained by the assumption of a checkerboard arrangement at $\rho = 0.5$. For this arrangement, all rodlike connections are interrupted so that $s = 0$ for $\rho = 0.5$. At $\rho = 1$, the capacitor is filled with a fully dense ferroelectric so that $s = 1$. The simplest description of the filling of the checkerboard is given by connecting these two limits with a linear function. Then the shape factor is $s(\rho) = 2\rho - 1$ and equation (6) finally becomes

$$P = P_t \, \rho(2\rho - 1) \tag{7}$$

It should be understood that the shape factor is merely a device to help us mix the two models of disks and rods. We have visibly included the results from the rod model, but also implicitly those of the disk model, because for it the polarization is zero except for $\rho \to 1$, where it rises steeply to the value of the fully dense body. Therefore, $s(\rho)$ can be said to represent the fraction of rodlike behavior and the remaining $(1 - s)$ the fraction of disklike behavior. The effect of equation (7) is to contract the polarization curve $P(E)$ along the P-axis. This is shown in Fig. 2(b) for a value of $\rho = 0.7$. For comparison, some hysteresis loops obtained with a 60-cycle alternating field are shown in Fig. 3. The density fraction is indicated in the figure. Although the loops for the low-density specimens were not driven to saturation, it is still evident that the model is in fair agreement with the experiment.

DIELECTRIC CONSTANT VS. DENSITY

Utilizing equation (7) with $E = E_t$, we can immediately write

$$\chi(\rho) = \chi_t \rho(2\rho - 1) \tag{8}$$

where $\chi(\rho)$ is the bulk susceptibility of the porous body and χ_t is the susceptibility of a fully dense ferroelectric. For the high suscepti-bilities of a ferroelectric, it also follows that the dielectric

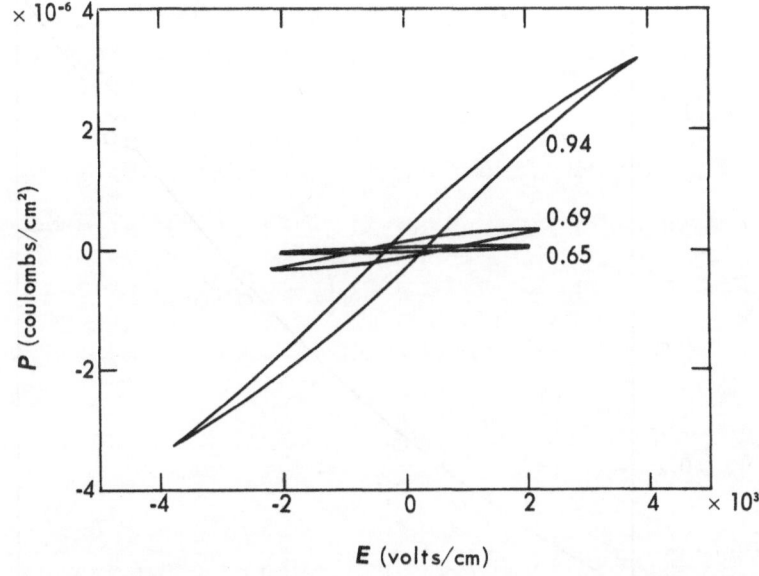

Fig. 3. Hysteresis loops of barium titanate of three different densities.

constant $\epsilon \approx \chi$, so the two can be used interchangeably. To test the validity of equation (8), the low-field (25 V/cm) dielectric constant at 1000 cps for sintered high-purity barium titanate of varying density is shown in Fig. 4. The agreement with the calculated curve is seen to be good. Since χ_i is not actually known, the experimental points had to be fitted to the curve. This was done at the high end of the scale. One obvious result of this analysis is that increasing the density from $\rho = 0.95$ (a typical experimental value) to the unattainable $\rho = 1$ yields only an additional 16% in ϵ. Therefore, no spectacular increase in dielectric constant can be expected from an attempt to push the density to the limit.

SPONTANEOUS STRAIN VS. DENSITY

Spontaneous strain is the mechanical manifestation of spontaneous polarization. In tetragonal barium titanate, it represents the difference between the axial ratios of the polarized and unpolarized crystal, or $(c/a) - 1$, where a and c are the lattice parameters. When the spontaneous strain is zero, the structure is cubic.

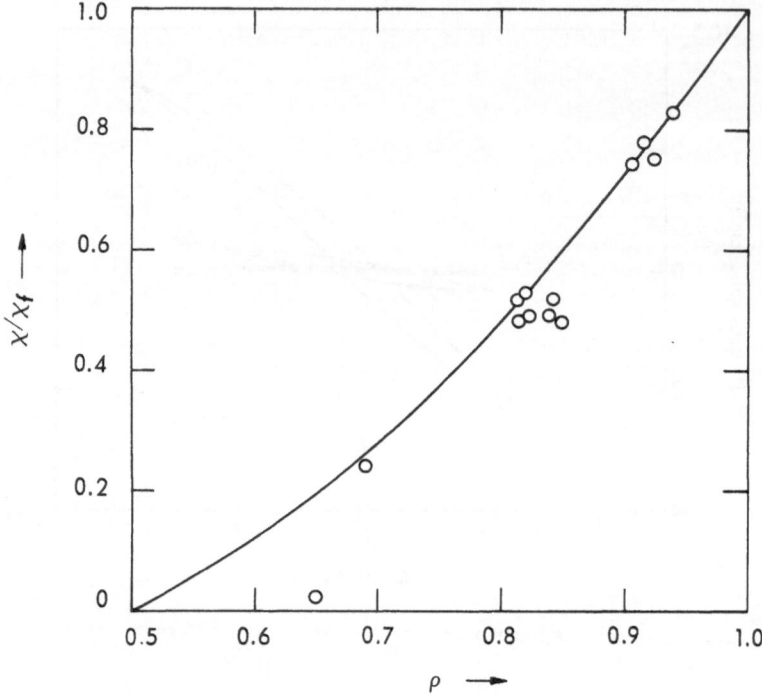

Fig. 4. Susceptibility versus density curve, calculated according to equation (8).

If the Devonshire thermodynamic model of a ferroelectric [3] is valid, it must be capable of explaining changes in spontaneous polarization, e.g., the loss of spontaneous polarization observed in small particles of barium titanate [4]. To understand this, consider a simplified Devonshire treatment of a ferroelectric crystal. Admitting only two directions of polarization, represented by positive and negative values of P, one obtains the free-energy curve shown in Fig. 5. The polarization curve also shown in Fig. 5 is derived by observing that the field is $E_f = d\Delta F(P)/dP$. If switching from negative to positive polarization is treated as a problem in heterogeneous nucleation, it will be seen that the branch 2 – 4 of the polarization curve is not stable, and that, ordinarily, the sequence 1 – 2 – 3 is followed leading to the familiar hysteresis loop. Shearing the polarization curve in a depolarizing field according to equation (3) could stabilize the central branch 2 – 4 and lead to zero

spontaneous polarization in a zero field.   In the presence of a
depolarizing field, which according to equation (3) is proportional
to the polarization of the ferroelectric, the energy will be increased
by a term that is proportional to the square of the polarization.
By adding this energy to the free energy of Fig. 5, we obtain
the dashed curve with its lower value of spontaneous polarization (P where ΔF has a minimum).   If the depolarizing field
is high enough, ΔF will have a single minimum at $P = 0$.   In view of
this tendency toward reduced spontaneous polarization, why do real
crystals show unreduced spontaneous strain in a depolarizing field?
The obvious answer is also found in Fig. 5 where it is seen that a
mixture of oppositely polarized domains (point on broken line) has
a lower energy than a single domain whose polarization is reduced
in a depolarizing field (minimum of dashed curve).   By using the
field rather than the free energy, this analysis becomes independent
of any details of the polarization curve.   This is shown next.

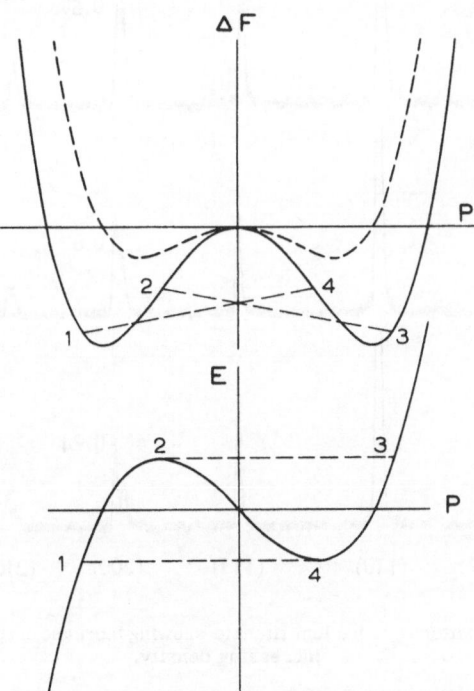

Fig. 5. Free energy versus polarization and field versus polarization.

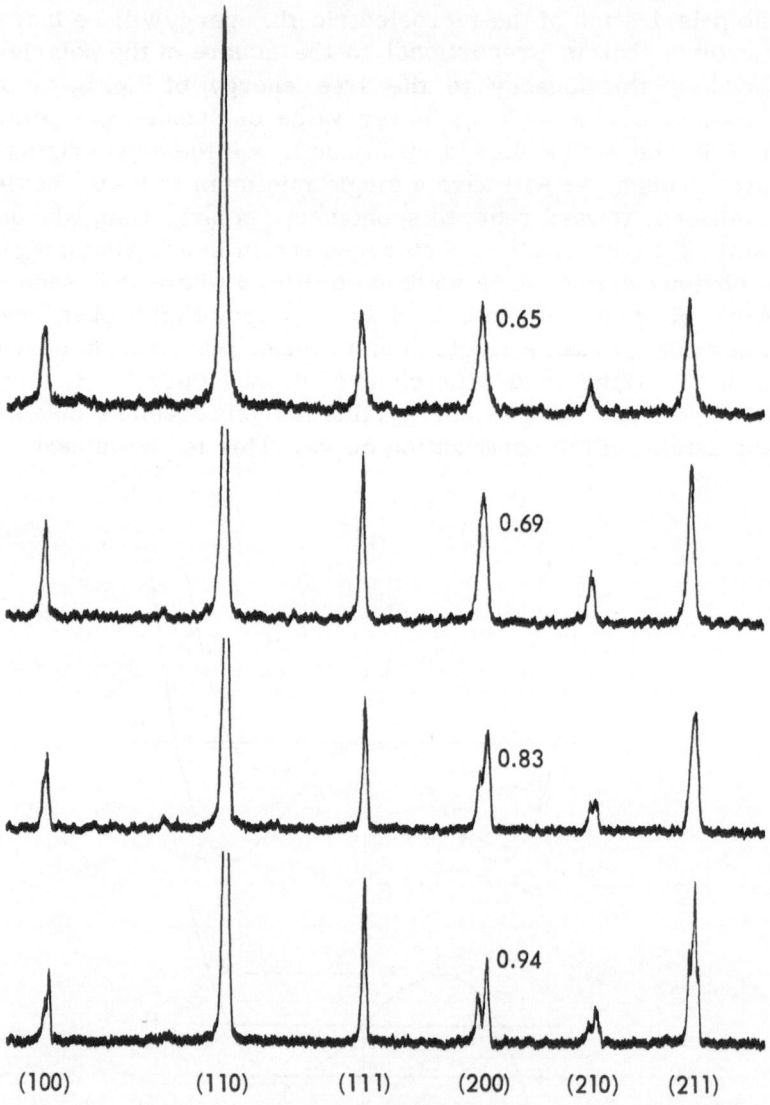

Fig. 6. Diffraction patterns of barium titanate showing increase of spontaneous strain with increasing density.

A disk-shaped ferroelectric subjected to a depolarizing field obeys equation (3) which, in the absence of an applied field E, is

$$0 = \epsilon_0 E_f + NP$$

or

$$P = -\epsilon_0 E_f / N \qquad (9)$$

The depolarization factor N is a fraction approaching 1 for a thin plate. In addition to equation (9), the field-free ferroelectric must also satisfy the equation of the unsheared polarization curve [unsheared in the sense discussed in connection with Fig. 1(b)]. It shall be written as

$$P = f(E_f) \qquad (10)$$

Equations (9) and (10) are satisfied simultaneously where the straight line of negative slope [equation (9)] intersects the polarization curve [equation (10)]. On the scale of Fig. 3, the line [equation (9)] coincides with the abscissa, and, consequently, it intersects the curve [equation (10)] at the two points of the coercive field where $P = 0$. In other words, the depolarizing field is practically equal to the coercive field. Thus, the depolarizing field tends to destroy polarization completely. This is achieved by a balance between domains of opposite polarization. There is a third solution of equations (9) and (10) on the metastable branch of the polarization curve. There polarization is destroyed by removing spontaneous polarization. Because this situation is not stable, it is found only in single-domain particles. Such particles are too small to have their (negative) volume energy of polarization support the (positive) interfacial energy associated with the domain boundaries. Therefore, there should be a cut-off particle size below which no spontaneous polarization exists. In reality, depolarization in barium titanate is only partial, even in the smallest particles, but this does not affect the present discussion.[*]

If barium titanate has a degree of purity and a particle size conducive to high depolarization, the maximum depolarization will be found in a loose powder. In a sintered body, the spontaneous polarization and strain will increase with particle contact and,

[*]A sharply defined critical size is not found in barium titanate. Instead, the spontaneous strain decreases monotonically over a range from 1 μ down to 100 A and extrapolates to zero for 0 A [4]. Suggested reasons are: (1) ease of flux closure due to the available six directions of polarization [5], (2) thin surface layer of high polarization providing flux closure [4], and (3) conduction in the ferroelectric [6]. The latter appears to us to be the most likely explanation because lower-purity particles generally show higher spontaneous strain.

hence, with density. It is easy to see how partial or complete flux closure can be obtained when there are several contacting single-domain particles, especially with the six directions of polarization available in barium titanate. For spherical particles of uniform size, the maximum number of contacts is achieved at $\rho = 0.74$; for the checkerboard arrangement discussed earlier, at $\rho = 1.0$. With this simple picture, we expect a noticeable rise in spontaneous polarization between $\rho = 0.74$ and $\rho = 1.0$. This assumes that only sintering and no recrystallization has taken place. Figure 6 shows the diffraction patterns of unrecrystallized ceramics sintered to densities between $\rho = 0.65$ and $\rho = 0.94$, all prepared from the same powder of low spontaneous strain. The change in spontaneous strain is best seen in the (200) cubic diffraction peak. In a truly cubic material, it should have the same inherent width as the neighboring (111) peak. Even in the low-density material ($\rho = 0.65$), it is seen to be broadened. It shows visible splitting in the material with $\rho = 0.83$ and greater splitting for $\rho = 0.94$. Thus, the increase in spontaneous strain with density is readily demonstrated.

It has been shown how the properties of the simplest heterogeneous ferroelectric can be understood. If the second phase consists of a dielectric with a low dielectric constant, the results are essentially the same. The presence of another ferroelectric or an electrically conducting phase as the second phase would change the details of the analysis, but not the principles on which it is based.

REFERENCES

1. H. H. Stadelmaier and S. W. Derbyshire, Materials Science Research, Vol. 1, Plenum Press (New York), 1963, p. 57.
2. G. Borelius, Handbuch der Metallphysik, Vol. 1, Akad. Verlagsges. (Leipzig), 1937.
3. A. F. Devonshire, Advan. Phys. 3:85 (1954).
4. M. Anliker, H. R. Brugger, and W. Känzig, Helv. Phys. Acta 27:99 (1954).
5. G. Shirane, F. Jona, and R. Pepinsky, Proc. IRE 43:1738 (1955).
6. W. Känzig, in: F. Seitz and D. Turnbull (eds.), Solid State Physics, Vol. 4, Academic Press (New York), 1957, p. 1.

Chapter 8

Interfacial Polarization Effects Associated with Surfaces and Interfaces

N. M. Tallan, H. C. Graham,
and J. M. Wimmer

Aeronautical Research Laboratories (U. S. Air Force)
Wright-Patterson Air Force Base, Ohio

The dispersion of the capacitance and conductance of a specimen with frequency, arising from the presence of interfaces between regions of differing dielectric properties, can be used to study the geometry and electrical properties of the more conducting regions. The measurement and interpretation of these interfacial polarization effects are discussed, with particular emphasis on their application to studies of surface effects in single crystals. The results of experimental studies of the variation in electrical properties of relatively thick (approximately 10μ) surface layers in sapphire produced by the adsorption and desorption of gases, incomplete equilibration of composition in various oxygen partial pressures, and the presence of impurities and preliminary results of interfacial polarization studies in sodium chloride crystals are presented.

THE INTERFACIAL POLARIZATION MECHANISM

When an electric field is applied to a material containing an interface between two regions of different electrical properties, more charge is delivered to or removed from the interfacial area, depending on the sense of the field, in the more conducting region than in the more insulating region. The polarization arising from this buildup of a net charge at the interface is reflected in a DC conductivity measurement as a time dependence of the current following application of the field and an "after-effect" current in the opposite direction after removal of the field. In an AC measurement as a function of temperature and frequency, this interfacial polarization may give rise to a frequency dependence, or dispersion, of both the permittivity (and, therefore, the observed capacitance) and the conductance of the specimen. At the same time, a peak will be present in a plot of both the absorption coefficient and dissipation factor of the specimen as a function of frequency or temperature.

Historically, interfacial polarization and space-charge effects, particularly the DC "after effect," were among the earliest observed in electrical measurements on insulating materials. Maxwell [1], in 1892, published a quantitative description of the dielectric properties of a material containing insulating and conducting layers, derived from considerations of the equivalent circuit of such a specimen. The treatment has since been extended by Wagner [2] to cases in which the conducting phase is distributed in spherical regions; cases of distribution in ellipsoidal regions were extended by Sillars [3] and Fricke [4].

Volger [5] has presented a particularly effective review of the dielectric behavior of a material containing very thin surface layers whose electrical properties differ from those of the bulk and some of the experimental observations of these interfacial polarization effects in polycrystalline materials. In the case of surface layer polarization,

$$\epsilon = \epsilon_\infty + \frac{\epsilon_s - \epsilon_\infty}{1 + \omega^2 \tau_\epsilon^2} \tag{1}$$

$$\sigma = \sigma_\infty - \frac{\sigma_\infty - \sigma_s}{1 + \omega^2 \tau_\sigma^2} \tag{2}$$

and

$$\tan \delta = \frac{\sigma_s}{\epsilon_s \omega} \cdot \frac{1}{1 + \omega^2 \tau_\delta^2} + \frac{\sigma_\infty \tau_\delta}{\epsilon_\infty} \frac{\omega \tau_\delta}{1 + \omega^2 \tau_\delta^2} \tag{3}$$

with

$$\left(\frac{\epsilon_s}{\epsilon_\infty} \right)^{1/2} \tau_\delta = \tau_\epsilon = \tau_\sigma = \frac{\epsilon_1 d_2 + \epsilon_2 d_1}{\sigma_1 d_2 + \sigma_2 d_1} \tag{4}$$

where ϵ and σ are the permittivity and conductivity, respectively, at any angular frequency ω, the subscripts s and ∞ on these quantities refer to low-frequency and high-frequency limiting values, the subscripts 1 and 2 refer to properties of the thin layer and bulk, respectively, τ represents effective relaxation time, and d is the thickness of a region. It is apparent from equations (1) – (4) that for given values of ϵ, σ, and d for each of the two regions, the permittivity and, therefore, the capacitance of the electroded specimen will fall from its low-frequency, or static, value to its high-frequency value in a range of frequencies centered about

TABLE I

Approximate Expressions for Various Quantities in the Two-Layer Model Under the Assumption $\epsilon_1 = \epsilon_2 = \epsilon_i$ and $d_1 \ll d_2$ (valid in the indicated ranges of ρ_1/ρ_2)

	$(\rho_1/\rho_2) \ll (d_1/d_2)$	$(d_1/d_2) \ll$ $(\rho_1/\rho_2) \ll$ $\sqrt{(d_2/d_1)}$	$\sqrt{(d_2/d_1)} \ll$ $(\rho_1/\rho_2) \ll (d_2/d_1)$	$(d_2/d_1) \ll (\rho_1/\rho_2)$
ϵ_s	ϵ_i	ϵ_i	$(d_1\rho_1^2/d_2\rho_2^2)\,\epsilon_i$	$(d_2/d_1)\,\epsilon_i$
ϵ_∞	ϵ_i	ϵ_i	ϵ_i	ϵ_i
ρ_s	ρ_2	ρ_2	ρ_2	$(d_1/d_2)\,\rho_1$
ρ_∞	$(d_2/d_1)\,\rho_1$	ρ_2	ρ_2	ρ_2
τ_δ	$\epsilon_i\rho_1$		$(d_2/d_1)\,\epsilon_i\rho_2$	$\sqrt{(d_2/d_1)}\,\epsilon_i\rho_2$
$\tan\delta$	Near peak $= \dfrac{d_1}{d_2}\dfrac{\omega\tau_\delta}{1+\omega^2\tau_\delta^2}$		Near peak $= \sqrt{\left(\dfrac{d_2}{d_1}\right)\dfrac{\omega\tau_\delta}{1+\omega^2\tau_\delta^2}}$	Near peak $= \sqrt{\left(\dfrac{d_2}{d_1}\right)\dfrac{\omega\tau_\delta}{1+\omega^2\tau_\delta^2}}$

After Volger [5].

$\omega = \tau_\epsilon^{-1}$; the conductivity will increase from its static to high-frequency value in the same frequency range; and the dissipation factor $\tan\delta$, which varies as ω^{-1} at low frequencies, will pass through a maximum at $\omega = \tau_\delta^{-1}$.

Volger [5] also shows that the results that may be encountered, if the permittivities of the two regions are equal, can be divided into four cases: (1) The surface layer is more conductive than the bulk; i.e., $\rho_1/\rho_2 \ll d_1/d_2$; (2) the resistances of the surface and bulk are not appreciably different, i.e., $d_1/d_2 \ll \rho_1/\rho_2 \ll (d_2/d_1)^{1/2}$; (3) the resistance of the surface layer is less than that of the bulk, but its resistivity is much greater than that of the bulk, i.e., $(d_2/d_1)^{1/2} \ll \rho_1/\rho_2 \ll d_2/d_1$; and (4) the resistance of the thin surface layer is much greater than that of the bulk, i.e., $d_2/d_1 \ll \rho_1/\rho_2$. The behavior for each of these cases is reproduced in Table I (after Volger [5]). In case 1, there would be essentially no permittivity dispersion, a large conductivity dispersion, and a small dissipation factor peak, with a maximum value of $d_1/2d_2$. In case 2, where the resistances of the regions are comparable, no interfacial polarization effects are observed. In case 3, there would be a very large permittivity dispersion and a very large dissipation factor peak, with a maximum value of $(d_2/4d_1)^{1/2}$, but essentially no conductivity dispersion. The behavior in case 4 would be similar to that for case 3, but there would also be a large conductivity dispersion. Between cases 1 and 2 and cases 2 and 3, one could include additional cases corresponding to transition ranges of ρ_1/ρ_2 near unity for which the dispersions and tan δ would be functions of both the thickness and resistivity ratios.

Whether the surface layer is more insulating or more conducting than the bulk, the thickness of the surface layer can easily be determined, by reference to Table I, from the magnitude of the dissipation factor maximum or the dispersion of either the permittivity or conductivity. In either case, too, since the relaxation time is inversely proportional to the conductivity of the more conductive region, this conductivity can be determined from the frequency corresponding to either the dissipation factor maximum or the midpoint of one of the dispersions. In addition, with the use of Table I and the observed peak height and the presence or absence of permittivity and conductivity dispersions to determine the applicable case, the static and high-frequency conductivities can be definitely identified with each of the regions of the material. This is particularly important in the many cases encountered where these separate conductivities cannot be determined readily by more direct techniques. Since the conductivity can generally be expressed by the relation

$$\sigma = \sigma_0 \exp\left(-\Delta H/kT\right) \tag{5}$$

the activation energy for conduction can be obtained from the frequency shift of the dissipation factor maximum or the change in static and high-frequency conductivity values with temperature. In practice, it is frequently easiest to use the frequency shift of the tan δ maximum. As in other conductivity measurements, this activation energy may include not only the energy for motion of the charge carrier, but also the energies for formation, ionization, and association or dissociation of defects if these processes affect the charge-carrier concentration.

Since the dissipation factor peaks expected for dipole-rotation processes are generally quite small (impurity-vacancy dipole concentrations of 1 ppm giving peak amplitudes of the order of 10^{-4}), the very large peaks and permittivity dispersions observed in the case of surface layers more insulating than the bulk are generally certain evidence of the interfacial polarization mechanism. However, when the surface layers are more conductive than the bulk, the peak height and dispersions obtained for interfacial polarization and dipole-rotation mechanisms are not significantly different. In this case, the mechanisms must be distinguished on the basis of the behavior of the relaxation time, which can be described by the relation

$$\tau = \tau_0 \exp\left(\Delta H / kT\right) \tag{6}$$

In the case of a dipole-rotation process, the activation energy involves only the energy for motion of the more mobile defect in the dipole, and the pre-exponential τ_0 (which involves a geometry factor, an entropy term, and a suitable value for the lattice-vibration frequency) is generally of the order of 10^{-13} to 10^{-14} sec and independent of sample treatment. In the case of an interfacial polarization process, on the other hand, the activation energy may differ from the known energy for motion since, being an energy for conduction, it may include the additional contributions mentioned earlier, and the pre-exponential (involving the permittivity, the pre-exponential of the conductivity, and perhaps the ratio of thicknesses) will often be quite different from the 10^{-13} to 10^{-14}-sec values and will be sensitive to sample treatment.

Although virtually all reported observations of these interfacial polarization effects have been concerned with polycrystalline materials, these studies can also be conducted on single crystals if the surface layer and bulk differ in charge-carrier type, concentration, or mobility. Such studies can provide significant information about the existence, origin, and properties of these surface layers, as well as a means of following the progress of such processes as surface adsorption or desorption, compositional changes due to interactions with the atmosphere, and impurity segregation at free surfaces or interfaces.

MEASUREMENT TECHNIQUES

The measurement of interfacial polarization effects, like any dielectric measurement, requires the use of a precision bridge capable of determining the capacitive and conductive components of the specimen impedance to the desired level of resolution. In addition, parallel capacitive and conductive paths in the sample holder and leads and long-range surface conduction paths on the specimen itself must be either compensated for or eliminated from the measurement. While these sources of error can be largely eliminated by the use of a guard ring, i.e., a three-terminal sample, as shown in Fig. 1a and 1b, and the addition of a Wagner ground or guard circuit to a conventional four-arm bridge, as in Fig. 1c, the balance of the resultant five-point bridge network is quite tedious. Recent advances in commercially available transformer ratio arm bridges have made their use in these measurements extremely attractive. The three-terminal sample and transformer bridge are shown schematically in Fig. 1d. Clearly, the electrode-to-guard ring surface impedances and the parallel path impedances in the sample holder and leads appear as shunting impedances across the transformer and detector. Until these shunt impedances fall to

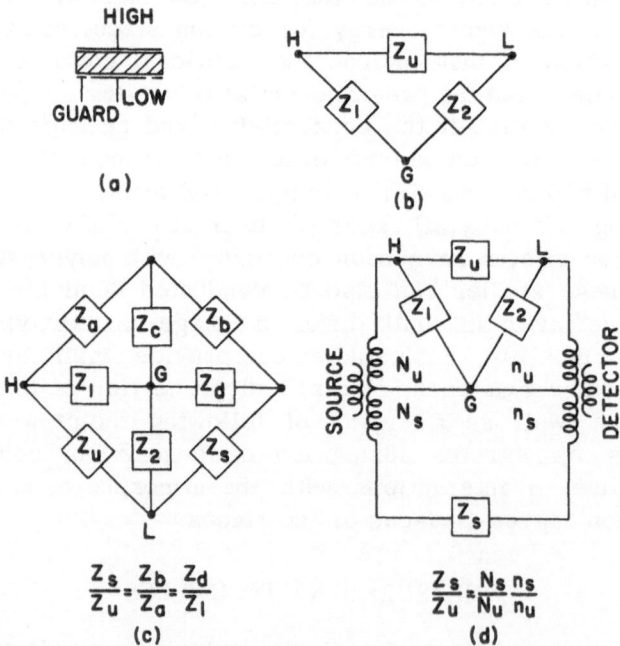

Fig. 1. Three-electrode specimen for dielectric measurements. (a) Electrode geometry; (b) equivalent three-terminal network; (c) application to a Schering bridge with a guard circuit; (d) application to a transformer ratio arm bridge.

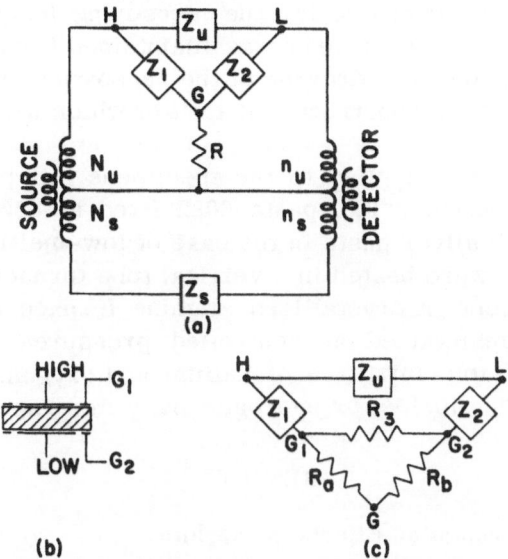

Fig. 2. The effect of series resistance in the guard lead. (a) Location of the series resistance using a three-electrode specimen; (b) four-electrode geometry; (c) equivalent three-terminal network, including the series resistances R_a and R_b between the guard rings and the guard point G at the bridge.

quite low values, they do not affect the accuracy of the determination of the sample impedance.

Unfortunately, the transformer bridge network shown in Fig. 1d has been found to be susceptible to large errors in the measurement of high-impedance specimens unless special precautions are taken to eliminate series resistance in the lead connecting the guard ring on the specimen to the bridge, designated as R in Fig. 2a. A dual guard ring, four-terminal specimen [6], shown in Fig. 2b and shown schematically in Fig. 2c, has been found to be very effective in reducing the magnitude of these errors.

To realize the freedom from the errors described to the extent permitted by the bridge and specimen electrode configurations, the sample-holder design must provide for either the total absence of parallel impedance paths or complete guarding of the measuring leads. The sample holder employed for the measurements to be reported utilized platinum-coated, single-bore ceramic rods to grip the specimen at the guard rings, thereby providing both sample support and contacts to the guard rings, and to simultaneously pro-

vide the required guarding for the measuring leads. The guard
shield continuity for each lead was maintained back to the bridge
terminals.   A schematic drawing of the "chopstick" sample holder
is given in Fig. 3, the construction details of which will be published
elsewhere.

Electrodes were applied to the specimens with platinum paste
(Engelhard Industries, Inc. paste 6082 fired at 825°C) whenever
possible or with silver paste in the case of low-melting materials.
The specimens were heated in a vertical tube furnace. The inside
of the impervious recrystallized alumina furnace tube could be
evacuated or maintained at controlled pressures of oxygen by
introducing suitable mixtures of helium and oxygen, carbon mon-
oxide and carbon dioxide, or hydrogen and water vapor.

RESULTS

Adsorption and Desorption Effects in Sapphire

In 1936, Hartmann [7] reported a several order-of-magnitude
increase in the DC conductivity of polycrystalline alumina compacts
heated in a vacuum of about 10^{-4} torr at 450°C. With continued
heating under these conditions, the measured activation energy for
conduction decreased from about 0.4 eV to a lower limit of about
0.25 eV.   Since the effects of the vacuum heat treatment could be
reversed by exposure to dry oxygen, Hartmann suggested that
alumina could be reversibly reduced under the conditions employed
and that the conductivity effects observed were associated with the
resultant nonstoichiometric compositions.

Recent studies [8] of the dielectric properties of single-crystal
alumina, under similar experimental conditions, reflected behavior
strikingly similar to that observed in Hartmann's earlier polycrystalline DC conductivity experiments. The analysis of the dielectric properties in terms of an interfacial polarization process,
however, suggests a rather different interpretation. The results of
initial measurements on sapphire specimens in vacuum were
generally similar to those reproduced in Fig. 4, in which are
shown the capacitance dispersion and broad dissipation factor peaks
observed at low temperatures.   The values of ΔH and τ_0 for the
polarization process responsible, computed from equation (6) and
the measured frequency shift of the tan δ maximum with temperature, were generally about 0.5 eV and 10^{-10} sec, respectively.
Extended heatings in vacuum at 400°C, however, significantly affected

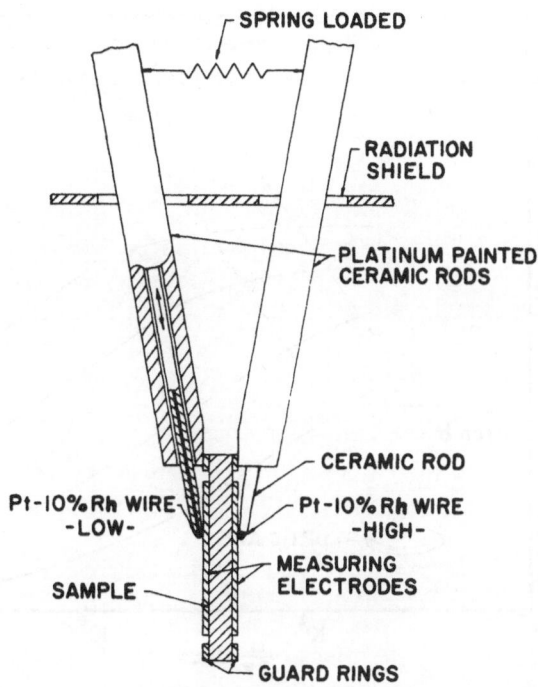

Fig. 3. Schematic drawing of four-terminal dielectric sample holder.

subsequent measurements of the peak; in all cases, the vacuum heat treatments shifted the peak, at a given temperature, toward higher frequencies, with corresponding changes in the computed magnitudes of ΔH and τ_0 toward apparently limiting values of about 0.2 eV and 10^{-8} sec. The effects of the vacuum heat treatments on ΔH and τ_0 could be readily reversed by exposure of the specimen to dry oxygen (as indicated in Fig. 5), dry hydrogen, or moist air.

The inconstancy of the ΔH and τ_0 values and the deviation of the τ_0 values from the 10^{-13} to 10^{-14}-sec range characteristic of dipole relaxation suggest an interpretation based on interfacial polarization. Since the magnitudes of the capacitance dispersion and dissipation factor maximum are small (of the order 10^{-2}), interpretation on the basis of surface layers would require that the surface layer be more conductive than the bulk of the crystal. Application of the relationships for this case ($\rho_1/\rho_2 \ll d_1/d_2$ in Table I) to the experimental data indicates the calculated thickness of the effec-

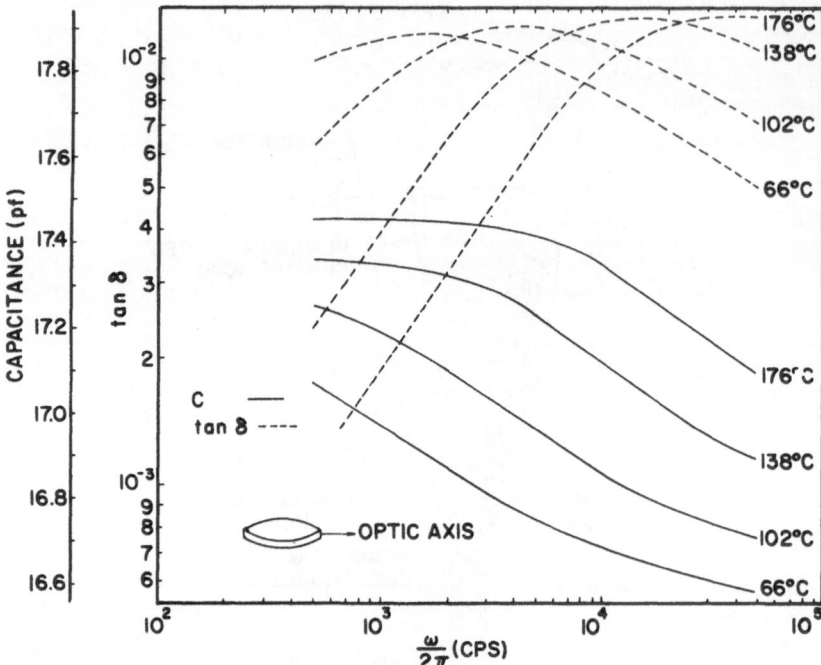

Fig. 4. Frequency dependence of the dissipation factor and capacitance of sapphire in vacuum before heat treatment.

tive surface layer would be about 10 μ since $d_1/d_2 = 2\,(\tan\delta)_{\max}$ and the overall sample thickness was 1 mm. From the observed magnitude of the background loss, the calculated bulk resistivity would be of the order of $10^{13}\,\Omega$-cm. From the observed values for τ_δ, the effective surface layer resistivities were of the order of $10^9\,\Omega$-cm, decreasing with vacuum heat treatment (shifting of the tan δ peak to higher frequencies, or smaller τ_δ values) to a value of about $10^8\,\Omega$-cm and increasing with exposure to various gases to a value of about $10^{10}\,\Omega$-cm.

The interpretation based on surface layers was suggested both by Hartmann's [7] earlier studies on polycrystalline alumina and by the extreme difficulty in making guard balances in the dielectric studies [8] after prolonged vacuum heat treatments. Direct-current conductivity measurements, using two electrodes on the same face of a sapphire specimen cut parallel to the optic axis, produced results remarkably similar to those obtained by Hartmann and supported the assumption of a conductive surface layer. The measured surface resistivity decreased markedly with time at 400°C in vacuum and increased upon exposure to dried oxygen or the atmosphere.   Activation energies computed from the temperature dependence of the DC surface resistance were similar to those observed in the dielectric studies.   These environmental surface conductivity effects were all readily reversible.

A reasonable explanation for the thick surface layer conduction effects observed may involve the surface double-charge layer that should exist in ionic materials.   It has long been recognized [9,10] that an electrical double layer, consisting of a net surface charge and a compensating space charge under the surface, exists at the external boundary of an ionic material because of an equilibrium concentration gradient of defects.   In the simple case of stoi-

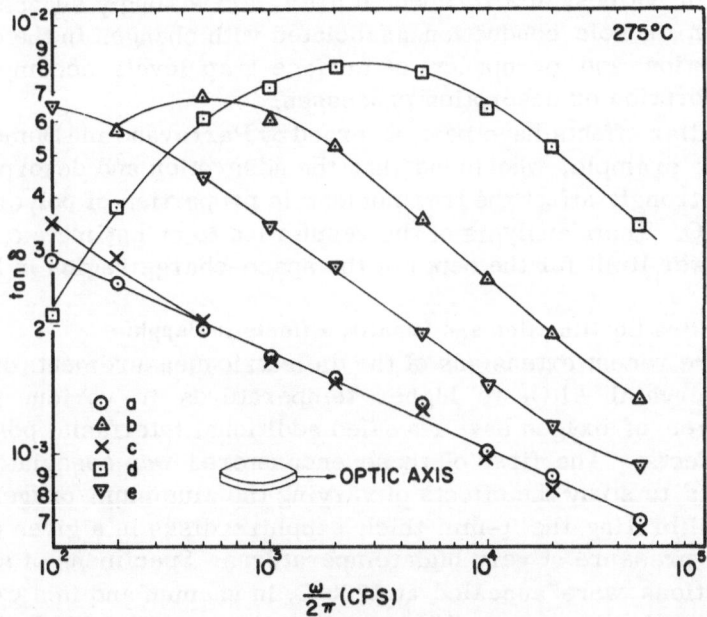

Fig. 5. Frequency dependence of the dissipation factor of sapphire as a function of heat treatment and exposure to dried oxygen. The curves represent (a) initial measurement; (b) after first vacuum heat treatment (400°C, ½ hr); (c) after exposure to dried oxygen (3 hr); (d) after second vacuum heat treatment (400°C, 1½ hr); and (e) after second exposure to dried oxygen (1 hr).

chiometric, pure material, if the energy of formation of a cation vacancy is less than that for an anion vacancy, an excess of cation vacancies will be emitted into the crystal and the surface will contain anion vacancies and cations in excess of their equilibrium bulk concentrations. The surface would, therefore, behave as an n-type, metal-excess material. Mass balance and electroneutrality would then require the existence of a p-type, metal-deficient region below the surface. Depending on the energies of formation of the defects and the permittivity of the material, this double-charge layer might be quite thick in the oxides.

The composition of this double-charge layer would immediately suggest that its electrical properties, and particularly its conductivity, should differ from those of the bulk. The adsorption or desorption of gases at such a surface would obviously modify the charge-carrier concentration within it, accounting for the observed variation in σ_0 and, therefore, in τ_0 during vacuum heat treatment or exposure to various gases. The changes in ΔH and the observed range of values, 0.2 to about 0.5 eV, are strongly suggestive of electron or hole conduction associated with changes in the energy distribution and occupancy of surface trap levels accompanying the adsorption or desorption processes.

Similar effects have been observed by Parravano and Domenicali [11], for example, who found that the adsorption and desorption of gases strongly affect the thermoelectric properties of polycrystalline NiO. Their analysis of the results led to an estimate of 600 A as a lower limit for the depth of the space-charge region in NiO.

Composition Equilibration and Impurity Effects in Sapphire

More recent extensions of the dielectric measurements on pure, single-crystal Al_2O_3 to higher temperatures in various partial pressures of oxygen have revealed additional interfacial polarization effects. The first of these encountered was associated with attempts to study the effects of varying the aluminum/oxygen ratio by equilibrating the 1-mm thick sapphire disks in a given oxygen partial pressure at very high temperatures. Specimens of several orientations were annealed at 1750°C, in vacuum and in a $CO-CO_2$ mixture, yielding a $p_{O_2} = 10^{-9}$ atm for times of 6 — 100 hr. In all cases, subsequent measurements of the dielectric properties indicated the existence of a region, assumed to be a surface layer, with a conductivity different from that of the bulk.

Representative results for two specimens are shown in Figs. 6

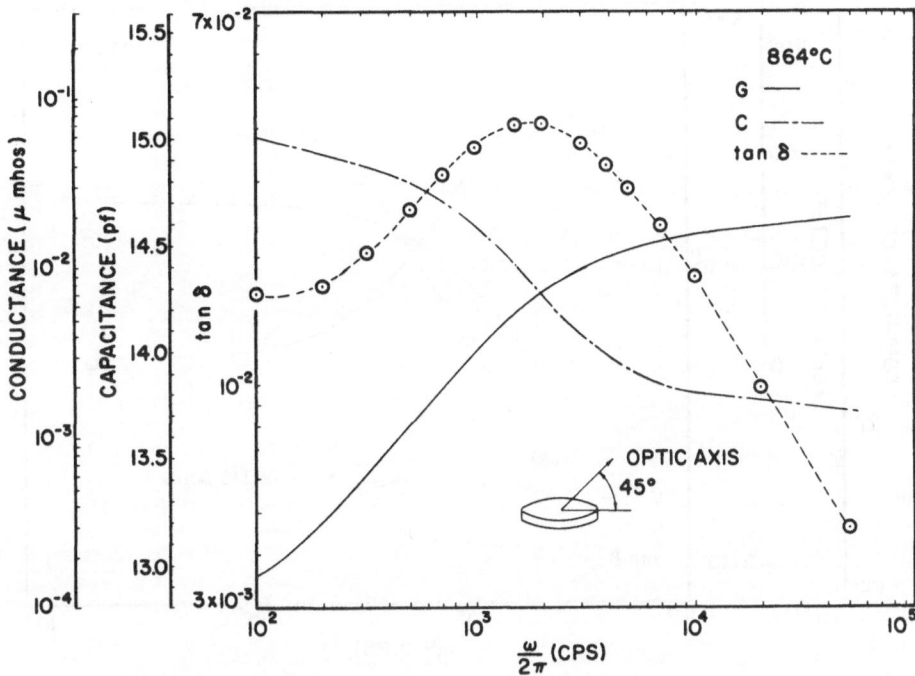

Fig. 6. Dissipation factor maximum and conductance and capacitance dispersions in sapphire produced by a relatively low-resistance surface layer.

and 7. The relatively small amplitude of the tan δ peak ($4 \cdot 10^{-2}$), the small amount of capacitance dispersion, and the large conductance dispersion, shown in Fig. 6 are, for an interfacial polarization mechanism, characteristic of a surface layer lower in resistance than the bulk (the case $\rho_1/\rho_2 \ll d_1/d_2$ in Table I). Based on the observed peak height and conductance dispersion, the effective thickness of the surface layer for this specimen is about 40μ, the bulk resistivity is about 10^{11} Ω-cm, and the surface layer resistivity is $10^7 - 10^8 \Omega$-cm. Values for ΔH, in this case the activation energy for conduction in the surface layer, and τ_0 of 2.4 eV and $2.5 \cdot 10^{-15}$ sec, respectively, were computed from the frequency shift of the peak with temperature using equation (6). Other cases of relatively conductive surface layers were encountered in which the computed ΔH and τ_0 values ranged from 1.5 to 1.8 eV and about $5 \cdot 10^{-9}$ to $2 \cdot 10^{-15}$ sec, respectively. The variation of the magnitudes of these

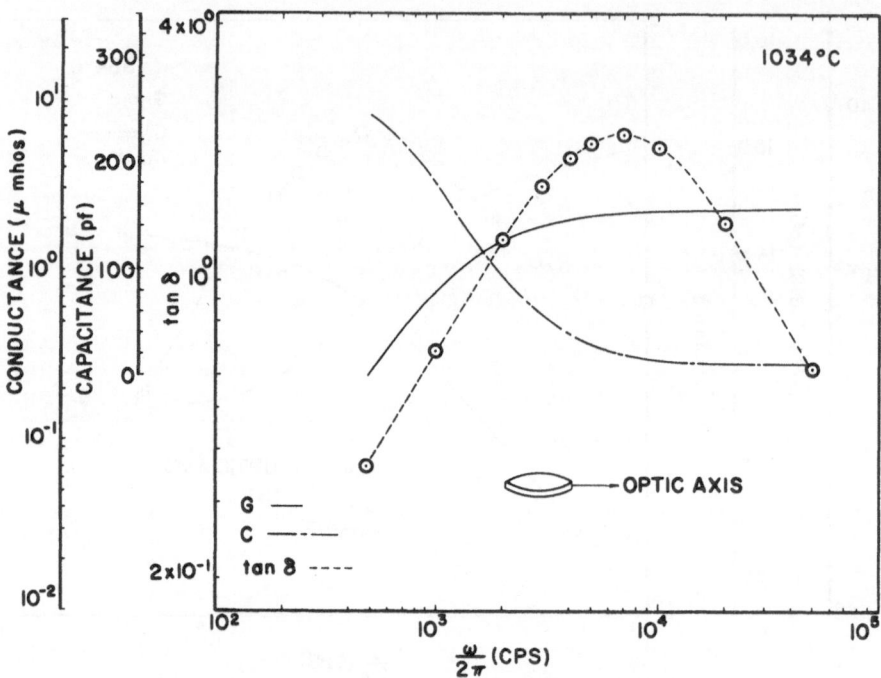

Fig. 7. Dissipation factor maximum and conductance and capacitance dispersions in
        sapphire produced by an insulating surface layer.

τ_0 values, compared to the 10^{-13} to 10^{-14}-sec values characteristic
of dipole rotations, supports the interpretation on the basis of interfacial polarization.

The very large amplitude of the tan δ peak (approximately 2) and
the large dispersions of both the capacitance and conductance, shown
in Fig. 7, are characteristic of interfacial polarization due to a surface layer higher in resistance than the bulk (the case $d_2/d_1 \ll \rho_1/\rho_2$
in Table I). The effective thickness of the surface layer, computed
from either the tan δ peak height or the capacitance dispersion, is
again about $40\,\mu$. The computed values for ΔH, in this case the
activation energy for conduction in the bulk, and τ_0 were 1.8 eV and
10^{-11} sec, respectively.

In all cases studied, the computed activation energies for conduction were either in the range $1.5 - 1.8$ eV or $2.3 - 2.5$ eV.
However, in several instances, the energy was found to change

from one range to the other between initial measurements after a very high-temperature anneal at a given oxygen pressure and subsequent measurements after prolonged heating at intermediate temperatures. The apparent source of this behavior and additional evidence for the assumption of a surface layer for the geometry of the nonbulk conducting region were found by varying the oxygen pressure in contact with a specimen for which a thinner surface layer was computed. The frequency shift, corresponding to a change in conductivity of the more conductive region (in this case, the surface layer) with oxygen pressure at constant temperature is shown in Fig. 8. This ability of the conductivity of a thin surface layer to respond rapidly to environmental changes suggests that even thicker surface layers might undergo changes in defect concentrations and predominant conduction mechanism with prolonged heating at higher temperatures.

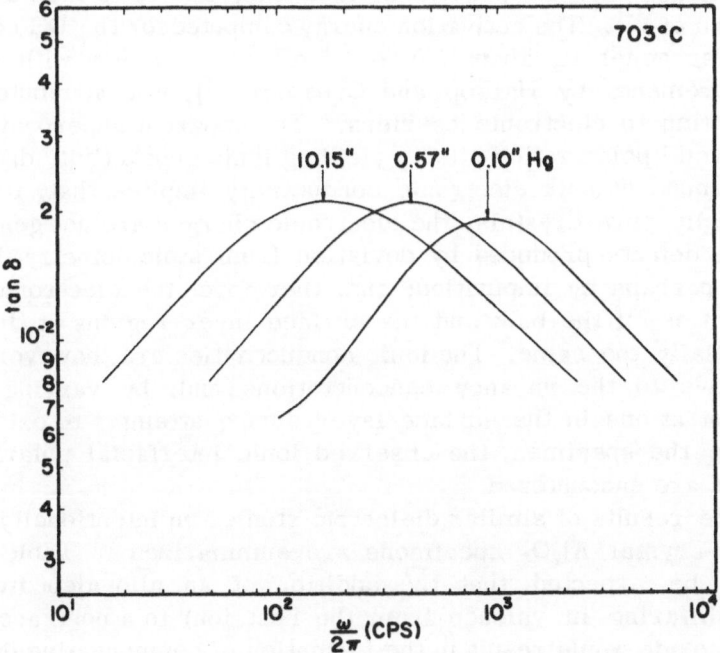

Fig. 8. Frequency shift of the dissipation factor maximum with oxygen pressure at constant temperature corresponding to changes in the conductivity of the surface layer on sapphire.

The results of all of the measurements on pure single-crystal specimens could be explained by the requirement of volume diffusion, and, therefore, extremely long times, for the equilibration of composition with temperature and oxygen partial pressure. This would naturally lead to the presence of a surface layer, of composition and conductivity different from that of bulk crystal, whose depth would depend on the length of the diffusion anneal. The observed values for the activation energies for conduction, 1.6 ± 0.2 eV and 2.4 ± 0.1 eV, computed from the interfacial polarization effects in the range 700 — 1200°C, are suggestive of ionic conductivity. The value 2.4 ± 0.1 eV is believed to be the energy required for oxygen-vacancy motion and is in excellent agreement with the value found for extrinsic oxygen diffusion [12]. The value 1.6 ± 0.2 eV is believed to be the energy required for aluminum-vacancy motion. In several specimens, the conductance and dissipation factor measurements at higher temperatures, in the range 1300 — 1550°C, were indicative of simple DC conduction, i.e., the conductance was independent of frequency and tan δ was a linear function of ω^{-1}. The activation energy computed for the DC conduction was generally about 4.0 to 4.5 eV, in agreement with earlier measurements by Harrop and Creamer [13], who attributed this conduction to electronic carriers. The apparent superposition of interfacial polarization effects yielding ionic conductivity data on a background of bulk electronic conductivity implies that, in these nominally pure crystals, the electronic charges are not generated by the defects produced by deviation from stoichiometry, but instead perhaps by impurities; and, therefore, the electronic conductivities of the bulk and the surface layer regions studied are essentially the same. The ionic conductivities are, however, very sensitive to the vacancy concentrations and, by varying these concentrations in the surface layer during attempts to oxidize or reduce the specimen, the observed ionic interfacial polarization effects are encountered.

The results of similar dielectric studies on intentionally doped single-crystal Al_2O_3 specimens are summarized in Table II. It might be expected that the addition of an aliovalent impurity (one differing in valence from the host ion) to a nontransitional metal oxide would result in the formation of compensating defects, either vacancies or interstitials, and perhaps the association of these defects and the aliovalent impurity to form permanent dipoles. While the amplitudes of the dissipation factor peaks observed

TABLE II

Dielectric Data for Doped Single-Crystal Al_2O_3

Dopant	ΔH(eV)	τ_0(sec)	$(\tan \delta)_{max}$	Probable defects	Atmosphere
Fe	2.3	$3.3 \cdot 10^{-17}$	$3 \cdot 10^{-2}$	Fe^{+2}, V_O	Air
	2.5	$2.7 \cdot 10^{-15}$	$5 \cdot 10^{-2}$	Fe^{+2}, V_O	Air
	2.2	$1.1 \cdot 10^{-15}$	$5 \cdot 10^{-2}$	Fe^{+2}, V_O	Air
Mn	1.8	$2.6 \cdot 10^{-14}$	$1 \cdot 10^{-2}$	Mn^{+4}, V_{Al}	Oxygen
	1.4	$1.2 \cdot 10^{-11}$	$2 \cdot 10^{-2}$	Mn^{+4}, V_{Al}	CO, heating
	2.7	$1.1 \cdot 10^{-16}$	$5 \cdot 10^{-2}$	Mn^{+2}, V_O	CO, cooling
Ni	0.85	$5.3 \cdot 10^{-9}$	$5 \cdot 10^{-2}$	Electronic	$CO-CO_2$
Mg	2.5	$2.7 \cdot 10^{-15}$	$4 \cdot 10^{-2}$	Mg^{+2}, V_O	Air
Si	1.36	$2.2 \cdot 10^{-15}$	$2 \cdot 10^{-1}$	Si^{+4}, V_{Al}	Two-terminal
	1.46	$1.2 \cdot 10^{-15}$	$2 \cdot 10^{-1}$	Si^{+4}, V_{Al}	measurements in air

would not be inconsistent with interpretation on the basis of a dipolar relaxation mechanism, the computed τ_0 values vary, in many cases, so widely from the anticipated 10^{-13} to 10^{-14}-sec range for dipolar rotation that most of the results would be more reasonably accounted for by an interfacial polarization mechanism. On the basis of an interfacial polarization model, the small peak amplitudes imply that the regions of nonbulk properties are more conductive than the bulk of the crystal.

The increased conductivity of these regions (which, if planar, either external surface or internal interface, would vary from about 5 to 50 μ in thickness for the impurities studied) might be induced either by a difference in valence or concentration of the impurity in these regions. The activation energies computed for the conductivity of these regions, with the exception of the nickel-doped specimen, were again indicative of ionic transport and agree very well with those observed in the pure crystals. The combinations of defects that might account for the increased conductivity in the more conductive regions are listed in the fifth column of Table II. The magnesium and silicon doping led to essentially invariant results within the temperature range studied which, if indeed

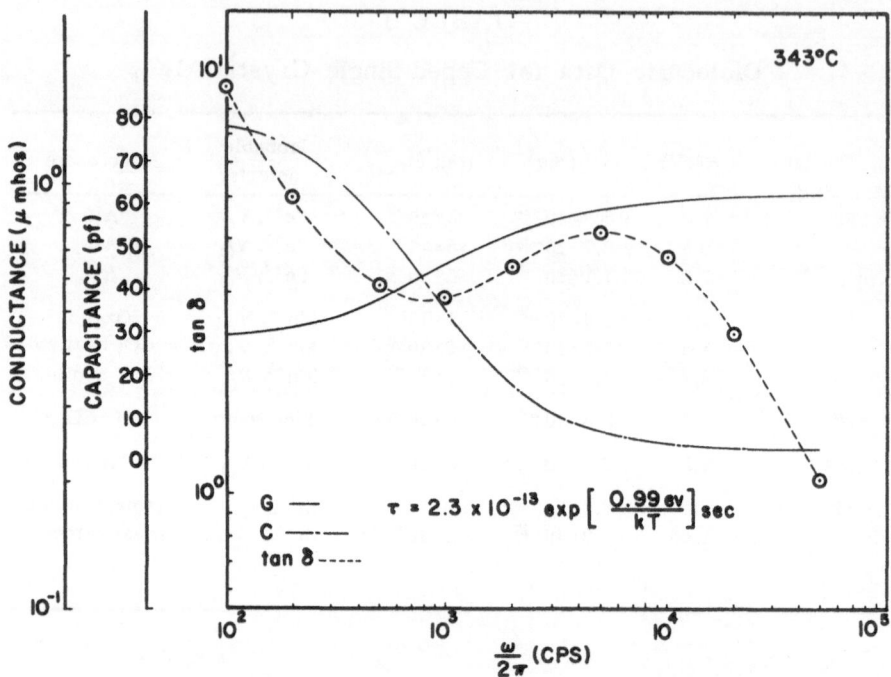

Fig. 9. Dissipation factor maximum and conductance and capacitance dispersions in single-crystal NaCl resulting from interfacial polarization.

indicative of interfacial polarization, would require a constant concentration of these impurities in the more conductive regions. The iron and manganese doping led to more variable results, and particularly some computed τ_0 values, almost certainly indicative of interfacial polarization, that may have been associated with either a different valence state or segregation in the more conductive regions. It may be noted that some of the computed τ_0 values listed would not be inconsistent with a dipolar relaxation mechanism, and, in these cases, the dielectric data alone cannot distinguish between the associated defects and interfacial polarization models.

Interfacial Polarization Effects in NaCl

Preliminary studies of the dielectric properties of both single-crystal and polycrystalline NaCl have indicated the existence of an easily reproducible interfacial polarization effect. Typical results for a single-crystal specimen, after annealing at 600°C for several hours, are shown in Fig. 9. Freshly grown crystals, pulled from the melt, which did not show a clearly resolved dissipation factor maximum upon initial measurement generally did after reducing the background conduction losses in the annealing process.

The magnitudes of the dissipation factor maximum and the observed dispersions in Fig. 9 could be accounted for, with the specimen thickness used, by the presence of a $10 - 20\ \mu$ insulating surface layer (probably, case $(d_2/d_1)^{1/2} \ll \rho_1/\rho_2 \ll d_2/d_1$ in Table I). Preliminary measurements of the potential profile across a crystal in a DC field were made using a very-small-diameter probe. The gradient was nonlinear near the electrodes, in support of a relatively insulating surface layer, but the majority of the potential drop occurred in the bulk of the crystal suggesting, as indicated by the dielectric data, that the resistivity of the surface layer is only one to two orders of magnitude greater than the bulk. The activation energy, 0.99 eV, computed from the frequency shift of the dissipation factor maximum with temperature, is consistent with the value obtained in this temperature range by Dreyfus and Nowick [14] using DC conductivity measurements.

Results essentially similar to those obtained in single crystals have been observed in dielectric measurements on pressed and sintered polycrystalline NaCl specimens. Preliminary measurements on copper-doped NaCl single crystals pulled from the melt, however, indicate that the presence of aliovalent impurities may greatly affect the observed behavior. Although the same interfacial polarization effects are observed immediately after annealing at about 600°C, they decrease markedly with time at $300 - 350$°C. Dipole relaxation effects, apparently associated with divalent copper — sodium vacancy pairs, were observed at much lower temperatures in these crystals.

Dielectric measurements and both conductivity and potential probe measurements on pure and copper-doped NaCl bicrystals pulled from the melt have so far shown no effects that could be directly attributed to the presence of the internal interface.

CONCLUSIONS

Interfacial polarization effects arising from the presence of surface layers with electrical properties different from those of the bulk may be observed in single-crystal, as well as in poly-crystalline, specimens. While the interfacial polarization studies described have been utilized primarily to determine the electrical properties of the material itself and perhaps the nature of the defects present, clearly they can also be applied to the study of any process which affects the conductivity of the material. It should be possible, for example, to continuously determine the rate of such processes as chemisorption, diffusion and corrosion, sintering, impurity segregation and precipitation, and oxidation or reduction in order to study the processes themselves.

ACKNOWLEDGMENTS

The authors are grateful to R. W. Vest for many helpful discussions and comments on the manuscript and to E. D. Wysong for assistance in the design and construction of the equipment used.

REFERENCES

1. J. C. Maxwell, Electricity and Magnetism, Vol. 1, Clarendon Press (Oxford), 1892, p. 452.
2. K. W. Wagner, "Erklarung der dielektrischen Nachwirkungsvorgange auf Grund Maxwellscher Vorstellungen," Arch. Elektrotechnik 2(9):371-387 (1914).
3. R. W. Sillars, "The Properties of a Dielectric Containing Semiconducting Particles of Various Shapes," J. Inst. Elec. Engrs. (London) 80(484):378-394 (1937).
4. H. Fricke, "The Maxwell-Wagner Dispersion in a Suspension of Ellipsoids," J. Phys. Chem. 57(9):934-937 (1953).
5. J. Volger, "Dielectric Properties of Solids in Relation to Imperfections," in: A. F. Gibson (ed.), Progress in Semiconductors, Vol. 4, John Wiley and Sons (New York), 1960, pp. 206-236.
6. N. M. Tallan, H. C. Graham, and R. W. Vest, "Origin of Apparent Negative Impedance in Three-Terminal Measurements," Rev. Sci. Instr. 33(10):1087-1088 (1962).
7. W. Hartmann, "Electrische Untersuchungen an Oxydischen Halbleitern," Z. Physik. 102(11):709-733 (1936).
8. N. M. Tallan and D. P. Detwiler, "An Anomalous Dissipation Factor Maximum in Sapphire," J. Appl. Phys. 34(6):1650-1656 (1963).
9. I. I. Frenkel, Kinetic Theory of Liquids, Clarendon Press (Oxford), 1946, pp. 36-40.
10. K. Lehovec, "Space-Charge Layer and Distribution of Lattice Defects at the Surface of Ionic Crystals," J. Chem. Phys. 21(7):1123-1128 (1953).
11. G. Parravano and C. A. Domenicali, "Thermoelectric Behavior of Solid Particulate Systems; Nickel Oxide," J. Chem. Phys. 26(2):359-366 (1957).
12. Y. Oishi and W. D. Kingery, "Self-Diffusion of Oxygen in Single-Crystal and Poly-crystalline Aluminum Oxide," J. Chem. Phys. 33(2):480-486 (1960).
13. P. J. Harrop and R. H. Creamer, "The High-Temperature Electrical Conductivity of Single-Crystal Alumina," Brit. J. Appl. Phys. 14(6):335-359 (1963).
14. R. W. Dreyfus and A. S. Nowick, "Ionic Conductivity of Doped NaCl Crystals," Phys. Rev. 126(4):1367-1377 (1962).

Chapter 9

The Effect of Interfaces on the Radiation
Production of Color Centers in Alkali Halides*

C. F. Gibbon and G. C. Kuczynski

University of Notre Dame
Notre Dame, Indiana

The changes in optical absorption spectra induced by gamma irradiation of polycrystalline pellets pressed from KCl, KBr, and KI powders have been studied and compared to the spectra of irradiated deformed single crystals. The results indicate that there is an accelerated rate of destruction of lattice defects in the presence of intercrystalline interfaces. In addition, an enhanced production of certain V-centers is observed, which is explained on the basis of the reaction of products of dissociation of water trapped in the interfaces with point defects existing in the interior of the grains.

INTRODUCTION

Color Centers in Alkali Halides

The study of color centers in alkali halides has provided considerable insight into the nature of and interaction between various crystalline imperfections in these simple ceramic systems. Color centers can be developed in halides in three different ways. First, an excess of one of the components can be introduced by heating the crystal in an atmosphere of either the appropriate alkali metal or halogen gas and quenching the material to preserve the non-stoichiometric condition. An excess of the metal produces a bell-shaped optical absorption band, usually in the visible, called the F-band. Successful addition of halogen to these compounds, by this additive coloration technique, has been achieved only in the case of iodides and bromides; a double band in the ultraviolet appears as a

*This paper is based in part on a thesis submitted by C. F. Gibbon to the Graduate School of the University of Notre Dame in partial fulfillment of the requirements for the degree of Doctor of Philosophy. The work was carried out in the Radiation Laboratory of the University of Notre Dame. This paper is AEC document number C00-38-351. (The Radiation Laboratory of the University of Notre Dame is operated under contract with the U. S. Atomic Energy Commission.)

result of this treatment. The shorter wavelength of the two bands is called the V_3-band and the longer, the V_2-band. A second method of introducing color centers into alkali halides is by exposing them to ionizing radiation. This technique produces the F-band plus one or more shorter wavelength bands usually designated V-bands; their exact nature and position depend on the irradiation temperature. The third method of coloration is injection of defects by a pointed electrode. The crystal is clamped between this electrode and a flat plate, and a voltage is impressed; if the pointed electrode is of negative polarity, F-coloration occurs, whereas, if it is positive, the V-band is introduced.

The generally accepted model of the defect giving rise to the F-band, i.e., the F-center, is that postulated by de Boer [1], namely, an electron trapped at a halide-ion vacancy. Under proper conditions, absorption bands can be induced at wavelengths longer than the F-band; these are the so-called M-, R-, and N-bands. Work by Van Doorn and Haven [2], Pick [3], Schnatterly and Compton [4], and others indicates that the centers responsible for these bands are aggregates of 2, 3, or 4 F-centers in next-nearest-neighbor positions.

Somewhat less is known about the nature of the V-type centers. During low-temperature irradiation of pure material only F-centers and V-type centers are created, and, since the F-center is known to be a donor-type defect, the V-bands are usually thought to be corresponding trapped hole (or acceptor) centers. Two centers—the V_K-center, which is produced by irradiation near 80°K, and a similar one, the H-center, which appears along with the F-center during irradiation at 4°K—are well understood due principally to the work of Känzig and his associates [5,6]. The V_K-center is apparently a hole trapped at two adjacent halide ions, i.e., an X_2^- molecule-ion on two lattice sites. The H-center can be thought of as an X_2^- molecule-ion on a single lattice site, or as an interstitial halogen atom. Both the V_K- and H-centers have a $\langle 110 \rangle$ axis of symmetry. The V-centers produced by irradiation at room temperature generally form a rather low broad optical band, which is sometimes very poorly resolved into two components, called V_2 and V_3, which may correspond to the double band formed by additive coloration. Although very little is known concerning the defects responsible for these bands in either the irradiated or additively colored material, Hersh [7] has pointed out the similarity between the spectra of halide crystals containing excess halogen and those of aqueous solutions of potassium halides containing dissolved halogens. He concludes that the V_2- and V_3-bands in KI and KBr and the V_2-band in KCl arise from a center which is a linear array of halogen molecule and halide ion in which the crystallinity of the material plays a minor role. This seems to be supported by Mollwo's observation [8] that the concentration of excess halogen in the crystal is proportional to the partial pressure of the halogen gas during additive coloration. The V_3-band in KCl, usually reported at 212 mμ (5.83 eV) lies very close to U-band at 214 mμ (5.79 eV); the U-band is attributed to the presence of a hydride ion in an anion vacancy. Lüty [9] has presented convincing evidence that the V_3-center is identical to the U-center. He has shown that, in the absence of OH$^-$ impurity (known by an absorption band in KCl at 204 mμ), no V_3-centers are produced by irradiation, only F and V_2. Zaleskiewicz and Christy [10] have denied this identification on the basis of differences in bleaching behavior between the V_3- and U-centers.

Acceptor-Type Centers and Interfaces

An interesting connection between acceptor (V-type) centers and interfaces in alkali halides was demonstrated by Hersh [11]. He showed that a large, fairly well-defined double band in the ultraviolet, comparable to that produced by additive coloration, can be produced in pressed transparent polycrystalline disks of alkali halides by irradiation. He noted that the V-bands were quite stable and remained almost fully developed even after total bleaching of the F-band; also, he found that the halogen was liberated when the X-rayed pellets were dissolved in water.

The purpose of this work was to compare the optical spectra of acceptor-type centers produced by irradiation in polycrystalline pellets obtained from pressed alkali halide powder, viz., produced in the presence of intercrystalline interfaces, with those generated in single-crystal disks which were deformed in the same, or in a similar, die with the same load under which the pellets were formed. It was hoped that the effect of the presence of interfaces on the coloration process might be better understood.

EXPERIMENTAL PROCEDURE

Specimen Preparation

The potassium halides used in this work were obtained in the form of unpolished disks about 2 mm thick and 13 mm in diameter.

These disks were cleaved in two (parallel to their faces) so that the specimens were about 1 mm thick.

The disks were subjected to a variety of treatments before irradiation; however, some were simply irradiated as cleaved. It was assumed that the surfaces of the crystals were sufficiently clean. Because some of them appeared scratched and had faint spots of unknown origin on them, about half the disks were etched to provide surfaces whose prior treatment was known. The etching procedure consisted of a 5-sec rinse in distilled water and a 30-sec quench in absolute alcohol. The specimens were then dried

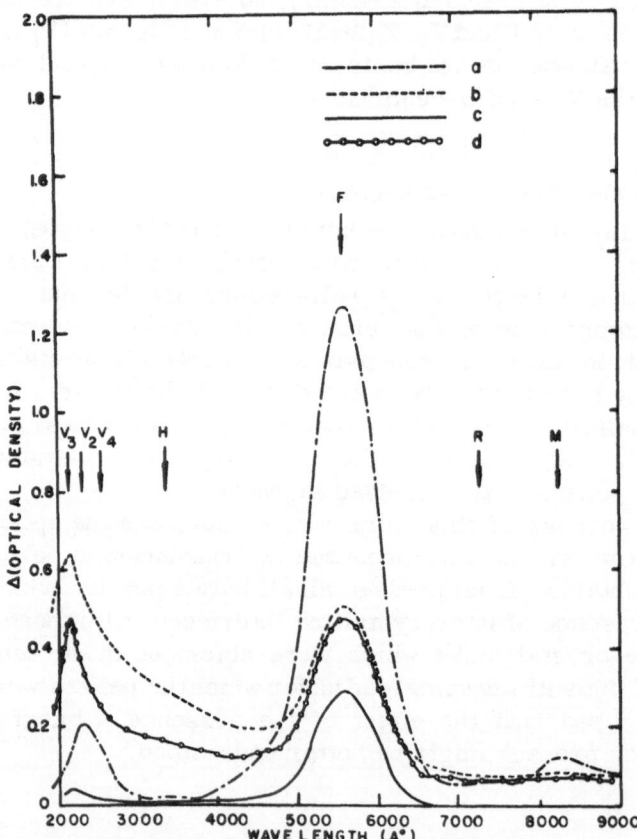

Fig. 1. Spectral changes produced by irradiation in KCl specimens prepared from non-heat-treated material; (a) deformed single crystal, specimen thickness $t = 0.094$ cm; (b) pellet, etched before grinding, $t = 0.091$ cm; (c) as-cleaved single crystal, $t = 0.114$ cm; (d) pellet, not etched before grinding, $t = 0.086$.

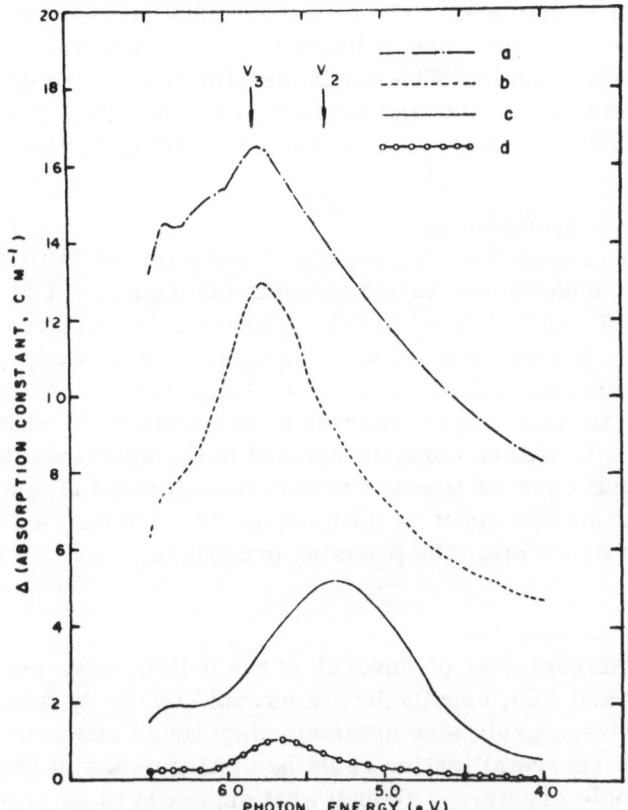

Fig. 2. V-bands in non-heat-treated KCl: (a) pellet, etched before grinding; (b) pellet, not etched before grinding; (c) deformed single crystal; (d) as-cleaved single crystal.

in the stream from a hot air gun for 60 sec. All the crystals which were to be simply deformed, and not ground and pressed into pellets, had to be etched to ensure their fitting into the die.

A number of specimens were then heat-treated at 560°C in a dry argon atmosphere for 12 hr and then furnace-cooled. The original purpose of this procedure was to remove any remaining moisture from the surface of the disks that were etched; some unetched crystals were also given the same treatment for comparison.

The transparent polycrystalline pellets were made by grinding the single-crystal specimens for about 5 min in an agate mortar and pressing the resulting powder to 14,000 kg/cm² in a standard evacuable die. The as-pressed specimens were 0.7–0.8 mm thick and 13 mm in diameter. The procedure was carried out in a room in which temperature and relative humidity were kept at 76°F and 40–50%, respectively. The single-crystal disks were deformed by placing them in the die and applying the same load in exactly the same manner as was applied to the polycrystalline material.

Irradiation and Spectroscopy

The specimens were irradiated at room temperature for 20 min in a 4-kCi underwater Co-60 gamma-ray source. The total dose was about $9 \cdot 10^{19}$ eV/cm^3 in KCl. The spectrum of each specimen was taken before and after irradiation on a Cary Model 14R recording spectrophotometer at room temperature. All the spectra presented in this paper represent the change in either optical density or absorption constant induced in the specimen by irradiation. In the case of pressed pellets or deformed single crystals, the irradiation and spectral measurements were carried out within one or two hours after the pressing procedure.

Grain-Size Measurements

Photomicrographs of several of the pellets were made. In the case of KI and KCl, usually the grains could be easily distinguished, and the average grain size determined by lineal analysis was about $6-9\,\mu$. No recrystallization could be distinguished in unetched deformed single crystals, although what appeared to be slip lines and some fragmented areas could be detected.

RESULTS

Potassium Chloride

The change in optical density (induced by irradiation) versus wavelength is plotted in Fig. 1 for KCl specimens produced from material which had not been heat-treated. The short-wavelength ends of the spectra are plotted in terms of absorption constant versus incident photon energy on an expanded scale in Fig. 2. Comparable results for specimens ground or pressed from single crystals previously heated at 560°C are presented in Figs. 3 and 4, respectively.

The peak height of all bands generated by irradiation is much enhanced in the deformed single crystals as compared to the as-cleaved disk. This has been observed by other workers. It should

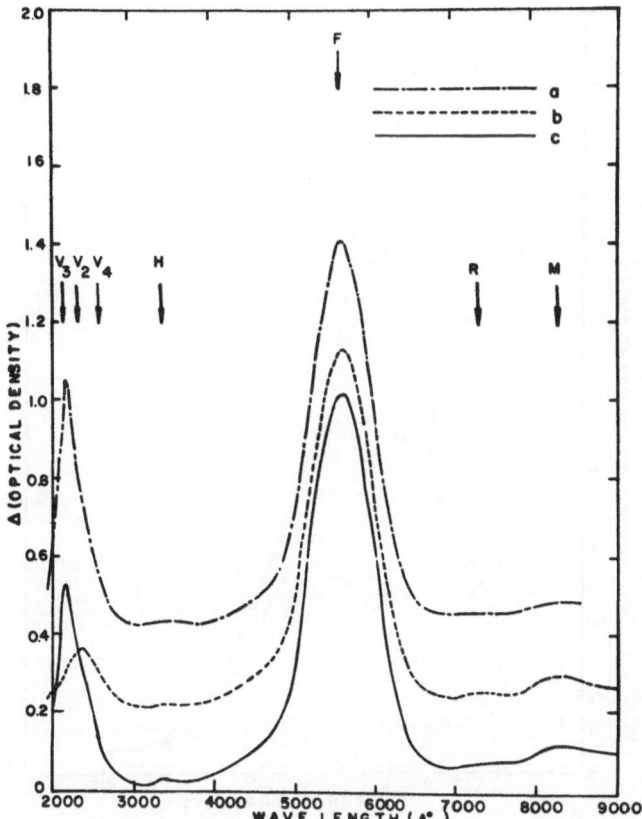

Fig. 3. Spectral changes produced by irradiation in KCl specimens prepared from heat-treated material: (a) pellet, not etched before heating, $t = 0.092$ cm, origin displaced up 0.4 OD units; (b) deformed single crystal, $t = 0.069$ cm, origin displaced up 0.2 OD units; (c) pellet, etched before heating, $t = 0.089$ cm.

be noted, however, that in both the as-received and heat-treated cases the particular V-band enhanced is the V_2-band.*

The pellets ground from as-received material, etched or unetched, showed much less enhancement in the F- and M-bands and a general elevation of optical density increasing in magnitude toward the blue. This opacity continued to increase after the end of the irradiation and was discovered by taking the spectra of

*This fact recently has been noted by other workers; see work by J. Z. Damm cited in J. Z. Damm and J. Kowalczyk, Phys. Stat. Sol. 6: 693 (1964); and see also W. A. Sibley, Phys. Rev. 133A: 1176 (1964).

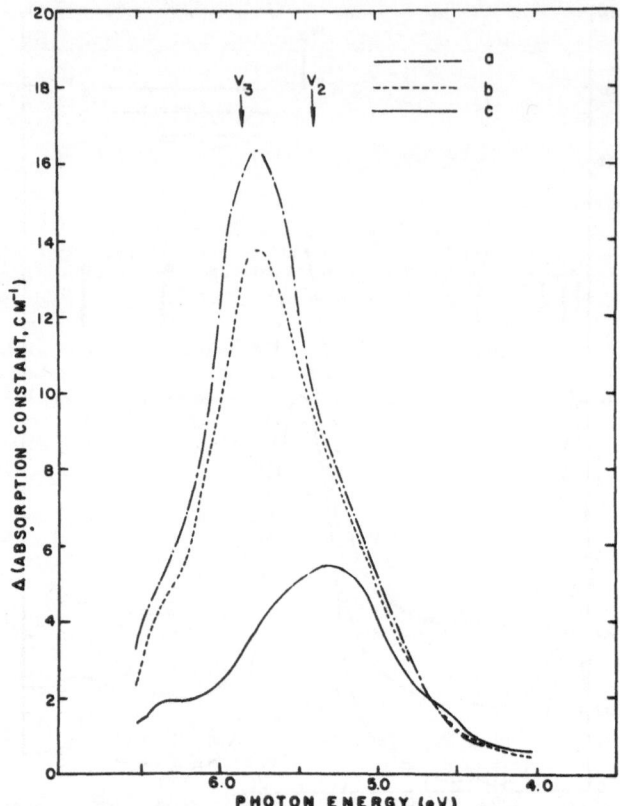

Fig. 4. V-bands in heat-treated KCl: (a) pellet, not etched before heating; (b) pellet, etched before heating; (c) deformed single crystal.

specimens several times consecutively.   This tendency of poly-crystalline pellets to "relax," or decrease in transparency, was eliminated by heat-treating the single-crystal disks from which the pellets were ground (see Figs. 3 and 4).   The spectra were then generally similar to those of deformed single crystals.   One common feature of the spectra taken from all the pellets is the great increase in intensity of the V_3-band.   This is especially noticeable in the specimens represented in Figs. 3 and 4.

The pre-irradiation spectra of all the KCl specimens were analyzed for the OH-band at 204 mμ.   No positive identification of this absorption could be made.   The as-cleaved single crystal and one of the deformed single crystals had spectra which indicated that this band might be present with an absorption constant of around 1 cm^{-1}. Two of the pellets also displayed faint traces of this absorption.

Potassium Bromide

The spectral changes caused by irradiation in specimens produced from non-heat-treated KBr are presented in Figs. 5 and 6. These data for pellets and deformed disks made from heat-treated source crystals are displayed in Figs. 7 and 8.

The deformed single crystals showed the typical, strongly enhanced F-band. Figures 6 and 8 reveal that the V-band peak in these deformed single-crystal disks appears to be at a somewhat lower energy than in some, but not all, of the pellets. The V_4-band is known to absorb at the approximate position found for the maximum in pressed single crystals (4.5 eV). However, this band is thought to be unstable at room temperature. Some enhancement

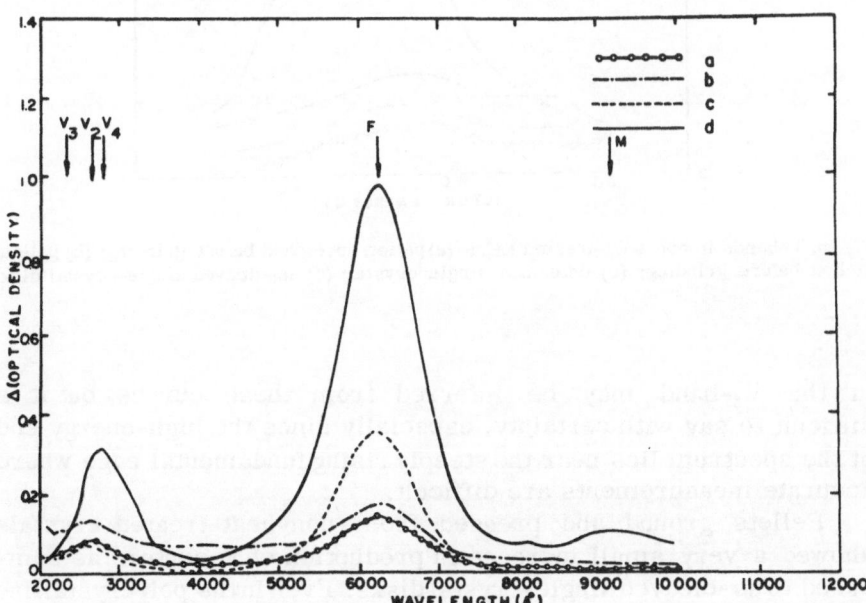

Fig. 5. Spectral changes produced by irradiation in KBr specimens prepared from non-heat-treated material: (a) pellet, etched before grinding, $t = 0.102$ cm; (b) pellet, not etched before grinding, $t = 0.117$ cm; (c) as-cleaved single crystal, $t = 0.129$ cm; (d) deformed single crystal, $t = 0.084$ cm.

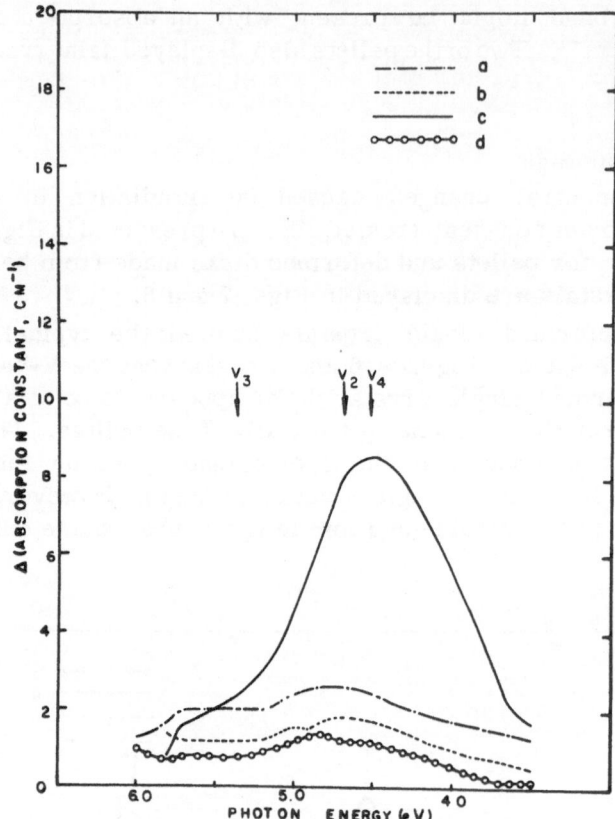

Fig. 6. V-bands in non-heat-treated KBr: (a) pellet, not etched before grinding; (b) pellet, etched before grinding; (c) deformed single crystal; (d) as-cleaved single-crystal disk.

in the V_3-band may be inferred from these curves, but it is difficult to say with certainty, especially since the high-energy end of the spectrum lies near the steeply rising fundamental edge where accurate measurements are difficult.

Pellets ground and pressed from non-heat-treated crystals showed a very small increase in production of V-centers as compared to as-cleaved single-crystal disks. Even in the polycrystalline samples made from heat-treated KBr, the V-band intensity is smaller than that of the deformed monocrystal; this is particularly evident if the general increase in background occurring during irradiation is subtracted from the spectra. This overall increase in opacity was observed in pellets ground from unetched material, heat-treated or not. In contrast to KCl, etching, rather than heat-treating, seemed to eliminate this effect in KBr. In agreement with the results obtained for the chloride, the more opaque pellets have a lower F-peak than the transparent ones if some approximate extrapolation of the background is subtracted from the observed value. No enhancement in the V_3-band could be discerned in the pellets.

Potassium Iodide

Figures 9 and 10 contain the optical density data collected on KI specimens prepared from single crystals which had not been heat-treated; Figs. 11 and 12 display the same information for pellets and deformed disks originating from heat-treated material.

A distinct enhancement in the V_2-band occurs in the deformed single crystals, while the V_3-band seems to be only mildly affected. A considerable intensity increase of the V_2-band was observed in the pellets, but, also, in contrast to the deformed single crystal, a

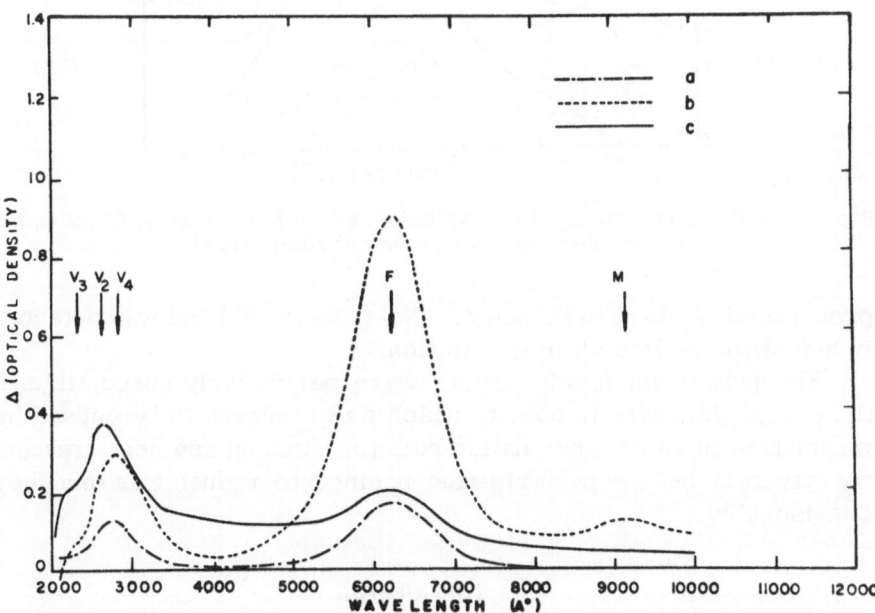

Fig. 7. Spectral changes produced by irradiation in KBr specimens prepared from heat-treated material: (a) pellet, etched before heating, $t = 0.104$ cm; (b) deformed single crystal, $t = 0.097$ cm; (c) not etched before heating, $t = 0.117$ cm.

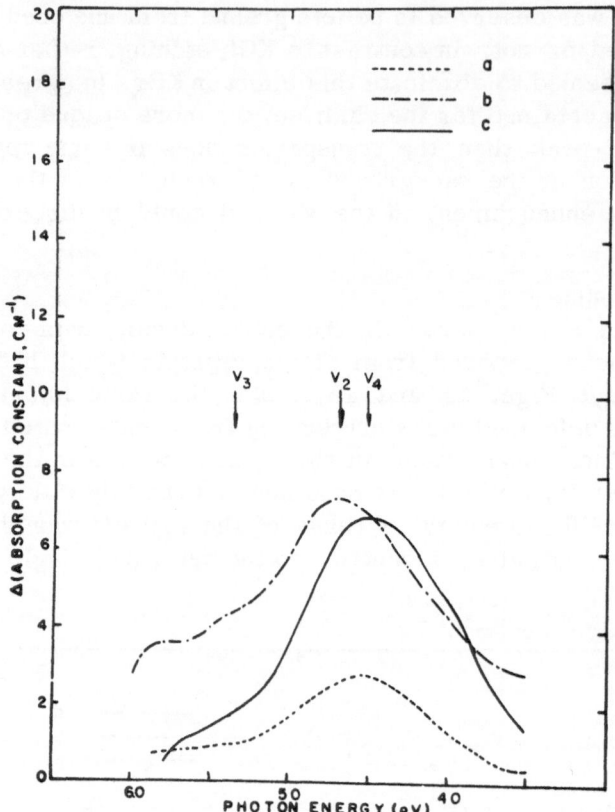

Fig. 8. V-bands in heat-treated KBr: (a) pellet, etched before heating; (b) pellet, not etched before heating; (c) deformed single crystal.

pronounced V_3-band was noted. No trace of F-band was detected in any of the pellets after irradiation.

The potassium iodide pellets were particularly susceptible to the overall increase in opacity which was observed to be caused by irradiation in some other halide pellets. Etching and heat-treating the crystals before pulverization seemed to reduce this tendency considerably.

DISCUSSION

Opacity

The development of opacity in the pellets during and after irradiation is an extrinsic effect due to the introduction of some foreign deposit from the source crystal surface during the grinding operation.   This is apparent because treatment of the source crystals before grinding and pressing eliminates or reduces the effect; these treatments include etching which removes surface layers of the crystal and heat-treating which should desorb or decompose foreign material present on the surface. The additional reduction in the colorability of the opaque pellets, as compared to those which remained transparent, also appears to be induced by this foreign substance.

Depression of the F and V_2 Colorability Increment

Deformation of all the potassium halide single crystals caused a sharp increase in the production of F- and V_2-centers, i.e., some protocenters were created by deformation which either are easily converted into the optical centers or increase the efficiency with which the color centers are produced.   In potassium chloride,

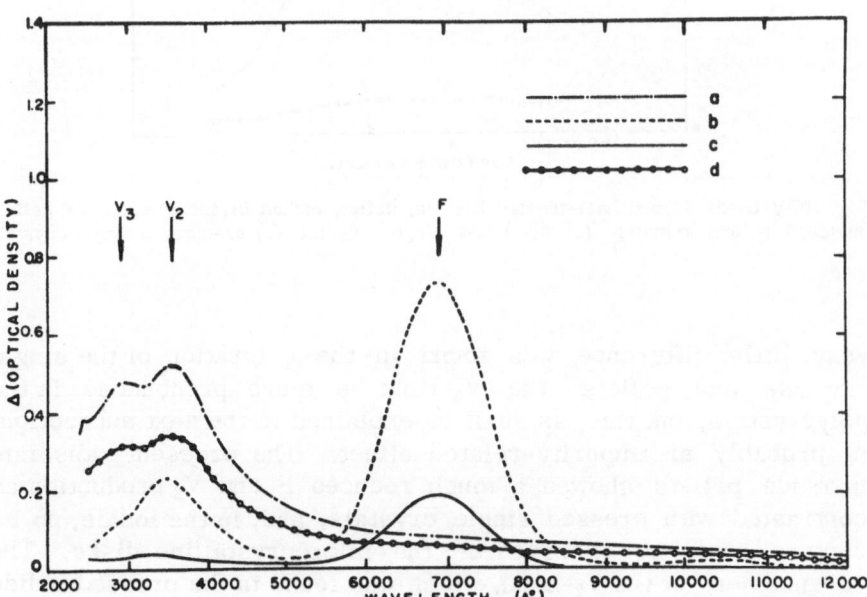

Fig. 9. Spectral changes produced by irradiation in KI specimens prepared from non-heat-treated material: (a) not etched before grinding, $t = 0.081$ cm; (b) deformed single-crystal disk, $t = 0.071$ cm; (c) as-cleaved single crystal, $t = 0.094$ cm; (d) pellet, etched before grinding, $t = 0.065$ cm.

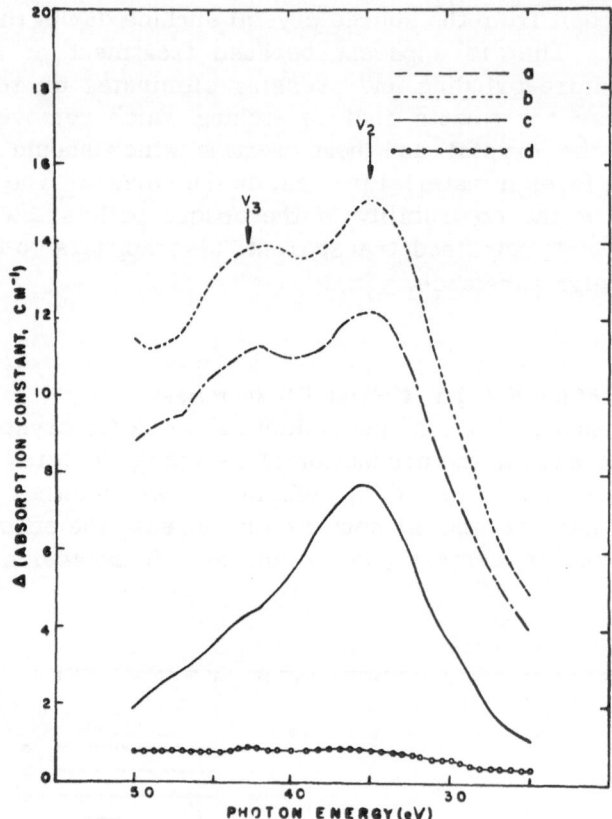

Fig. 10. V-bands in non-heat-treated KI: (a) pellet, etched before grinding; (b) pellet, unetched before grinding; (c) deformed single crystal; (d) as-cleaved single crystal.

very little difference was found in the coloration of the single crystals and pellets; the V_3-band is more pronounced in the polycrystals, but this, as shall be explained in the next subsection, is probably an impurity-related effect.   The pressed potassium bromide pellets showed a much reduced F and V_2 production as contrasted with pressed single crystals, and, in the iodide, no F-centers could be observed in the microcrystalline disks.   The enhancement of the V_2-band, which was found in the pressed iodide single crystals accompanying the F-center increment, may also be absent in the pellets and may simply be masked by large V_2- and V_3-bands from another source.   These bands could be produced in the presence of moisture adsorbed in the interfaces in a photolytic process which will be described later. In summary, there definitely is a depression of the F-band and (except in potassium iodide, where the effect is not clear) of the V_2-bands in the pellets, as compared to deformed single crystals. The decrement observed increases with increasing anion size.

This phenomenon could be due to two factors: First, the proto-centers may be more rapidly destroyed in the pellets than in the monocrystals during the one to two hours between pressing and irradiation; and, second, the centers may be quickly bleached during irradiation in the pellets. It is difficult to decide which factor is more important on the basis of the present data, and, indeed, both alternatives may hold. In this regard, it is interesting to note the results of Laurent and Bénard [12], who studied the diffusion of ions in single-crystal alkali halides. In the potassium halide series, they found a decreasing activation energy for diffusion of both the potassium cation and halide anion as the halide ion becomes larger.

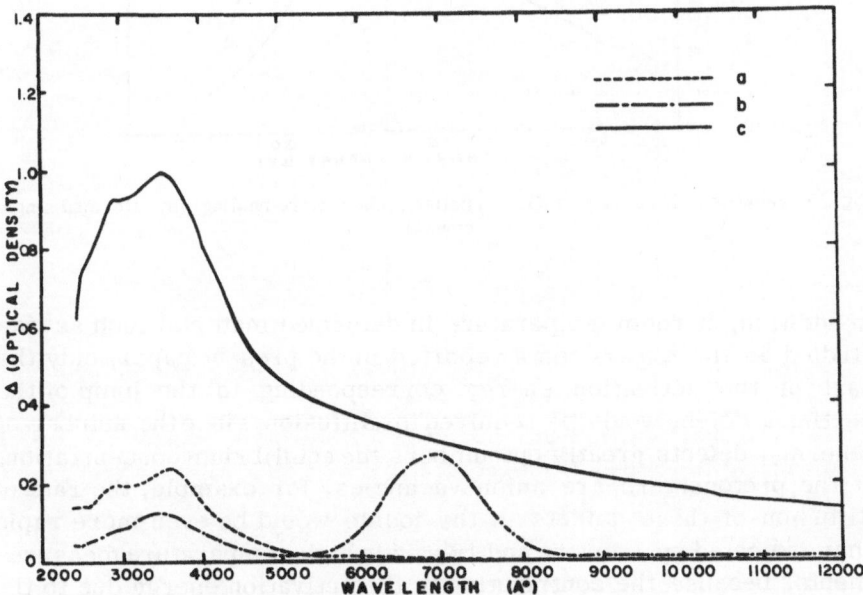

Fig. 11. Spectral changes produced by irradiation in KI specimens prepared from heat-treated material: (a) pellet, etched before heating, $t = 0.043$ cm; (b) deformed single crystal, $t = 0.043$ cm; (c) pellet, not etched before heating, $t = 0.064$ cm.

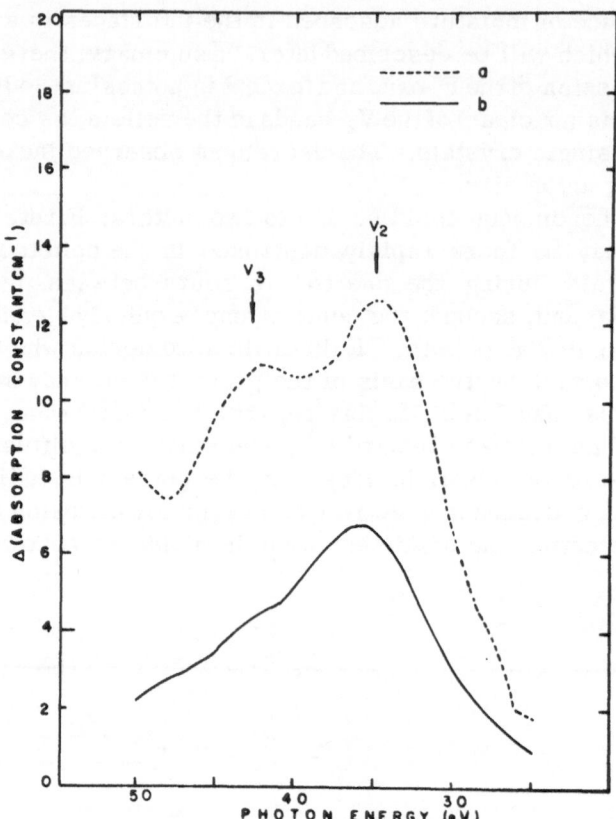

Fig. 12. V-bands in heat-treated KI: (a) pellet, etched before heating; (b) deformed single crystal.

In addition, at room temperature in deformed material such as was studied in the experiments reported in the present paper, only the part of the activation energy corresponding to the jump of the pertinent defect would be required for diffusion, since the number of athermal defects greatly outnumbers the equilibrium concentration. If the protocenters are anion vacancies, for example, the rate of diffusion of these defects in the iodide would be even more rapid than indicated by Laurent and Bénard's high-temperature measurements, because the contribution to the activation energy due to the energy of formation of vacancies should be greater, the larger the ion. Thus, the activation energies for diffusion at room temperature should be even smaller for vacancies of the larger anions.

V-Band Enhancement in the Pellets

For potassium chloride, a large enhancement in the V_3-band occurs in the pellets. As previously discussed, Lüty [9] found that the V_3-band could be developed by irradiation only in single crystals grown in air, which, as could be ascertained from pre-irradiation measurements of the OH-band, contained water-related impurities. The close proximity of the V_3- and U-bands led Lüty to suggest that they are identical, i.e., both are due to a hydride ion substitutionally located on an anion vacancy. It seems reasonable to conclude that, in the pellets, moisture adsorbed on the powder surfaces during grinding, and, thus, trapped in the interfaces, is decomposed during radiation according to

$$H_2O + h\nu \rightarrow H^+ + OH^- \tag{1}$$

One or both of these species can then rapidly diffuse into the crystal and participate in whatever reactions lead to the production of the V_3- or U-centers.

The potassium halides may be photolytically decomposed. In the pellets, the presence of the adsorbed water in the interface may cause the end products of this photolysis, if it occurs, to be optically observable as the large double V-band typical of additively colored halides. During irradiation, this photolytic (radiolytic) reaction proceeds as

$$X^- + h\nu \rightarrow X^0 + e^- \tag{2}$$

where X^0 and X^- stand for the halogen atom and ion, respectively. It will be recalled that Hersh [7] has noted the similarity between the ultraviolet spectra of halogens dissolved in aqueous solutions of their salts and those of potassium halides additively colored with halogen or X-rayed potassium halide pellets. He noted that the double-band characteristic of the solution is attributed to the poly-halide ion X_3^-, and, thus, the bands observed in the crystalline material may be due to a similar entity in which the optical transitions are not appreciably perturbed by the surroundings. In the pellets near the interface, the electrons produced in reaction (2) may be consumed by

$$H^+ + e^- \rightleftharpoons \tfrac{1}{2}H_2 \tag{3}$$

The halogen atoms can then react according to

$$X^0 + X^0 \rightleftharpoons X_2 \tag{4}$$

and

$$X_2 + X^- \rightleftharpoons X_3^-$$ (5)

The ion X_3^- would then give rise to ultraviolet bands. However, there will be a competing reaction, i.e., the formation of the hydrohalogenic acids by one or both of the following reactions:

$$H_2 + X_2 \rightleftharpoons 2HX$$ (6)

$$X_3^- + H_2 \rightleftharpoons 2HX + X^-$$ (7)

As the solvation energies of the ions in the aqueous solution and in the crystalline environment are not very different, we can analyze this problem on the basis of the behavior of these reactions in aqueous solution. It is well known that, except for the case of iodine, reactions (6) and (7) always proceed to the right. Thus, in KCl and KBr hydrogen reacts with the halogen and suppresses the formation of the optically active halide molecule-ion. Only in KI is this species stable enough to appear in the spectra.

REFERENCES

1. J. H. de Boer, Rec. Trav. Chim. Pays-Bas 56: 301 (1937).
2. C. S. Van Doorn and Y. Haven, Philips Research Repts. 11: 479 (1956).
3. H. Pick, Z. Physik. 159: 69 (1960).
4. S. Schnatterly and W. D. Compton, Phys. Rev. 135: A227 (1964).
5. T. G. Castner and W. Känzig, J. Phys. Chem. Solids 3: 178 (1957).
6. W. Känzig and T. O. Woodruff, J. Phys. Chem. Solids 9: 70 (1958).
7. H. N. Hersh, Phys. Rev. 105: 1410 (1957).
8. W. Mollwo, Ann. Physik. 29: 394 (1937).
9. F. Lüty, J. Phys. Chem. Solids 23: 677 (1962).
10. T. P. Zaleskiewicz and R. W. Christy, Phys. Rev. 135: A194 (1964).
11. H. N. Hersh, J. Chem. Phys. 27: 1330 (1957).
12. J. F. Laurent and J. Bénard, J. Phys. Chem. Solids 3: 7 (1957).

Chapter 10

Grain Boundary Effects in Relation to the Dielectric Properties of Infrared Transmitting Materials

T. J. Gray and D. Guile*

State University of New York College of Ceramics
Alfred University
Alfred, New York

Dielectric and infrared transmitting properties of single-crystal and hot-pressed polycrystalline specimens of MgO, MgF_2, and other infrared transmitting materials have been investigated. The frequency dispersion of the dielectric loss has been determined from 100 cps to 100 kcps for the temperature range 25–550°C. DC conductivity has also been determined. Initial attempts to separate the grain boundary contribution from that of the bulk by a Sillars–Koops-type analysis were not entirely satisfactory. An alternate qualitative analysis is proposed. The apparent activation energy for the dielectric-loss mechanism has been determined.

INTRODUCTION

As early as 1955 it was demonstrated in this laboratory that semi-conductor materials could readily be fabricated by a hot-pressing technique to yield specimens having essentially single-crystal characteristics. This technique was demonstrated with respect to several infrared transmitting materials and was subsequently followed by commercial development of these materials. From the start, it was appreciated that the factors of initial purity and use of a vacuum or a controlled atmosphere were of the greatest importance. Although materials fabricated by this technique are now available from many sources, detailed and independent information correlating transmission and dielectric properties over a wide temperature range has not been available. Some of these data will be presented at this time and further information in the near future.

*Present address: Ceramic Research Laboratory, Corning Glass Works, Corning, New York.

In dealing with polycrystalline specimens, it would be advantageous to correlate physical properties with grain boundary constitution.   One mode of effecting this correlation appears to be the application of a Sillars–Koops-type analysis [1, 2] to the dispersion curves for dielectric loss and permittivity plotted against frequency. It will be indicated, however, that the model employed in these analyses is unsatisfactory and can be used only as a guide for improved analysis on a somewhat different model.

It has already been shown in this laboratory that pure alumina hot-pressed materials of extremely fine grain and theoretical density exhibit essentially the same elastic properties as single-crystal materials.   However, internal friction measurements disclose significant differences and suggest that these, in correlation with dielectric measurements, may lead to an improved understanding of grain boundary conditions.

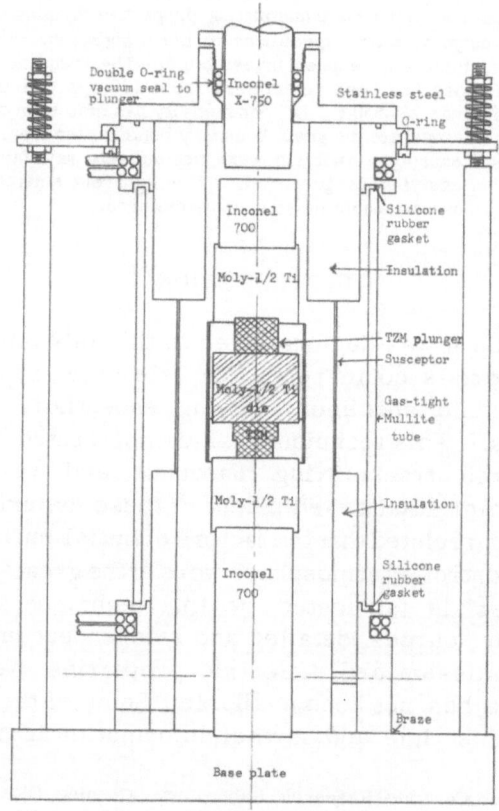

Fig. 1. Vacuum hot-press.

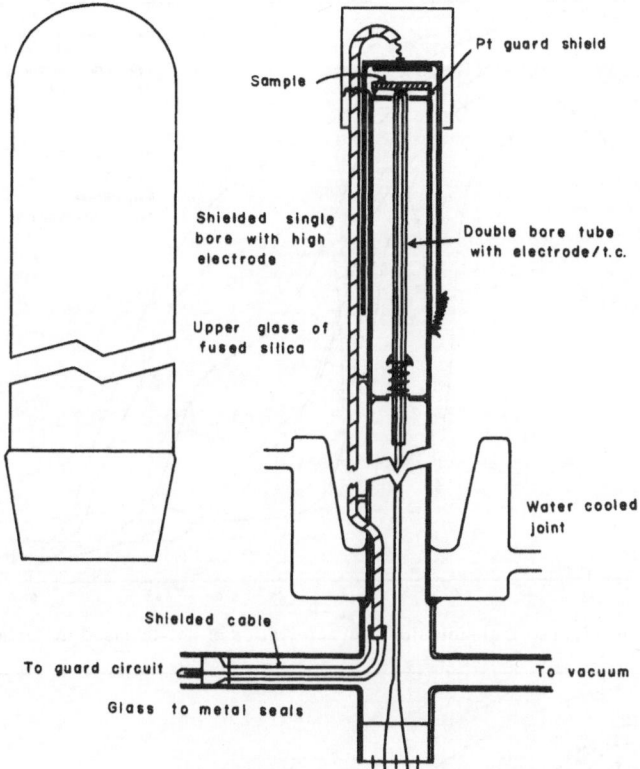

Fig. 2. Measuring-tube assembly.

EXPERIMENTAL

Specimens were fabricated from ultrapure materials in the form of submicron powders by the technique of hot-pressing. The carefully dried powders were handled, when necessary, in a helium-filled dry box and compacted either directly into the die cavity or into a tantalum-foil capsule. In all cases, the dies were constructed of titanium—molybdenum alloy with thin-walled boron nitride sleeves to minimize contamination. TZM-alloy push rods were tipped with boron nitride disks. A separate susceptor of molybdenum sheet was used to optimize coupling to the high-frequency coil for increased speed of heating. Pressures of 25,000 to 30,000 psi were employed in the temperature range 650–1200°C.

The press assembly is shown in Fig. 1. Provision is included for initial or continuous evacuation to 10^{-6} mm Hg, after which

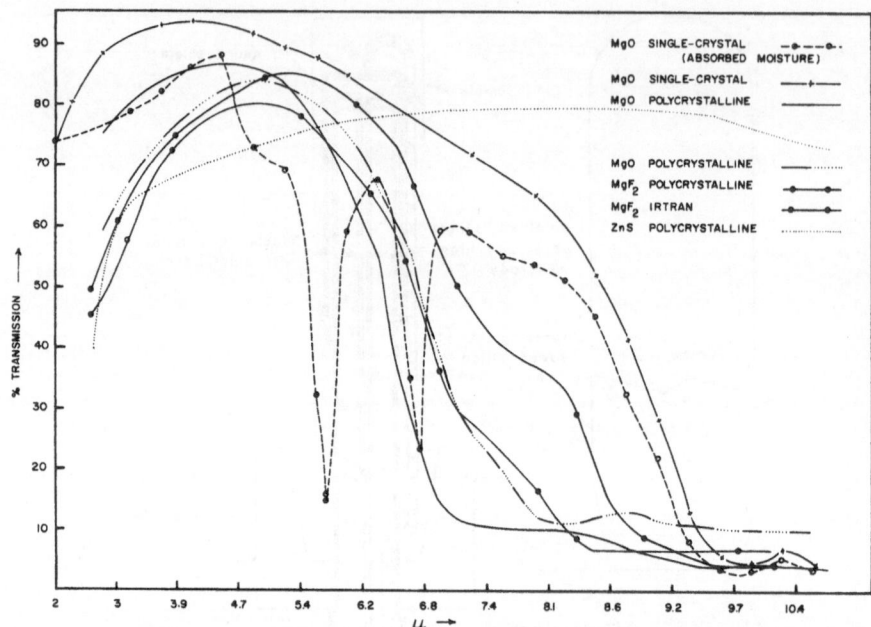

Fig. 3. Infrared transmission characteristics of hot-pressed materials.

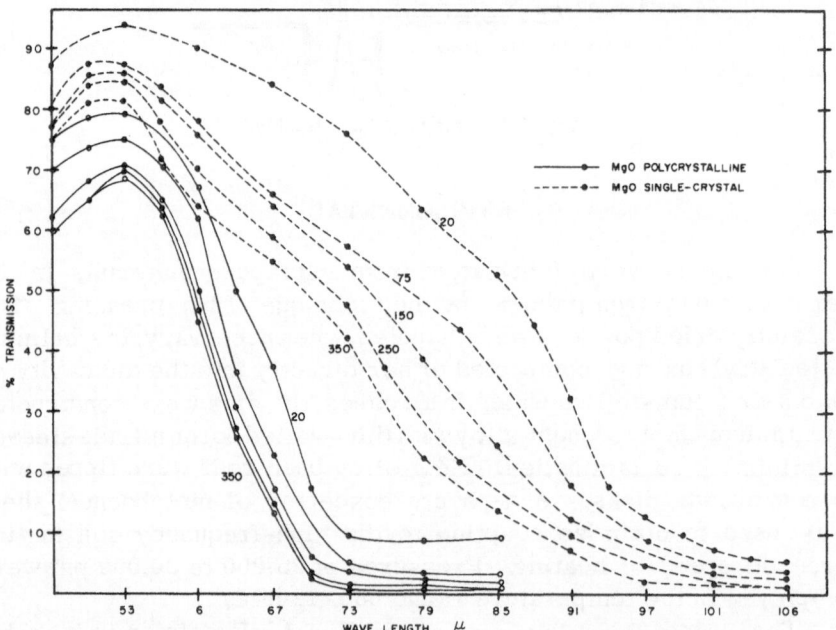

Fig. 4. Comparison of infrared transmission of single-crystal and polycrystalline (hot-pressed) MgO over the temperature range 20–350°C.

inert gases can be introduced to constant pressure. Power for the induction coil is provided by a 35-kVA Ajax arc-type unit. The upper and lower plates are water-cooled, as is the sliding O-ring seal on the main push rod. The outer mullite tube is protected by radiation shields and sealed top and bottom by spring-loading against silicone-rubber gaskets.

After fabrication, all disks were optically polished, ultrasonically cleaned, and thoroughly dried before being examined initially for infrared transmission in the range $1 - 25\,\mu$ at temperatures from below room temperature to 500°C. Platinum electrodes, including a guard electrode, were then sputtered onto the surfaces of the specimens, and the dielectric properties were measured by a General Radio 716C bridge with Wagner ground for frequencies from 10^2 to 10^5 cycles. An oscilloscope with high-gain preamplifier was employed as a null detector. For measurements of DC conductance, a General Radio megohm bridge was employed. All measurements were conducted either in vacuum or in an inert atmosphere after extended evacuation at a moderately elevated temperature.

The measuring-tube assembly is shown in Fig. 2, where it can be noted that very extensive shielding with platinum foil and platinum tubes has been employed to ensure accurate measurements. This is particularly necessary in view of effects which might otherwise be experienced.

RESULTS

Infrared transmission data have been obtained over the range $1 - 25\,\mu$ using a Perkin-Elmer Model 112 with KBr optics modified for the incorporation of a hot stage. Typical infrared transmission characteristics are presented in Figs. 3 and 4, the latter giving the results for various temperatures. Additional data on this aspect have been presented elsewhere [3]. One noteworthy aspect of Fig. 3 concerns single-crystal MgO (periclase), for which two large absorption peaks develop simply by exposure to atmospheric humidity. These peaks can be entirely removed by repolishing the specimen under absolute alcohol and examining in a dry, controlled atmosphere.

The dielectric properties of MgO are illustrated in Figs. 5, 6, and 7. It should be noted that the loss curves for single-crystal MgO (Fig. 6), after careful drying and evacuation at moderately

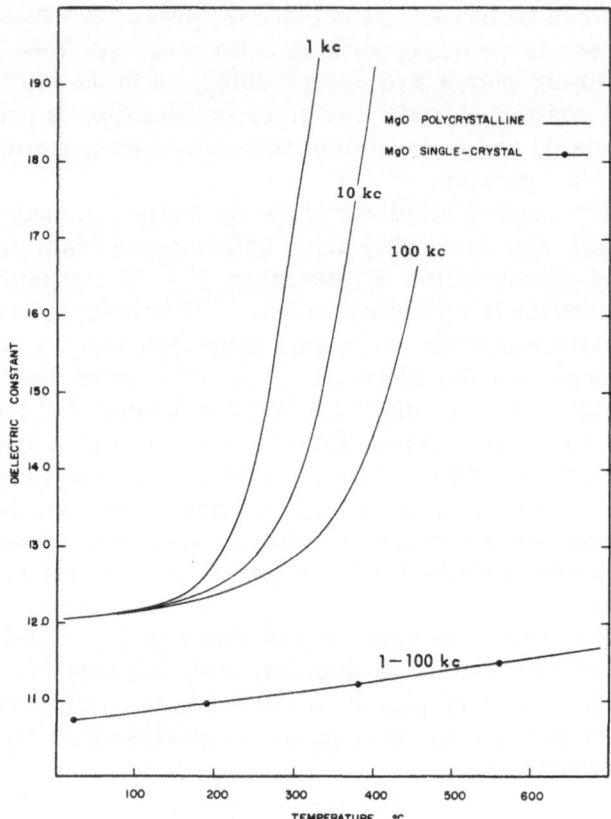

Fig. 5. Variation of dielectric constant with temperature for various frequencies for single–crystal and hot–pressed polycrystalline MgO.

elevated temperatures, are several orders of magnitude lower than those for the corresponding polycrystalline specimens. Exposure to the atmosphere raises the loss very significantly.

Corresponding information is presented in Figs. 8, 9, and 10 for magnesium fluoride and in Figs. 11 and 12 for zinc sulfide. In these cases, no satisfactory single-crystal specimens were currently available. The zinc sulfide specimens, while exhibiting good infrared transmission characteristics to 15μ, developed high dielectric loss at temperatures above 150°C.

Electron micrographs of typical hot-pressed MgO and MgF_2 specimens are given in Figs. 13 and 14, respectively.

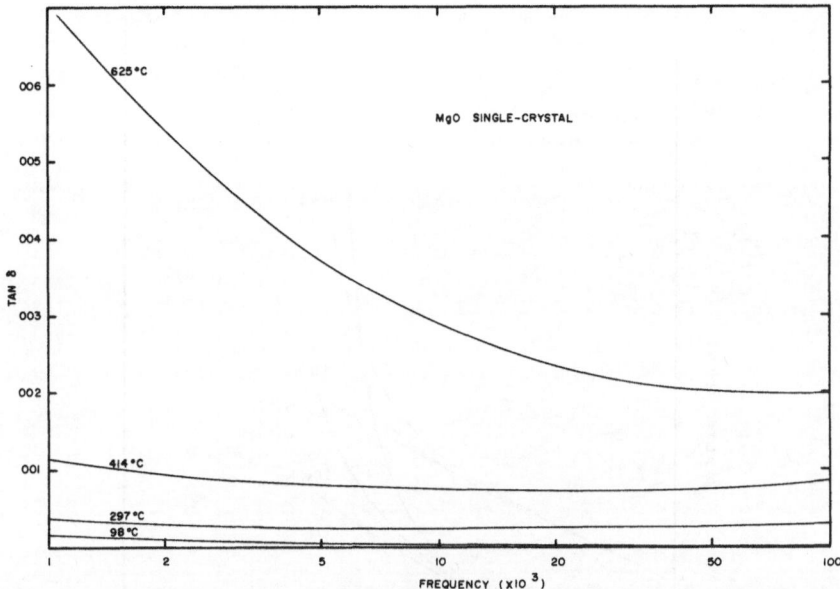

Fig. 6. Dielectric-loss frequency-dispersion curves for single-crystal MgO from 98 to 625°C.

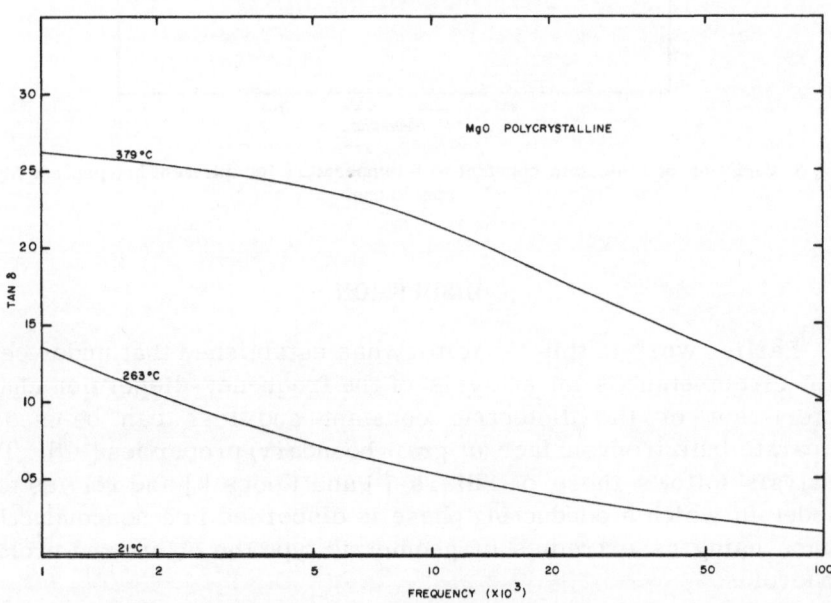

Fig. 7. Dielectric-loss frequency-dispersion curves for hot-pressed polycrystalline MgO from room temperature to 379°C.

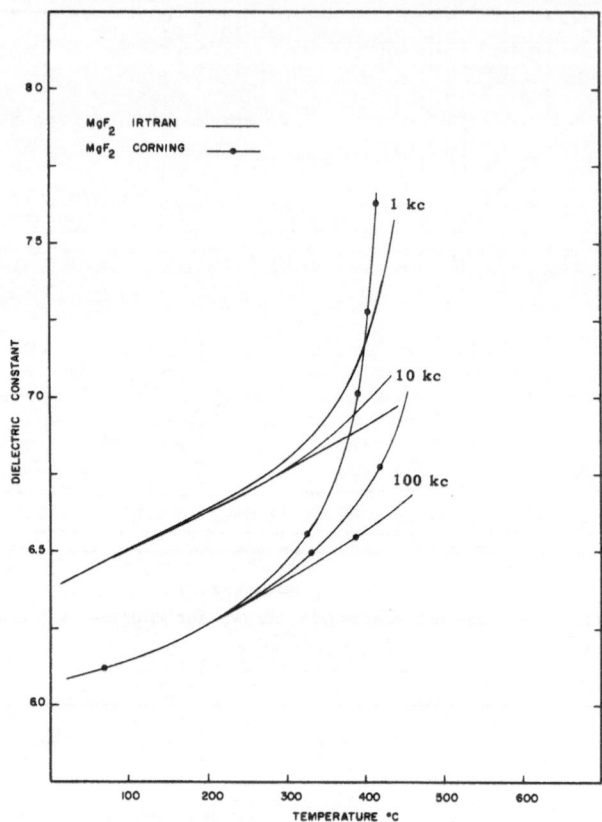

Fig. 8. Variation of dielectric constant with temperature for different hot–pressed MgF$_2$ specimens.

DISCUSSION

Earlier work in this laboratory has established that under certain circumstances an analysis of the frequency-dispersion characteristics of the dielectric constant and loss can be used to separate bulk from surface (or grain boundary) properties [4, 5]. The analysis follows those of Sillars [1] and Koops [2] and relates to a model in which a conducting phase is dispersed in a nonconducting phase using as extremes of geometric type the oblate and prolate spheroids.

With a model of a stratified polycrystalline dielectric system similar to the one displayed in Fig. 15, Volger [6] determined that the dielectric properties could be represented, under certain specific conditions of analysis by the general expressions shown in Table I.

Providing ϵ_1 is approximately equal to ϵ_2, the ratio of the thickness of the insulating phase to that of the conducting phase X is very much less than unity and the conductivity of the one phase is very much greater than that of the other. It has been shown [4] that

$$\rho_{\infty} = \rho_2 \tag{1}$$

$$\rho_0 = X\rho_2 \tag{2}$$

$$\epsilon_{\infty} = \epsilon^1 \tag{3}$$

$$\epsilon_0 = \epsilon^1/X \tag{4}$$

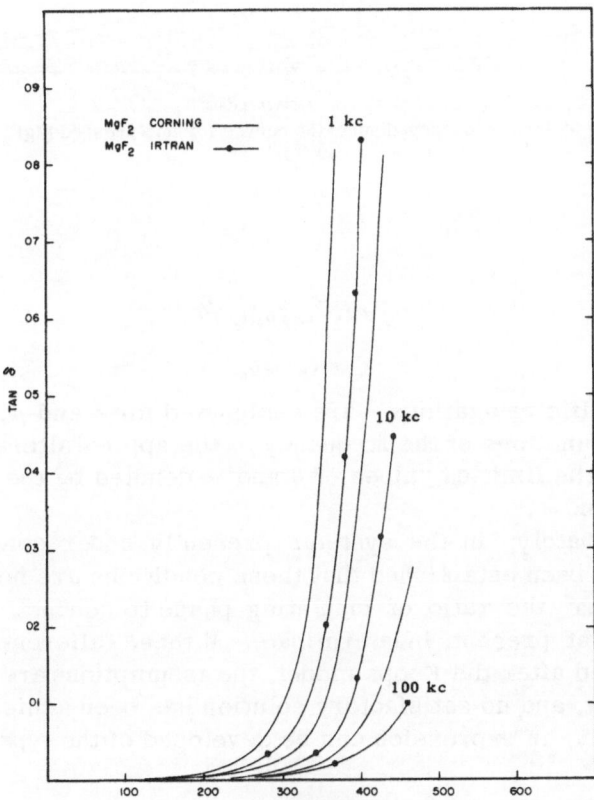

Fig. 9. Variation of dielectric loss with temperature for different hot-pressed MgF$_2$ specimens.

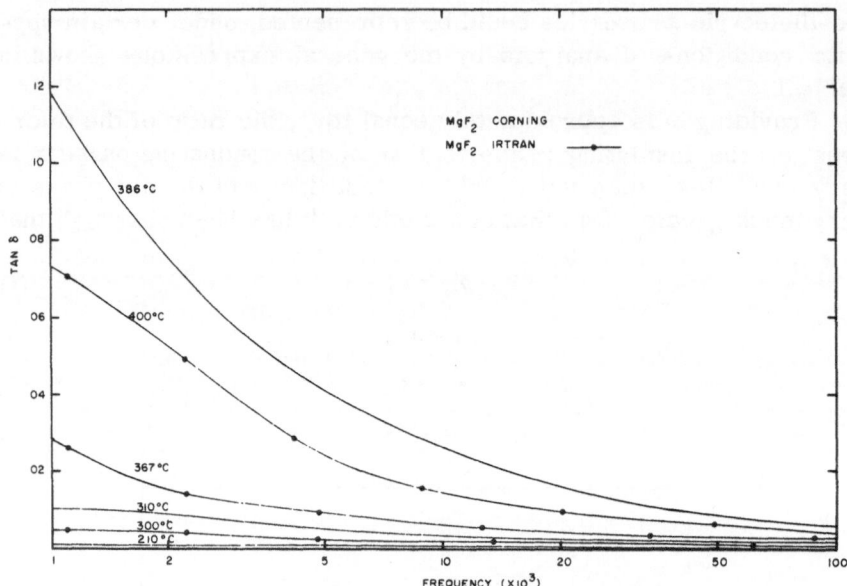

Fig. 10. Dielectric-loss frequency-dispersion curves for hot-pressed MgF$_2$ specimens to 400°C.

and that

$$\tau_\rho = \epsilon_S \cdot \epsilon_0 \left(\rho_0 \rho_\infty\right)^{1/2} \tag{5}$$

$$\tau_\epsilon = \epsilon_S \cdot \epsilon_0 \rho_\infty \tag{6}$$

where specific relaxations τ are designated for ϵ and ρ, which are monotonic functions of the frequency of the applied alternating field ω between the limiting values of 0 and ∞ denoted by the subscripts S (static) and ∞.

Unfortunately, in the systems presently under consideration, it has now been established that these conditions are not obtained; in particular, the ratio of insulating phase to conductive phase is large and, at present, indeterminate. If these different conditions are analyzed after the Koops model, the assumptions are seen to be inconsistent, and no satisfactory solution has been achieved so far. Qualitatively, an expression can be developed of the type

$$\rho_p^\infty = \frac{\rho_1 \rho_2}{\rho_1 + X \rho_2} \tag{7}$$

Since $\rho_2 > \rho_1$, this shows that, if the resistivity of the grain boundary decreases, the term ρ_p^∞ should decrease; and, if the fractional proportion of grain boundaries increases, then again ρ_p^∞ should decrease. This is, in fact, observed, but since it is not at present possible to determine X, the individual values for ρ_1 and ρ_2 cannot be determined exactly.

Attempts to fit the dispersion curves to a Koops-type analysis have been only partially successful at higher frequencies, and the absence of relaxation (absorption) peaks, as observed in single-crystal periclase, suggests that the grain boundary conduction processes are complex—very probably mixed electronic and ionic conduction. It is considered likely that an improved mathematical

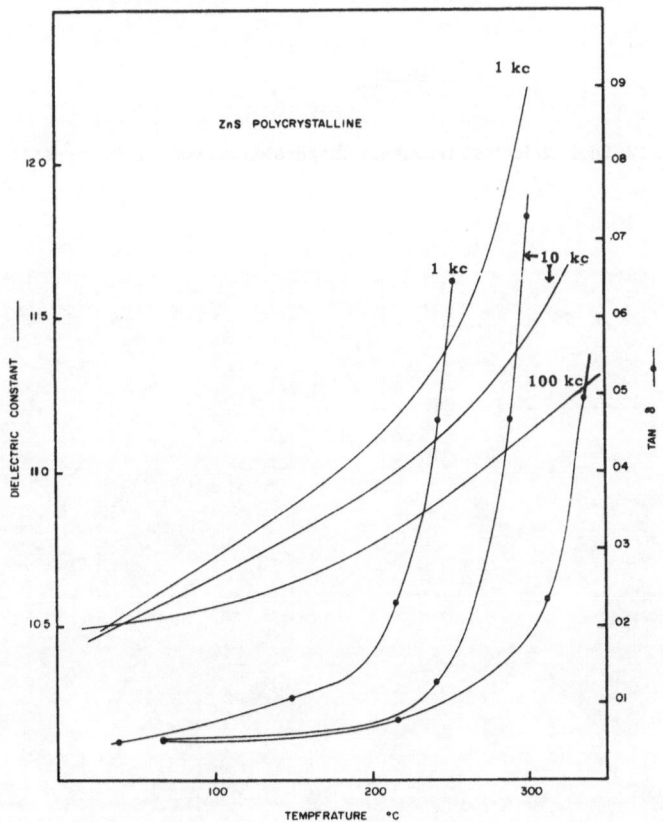

Fig. 11. Variation of dielectric constant with temperature for two different specimens of ZnS.

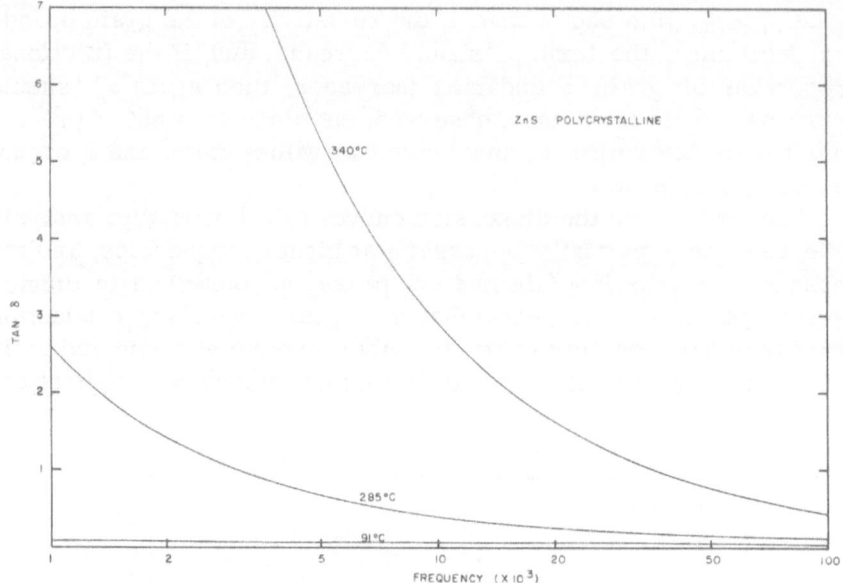

Fig. 12. Dielectric-loss frequency-dispersion curves for hot-pressed ZnS.

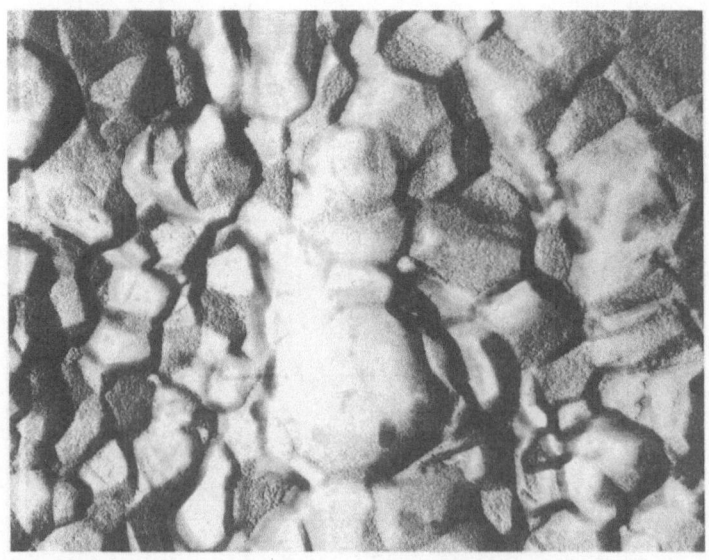

Fig. 13. Electron micrograph of MgO. Grain size, $0.5-1\mu$.

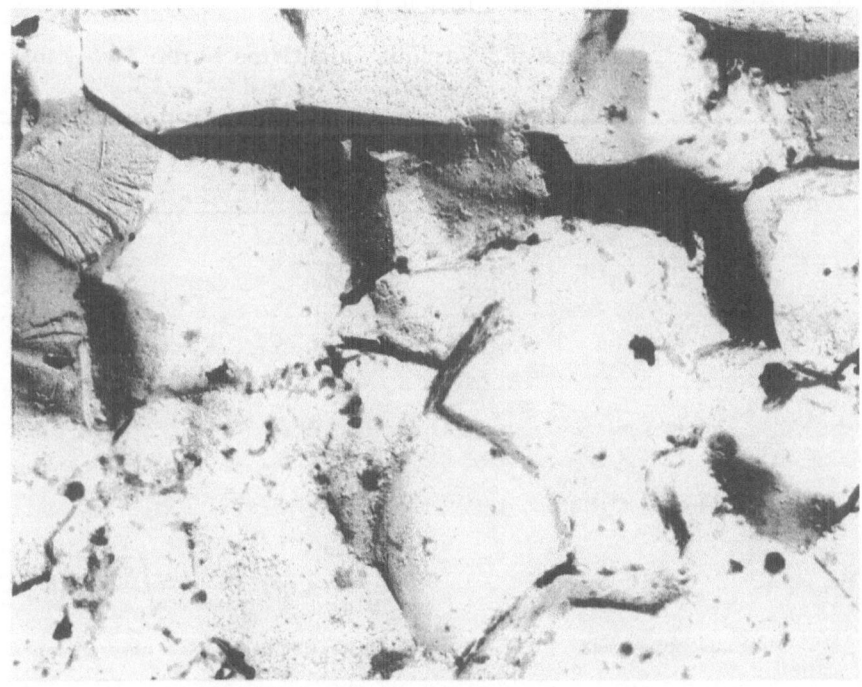

Fig. 14. Electron micrograph of MgF_2. Grain size, $3-4\,\mu$.

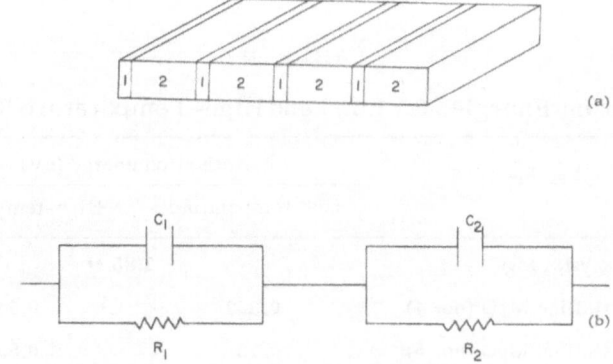

Fig. 15. Stratified polycrystalline dielectric specimen: (a) stratified dielectric and (b) electric equivalent.

TABLE I

Approximate Expressions for Various Quantities in the Two-Layer Model [*]

	$\dfrac{\rho_1}{\rho_2} \ll \dfrac{d_1}{d_2}$	$\dfrac{d_1}{d_2} \ll \dfrac{\rho_1}{\rho_2} \ll \sqrt{\dfrac{d_2}{d_1}}$	$\sqrt{\dfrac{d_2}{d_1}} \ll \dfrac{\rho_1}{\rho_2} \ll \dfrac{d_2}{d_1}$	$\dfrac{d_2}{d_1} \ll \dfrac{\rho_1}{\rho_2}$
ϵ_S	ϵ_i	ϵ_i	$\dfrac{d_1 \rho_1^2}{d_2 \rho_2^2} \cdot \epsilon_i$	$\dfrac{d_2}{d_1} \cdot \epsilon_i$
ϵ_∞	ϵ_i	ϵ_i	ϵ_i	ϵ_i
ρ_S	ρ_2	ρ_2	ρ_2	$\dfrac{d_1}{d_2} \cdot \rho_1$
ρ_∞	$\dfrac{d_2}{d_1} \cdot \rho_1$	ρ_2	ρ_2	ρ_2
τ_δ	$\epsilon_0 \epsilon_i \rho_1$		$\sqrt{\dfrac{d_2}{d_1}} \cdot \epsilon_0 \epsilon_i \rho_2$	$\sqrt{\dfrac{d_2}{d_1}} \cdot \epsilon_0 \epsilon_i \rho_2$
		Near absorption peak	Near absorption peak	Near absorption peak
$\tan \delta$	$\tan \delta = \dfrac{d_1}{d_2} \cdot \dfrac{\omega \tau_\delta}{1 + \omega^2 \tau_\delta^2}$		$\tan \delta = \sqrt{\dfrac{d_2}{d_1}} \cdot \dfrac{\omega \tau_\delta}{1 + \omega^2 \tau_\delta^2}$	$\tan \delta = \sqrt{\dfrac{d_2}{d_1}} \cdot \dfrac{\omega \tau_\delta}{1 + \omega^2 \tau_\delta^2}$

[*]Assumptions are $\epsilon_1 = \epsilon_2 = \epsilon_i$ and $d_1 \ll d_2$. Valid in the indicated ranges of ρ_1/ρ_2. Dielectric constant, ϵ; thickness, d; resistivity, ρ; and specific relaxation time, τ.

TABLE II

Activation Energies for Low- and High-Temperature Regions

Sample	Activation energy (eV)	
	Low-temperature	High-temperature
Single-crystal MgO	$\leftarrow$ 1.85 $\rightarrow$	
Polycrystalline MgO (no. 4)	0.343	0.503
Polycrystalline MgO (no. 5)	0.43	0.509
MgF_2	0.889	1.265
ZnS	1.30	1.415

analysis following a modification to the Sillars model may ultimately prove satisfactory.

If the loss mechanism is viewed as a mixed-conduction process at the grain boundary, it is permissible to analyze the results on the basis of an activated process following an Arrhenius expression. If this analysis is conducted, the curves fall into two straight portions indicative of two conduction processes with different activation energies in which it is reasonable to attribute the high-temperature process to the ionic conduction and the low-temperature process to the electronic component. Evidence for the ionic contribution was obtained by measuring the DC resistance with a General Radio megohm bridge when the initial resistance increased within $10 - 30$ sec by $2 - 3$ orders of magnitude attributable to interfacial polarization. Values for the activation energy for the two regions of conduction are given in Table II.

It can be concluded from these data that, while the detailed investigation of the dielectric properties of polycrystalline aggregates leads to important phenomenological data, it is difficult at present to translate this into a complete and detailed fundamental analysis. In order to achieve this, it will be necessary to re-examine these dielectric properties, possibly under the influence of an applied bias and to determine the individual electronic and ionic contributions. Any realistic model must, in addition, make provision for interfacial polarization and indeed calls for a complete re-analysis of the system.

ACKNOWLEDGMENTS

The authors wish to express their appreciation to Dr. J. K. Zope for his assistance in performing the detailed study of infrared transmission. The support of the Office of Naval Research under Contract Nonr 1503(01) and the Corning Glass Works is gratefully acknowledged.

REFERENCES

1. R. W. Sillars, J. Inst. Elec. Engrs. (London) 80:378 (1937).
2. C. G. Koops, Phys. Rev. 83:121 (1951).
3. T. J. Gray and J. K. Zope, "Infrared Transmitting Materials," presented at the San Francisco Meeting of the American Ceramic Society, October 1964.
4. T. J. Gray, Defect Solid State, Interscience Publishers, Inc. (New York), 1957.
5. N. Tallan, Ph.D. Thesis, State University of New York College of Ceramics, Alfred University, Alfred, New York, 1959.
6. J. Volger, Progress in Semiconductors, Vol. 4, John Wiley and Sons (New York), 1960.

Chapter 11

Optical Studies on Hot-Pressed Polycrystalline CaF$_2$ With Clean Grain Boundaries

E. Carnall, Jr., S. E. Hatch, and W. F. Parsons

Eastman Kodak Company
Rochester, New York

It has been demonstrated that with careful powder preparation and vacuum hot-pressing techniques, optical-quality polycrystalline CaF$_2$ can be made, the density of which is 99.998% of theoretical. The transmittance, refractive index, and optical homogeneity of hot-pressed and single-crystal CaF$_2$ are compared and shown to be virtually identical. Dysprosium-doped CaF$_2$ lasers have been operated in both the hot-pressed polycrystal and single-crystal forms. The thresholds for laser oscillation in the two forms are compared. It can be inferred from the data presented that clean grain boundaries in an isotropic material do not measurably affect the ordinary optical properties.

INTRODUCTION

In unpublished remarks, a number of people have speculated that a theoretically dense polycrystal of an isotropic, pure material would be optically indistinguishable from a single crystal of the same material. Data on the refractive index and interferometric, transmittance, and laser data that have been obtained on our hot-pressed CaF$_2$ very strongly support this contention.

Calcium fluoride is a convenient material to use in comparing the optical properties of single-crystal and polycrystalline materials. Large, high-quality single crystals are readily available, and the optical properties are well established. Therefore, they serve as critical standards against which the optical quality of our polycrystal can be evaluated. Furthermore, the properties of calcium fluoride are amenable to fairly standard hot-pressing techniques.

EXPERIMENTAL

As a raw material for these studies, we have used commercially available, single-crystal calcium fluoride. The large pieces of

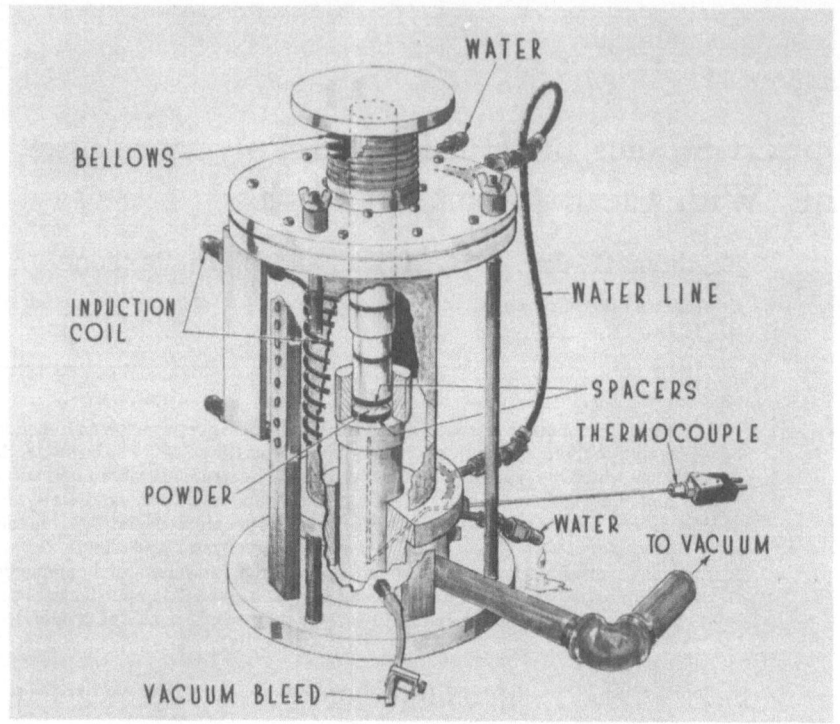

Fig. 1. Vacuum hot-pressing apparatus.

single crystal are pulverized to form a powder of approximately $180\,\mu$ average particle size. Extreme care must be taken to remove any contaminants introduced during pulverization. The powder is then inserted in the vacuum hot-pressing chamber shown in Fig. 1. The charge is enclosed between two pyrolytic graphite spacers in a metal cylinder. Pressure is applied through a single-acting plunger, and the mold is heated by a 10-kc induction coil surrounding the quartz cylinder. The material is hot-pressed in a vacuum atmosphere (maintained by a roughing pump) at 900°C at a pressure of 40,000 psi for approximately 30 min. An example of the resulting polycrystal is shown in Fig. 2. The sample shown at the left is a polished piece of commercial single crystal and that on the right a polished hot-pressed polycrystal. To the naked eye, the hot-pressed sample is as colorless and transparent as the single crystal.

A photomicrograph of an etched surface of the polycrystal is shown in Fig. 3. As one would expect from the visual clarity of the material, no pores are visible at this magnification. It is apparent

Fig. 2. Single–crystal CaF$_2$ (0.200 in.) and hot-pressed polycrystalline CaF$_2$ (0.250 in.).

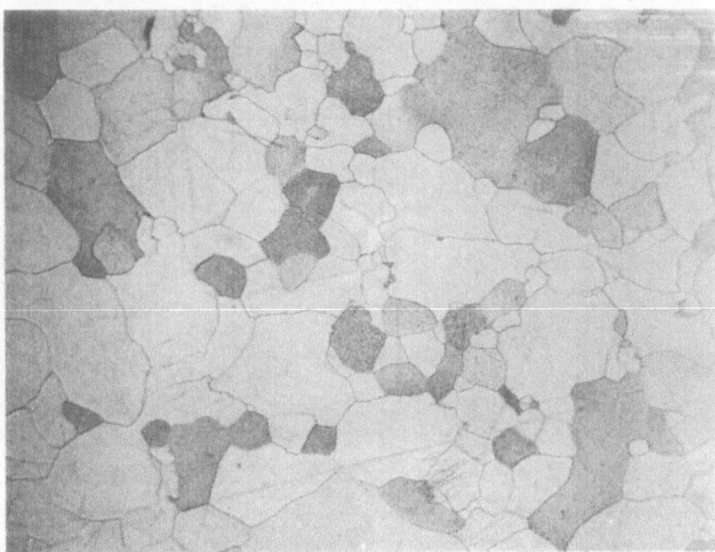

Fig. 3. Photomicrograph of hot-pressed polycrystalline CaF$_2$. Average particle size, 180 μ.

TABLE I

Density of Single-Crystal and Hot-Pressed
Polycrystalline CaF_2

	Density (g/cm³)
Theoretical (computed X-ray)	3.17935
Single crystal	3.17934
Hot-pressed polycrystal	3.17928
Polycrystal/single crystal	99.998%

Accuracy, $\pm 1 \cdot 10^{-5}$.

that the pulverization and subsequent hot-pressing processes result
in a wide range of grain sizes whose approximate average is 180 μ
with extremes of 35 and 500 μ.

The density of the hot-pressed calcium fluoride is compared to
that of the single crystal in Table I. The measurements, having an
accuracy of $\pm 1 \cdot 10^{-5}$, were made by Professor Smakula's group at

200 μ

Fig. 4. Ultramicrograph of hot-pressed polycrystalline CaF_2.

TABLE II

Refractive Index of Single-Crystal and Hot-Pressed Polycrystalline CaF$_2$

Wavelength (mμ)	Refractive index	
	Polycrystal	Single crystal
486	1.43698	1.43698
546	1.43490	1.43491
589	1.43380	1.43378
656	1.43244	1.43243

Temperature, 24°C; accuracy, $\pm 1 \cdot 10^{-5}$.

the Massachusetts Institute of Technology. It can be seen that a very small but measurable difference in density exists between the single crystal and the polycrystal. We feel that the slight difference is due to small, unidentified second-phase inclusions which may be trace quantities of CaO, or possibly pores or microcracks.

These inclusions can be seen with the aid of ultramicroscope techniques. With ultramicroscopy, the intent, of course, is to illuminate a very thin plane of the sample perpendicular to the viewing axis of the microscope. One then focuses the microscope at this illuminated plane and can view any second-phase inclusions as small bright dots of scattered light against a dark background. Any particle with a size of the order of 0.01 μ or larger can be seen when this technique is properly employed. The size is, of course, indeterminate at this small scale, because one is seeing only the diffraction image of the particle. An example of the appearance of hot-pressed calcium fluoride under ultramicroscope illumination is shown in Fig. 4. Unfortunately, our apparatus illuminated somewhat more than a true plane, and particles which are out of focus tend to clutter up the photograph. A much clearer indication of what is actually present can be gotten by observation with a stereomicroscope while moving the focus through the illuminated plane. When viewed in this manner, the contaminant is seen as small, physically isolated scattering centers.

In spite of this small remaining contamination, the optical properties of the polycrystalline material are found to be very close to those of the single crystal. Table II shows the refractive index of the starting single crystal and some of the same material which has been pulverized and hot-pressed. The data were taken on a

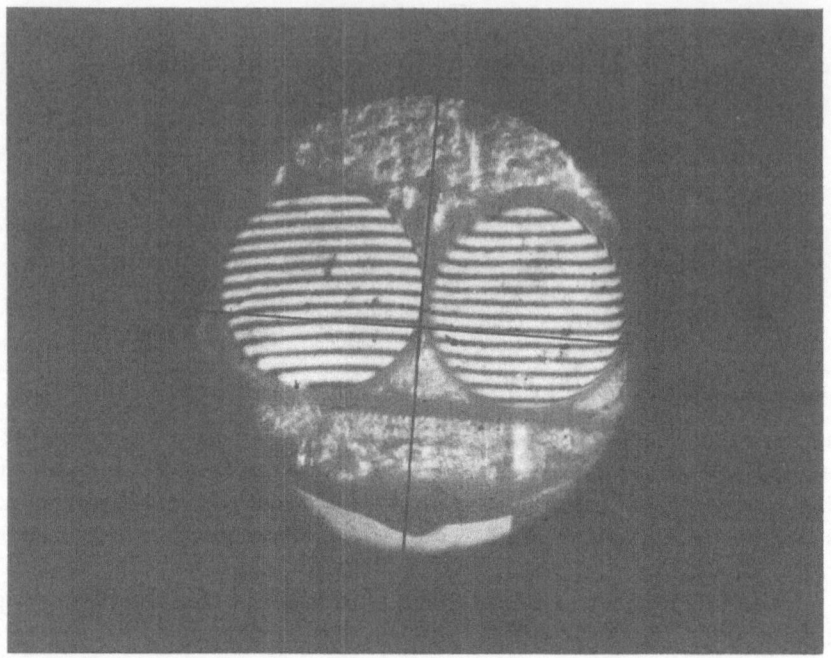

Fig. 5. Interferogram of single-crystal CaF$_2$ (left) and hot-pressed polycrystalline CaF$_2$ (right).

Gaertner scientific spectrometer using all four microscopes. The accuracy is ±0.00001. As can be seen, the refractive indices agree within the experimental error.

A comparison of the optical homogeneity of the single crystal and polycrystal is shown in Fig. 5. The polycrystalline sample is on the right and the single crystal on the left. The picture was taken on a Twyman-Green interferometer using samples of identical thickness between two large oiled-on optical flats. The slight bow in the fringe pattern is due to a small amount of power in one of the oiled-on flats. This picture was taken merely to provide a visual comparison between the two materials. More accurate data were obtained using a hot-pressed sample approximately $\frac{3}{4}$ in. thick whose faces were polished flat to less than one-tenth wave. Using a helium–neon gas laser, an interference pattern was observed between light reflected from the front and back surfaces of the sample. In this manner, one of the beams being interfered has passed through twice the thickness of the sample, while the other has been reflected from the front surface only. If the two surfaces are extremely flat, any distortion seen in the fringe pattern must be due to internal inhomogeneity in the material. Refractive index variation of about $5 \cdot 10^{-6}$ should be detectable in this manner. The hot-pressed polycrystal showed no detectable refractive index inhomogeneity in this examination.

It is clear from the last two figures that neither remanent second phases nor the presence of randomized internal grain boundaries had any measurable effect on the refractive index. One optical property which is measurably affected is the specular transmittance, as shown in Fig. 6; however, the difference is not measurable in a thickness of 0.4 in. at wavelengths greater than $1.5\,\mu$. We feel that the small difference observed at the shorter wavelengths is due to scattering by the second-phase contamination, rather than scatter by the pure, uncontaminated grain boundaries, but, of course, this cannot be proved until the contamination has been completely eliminated.

As a final point in this attempt to show that clean grain boundaries are optically inactive, we will briefly cite results obtained with the successful operation of a hot-pressed polycrystalline CaF$_2$: Dy^{+2} laser. In this case, the starting material was prepared by vacuum-melting single-crystal CaF$_2$ together with the desired

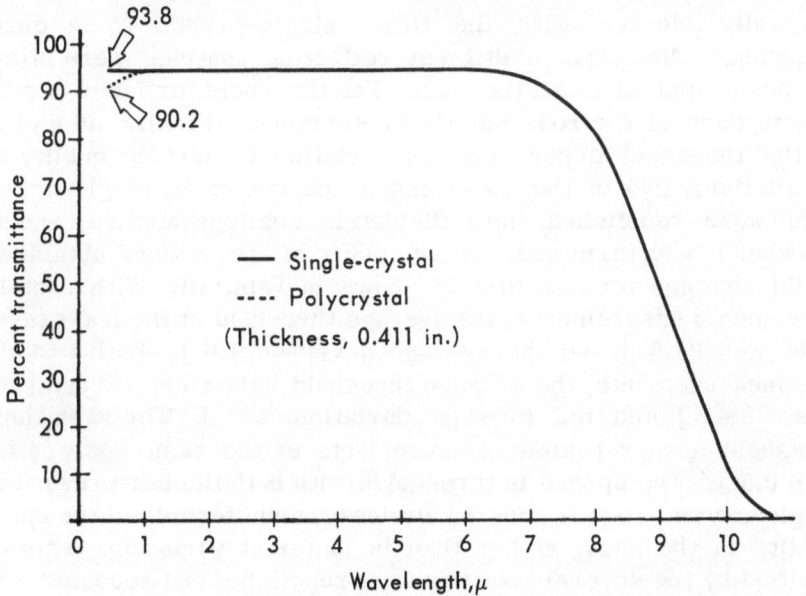

Fig. 6. Transmittance of single-crystal and hot-pressed polycrystalline CaF$_2$.

TABLE III

Threshold for Laser Action in Single-Crystal and Hot-Pressed Polycrystalline $CaF_2:Dy^{+2}$

Sample	Number of measurements	Average (J)	Average deviation (J)
Hot-pressed	17	24.6	1.6
Single crystal	5	25.4	4.7

Temperature, 77°K.

quantity of anhydrous DyF_3. The solidified multicrystalline boule was then pulverized and hot-pressed in the same manner as with undoped material. The dysprosium enters the CaF_2 lattice in the trivalent state and is subsequently partially reduced to the divalent state by $\frac{1}{4}$-MeV X-irradiation. Fifteen different rods $\frac{1}{8}$ in. in diameter and $\frac{7}{8}$ in. long were cut from different hot-pressed samples. For comparison, three $CaF_2:Dy^{+2}$ single crystals of the same size and similar doping were purchased from Optovac, Inc. The faces of the rods were polished flat to one-tenth wave and parallel to within 5 sec of arc. To ensure at least a partial overlap of surface quality and parallelism, two of the hot-pressed rods were integrally blocked with the three single-crystal rods during polishing. Dielectric multilayer reflecting coatings were simultaneously applied to all the rods. The threshold for laser oscillation in each of the rods was then determined. Finally, as a check of the threshold dependence on variation in surface quality and parallelism, two of the hot-pressed and two of the single-crystal rods were repolished, new dielectric coatings applied, and the thresholds redetermined. A summary of the results obtained at liquid-nitrogen temperature is shown in Table III. With a total of seventeen measurements, the average threshold of the hot-pressed rods was 24.6 J and the average deviation 1.6 J. With a total of five measurements, the average threshold of the single-crystal rods was 25.4 J and the average deviation 4.7 J. The variation in threshold upon repeated measurements of the same rod was less than 0.5 J. The spread in thresholds with both the hot-pressed and single-crystal rods is most likely due to nonuniformity in the optical quality of the ends, rather than in material variation. This was verified by the several rods that were repolished and recoated. The average threshold and the average deviation of the group did not change significantly; however, the actual deviation from the mean threshold of any given rod after repolishing appeared completely unrelated to its deviation from the mean before repolishing. These data do not allow us to make any quantitative claims concerning the optical quality of the hot-pressed polycrystal. They do, however, demonstrate that, at the 2.36-μ Dy^{+2} fluorescence, the optical quality of the hot-pressed polycrystal must be very close to that of a single crystal.

CONCLUSION

The data presented have demonstrated that clean grain boundaries in an isotropic material do not measurably affect the refractive index or index homogeneity. Furthermore, the data strongly suggest that such grain boundaries are also optically inactive in specular transmittance and the establishment of laser oscillation.

Chapter 12

Application of Transmission Electron Microscopy to Studies of Relative Interfacial Energies in Solids

M. C. Inman

Pennsylvania State University
University Park, Pennsylvania

In the transmission electron microscope, interface junctions were observed at high magnification with the additional advantage of great depth of focus. The resulting three-dimensional nature of electron micrographs, combined with appropriate procedures of selected area diffraction, permits precise determination of relative interfacial energies. At the present time, the techniques described in this paper have been applied only to metallic solids. However, the results already obtained are so encouraging that similar studies of ceramic interfaces are very desirable.

INTRODUCTION

The techniques of transmission electron microscopy of thin films provide a powerful general experimental method for study of the relative energies of interfaces in solids, an area of application which has been completely neglected in the past. It is, therefore, the purpose of this paper to review briefly some encouraging results which the author has obtained with metallic solids, and to draw attention to the desirability of similar studies in the field of ceramics. Much new and useful information concerning such matters as relative interfacial energy and associated crystallographic misfit is now obtainable by electron transmission techniques. It also appears that the dependence of interfacial energy upon temperature and impurity content may be elucidated. Apart from their intrinsic interest, these parameters have an important, but as yet ill-defined, role in the determination of bulk properties, with special reference to the mechanical bulk properties of ceramic materials.

REVIEW OF EXPERIMENTAL TECHNIQUES

The absolute energy of a solid–solid interface, such as a grain boundary, is not a quantity amenable to direct measurement, although in a few special instances absolute measurements have been made [1-3]. In general, the energy of a given grain boundary is measured relative to that of some other grain boundary, a twin boundary, or a free surface [4]. Relative energy measurements of this nature depend essentially upon the precise determination of dihedral angles between intersecting interfaces, which are assumed to lie in equilibrium configuration at their common line of intersection. With simplifying assumptions (neglect of torque terms), the situation corresponds to a simple equilibration of surface tension forces γ so that, in the case of a grain boundary triple junction, for example,

$$\gamma_1/\sin a_1 = \gamma_2/\sin a_2 = \gamma_3/\sin a_3 \tag{1}$$

where $a_1, a_2,$ and a_3 are the three dihedral angles and $\gamma_1, \gamma_2,$ and γ_3 are the three grain boundary tensions. If the grain boundaries have been equilibrated at a temperature such that the atomic mobility is large, the grain boundary tensions γ_{gb} are assumed to be numerically equivalent to the grain boundary free energies F_{gb}. Similar arguments are used in the case of grain boundary intersections with twin boundaries and with free surfaces.

Prior to the application of electron transmission techniques [4,5], interfacial energy ratios were usually determined by measurement of surface trace angles ψ, that is, the angles between the traces of intersecting boundary planes in a suitably prepared surface, using light-microscopy techniques. Light microscopy suffers from two major disadvantages in this area of application. First, the resolution is not better than about one micron so that measured surface trace angles will be in serious error if the boundaries possess strong curvature at distances less than one micron from the line of intersection. Second, surface trace angles are only equivalent to dihedral angles when all the intersecting boundary planes lie normal to the prepared surface plane. Since this is rarely the case, a second serious source of error is introduced. Both of the above major disadvantages can be overcome by the application of electron transmission techniques to studies of interface intersections. The method of application is illustrated in the next section by reference to grain boundary–twin boundary intersections and grain boundary triple junctions.

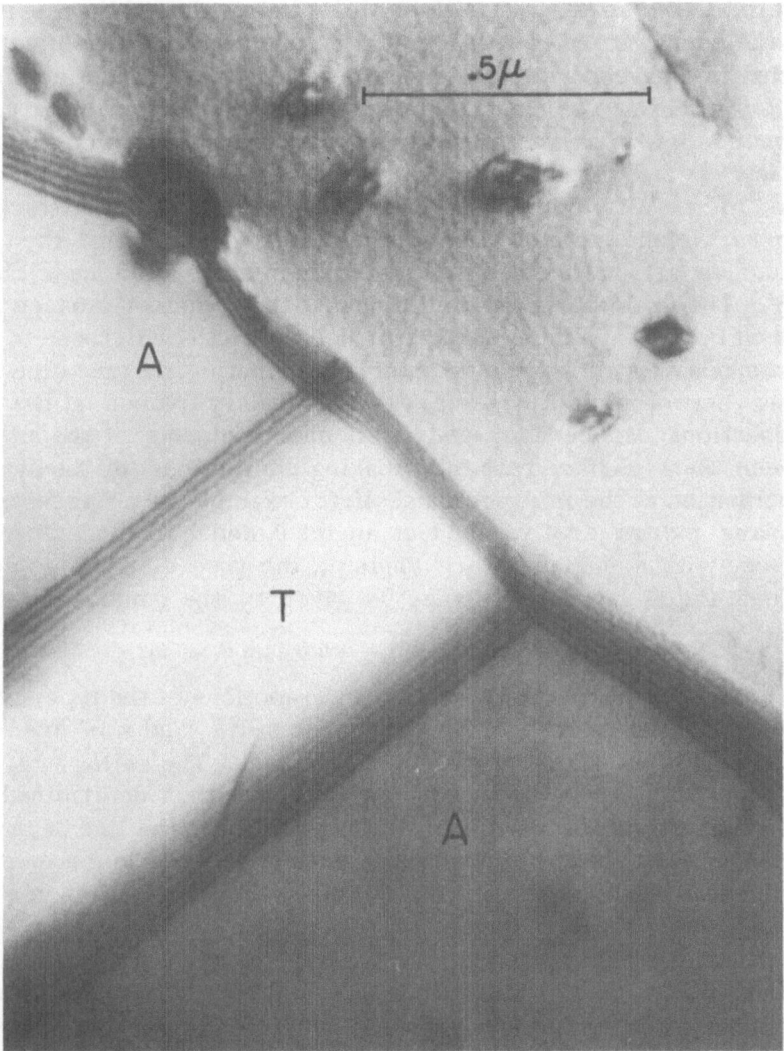

Fig. 1. Electron micrograph of the intersection of a twin band with a grain boundary in a face-centered cubic metal crystal (stainless steel). Grain surface A close to (117); 100,000×, reduced for reproduction by 30%.

TRANSMISSION STUDIES OF GRAIN BOUNDARY-TWIN BOUNDARY INTERSECTIONS AND GRAIN BOUNDARY TRIPLE JUNCTIONS

Figure 1 is an electron transmission micrograph of a grain boundary intersecting two parallel {111} twin boundaries in a face-

centered cubic metal crystal (stainless steel). It will be observed
that the combined three-dimensional nature and high magnification
of this micrograph give a most detailed view of the interface
configurations at the lines of intersection. In the particular case
shown in Fig. 1, it was found by selected area diffraction that the
surface of the parent grain A on each side of the twin band T is
close to a (117) plane. A detailed analysis of the diffraction
patterns which exhibit sharp Kikuchi lines showed that the twin
boundaries are inclined to the specimen surface at an angle close
to 68°. The grain boundary inclinations to the specimen surface are
deduced by a comparison of their projected widths with those of the
twin boundaries. With this geometric information, the true dihedral
angles between the twin and grain boundary planes at the two
intersections can be computed from measurements of the angles
between their surface traces by making corrections for the angles
of inclination at the intersections. If, for example, two intersecting
boundary planes are inclined at angles θ_1 and θ_2 to the specimen
surface, with a surface trace angle ψ_1, the true dihedral angle a_1
between these boundary planes is given by the general relation

$$\cos a_1 = -\cos \theta_1 \cos \theta_2 + \cos \psi_1 \sin \theta_1 \sin \theta_2 \qquad (2)$$

For grain boundary–twin boundary intersections of the type shown
in Fig. 1, the pairs of dihedral angles a_1, a_2 and a_3, a_4 are determined for each of the two intersections. The ratios γ_t/γ_{gb} of
twin boundary–grain boundary tensions are then determined by
resolving tension forces in a direction parallel to the line of intersection of the twin planes with the specimen surface in the manner
first proposed by Fullman [6]. This procedure eliminates the
unknown and probably large torque term $\partial \gamma_t / \partial a$ associated with a
twin boundary because of the very small crystallographic misfit
across this boundary. Since the specimen material is equilibrated
at temperatures near the melting point, it is assumed that the
boundary tensions and boundary free energies are numerically
equivalent so that $\gamma_t / \gamma_{gb} = F_t / F_{gb}$.

The typical appearance of a grain boundary triple junction in a
transmission electron micrograph is illustrated in Fig. 2. As in the
case of grain boundary–twin boundary intersections, the important
features to note here are the high resolution of the surface trace
angles and the different inclinations of the three grain boundary
planes with respect to the specimen surface. In the case of a grain
boundary triple junction, there is no convenient specimen-thickness
scale such as that provided by {111} twin planes for precise

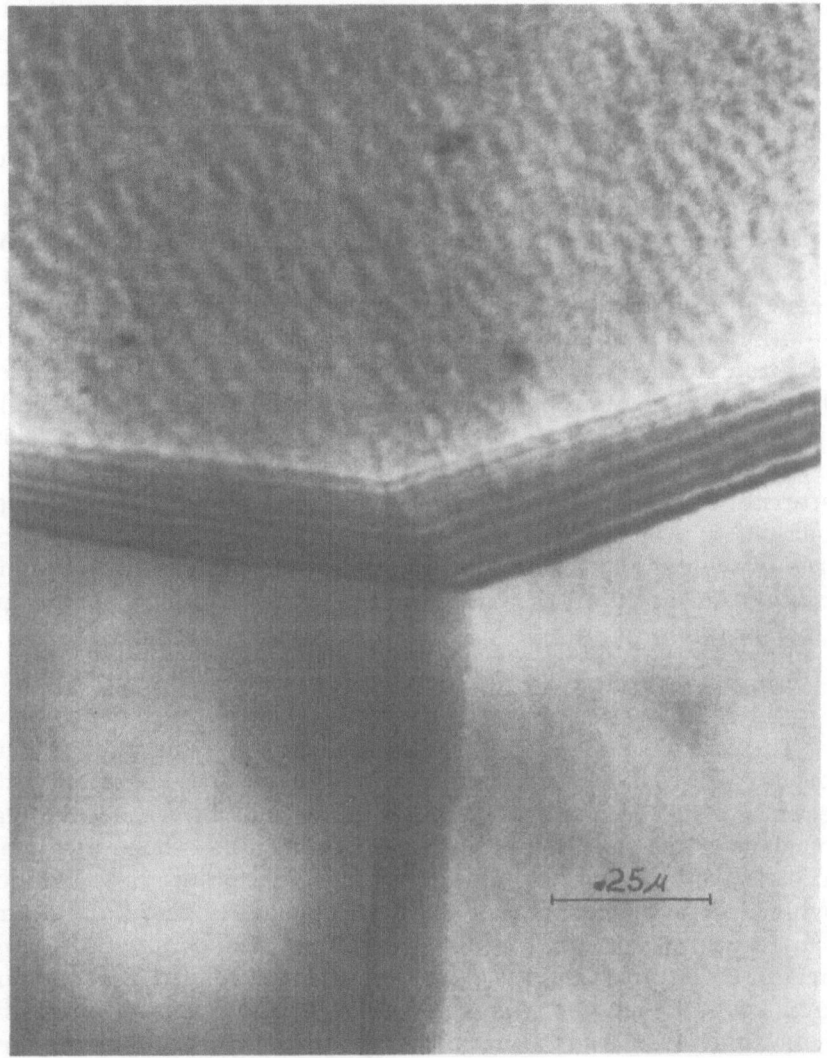

Fig. 2. Transmission electron micrograph of a grain boundary triple junction in Inconel alloy; 110,000 ×, reduced for reproduction by 30%.

absolute determination of the boundary inclination angles θ. However, since most modern electron microscopes are fitted with a specimen-tilting stage, these inclination angles can be found by observing the change Δx in the boundary projection width x on the image screen as a function of the angle ϕ through which the specimen is tilted. If, for example, a given grain boundary plane is inclined at an angle θ to the specimen surface, the following relation gives θ in terms of x, Δx, and ϕ for the case where the axis of tilt coincides with the surface trace of the particular grain boundary plane:

$$\theta = \tan^{-1}\left(\frac{1 + \Delta x/x - \cos\phi}{\sin\phi}\right) \tag{3}$$

If the axis of tilt does not coincide with the surface trace direction of the grain boundary plane, a more complicated relation than equation (3) is obtained. It is also worth noting here that since the "sense" of the tilt angle ϕ (clockwise or counterclockwise) is known, it is possible to determine which grain boundary traces lie in the top surface and which in the bottom surface of the specimen. The sense of the inclination θ to the specimen surface is then also known, so that the crystallographic misfit across each grain boundary can be unambiguously determined from selected area diffraction patterns taken in each of the three grains which border the triple junction. It is possible, therefore, to determine the relative tensions (free energies) and also the orientation with respect to the adjoining crystals of each of the intersecting grain boundary planes at a triple junction.

GENERAL REVIEW OF RESULTS AND FUTURE APPLICATIONS

To date, this author [7] has examined some hundreds of grain boundary–twin boundary intersections using the techniques just described and some initial studies have also been made on grain boundary triple junctions. All of this work has been done with metal crystals, but, from the foregoing description of the techniques involved, it is clear that they are quite general in scope and, therefore, equally applicable to solid–solid interfaces in ceramic materials. To illustrate the precision attainable with transmission microscopy, Fig. 3 shows schematically the observed spread of measurement of the twin boundary–grain boundary free energy ratios F_t/F_{gb} associated with light microscopy and electron-microscopy techniques of varying degrees of sophistication. This diagram reveals that measurement errors associated with the ratio F_t/F_{gb} decrease in striking fashion as one proceeds from light-microscopy studies of surface trace angles to electron-microscope studies of surface trace angles, and finally to a three-dimensional analysis of interface intersections along the lines outlined in the previous section. The measurement errors are taken

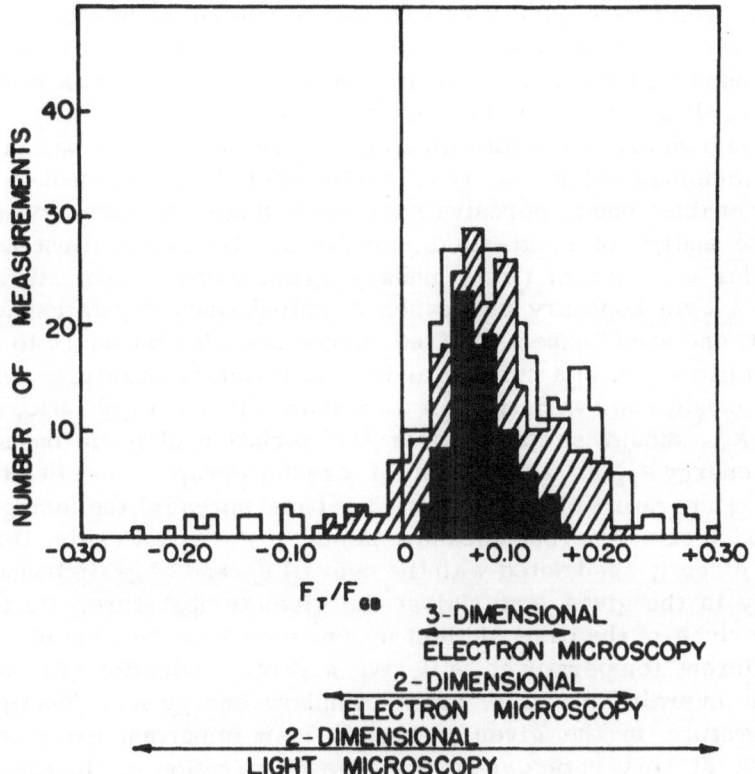

Fig. 3. Schematic diagram showing errors associated with various techniques for measurement of relative interfacial energies.

to be directly related to the base width of the measurement histograms in Fig. 3. It has been found that the width of these histograms exhibits a further significant contraction if three-dimensional analyses are based upon electron micrographs with selected area diffraction patterns containing sharp Kikuchi lines, because inter- pretation of these lines ensures that the specimen plane normal to the electron beam is defined as accurately as possible. One final source of measurement error peculiar to three-dimensional electron transmission studies of interface intersections arises from the possibility that the specimen surface plane is not coincident with the crystallographic plane normal to the electron beam. If the electron micrographs contain sufficient visual information in the form of slip plane projections, measurement of the angles between the traces of these planes enables one to determine whether or not an appreciable surface error exists in any given case. Experience has clearly shown that elimination of doubtful cases again reduces the overall spread of the histograms shown in Fig. 3.

The high precision with which interfacial energy ratios can now be determined using the transmission techniques outlined in this paper enables one to perceive significant general features concerning the energy of solid–solid interfaces. By way of illustration, consider the case of twin boundary–grain boundary intersections. Since a twin boundary possesses a well-defined crystallographic misfit, one would expect the free energy of such a boundary to have a specific value in a pure material at a given temperature. Under these conditions, variations of a measured free energy ratio, such as F_t/F_{gb}, should simply reflect the variation of grain boundary free energy F_{gb} as a function of crystallographic misfit at the given temperature of observation. The basal spread of the histogram formed from numerous measurements of F_t/F_{gb} ratios is, therefore, directly associated with the overall spread of grain boundary energy in the given material at the given temperature. Further, comparison of the base width of measurement histograms obtained at different temperatures will give a direct indication of changes in the overall spread of grain boundary energy as a function of temperature in the given material. An important extension of studies of this nature concerns the observation of the effect of impurity additions upon the spread of grain boundary energy at a given temperature. The present author has already obtained some preliminary results [7] in this area with a metallic system (antimony impurity additions to copper) which appears to show that the spread of grain boundary energy at the temperature of observation (950°C) is increased by minor impurity additions. Further studies are required, however, to establish this result with certainty. Investigations in this area have important implications concerning the incidence of brittle grain boundary fracture in metallic systems as a function of temperature and impurity content. For this reason, it is believed that analogous studies of ceramic materials where brittle fracture can be a much more severe problem would be very revealing. In the case of grain boundary triple junctions, there appears to be good reason to suppose that transmission studies will produce new and useful information relating to the dependence of grain boundary energy upon crystallographic misfit. A direct comparison of metallic and ceramic systems in this respect would be of great interest.

CONCLUSION

To summarize this brief review of the application of electron transmission techniques to interfacial energy problems, the results already obtained with metallic solids using the techniques outlined in this paper are sufficiently encouraging that analogous studies of ceramic materials seem very desirable. More precise information concerning the dependence of interface energy upon crystallographic misfit, temperature, and impurity additions can be obtained, which may help to elucidate the bulk properties of ceramics with special reference to the important problem of brittle fracture in these materials.

REFERENCES

1. H. U. Astrom, Acta Met. 4: 562 (1956).
2. W. W. Mullins, Acta Met. 4: 421 (1956).
3. J. Washburn, R. B. Shaw, T. L. Johnston, R. J. Stokes, and E. R. Parker, Univ. of California 14th Tech. Rept. No. 27, 1956.
4. M. C. Inman and H. R. Tipler, Met. Rev. Inst. Metals 8: 105 (1963).
5. M. C. Inman and A. R. Khan, Phil. Mag. 6:937 (1961).
6. R. L. Fullman, J. Appl. Phys. 22: 448 (1951).
7. M. C. Inman, L. E. Murr, and A. R. Khan, (to be published).

SUPPLEMENTAL NOTE*

I wish to point out that we have studied a device for thinning samples of bulk materials meant for direct examination by transmission electron microscopy.† In this apparatus, two multibeam ion guns simultaneously bombard the sample on the two opposite faces at incident angles of 75°. The samples are first thinned to a thickness of $30-50\mu$ by mechanical polishing and, thereafter, by ionic bombardment to electronic transparency. The electron micrographs of Figs. A–E show some of the results obtained on ceramics and metals.

*This section contributed by Max Paulus, Centre National de la Recherche Scientifique, Bellevue, France.
†See M. Paulus and F. Reverchon, J. Phys. Radium 22 (S.6): 103A–107A (1961) and see also M. Paulus and F. Reverchon, Colloq. Intern. Centre Natl. Rech. Sci. (Paris) sur le Bombardement Ionique, No. 113: 223–232 (1962).

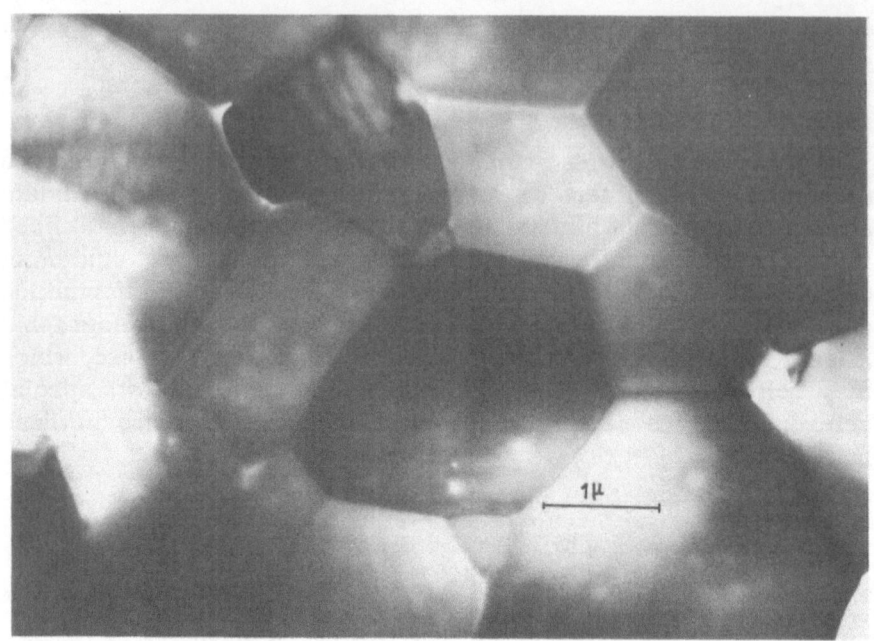

Fig. A. Polycrystalline manganese zinc ferrite (spinel structure) at low magnification.

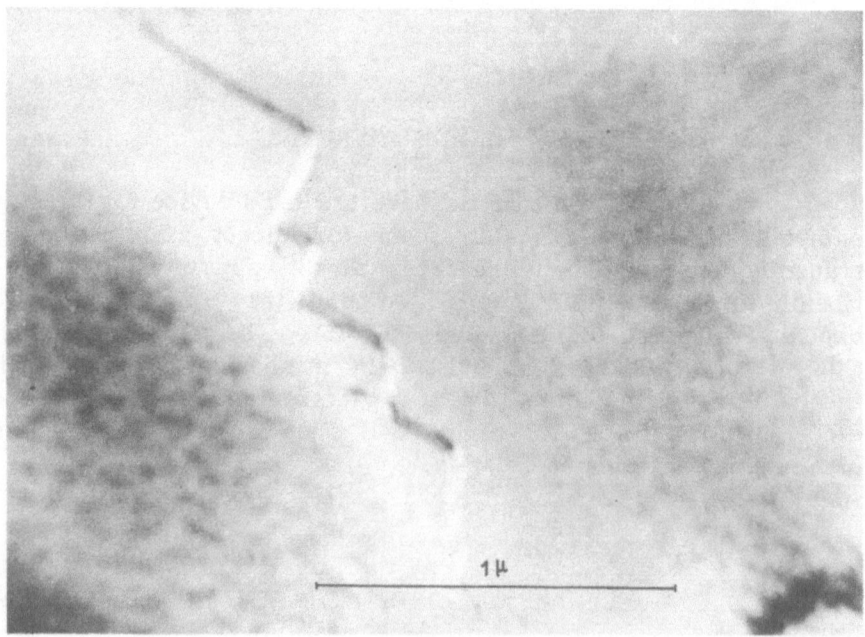

Fig. B. Details of a grain boundary in a polycrystalline manganese zinc ferrite (spinel structure).

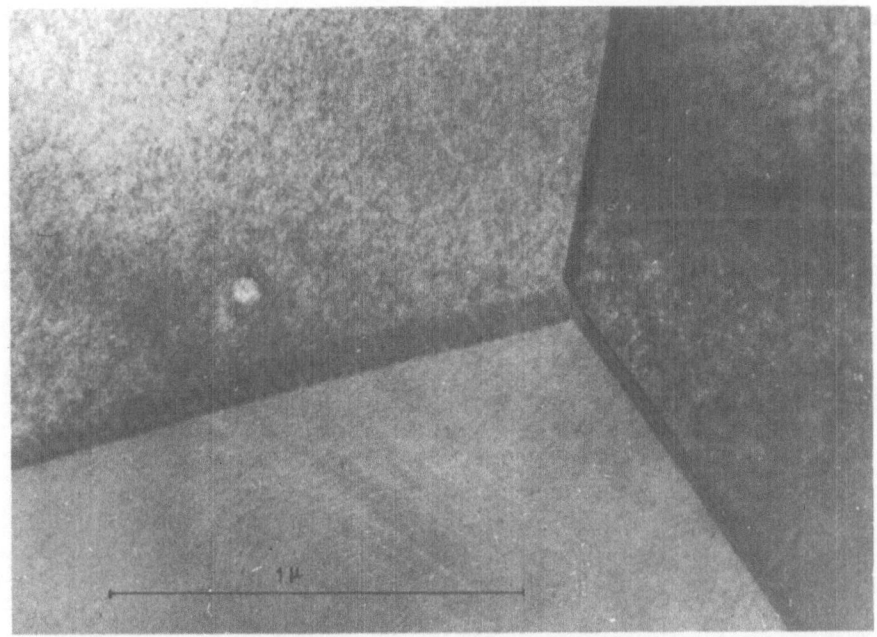

Fig. C. Details of a low-angle twin boundary in a polycrystalline manganese zinc ferrite.

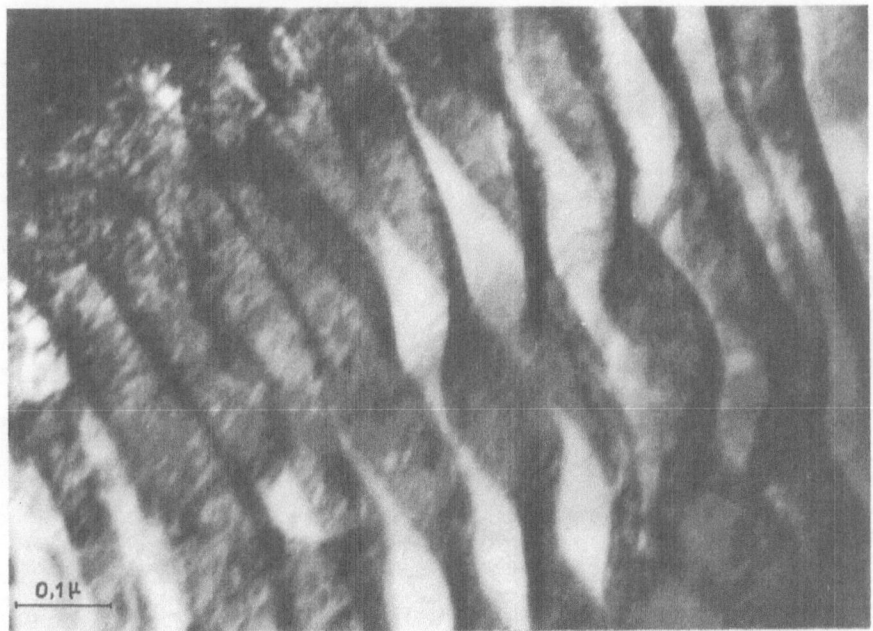

Fig. D. Pearlite in a 0.7% carbon steel. The lamellae of cementite are twisted by rolling.

Fig. E. Stacking faults and dislocations in a stainless steel after rolling.

PART III. Grain Boundary Contributions to Deformation

R. J. STOKES, Presiding

Honeywell Research Center
Hopkins, Minnesota

Chapter 13

Deformation of Polycrystalline Ceramics

Stephen M. Copley

Pratt & Whitney Aircraft
North Haven, Connecticut

and Joseph A. Pask

University of California
Berkeley, California

The mechanical behavior of nominally dense, pure, polycrystalline, ceramic materials in which dislocations are mobile is discussed. Deformation by crystallographic slip is first considered. The von Misés analysis is developed for a two-dimensional polycrystalline aggregate, and a generalization is then made to three dimensions and to simple ceramic structures. Deformation by grain boundary sliding is also considered. Stress–strain curves and photomicrographs are presented for two types of polycrystalline MgO at temperatures up to 1500°C that illustrate both deformation modes. Problems encountered in realizing deformation of polycrystalline ceramic materials are reviewed, and methods for modifying the behavior of ceramic materials in which dislocations are mobile are discussed.

INTRODUCTION

If a polycrystalline material is to exhibit appreciable ductility, it must deform primarily by crystallographic slip. Each grain must contain segments of dislocation line that can move and multiply, thereby spreading slip throughout its volume. To maintain continuity of material during deformation, each grain must also be able to undergo general changes in shape. This is possible, however, only if slip occurs on certain families of slip systems. Finally, for ductility, the grain boundaries must be strong enough to transmit stresses between grains without shearing or parting.

Although a number of ceramic materials exhibit ductility as single crystals, polycrystalline ceramics are generally brittle. In this report, it is shown that such behavior can generally be attributed to failure to satisfy one or more of the preceding conditions. The discussion is restricted to dense, single-phase, polycrystalline

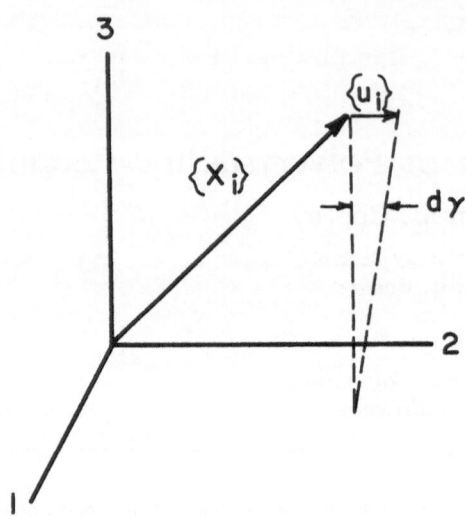

Fig. 1. The displacement of a point resulting from shear parallel to the (001) plane in the [010] direction, i.e., $\{n_i\} = (0, 0, 1)$ and $\{l_i\} = (0, 1, 0)$.

materials. Topics treated include crystallographic slip, grain boundary sliding, and experimental data on polycrystalline MgO.

CRYSTALLOGRAPHIC SLIP

The von Misés Analysis

Slip produces displacements parallel to certain crystallographic planes and directions. Each slip plane and corresponding slip direction is called a slip system. In metals, the slip plane is generally that of densest packing and widest spacing; the slip direction, that of the densest row of atoms. In ceramic materials, the choice of slip plane and slip direction is affected also by the presence of electrostatic forces and directional bonds.

Under consideration is a volume element in a polycrystalline material which is small enough so that the distribution of shear is approximately uniform and yet large enough so that the inhomogeneities inherent in slip itself are not resolved. The displacement $\{u_i\}$ of an arbitrary point $\{x_i\}$ due to a shear strain $d\gamma$ parallel to the $\overline{1}2$ plane and the 2 direction is shown in Fig. 1. The magnitude of the displacement is given by the product of the shear strain and the perpendicular distance from the point to the slip plane containing the origin. The direction of the displacement is the shear direction. The ith component of the displacement vector can be obtained by multiplying its magnitude by the cosine of the angle between the slip direction and the ith axis. Thus, the displacement of a point due to a uniform shear parallel to a plane whose normal is $\{n_j^\alpha\}$ and in a direction $\{l_i^\alpha\}$ is given by

$$u_i^\alpha = \sum_{j=1} dy^\alpha n_j^\alpha x_j \, l_i^\alpha \qquad (1)$$

where the index α is used to label slip systems. The total displacement of a point due to shear on n slip systems is

$$u_i = \sum_{\alpha=1}^{n} u_i^\alpha \qquad (2)$$

The strain components can be obtained from the total displacement by differentiating with respect to position coordinates.

$$\epsilon_{ij} = \tfrac{1}{2}\left(\frac{\delta u_i}{\delta x_j} + \frac{\delta u_j}{\delta x_i}\right) \qquad (3)$$

Since $\epsilon_{ij} = \epsilon_{ji}$, equation (3) represents a set of six linear equations relating the strain components to the shears. In shear deformation, no volume change occurs and, thus, $\epsilon_{11} + \epsilon_{22} + \epsilon_{33} = 0$. It follows that the most general uniform strain that can be produced by shear deformation can be represented by five independent strain components. The following cases result from omitting one of the equations for the normal strains and setting E equal to the matrix of coefficients of dy^α:

1. If the number of shears is less than five $(n < 5)$, it is not possible to fix the five strain components independently.
2. If $n = 5$, it is possible to fix the five strain components independently, providing that the determinant of $E \neq 0$.
3. If $n > 5$, it is possible to fix the five strain components independently, providing that a 5×5 minor of $E \neq 0$. In this case, $n - 5$ shears may be assigned arbitrary values.

Thus, five shears, independent in the sense that they can independently fix five components of the strain tensor, are required to produce the most general uniform strain. This result was first derived by von Misés [1]. The analysis here follows that of Bishop [2]. Although it is valid only for infinitesimally small strains, this analysis can be applied to large strains by viewing the components of the strain tensor as strain rates multiplied by an infinitesimal length of time.

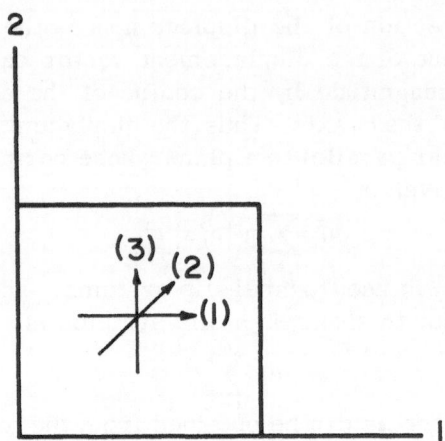

Fig. 2. A two-dimensional grain with [10], [11], and [01] slip directions.

Because of Taylor's analysis, it has been generally assumed that a polycrystalline material could maintain continuity during plastic deformation only if its grains could undergo a general uniform strain [3]. On the basis of the von Misés analysis, this condition is equivalent to requiring that a material have five independent slip systems. A clearer picture of why slip systems are independent or dependent and why a certain number of slip systems is necessary to maintain continuity during deformation can be obtained by considering a two-dimensional polycrystalline material.

Deformation of a Two-Dimensional Polycrystalline Aggregate

In two dimensions, there are three strain components—$\epsilon_{11}, \epsilon_{22},$ and ϵ_{12}. Of these, however, only two can be independently fixed

TABLE I

Values of $\{n_i^\alpha\}$ and $\{l_i^\alpha\}$

α	Slip system	n_1	n_2	l_1	l_2
1	[10]	0	1	1	0
2	[11]	$-1/\sqrt{2}$	$1/\sqrt{2}$	$1/\sqrt{2}$	$1/\sqrt{2}$
3	[01]	1	0	0	1

because area is conserved in shear deformation so that $\epsilon_{11} + \epsilon_{22} = 0$. Consequently, only two independent slip directions are required for a general change in shape.

Figure 2 shows a two-dimensional crystal containing three slip directions—[10], [11], and [01]. The total displacement of a point due to shear parallel to these slip directions may be calculated from equations (1) and (2) with the appropriate values of $\{n_i^\alpha\}$ and $\{l_i^\alpha\}$ as indicated in Table I. By substitution and regrouping, the total displacements become

$$u_1 = -\tfrac{1}{2}\, d\gamma^2\, x_1 + (d\gamma^1 + \tfrac{1}{2}\, d\gamma^2)x_2 \qquad (4a)$$

$$u_2 = (-\tfrac{1}{2}\, d\gamma^2 + d\gamma^3)x_1 + \tfrac{1}{2}\, d\gamma^2\, x_2 \qquad (4b)$$

Differentiation of the total displacements with respect to position coordinates, as in equation (3), yields the following strain components:

$$\epsilon_{11} = -\tfrac{1}{2}\, d\gamma^2 \qquad (5a)$$

$$\epsilon_{22} = \tfrac{1}{2}\, d\gamma^2 \qquad (5b)$$

$$\epsilon_{12} = \epsilon_{21} = \tfrac{1}{2}\, d\gamma^1 + \tfrac{1}{2}\, d\gamma^3 \qquad (5c)$$

By omitting one of the normal strain equations [equation (5a)], an E-matrix with a nonvanishing 2×2 minor is obtained. Thus, slip directions 1, 2, and 3 constitute two independent slip directions. It can be seen that directions 1 and 2 or 2 and 3 also constitute two independent slip directions, but directions 1 and 3 provide only one.

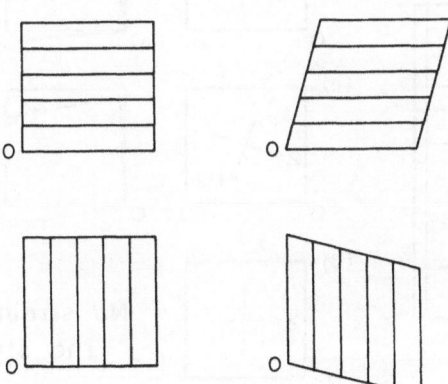

Fig. 3. The change in shape of a two-dimensional grain resulting from shear parallel to two orthogonal slip directions.

In physical terms, a slip system (or, in two dimensions, a slip direction) is independent if shear on it produces a change in shape which cannot be produced by a superposition of shears on the other slip systems. Slip systems fail to be independent if, because of crystal symmetry, shear on one produces a change in shape which differs by only a rigid body rotation from that produced by shear on the other. In two dimensions, orthogonal slip systems such as 1 and 3 (Fig. 3) provide a good example of this case. Slip parallel to [10] produces a shape that differs by only a 90° rotation from the shape produced by slip parallel to [0$\bar{1}$]. Another case in which slip systems fail to be independent is found in fcc metals where slip occurs on the {111} <1$\bar{1}$0> slip systems. Although each slip plane contains three slip directions, only two of the resulting three slip systems are independent. A change in shape due to shearing on one slip system can always be produced by a superposition of shearing parallel to the other two.

Figure 4(a) shows the grains of a two-dimensional polycrystalline specimen, before and after uniaxial compression. Continuity is maintained because each grain has taken on the overall strain

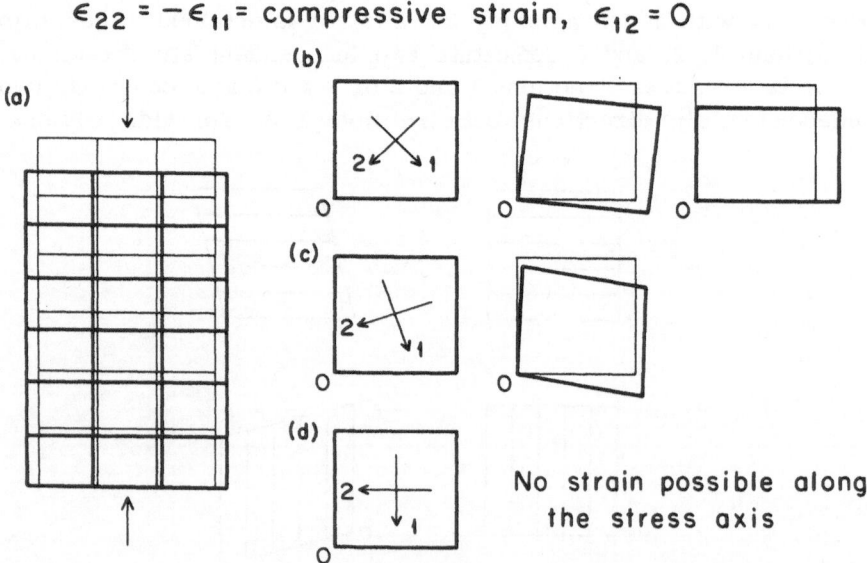

$\epsilon_{22} = -\epsilon_{11} =$ compressive strain, $\epsilon_{12} = 0$

No strain possible along the stress axis

Fig. 4. The possible changes of shape of a two-dimensional grain with only one independent slip system.

of the specimen (i.e., in each grain ϵ_{22} equals the compressive strain of the specimen, $\epsilon_{11} = -\epsilon_{22}$ to conserve area, and $\epsilon_{12} = 0$, where the strain components are referred to axes parallel and perpendicular to the compression axis). A grain oriented at random with respect to the compression axis can change shape in the required manner, however, only if it has two independent slip directions. The possible shape changes for a grain with two orthogonal slip directions and, therefore, only one independent slip direction are shown in (b), (c), and (d) of Fig. 4. The required change of shape can be obtained only if the grain is oriented with its slip systems at 45° to the compression axis, as shown in Fig. 4(b). In this orientation, shear parallel to one slip direction produces the required change of shape accompanied by a rotation. Equal amounts of shear parallel to both slip directions produce the same change of shape with no rotation. Grains in other orientations, however, cannot change shape as required. The grain in Fig. 4(c) can take on the required uniaxial strain ϵ_{22}, but will not remain rectangular regardless of how shear is distributed between its two slip directions. The grain in Fig. 4(d) cannot strain along the specimen axis. Thus, it can be seen that continuity cannot be maintained in a polycrystalline material with randomly oriented grains if deformation occurs on an insufficient number of independent slip systems.

Application of the von Misés Analysis to Ceramic Structures

Groves and Kelly have recently examined a number of simple ceramic structures to see if they possess the five independent slip systems that are required for polycrystalline ductility [4]. Their results are summarized in Table II.

In crystals with the NaCl structure, slip can occur on two families of slip systems. In those crystals with ionic bonding, at low temperatures, dislocations move on $\{110\}$ $<1\bar{1}0>$ slip systems, producing straight slip steps; at somewhat higher temperatures, dislocations can also move on $\{001\}$ $<1\bar{1}0>$ slip systems, producing slip steps that are wavy [5]. According to Table II, slip must occur on both the $\{110\}$ $<1\bar{1}0>$ and the $\{001\}$ $<1\bar{1}0>$ families to provide five independent slip systems; the $\{110\}$ $<1\bar{1}0>$ family alone provides only two. As predicted, polycrystalline NaCl [6], AgCl [7], KCl [8], and LiF [9] are brittle at low temperature, but become ductile at temperatures where slip on both $\{110\}$ $<1\bar{1}0>$ and $\{001\}$ $<1\bar{1}0>$ families is possible. Similar behavior has been observed by the authors in polycrystalline MgO and will be described later.

TABLE II

Families of Slip Systems for Simple Ceramic Structures

Structure	Slip systems	Number of independent systems
NaCl	$\{110\}$ <1$\bar{1}$0>	2
	$\{110\}$ <1$\bar{1}$0> + $\{001\}$ <1$\bar{1}$0>	5
Al_2O_3	(0001) <11$\bar{2}$0>	2
	(0001) <11$\bar{2}$0> + $\{10\bar{1}0\}$ <11$\bar{2}$0>	4
CsCl	$\{110\}$ <001>	3
CaF_2	$\{001\}$ <1$\bar{1}$0>	3
	$\{001\}$ <1$\bar{1}$0> + $\{110\}$ <1$\bar{1}$0>	5
TiO_2 (rutile)	$\{101\}$ <10$\bar{1}$> + $\{110\}$ <001>	4
$MgO \cdot Al_2O_3$, TiC	$\{111\}$ <1$\bar{1}$0>	5

In alumina, slip occurs on (0001) <11$\bar{2}$0>; however, there is a lack of agreement as to the second slip system [4, 10]. Groves and Kelly, using the (0001) <11$\bar{2}$0> and (10$\bar{1}$0) <11$\bar{2}$0> slip systems, showed that they provided only four independent slip systems. Although single crystals of sapphire deform in tension above 1270°C (the transition temperature is strain-rate sensitive) [11], polycrystalline Al_2O_3 in the form of Lucalox exhibits no ductility up to 1900°C [12]. In addition to its lack of sufficient independent slip systems, the ductility of polycrystalline Al_2O_3 is also limited at low temperatures by a high frictional resistance to dislocation motion. CsCl, on the other hand, has a low resistance to dislocation motion at room temperature and, thus, is extremely soft. The $\{110\}$ <001> slip systems provide only three independent slip systems [13], however; and, as predicted, this material lacks ductility in polycrystalline form [14].

CaF_2 should behave in a manner similar to MgO. At low temperatures, slip occurs on the (001) <1$\bar{1}$0> slip systems which provide three independent slip systems. At elevated temperatures, slip also occurs on the (110) <1$\bar{1}$0> slip systems, producing a total of five independent slip systems [15]. The mechanical behavior of single-crystal and polycrystalline CaF_2 is discussed by Pratt et al.* In TiO_2 (rutile), on the other hand, slip occurs on the $\{101\}$ $<10\bar{1}>$ and the $\{110\}$ $<001>$ slip systems, producing a total of four independent slip systems [16]. Thus, TiO_2 should be brittle in polycrystalline form. The $\{101\}$ $<10\bar{1}>$ slip systems in TiO_2 are not equivalent to the $\{110\}$ $<1\bar{1}0>$ slip systems in NaCl because TiO_2 is tetragonal.

The family of slip systems for $MgAl_2O_4$ (spinel) and TiC is $\{111\}$ $<1\bar{1}0>$, which provides five independent slip systems; however, deformation occurs by slip only at elevated temperatures [17, 18].† An interesting question with regard to both of these materials is the effect of stoichiometry on their flow stress. Both materials exist over a finite composition range, and, thus, it is possible that a large component of their flow stress is due to lattice defects resulting from nonstoichiometry. Recent results by Williams, however, show that in TiC the flow stress increases as the composition approaches that of the stoichiometric compound [19].

GRAIN BOUNDARY SHEARING

At temperatures in the neighborhood of $T_m/2$, metals and probably most ceramics begin to exhibit some grain boundary sliding. Early investigations of MgO and Al_2O_3 by Wachtman and Lam [20] and by Chang [21] reported changes in the elastic modulus. Recently, more direct evidence of grain boundary sliding in ceramic materials was obtained by Adams and Murray [22], who studied its occurrence in bicrystals of MgO. They found that grain boundary sliding rates increased rapidly with increasing temperature and increasing stress and that the resistance to grain boundary sliding depends on boundary orientation. In general, the more complex the boundary, the more readily sliding occurs.

Effect of Orientation and Temperature

Murray and Mountvala have made extensive measurements of the effect of temperature and orientation on grain boundary sliding in MgO [23]. Bicrystals were cut in the shape of rectangular parallelepipeds with the boundary lying at approximately 45° to the

---

*See P. L. Pratt, C. Roy, and A. G. Evans, "The Role of Grain Boundaries in the Plastic Deformation of Calcium Fluoride," this volume, Chapter 14.
†See Choi and Palmour, this volume, Chapter 25.

tensile axis. The boundary orientations were specified as illustrated in Fig. 5. The (1, 2, 3) and the (1', 2', 3') axes correspond to <100> directions in each crystal. The axis closest to the boundary normal is labeled 3. The axis closest to 3 is labeled 3'. The angle a between axis 3 and axis 3' is defined as the angle of tilt. To obtain the angle of twist β, 3' is rotated to coincide with 3 so that (1, 2) and (1', 2') are coplanar. The angle between axis 1 and axis 1' is then defined as the angle of twist. It should be noted that this scheme gives atomic misorientations only for pure twist and pure tilt boundaries.

After determining their boundary orientations, the bicrystal specimens were loaded in compression and their fracture stress measured. In all cases, the specimens were observed to fracture by shearing along the boundary.

Figure 6 shows the effect of boundary orientation on fracture stress at 1400°C [23]. When the angle of tilt was less than 20°, the fracture stress decreased rapidly with increasing twist. When the angle of tilt was greater than 30°, the decrease was less pronounced.

Figure 7 shows the effect of temperature on fracture for several boundary orientations. It can be seen that the fracture stress

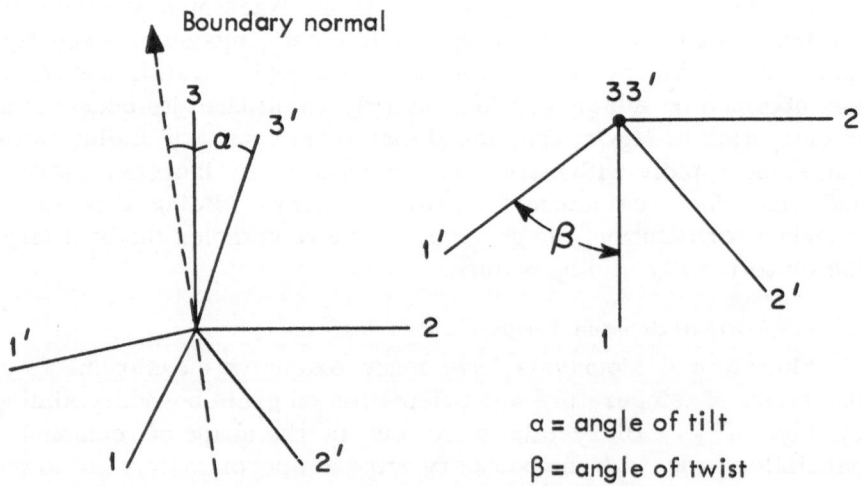

Fig. 5. Scheme for describing grain boundary misorientation in bicrystals (after Murray and Mountvala [23]).

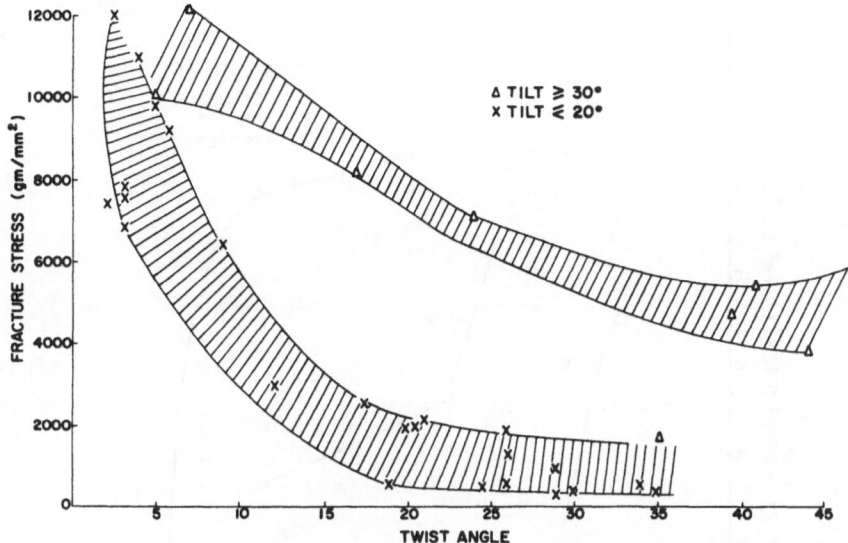

Fig. 6. The effect of grain boundary misorientation on strength of MgO bicrystals (after Murray and Mountvala [23]).

decreases rapidly at a critical temperature which depends on boundary orientation. This critical temperature decreases with increasing twist and decreasing tilt.

From these results, it can be predicted that grain boundary sliding should start contributing to the deformation of polycrystalline MgO at about 1300°C. It should be emphasized, however, that such sliding at normal strain rates does not constitute a ductile mode of deformation because it is necessarily accompanied by the formation of accommodation cracks at three-grain junctions.

DEFORMATION OF POLYCRYSTALLINE MgO*

Because of the extensive information available on mechanical behavior in single-crystal and bicrystal form, MgO was chosen for polycrystalline deformation studies [22–27]. MgO has the NaCl crystal structure. At low temperatures, it is a material in which dislocations are mobile, but which possesses an insufficient num-

---

*This section, including the experimental results and discussion, is based in part on the authors' paper, "Deformation of Polycrystalline MgO at Elevated Temperatures," J. Am. Ceram. Soc. 48:636–642 (1965).

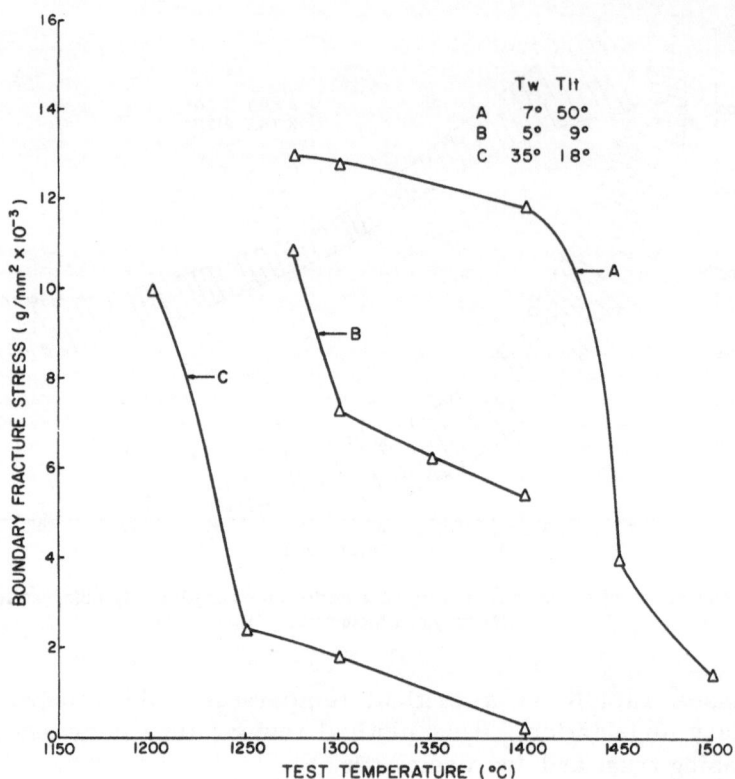

Fig. 7. The effect of temperature on grain boundary strength in MgO bicrystals (after Murray and Mountvala [23]).

ber of independent slip systems for ductility. At elevated temperatures, MgO has five independent slip systems; however, ductility comparable to that observed in polycrystalline NaCl, AgCl, KCl, or LiF has not yet been reported [12, 25]. On the basis of the observations of Murray and Mountvala [23], grain boundary sliding should appear in polycrystalline MgO above 1300°C. In this section, the results of stress–strain experiments on two different types of polycrystalline MgO are presented. Both types were nominally dense and pure, but they deformed differently, thus emphasizing the importance of subtle differences in structure in determining mechanical behavior. (Structure is used in the broad sense to include composition, microstructure, flaws, etc. Microstructure specifically refers to the geometric distribution of identifiable grains

TABLE III

Spectroscopic Analysis of Polycrystalline MgO

| Element | Impurities, % | |
	Type 1	Type 2
Mg	Principal	Principal
Fe	–	–
Ba	–	–
Si	0.01	0.015
Mn	–	–
Al	<0.005	<0.005
Ca	0.003	0.02
Cu	0.0008	0.0008
Ti	0.006	–
Li	0.0075	–
Ni	–	–
Cr	0.004	–
B	–	–

Analyses made by the American Spectroscopic Laboratories, San Francisco, California. Constituents reported as oxides of the elements indicated.

and phases, but frequently microstructure is also used in the broad sense.)

Experimental Procedure

Specimens. The composition of each type of polycrystalline MgO was determined by spectrographic analysis. The constituents are reported in Table III as oxides of the elements.

Type 1 was made in the authors' laboratory using a technique based on one described by Rice [28]. In brief, a mixture of powdered MgO with 3% LiF was hot-pressed at 1000°C at a pressure of 3000 psi for about 30 min in a graphite die in vacuo (10^{-3} mm Hg). The resulting compact was then fired at 1300°C for 4 hr in air. This procedure resulted in transparent polycrystalline MgO containing approximately 75 ppm of lithium* with a density of 3.579 g/cm³ † and with the microstructure shown in Fig. 8. The grain

*Private communication from Roy W. Rice states that radio-tracer techniques indicate that such specimens contain approximately 0.08% fluorine.

†All reported density values were obtained by use of Archimedes principle and, thus, are comparable; the X-ray density value for MgO single crystals is 3.5733 g/cm³.

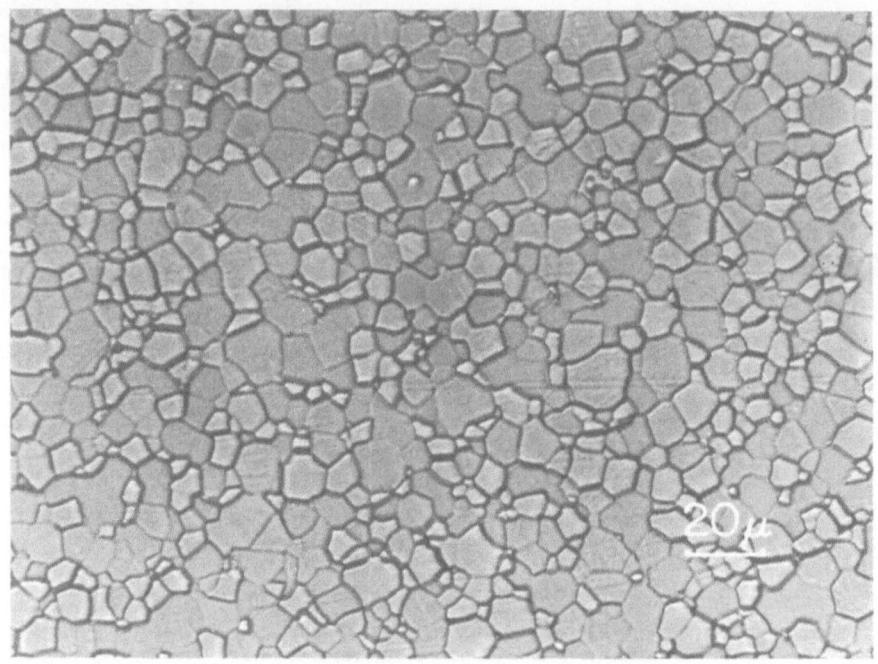

Fig. 8. Microstructure of Type 1 polycrystalline MgO.

size was uniform, the largest cross sections being about $15\,\mu$. These specimens were called Type 1 (S.G.). Some were given a high-temperature anneal to cause grain growth. After this anneal, the grains varied in size from region to region, and grain cross sections up to $100\,\mu$ were observed. These specimens were called Type 1 (L.G.).

Type 2 specimens were obtained from the Honeywell Research Center, Hopkins, Minnesota, and were formed by an isostatic pressing and sintering method. They were white and translucent, had a density of 3.546 g/cm^3, and had the microstructure shown in Fig. 9. A number of very small pores can be seen within the grains and on the grain boundaries. The largest grain cross sections are about $50\,\mu$.

Compression specimens of each type of polycrystalline MgO were cut with a diamond saw. The specimens were prisms with approximately square cross sections about $0.2 \times 0.2 \times 1.0$ in. Coarse abrasive paper (220A) was used to grind specimen ends parallel.

Stress-Strain Experiments. Stress—strain curves were obtained for each type of MgO at a number of constant temperatures. Specimens were loaded in compression at a constant force rate such that the initial stress rate was 20 psi/sec. The stresses were calculated from initial specimen cross sections. Strains were determined by measuring the displacement of two small divot holes initially 0.5 in. apart on a side face of the specimen. All reported strains are true strains; a true strain is equal to the natural logarithm of one plus the engineering strain.

Specimens were heated in air in a furnace with $MoSi_2$ heating elements. Alumina buttons were placed between specimen ends and the loading rams, which were also made of alumina. Thin platinum sheets were placed between the specimen ends and the buttons as reaction barriers. A detailed description of the overall apparatus used in this investigation will be published separately [29].

Examination of Specimens. Several specimens were carefully polished on one face to allow microscopic examination before and

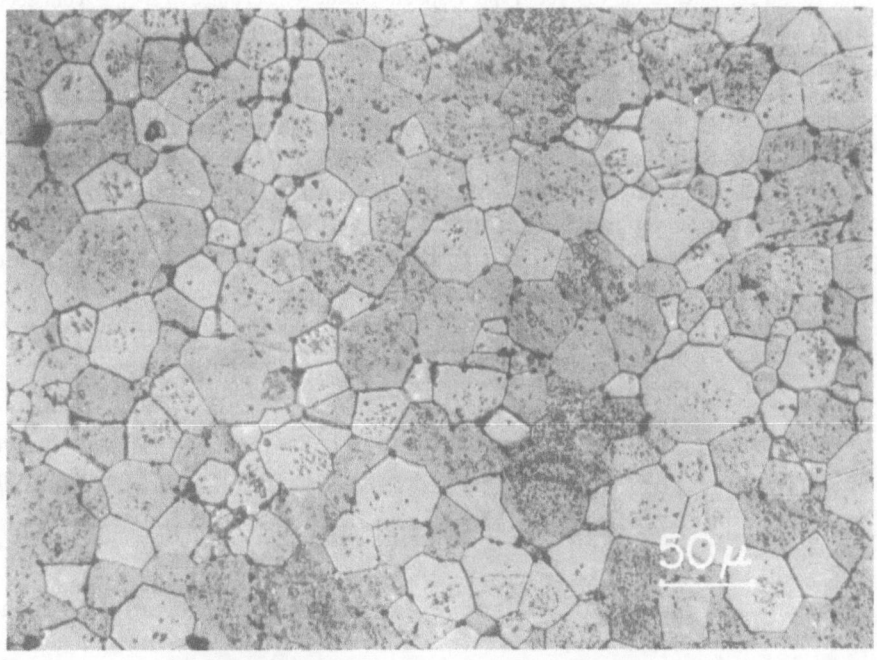

Fig. 9. Microstructure of Type 2 polycrystalline MgO.

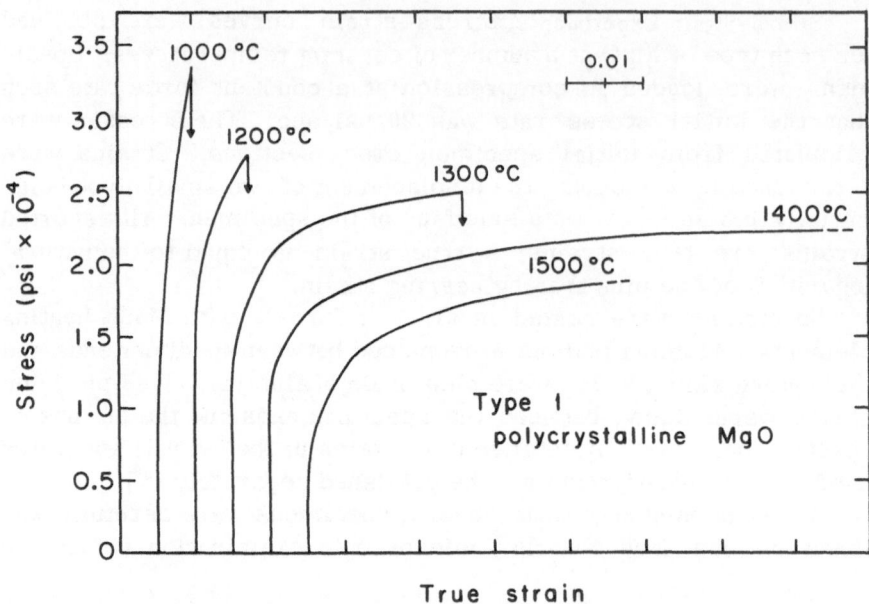

Fig. 10. Stress–strain curves for Type 1 specimens at temperatures ranging from 1000 to 1500°C.

after deformation.   The faces were (1) flattened by grinding with 600-grit SiC, (2) lapped with 1–2 μ diamond grit, and (3) lapped with a high-speed wheel using Linde A on Politex polishing disks (Style PA-K).* All specimens were chemically polished for 1 min in 85% orthophosphoric acid and 1 part sulfuric acid. MgO single-crystal specimens immersed in this etch for 1 min at 26°C showed pits only on the {100} and {110} faces.

All photographs were taken with a Leitz metallographic microscope (MM5) using 4 × 5 in. Polaroid 55P/N film.

Results

In a force rate experiment, the slope of the stress–strain curve is related to the strain rate at small strains by the equation

$$\frac{d\sigma}{d\epsilon} = \frac{d\sigma}{dt}\left(\frac{d\epsilon}{dt}\right)^{-1} \tag{6}$$

where σ is the applied force divided by the cross section of the undeformed specimen, ϵ is the strain, and t is the time. It can

*Geoscience Instrument Corp.

be seen that, for a constant force rate, a small slope of the curve indicates a high strain rate; and a high slope, a low strain rate.

Type 1 Specimens. Figure 10 shows typical stress–strain curves for Type 1 (S.G.) specimens at temperatures from 1000 to 1500°C. At 1000°C, fracture occurred with little bulk strain, and the basic shape of the curve is different from those obtained at higher temperatures. At 1400 and 1500°C, after initial yielding, the stress–strain curves rapidly became flat, indicating high strain rates; and the experiments were terminated by fracture of the alumina supporting buttons. At 1200 and 1300°C, stress–strain curves are intermediate and terminated by fracture of the specimen.

After deformation, it was noticed that the Type 1 (S.G.) specimens had lost their transparency, the 1400 and 1500°C specimens being appreciably opaque. These observations suggested that deformation might be occurring by grain boundary shearing and separation. To check this hypothesis, lines approximately $1\,\mu$ wide

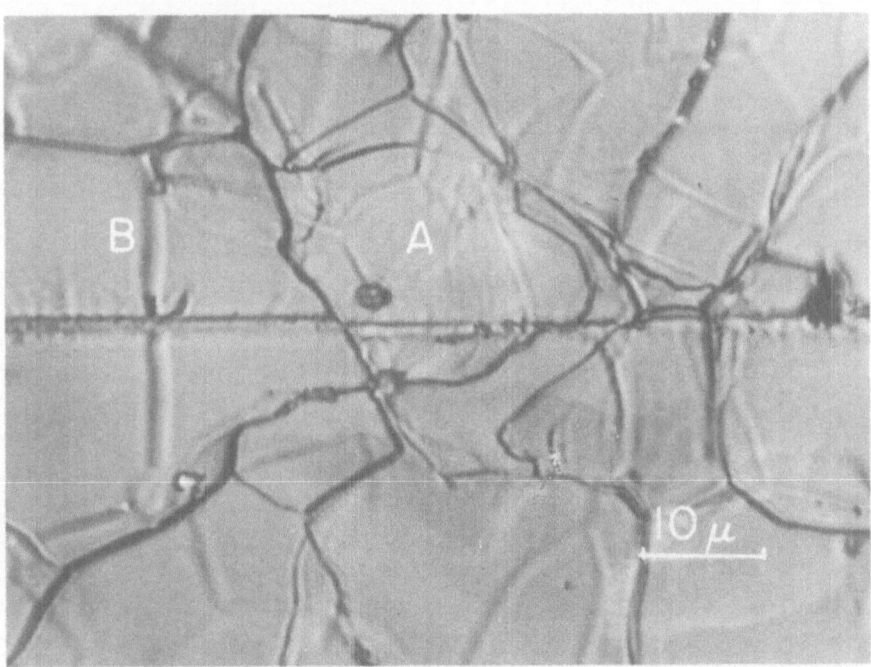

Fig. 11. The displacement of a scribed line indicating grain boundary sliding in a Type 1 specimen.

were scribed with a diamond point on the mechanically and chemically polished surface of a Type 1 (S.G.) specimen. Figure 11
shows the surface of this specimen after it was strained 3% at
1400°C. The displacement of the scribed lines between grains A
and B is clear evidence of grain boundary shearing. Of particular
interest is the cleavage fracture developing in grain A. Such fracture is a necessary consequence of grain boundary shearing and
allows for accommodation at three-grain junctions. Although
occasionally occurring within grains (as in grain A), such localized
fracture occurred most frequently along the grain boundaries.

A second feature of interest in Fig. 11 is the ghost lines, which
were not present before the deformation experiment. These lines
are the result of thermal etching and correspond to the positions
where the grain boundaries were stationary for a period of time at
1400°C. The final grain boundary positions are darker than the
ghost lines. They are sharp-bottomed due to the presence of the

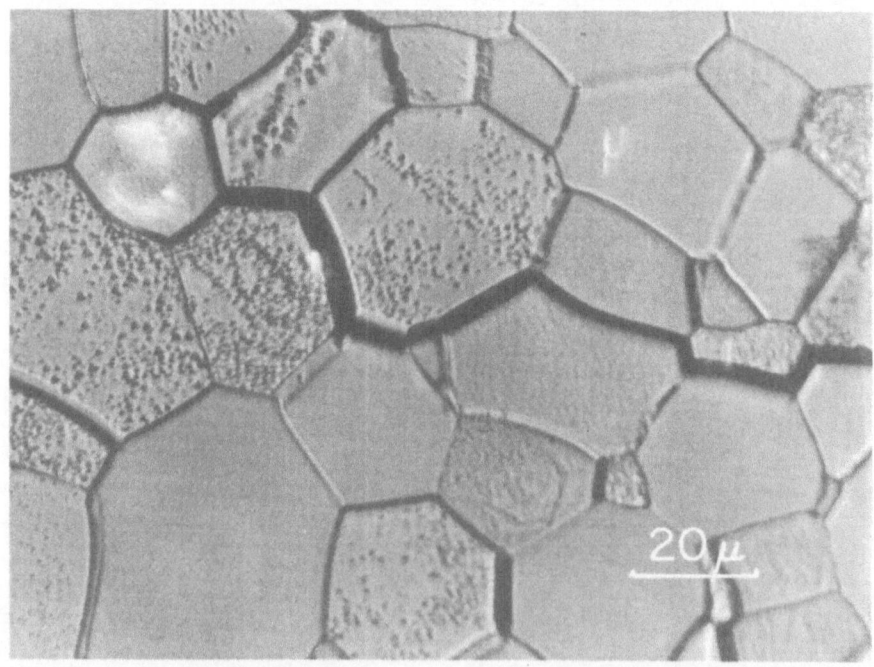

Fig. 12. Intergranular cracking in a Type 1 specimen deformed at 1300°C.

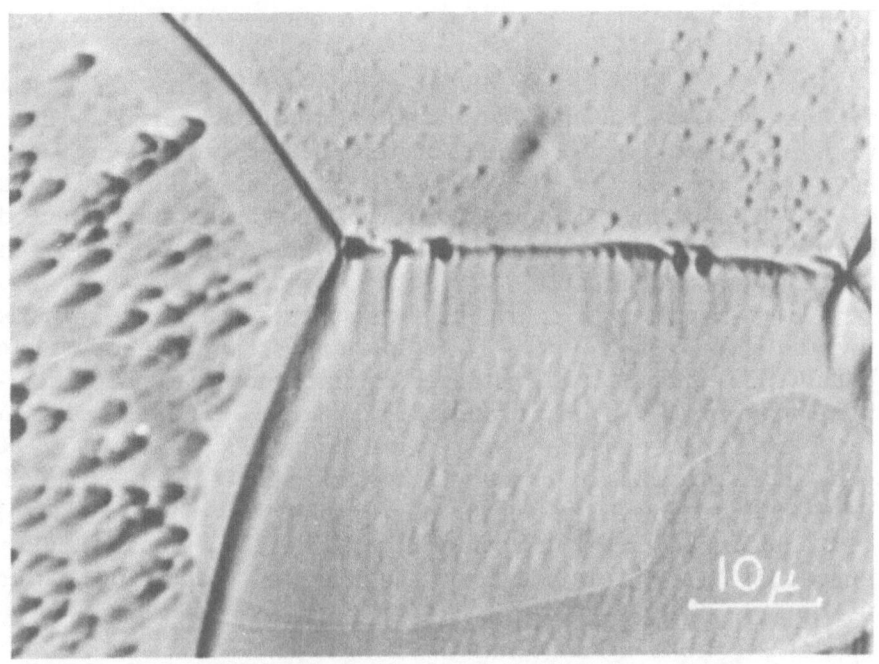

Fig. 13. Cylindrical voids marking boundaries where sliding has occurred in a Type 1 specimen.

boundary, while the ghost lines are rounded due to additional thermal polishing after the boundary moved.

The translucency developed by the deformed Type 1 (S.G.) specimens is, thus, accounted for by localized fracture and grain boundary separations. These discontinuities also account for the high strain rates shortly after yielding because they reduce the cross-sectional area resisting deformation. This effect is emphasized with increasing strain because, as the amount of localized fracture increases, the gaps between grains join together to form long intergranular cracks roughly parallel to the loading axis.

Figure 12 shows an intergranular crack in a Type 1 (L.G.) specimen deformed to fracture at 1300°C. Although little bulk straining by slip actually occurred, dislocation etch pits can be seen on the favorably oriented grains. In a large grain specimen, the gaps resulting from grain boundary shearing are quite large and can spread easily, causing bulk fracture. Thus, from 1000 to

1500°C, the Type 1 (L.G.) specimens were observed to fracture with little plastic strain.

Figure 13 shows cylindrical voids which were frequently observed at grain boundaries on which sliding had occurred. These voids probably mark the position of steps in the boundary which may have been produced by slip. The cylindrical shape of these voids can be seen because grains of different orientation were attacked by the polish at different rates, thus revealing the boundaries of the slowly attacked grains.

Type 2 Specimens. Figure 14 shows typical stress–strain curves for the Type 2 specimens at temperatures ranging from 400 to 1400°C. Experiments were stopped prior to fracture to obtain complete specimens for microscopic examination. It can be seen that the Type 2 specimens were ductile well below 1200°C. Between 800 and 1200°C, parabolic-shaped stress–strain curves were obtained. At 1400°C, however, the curve was similar to that for the Type 1 (S.G.) specimen, in that it rapidly became flat after initial yielding, suggesting that grain boundary shearing was occurring at this

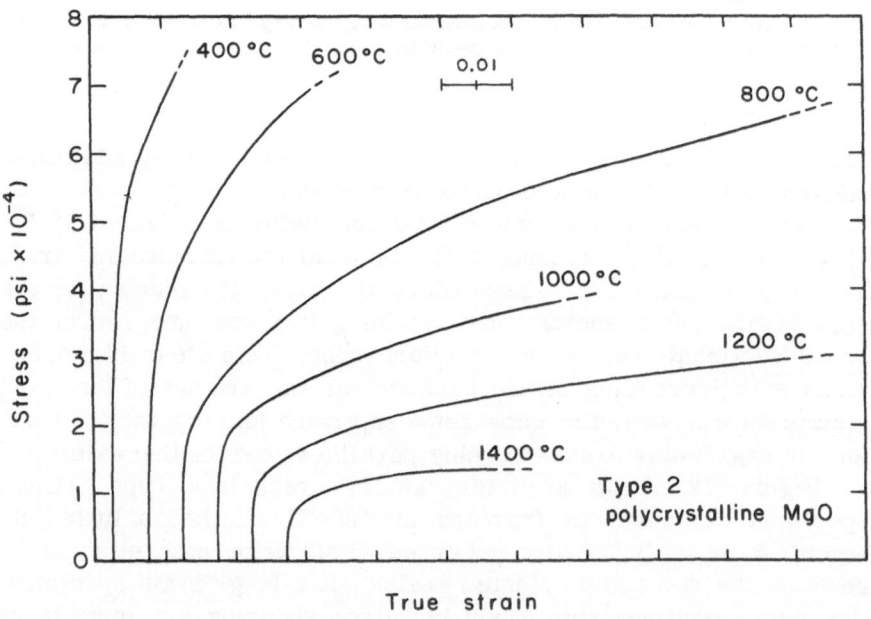

Fig. 14. Stress–strain curves for Type 2 specimens at temperatures from 400 to 1400°C.

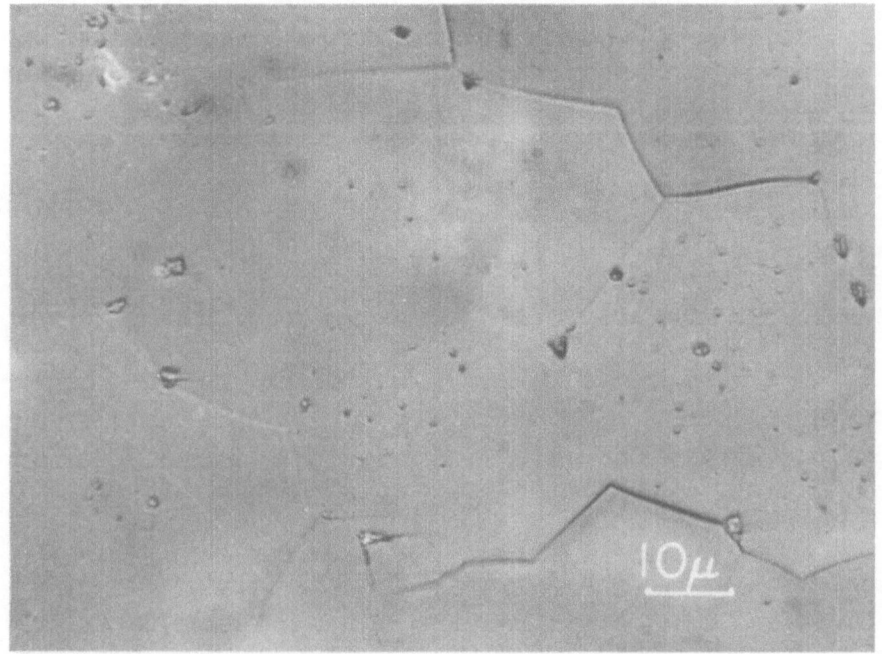

Fig. 15. Microstructure of a Type 2 specimen deformed at 400°C.

temperature. At 400°C, the stress–strain curve was not parabolic and little bulk strain occurred, while, at 600°C, it was intermediate in nature. Figure 15 shows the surface of a specimen deformed at 400°C. Many grain boundaries were cracked; however, little intra-granular cracking was observed.

Figure 16 shows the surface of a Type 2 specimen deformed about 5% at 1000°C. This surface had been mechanically and chemically polished prior to deformation, but was not treated after deformation. The dark spots were caused by chemical polish and were visible on the surface before deformation. The wavy lines appeared after deformation and are similar to the wavy slip steps observed previously in NaCl, AgCl, KCl, and LiF. These steps were quite pronounced and appeared in almost all of the grains. They occasionally crossed grain boundaries and often became forked as they approached grain boundaries. Similar steps were observed on specimens deformed at 800°C, but not at 600°C. The 1200 and 1400°C specimens were not prepared for this observation.

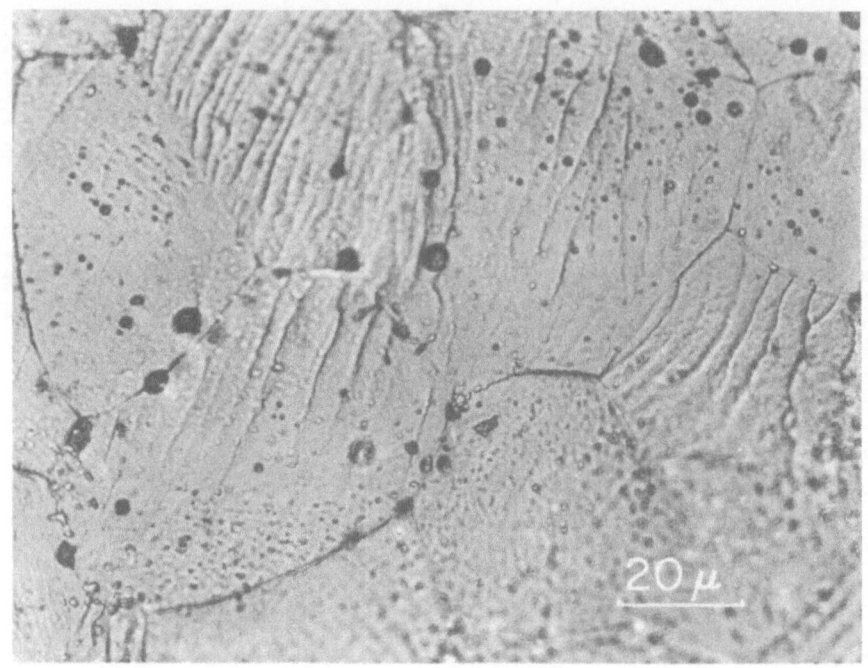

Fig. 16. Wavy slip in a Type 2 specimen deformed at 1000°C.

Discussion

Of the two types of specimens studied in this investigation, only the Type 2 exhibited behavior comparable to that previously reported for AgCl, NaCl, KCl, and LiF. At 600°C and below, these specimens were brittle and their deformation was accompanied by a pulling apart of grain boundaries. At 800°C and above, they were ductile and wavy slip steps were observed. A transition from brittle to ductile behavior thus occurred between $0.28\ T_m$ and $0.35\ T_m$. In AgCl, NaCl, and KCl, similar transitions were observed in tension at about $0.24\ T_m$, $0.44\ T_m$, and $0.40\ T_m$, respectively [6-8]. In LiF, a similar transition was observed in compression at about $0.67\ T_m$[9].

Before comparing the behavior of the two types of polycrystalline MgO, the behavior of the Type 2 specimens will be considered in more detail. The deformation of these specimens can be further understood by comparing their stress–strain behavior to that of uniaxially stressed single crystals. Figure 17 gives yield stress values for MgO single crystals with <100> and <111> stress axes as a function of temperature [25, 26]. A <100> stress axis results in equal, nonzero resolved shear stresses on four {110} <1$\bar{1}$0> slip systems. A <111> stress axis results in equal, nonzero resolved shear stresses on three {001} <1$\bar{1}$0> slip systems. In both orientations, the resolved shear stress on all other {110} <1$\bar{1}$0> and {001} <1$\bar{1}$0> slip systems is zero. The ratio of yield stresses for single crystals with a <111> stress axis to those with a <100> stress axis provides a measure of the relative mobility of dislocations on the two families of slip systems. This ratio is 13/1 at 350°C, 4.5/1 at 600°C, and 2.9/1 at 1600°C. Because slip on both families of slip systems is necessary to satisfy the von Misés requirement, an increase in ductility with increasing temperature would be predicted.

Also shown in Fig. 17 are the yield stress values for the Types 1 (S.G.) and 2 polycrystalline specimens. It can be seen that the

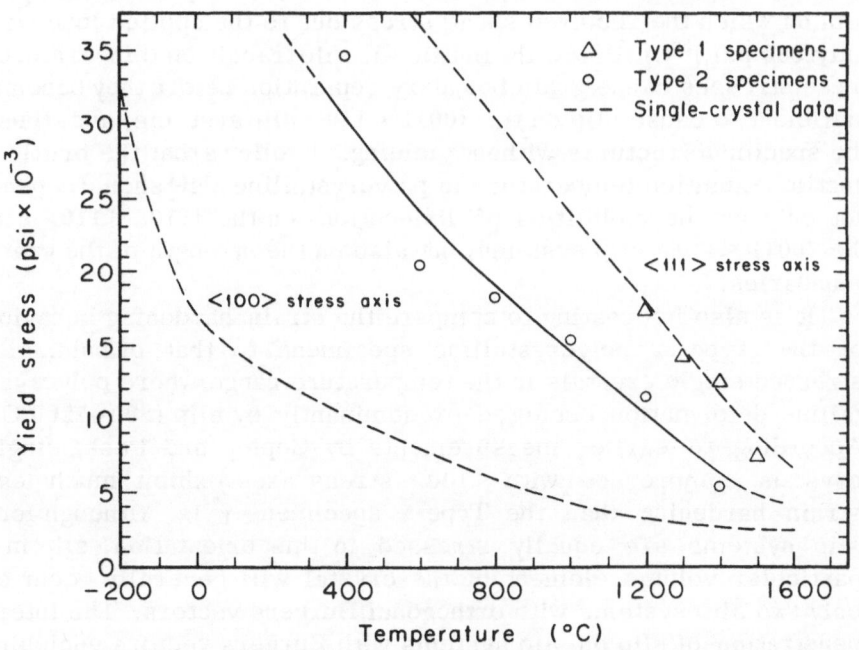

Fig. 17. Yield stress vs. temperature curves for <100> and <111> oriented single crystals and for Type 1 and Type 2 polycrystalline specimens.

Type 2 specimens yielded at stresses considerably below those necessary to yield a single crystal with a <111> stress axis. This fact is particularly significant because a <111> stress axis produces a near maximum resolved shear stress of 0.472 times the uniaxial stress on the three {001} <1$\bar{1}$0> slip systems. Thus, slip on these slip systems in the randomly oriented grains of the Type 2 specimens could not have resulted from the uniaxial stress alone, but is evidence of additional stresses due to interactions between grains.

These observations suggest the following explanation for the yielding behavior of the Type 2 specimens. With the application of stress, slip occurs first on the {110} <1$\bar{1}$0> slip systems because of their lower resistance to dislocation motion. Slip on these slip systems can provide, however, only two independent slip systems so that long-range stresses build up in each grain. Above the brittle–ductile transition temperature, these stresses force slip on the {001} <1$\bar{1}$0> slip systems and yielding occurs in the polycrystalline piece. (This phenomenon was quantitatively demonstrated for large grain polycrystalline LiF by Scott and Pask wherein a localized shear stress of 9000 psi was shown to develop on a slip system on which the resolved shear stress due to the applied load was only 880 psi [30].) Below the brittle–ductile transition temperature, these stresses cause grain boundary separation before they become sufficient to cause slip on the {001} <1$\bar{1}$0> slip systems, and, thus, the specimen fractures without yielding. It follows that the brittle–ductile transition temperature in polycrystalline MgO should depend not only on the mobilities of dislocations on the {110} <1$\bar{1}$0> and the {001} <1$\bar{1}$0> slip systems, but also on the strength of the grain boundaries.

It is also interesting to compare the strain-hardening behavior of the Type 2 polycrystalline specimens to that of uniaxially stressed single crystals in the temperature range where polycrystalline deformation occurred predominantly by slip (800–1200°C). According to earlier measurements by Copley and Pask, single crystals compressed with <100> stress axes exhibit much less strain hardening than the Type 2 specimens [26]. Although four slip systems are equally stressed in this orientation, slip in a particular volume element of the crystal will generally occur on only two slip systems with orthogonal Burgers vectors. The interpenetration of slip on slip systems with Burgers vectors enclosing an angle of 60° (or 120°) is observed only where a transition occurs from one set of orthogonal slip systems to the other. Long-range

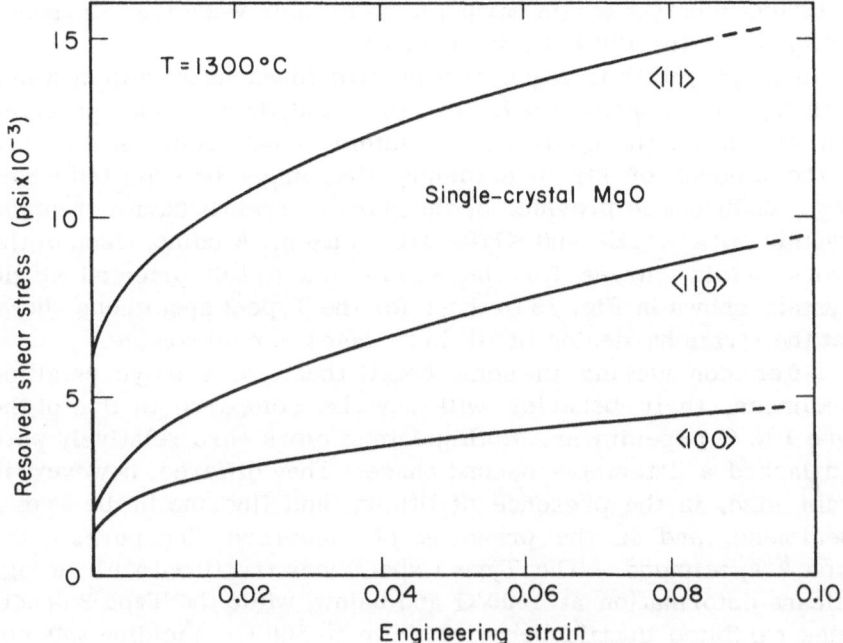

Fig. 18. Resolved shear stress–strain curves for single crystals with <100>, <110 >, and <111> stress axes, deformed at 1300°C.

stress interactions exist between dislocations on these slip systems, and dislocations with Burgers vectors at 120° can react to form sessile dislocation networks, which are observed where such transitions occur [26, 27, 31]. These observations suggest that greater strain hardening should result from interpenetrating slip on the oblique slip systems than on the orthogonal ones.

Figure 18 shows stress–strain curves obtained by compressing single crystals of MgO with <100>, <111>, and <110> stress axes at 1300°C. In the <111> orientation, slip occurs on three {001} <1̄10> slip systems, providing that the ends are constrained from lateral movement. The Burgers vector of each of these slip systems is at 60° (or 120°) relative to the other two. In the <110> orientation, slip occurs on four {110} <1̄10> slip systems. The interpenetration of slip on slip systems with Burgers vectors enclosing an angle of 60° (or 120°) is again necessary if the specimen ends are constrained. It can be seen that the critical resolved shear stress is greater for the {001} <1̄10> than for the {110} <1̄10> slip systems at 1300°C and that strain hardening is greater when the interpenetration of 60° (or 120°) slip is required.

In polycrystalline MgO, slip on five independent slip systems in each grain is necessary for ductility, and, thus, an interpenetration of slip on the oblique slip systems must occur. A measure of the amount of strain hardening that might be expected under these conditions is provided by the stress–strain behavior of single crystals with <111> and <110> stress axes. A comparison of the stress–strain curves for the <111> and <110> oriented single crystals shown in Fig. 18 to those for the Type 2 specimens shows that the strain hardening in all three cases is comparable.

After considering in some detail the Type 2 polycrystalline specimens, their behavior will now be compared to that of the Type 1 (S.G.) specimens. Both specimen types were relatively pure and lacked a detectable second phase. They differed, however, in grain size, in the presence of lithium and fluorine in the Type 1 specimens, and in the presence of dispersed fine pores in the Type 2 specimens. The Type 1 specimens fractured without significant deformation at 1000°C and below, while the Type 2 specimens exhibited ductile behavior down to 800°C. Yielding was observed in the Type 1 (S.G.) specimens only at 1200°C and above where they deformed primarily by grain boundary shearing. As shown in Fig. 17, the Type 1 (S.G.) specimens generally supported greater stresses without yielding than those required to yield the Type 2 specimens.

The less-ductile behavior of the Type 1 specimens may be caused by (1) lower dislocation mobility, (2) greater difficulties in initiating slip, or (3) grain boundary weakness. A reduction of the mobility of dislocations may be caused by the presence of lithium and fluorine. Gorum, Luhman, and Pask have shown that small solute additions can increase the yield stress of MgO single crystals by as much as four times [32]. Microhardness measurements on both types of specimens, however, detected no difference in hardness.

If dislocations and slip are initiated at only a few points in a grain, they will cause nonhomogeneous stress concentrations and fracture. Stokes and Li on the basis of room-temperature tensile experiments on polycrystalline MgO concluded that mobile dislocations can be introduced through stress concentrations associated with pores [33]. Thus, it is possible that the presence of the small amount of pores in the Type 2 specimens may play a significant role in initiating slip.

Lastly, the presence of lithium and fluorine in the Type 1 specimens, particularly at grain boundaries, can be expected to have a significant effect on the structure, strength, and behavior of the grain boundaries. Modifications in their structure can increase the difficulties in forcing slip on {001} <1$\bar{1}$0> slip systems in adjoining grains. On this basis, the Type 2 specimens have grain boundaries whose strength and characteristics are considerably more favorable for allowing the forcing of slip in {001} <1$\bar{1}$0> slip systems and the development of wavy slip. The greater ductility of the Type 2 specimens as compared to that of the Type 1 (S.G.) does not appear to be attributable to some unknown favorable role of their larger grain size, since the Type 1 (L.G.) specimen, the grain size of which is still larger, showed no ductility at 1000°C.

DUCTILITY IN POLYCRYSTALLINE CERAMICS

Crack Formation

The demonstration by Gorum, Parker, and Pask that MgO single crystals exhibited substantial ductility at room temperature stimulated considerable interest in developing a ductile, refractory, polycrystalline ceramic material [34]. Initial studies were directed toward elucidating the process of crack nucleation in single crystals of MgO. Stokes, Johnston, and Li showed that cracks form at slip band intersections parallel to {110} planes in both compression and bend specimens [35-37]. Once formed, the behavior of these cracks was found to depend on the distribution of slip in the crystal [38]. If slip was confined to only a few slip bands, the cracks were found to spread rapidly causing fracture. If many slip bands were present, fewer cracks were nucleated and the growth of these cracks was frequently observed to cease when they crossed neighboring slip bands, even though their length exceeded that necessary for propagation as a Griffith crack. Such stabilization was attributed to the additional energy necessary to propagate a crack across a slip band. As pointed out by Gilman, the surface area of a crack is increased in such a region due to formation of steps where screw dislocations are cut, and, thus, more strain energy must be expended [39].

The nucleation of cleavage cracks in MgO was at first believed to result from the coalescence of edge dislocations piling up at a barrier, a mechanism first proposed by Zener [40] and analyzed mathematically by Stroh [41, 42]. It soon became apparent, however, that this mechanism alone could not completely explain the observed behavior. It could not explain, for example, why cracks were formed only at intersections where complete interpenetration of slip bands occurred. Recently, many such details of crack formation in MgO single crystals have been clarified by Argon and Orowan [43]. They show that high stresses of a macroscopic nature exist where dislocation bands intersect and that these stresses, although often reduced by plastic deformation in the form of kinking, play a dominant role in the formation of the observed cracks.

Since cleavage cracks were observed to form at slip band intersections, it was expected that they would also form in polycrystalline MgO where slip bands would almost certainly be blocked by grain boundaries. This behavior was confirmed in experiments on MgO bicrystals by Johnston, Stokes, and Li [44] and by Westwood [45]. Johnston et al. showed, in fact, that a direct correlation existed between the fracture stress of MgO bicrystals and the stress required to form slip bands in them. Ku and Johnston [46] (see also Johnston and Parker [47]) have recently studied the stress necessary to form a crack at the intersection of a slip band and a grain boundary in MgO in greater detail. Slip bands were initiated at microhardness indentations on the surfaces of bicrystals which had been extensively polished to remove all other dislocation sources. It was found that the fracture stress of the bicrystals σ_f obeyed the well-known Petch equation

$$\sigma_f = \sigma_0 + AL^{-\frac{1}{2}} \tag{7}$$

where, in this case L was equal to the distance from the microhardness indentation to the boundary. It was also found that a physical significance could be attached to σ_0 and A. The constant σ_0, for all boundaries studied, was found to equal the stress required to form a slip band. The constant A could be correlated with the strength of the grain boundary. This result suggests that crack nucleation might be suppressed in polycrystalline MgO by reducing the grain size as proposed by Parker several years ago, and that the effect of reducing the grain size will be greatest in the material with the strongest grain boundaries [48].

Dislocation Sources and Distribution of Slip

Although much progress has been made in understanding the relationship of slip to crack formation in single crystals and bicrystals of MgO, several important facts remain to be established in the case of polycrystalline MgO. One of the more important of these is the distribution of slip in the grains of polycrystalline MgO deformed at various temperatures. If slip occurs inhomogeneously, there is little chance of ductility. If it occurs homogeneously, however, the probability of nucleating cracks is reduced, and the probability of stabilizing cracks, once formed, is greater.

The distribution of slip depends on the number of sources (i.e., segments of mobile dislocation line) present in each grain. The number of such sources present in a single-phase ceramic material should ordinarily be small. Although they are often deformed at some stage of their preparation (hot-pressing of the Type 1 specimens), ceramic materials are generally annealed at high temperatures to obtain maximum density. The grain growth resulting from this anneal generally removes all dislocation substructure from within the grains. Thus, the distribution of slip in such materials should depend on the capacity of the grain boundaries to act as sources of dislocations.

Although grain boundaries in metals are known to act as dislocation sources, there is no direct evidence that they act as sources in ceramic materials. In metals, the energy of a boundary joining grains with a substantial mismatch should vary only slightly with the change in orientation accompanying the emission of a dislocation. On the other hand, the same change in orientation at a ceramic grain boundary might cause a large change in energy by causing like charges to approach each other. Another factor which may limit the capacity of a ceramic grain boundary to act as a source is its affinity for foreign atoms. A concentration of foreign atoms in the boundary region could reduce its effectiveness as a source of dislocations by raising the local friction stress.

Further, the nature and amount of impurities at the boundaries can also be of great significance in other ways. Impurities can lead to the presence of essentially another phase as a film of varying thickness between the grains. Under these conditions, the strength of the grain boundaries may not be adequate to transmit stresses and strains. Deformation by shearing then can occur along grain boundaries leading to separation. The presence of lithium and fluorine in the Type 1 specimens may be responsible in this way for their grain boundary shearing described earlier. The behavior of the Type 2 specimens then may be attributed to the presence of stronger and more coherent grain boundaries. Research efforts to correlate well-characterized grain boundaries with the behavior of polycrystalline pieces are still quite limited.

Carnahan has attempted to increase the strength of polycrystalline MgO at room temperature by adding a second dispersed phase of tungsten spheres [49]. It was reasoned that such spheres might produce mobile dislocations when the specimen was stressed, as observed by Jones and Mitchell in the case of glass spheres in AgCl [50]. The increased number of sources would be expected to produce a more homogeneous distribution of slip and, thus, raise the strength; however, no increase in strength was observed. No effort was made to directly establish the distribution of slip in the grains either with or without the tungsten spheres.

Another approach for producing a more homogeneous distribution of slip in a polycrystalline ceramic is suggested by the recent experiments of Bullen and co-workers on recrystallized chromium [51]. They report that recrystallized chromium, which is ordinarily quite brittle, exhibits considerable ductility if subjected to a hydrostatic pressure prior to, but not during, deformation. They attribute this behavior to the formation of mobile dislocation segments at internal inhomogeneities, such as precipitates or inclusions. If this interpretation is correct, then it suggests that a homogeneous distribution of slip might be produced in a polycrystalline ceramic material by alloying followed by a heat treatment to form a suitably dispersed precipitate and finally by application of hydrostatic pressure.

Movement of Slip Systems

Another fact which needs to be established in the case of polycrystalline MgO is the role of cross-slip in relieving stress concentrations where slip bands intersect grain boundaries. (The term cross-slip, in this case, refers to the movement of screw dislocations with $a/2$ {110} Burgers vectors for short distances on {100} planes.) Such relief has been postulated by Stokes and Li to explain the slight ductility exhibited by NaCl bicrystals at room temperature and by MgO bicrystals at 450°C [6]. Such relief would explain the lack of cleavage in the grains of the Type 2 specimen deformed at 400°C. It might be argued that the formation of cleavage cracks in this specimen was suppressed by the compression loading; however, as discussed previously, high stresses were generated between grains in addition to the applied load. Since these stresses were sufficient to cause grain boundary parting, it would seem that cleavage cracks, if nucleated, should have been observed. Further information about stress relief by cross-slip might be obtained by careful study of crack nucleation by individual slip bands in bicrystals of MgO as a function of temperature.

The greatest barrier to promoting ductility in ceramic materials is their failure to satisfy the von Misés requirement for movement on five independent slip systems. This failure is intrinsic and generally can be related to the atomic structure at the core of the dislocation. In MgO and other predominantly ionic materials with the NaCl-type structure, Gilman has pointed out that the high stress required for slip on {001} <1$\bar{1}$0> slip systems is due to the close approach of anions at the dislocation core when it is in mid-glide position [5]. Such slip is less difficult in compounds with more polarizable anions or at higher temperatures where the lattice is expanded.

Kelly and Clarke [52] have suggested that some form of alloying might be used to make the stress required for slip on the {001} <1$\bar{1}$0> slip systems more comparable to that required for {110} <1$\bar{1}$0> slip in MgO. The results of several investigations of the effect of alloying on NaCl-type compounds have not been encouraging, however, in this regard. Stoloff, Lezius, and Johnston studied the behavior of KCl—KBr alloys [8]. Since the bromide ion is more polarizable than the chloride ion, a lowering of the stress required for {001} <1$\bar{1}$0> slip would be predicted with increasing bromide-ion additions. Instead, it was found that the bromide-ion additions suppressed slip on these slip systems. Liu, Stokes, and Li have investigated the mechanical behavior of MgO—NiO and MgO—MnO alloys, but they observed no change in slip mode at room temperature [53].

In addition to the problem of realizing movement on five independent slip systems to satisfy the von Misés requirement, there is the additional problem of intersections of slip systems which also must occur or be accommodated in some manner to obtain ductility. Resistance to such intersections has been discussed in the previous section. Efforts to minimize such problems and related effects by alloying or other means have not been reported.

Polycrystalline Material of Preferred Orientation

Although failure to satisfy the von Misés requirement precludes continuity in a deformed polycrystalline material with randomly oriented grains, it does not preclude continuity in a polycrystalline material with a preferred orientation which has undergone a spe-

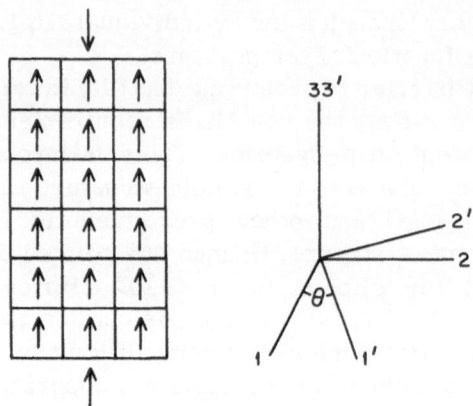

Fig. 19. Schematic diagram of an isoaxial compact.

cific change in shape. Consider, for example, the capacity of a polycrystalline specimen of MgO to undergo a uniaxial strain, if each grain is oriented so that a $<100>$ axis is parallel to the stress axis, as shown in Fig. 19. Continuity can be maintained if each grain takes on the overall strain of the specimen, i.e., if in each grain

$$\epsilon_{33} = \text{uniaxial strain} \tag{8a}$$

$$\epsilon_{11} = \epsilon_{22} = -\tfrac{1}{2}\epsilon_{33} \tag{8b}$$

and

$$\epsilon_{12} = \epsilon_{13} = \epsilon_{23} = 0 \tag{8c}$$

where the strain components are referred to axes 1, 2, and 3 (Fig. 19). If a grain is oriented so that its $<100>$ axes lie parallel to the (1,2,3) axes, equal amounts of slip parallel to the four $<110>$ planes lying at 45° to stress axis can produce the required strains. [This grain is the three-dimensional analog of the grain shown in Fig. 4(b).] If a grain is oriented so that its $<100>$ axes 2' and 3' are rotated by an angle θ about the stress axis, it can also change shape in the required manner. This result can be seen by expressing the strain components referred to the primed axes in terms of strain components referred to the unprimed axes. The two sets of strain components are related by the linear equation

$$\epsilon'_{lm} = a_{li}a_{mj}\epsilon_{ij} \tag{9}$$

where

$$a_{li} = \mathbf{e}'_l \cdot \mathbf{e}_i \tag{10}$$

and $\{\mathbf{e}'_l\}$ and $\{\mathbf{e}_i\}$ are unit vectors for the primed and unprimed systems, and the summation convention is to be applied. Expansion of this equation yields the following:

$$\epsilon'_{11} = \cos^2\theta\epsilon_{11} + \sin^2\theta\epsilon_{22} + 2\sin\theta\cos\theta\epsilon_{12} \tag{11a}$$

$$\epsilon'_{22} = \sin^2\theta\epsilon_{11} + \cos^2\theta\epsilon_{22} - 2\sin\theta\cos\theta\epsilon_{12} \tag{11b}$$

$$\epsilon'_{33} = \epsilon_{33} \tag{11c}$$

$$\epsilon'_{12} = -\cos\theta\sin\theta\epsilon_{11} + \sin\theta\cos\theta\epsilon_{22} + (\cos^2\theta - \sin^2\theta)\epsilon_{12} \tag{11d}$$

$$\epsilon'_{13} = \cos\theta\epsilon_{13} + \sin\theta\epsilon_{23} \tag{11e}$$

$$\epsilon'_{23} = -\sin\theta\epsilon_{13} + \cos\theta\epsilon_{23} \tag{11f}$$

The strain components referred to the primed system which correspond to the required strains in the unprimed system may be found by substituting conditions (8a)–(8c) into the preceding equations. The resulting strain components are

$$\epsilon'_{33} = \text{compressive strain} \tag{12a}$$

$$\epsilon'_{11} = \epsilon'_{22} = -\tfrac{1}{2}\epsilon'_{33} \tag{12b}$$

$$\epsilon'_{12} = \epsilon'_{13} = \epsilon'_{23} = 0 \tag{12c}$$

which, as already discussed, can be produced by equal amounts of slip parallel to the four {110} planes at 45° to the stress axis, thus proving the assertion.

Finally, it is interesting to note that an isoaxial compact, such as that shown in Fig. 19, will exhibit an unusual anisotropy in grain boundary strength. If a uniaxial stress is applied perpendicular to that indicated in Fig. 19 and at 45° to the plane of the page, the maximum resolved shear stress falls entirely on tilt boundaries, which, according to the data of Murray and Mountvala presented previously, are the most resistant to sliding. The resolved shear stress on the weaker twist boundaries is zero.

ACKNOWLEDGMENTS

The authors would like to express their thanks to N. E. Olson and M. Mendelson who assisted with the experimental phase of this investigation. The authors are also grateful to the Advanced Materials Research and Development Laboratory of Pratt & Whitney Aircraft for their cooperation in completion of the manuscript. This work was done under the auspices of the U. S. Atomic Energy Commission.

REFERENCES

1. R. von Misés, "Mechanik der plastichen Formänderung von Kristallen (Mechanics of Plastic Deformation of Crystals)," Z. Angew. Math. Mech. 8:161 (1928).
2. J. F. W. Bishop, "A Theoretical Examination of the Plastic Deformation of Crystals by Glide," Phil. Mag. 44:51–64 (1953).
3. G. I. Taylor, "Plastic Strain in Metals," J. Inst. Metals 62:307–324 (1938).
4. G. W. Groves and A. Kelly, "Independent Slip Systems in Crystals," Phil. Mag. 8:877–887 (1963).
5. J. J. Gilman, "Plastic Anisotropy of LiF and Other Rocksalt-Type Crystals," Acta Met. 7:608–613 (1959).
6. R. J. Stokes and C. H. Li, "Dislocations and the Strength of Polycrystalline Ceramics," in: H. H. Stadelmaier and W. W. Austin (eds.), Materials Science Research, Vol. 1, Plenum Press (New York), 1963, pp. 133–157.
7. R. D. Carnahan, T. L. Johnston, R. J. Stokes, and C. H. Li, "Effect of Grain Size on the Deformation of Polycrystalline Silver Chloride at Various Temperatures," Trans. AIME 221:45–49 (1961).
8. N. S. Stoloff, D. K. Lezius, and T. L. Johnston, "Effect of Temperature on the Deformation of KCl–KBr Alloys," J. Appl. Phys. 34:3315–3322 (1963).
9. D. W. Budworth and J. A. Pask, "Flow Stress on the {100} and {110} Planes in LiF, and the Plasticity of Polycrystals," J. Am. Ceram. Soc. 46:560–561 (1963).
10. R. Scheuplein and P. Gibbs, "Surface Structure in Corundum: II, Dislocation Structure and Fracture of Deformed Single Crystals," J. Am. Ceram. Soc. 45:439–452 (1962).
11. M. K. Kronberg, "Dynamical Flow Properties of Single Crystals of Sapphire," J. Am. Ceram. Soc. 45:274–279 (1962).
12. P. R. V. Evans, "Effect of Microstructure," in: N. A. Weil (ed.), Studies of the Brittle Behavior of Ceramic Materials, U. S. Air Force Report ASD-TR-61-628, Part II, 1963, pp. 164–202.
13. S. M. Copley, "Independent Slip Systems in CsCl-Type Crystals," Phil. Mag. 8:1599 (1963).
14. L. D. Johnson and J. A. Pask, "Mechanical Behavior of Single-Crystal and Polycrystalline Cesium Bromide," J. Am. Ceram. Soc. 47:437–444 (1964).
15. C. Roy, Ph.D. Thesis, Imperial College of Sciences and Technology, London, 1962.
16. K. H. G. Ashbee and R. E. Smallman, "Stress–Strain Behavior of Titanium Dioxide (Rutile) Single Crystals," J. Am. Ceram. Soc. 46:211–214 (1963).
17. J. Hornstra, "Dislocations in Spinels and Related Structures," in: H. H. Stadelmaier and W. W. Austin (eds.), Materials Science Research, Vol. 1, Plenum Press (New York), 1963, pp. 88–97.
18. W. S. Williams and R. D. Schaal, "Elastic Deformation, Plastic Flow, and Dislocations in Single Crystals of Titanium Carbide," J. Appl. Phys. 33:955–962 (1962).
19. W. S. Williams, "Influence of Temperature, Strain Rate, Surface Condition, and Composition on the Plasticity of Transition Metal Carbide Crystals," J. Appl. Phys. 35:1329–1338 (1964).
20. J. B. Wachtman and D. G. Lam, "Young's Modulus of Various Refractory Materials as a Function of Temperature," J. Am. Ceram. Soc. 42:254–260 (1959).
21. R. Chang, "High-Temperature Creep and Anelastic Phenomena in Polycrystalline Refractory Oxides," J. Nucl. Mater. 1:174–181 (1959).

22. M. A. Adams and G. T. Murray, "Direct Observations of Grain-Boundary Sliding in Bi-Crystals of Sodium Chloride and Magnesia," J. Appl. Phys. 33:2126-2131 (1962).
23. G. T. Murray and A. J. Mountvala, "The Role of the Grain Boundary in the Deformation of Ceramic Materials," U. S. Air Force Report ASD-TDR-62-225, Part II (Contract AF 33(616)-7961), March 1963, p. 24.
24. C. O. Hulse and J. A. Pask, "Mechanical Properties of Magnesia Single Crystals in Compression," J. Am. Ceram. Soc. 43:373-378 (1960).
25. C. O. Hulse, S. M. Copley, and J. A. Pask, "Effect of Crystal Orientation on Plastic Deformation of Magnesium Oxide," J. Am. Ceram. Soc. 46:317-323 (1963).
26. S. M. Copley and J. A. Pask, "Plastic Deformation of MgO Single Crystals up to 1600°C," J. Am. Ceram. Soc. 48: 139-146 (1965).
27. R. B. Day and R. J. Stokes, "Mechanical Behavior of Magnesium Oxide at High Temperatures," J. Am. Ceram. Soc. 47:493-503 (1964).
28. R. W. Rice, "Production of Transparent MgO at Moderate Temperatures and Pressures," presented at the Sixty-Fourth Annual Meeting Am. Ceram. Soc. (New York) (White Wares Division, No. 5), April 30, 1962.
29. C. O. Hulse and S. M. Copley, "High-Temperature Compressive Deformation Equipment," to be published.
30. W. D. Scott and J. A. Pask, "Deformation and Fracture of Polycrystalline LiF," J. Am. Ceram. Soc. 46:284-293 (1963).
31. B. H. Kear, A. Taylor, and P. L. Pratt, "Some Dislocation Interactions in Simple Ionic Crystals," Phil. Mag. 4:665-672 (1959).
32. A. E. Gorum, W. J. Luhman, and J. A. Pask, "Effect of Impurities and Heat Treatment on Ductility of MgO," J. Am. Ceram. Soc. 43:241-245 (1960).
33. R. J. Stokes and C. H. Li, "Dislocations and the Tensile Strength of Magnesium Oxide," J. Am. Ceram. Soc. 46:423-434 (1963).
34. A. E. Gorum, E. R. Parker, and J. A. Pask, "Effect of Surface Conditions on Room-Temperature Ductility on Ionic Crystals," J. Am. Ceram. Soc. 41:161-164 (1958).
35. R. J. Stokes, T. L. Johnston, and C. H. Li, "Crack Formation in MgO Single Crystals," Phil. Mag. 3:718-725 (1958).
36. R. J. Stokes, T. L. Johnston, and C. H. Li, "Further Observations of Stroh Cracks in MgO Single Crystals," Phil. Mag. 4:137 (1959).
37. R. J. Stokes, T. L. Johnston, and C. H. Li, "The Relationship Between Plastic Flow and Fracture Mechanism in MgO Single Crystals," Phil. Mag. 4:920-932 (1959).
38. R. J. Stokes, T. L. Johnston, and C. H. Li, "Effect of Surface Condition on the Initiation of Plastic Flow in MgO," Trans. AIME 215:437-444 (1959).
39. J. J. Gilman, "Creation of Cleavage Steps by Dislocations," Trans. AIME 212:310-315 (1958).
40. C. Zener, "The Micromechanisms of Fracture," in: F. Jonassin, W. P. Roop, and R. T. Bayless (eds.), Fracturing of Metals, American Society of Metals (Cleveland), 1948, pp. 3-31.
41. A. N. Stroh, "The Formation of Cracks as a Result of Plastic Flow," Proc. Roy. Soc. (London) A223:404 (1954).
42. A. N. Stroh, "The Formation of Cracks in Plastic Flow, II," Proc. Roy. Soc. (London) A232:548 (1955).
43. A. S. Argon and E. Orowan, "Crack Nucleation in MgO Single Crystals," Phil. Mag. 9:1023-1039 (1964).
44. T. L. Johnston, R. J. Stokes, and C. H. Li, "Crack Nucleation in Magnesium Oxide Bi-Crystals Under Compression," Phil. Mag. 7:23-34 (1962).
45. A. R. C. Westwood, "On the Fracture of Magnesium Oxide Bi-Crystals," Phil. Mag. 6:195 (1961).
46. R. C. Ku and T. L. Johnston, "Fracture Strength of MgO Bicrystals," Ford Motor Company Scientific Laboratory Report, Dearborn, Michigan, 1963.
47. T. L. Johnston and E. R. Parker, "Fracture of Nonmetallic Crystals," in: D. C. Drucker and J. J. Gilman (eds.), Fracture of Solids, Interscience Publishers (New York), 1963, pp. 267-287.

48. E. R. Parker, "Fracture of Ceramic Materials," in: B. L. Averbach, D. K. Felbeck, G. T. Hahn, and D. A. Thomas (eds.), M. I. T. Press and John Wiley and Sons, Inc. (New York), 1959, pp. 181-192.

49. R. D. Carnahan, "Mechanical Behavior of Hot-Pressed MgO Containing a Dispersed Phase," J. Am. Ceram. Soc. 47:305-306 (1954).

50. D. A. Jones and J. W. Mitchell, "Observations on Helical Dislocations in Crystals of Silver Chloride," Phil. Mag. 3:1 (1958).

51. F. P. Bullen, F. Henderson, and H. L. Wain, "The Effect of Hydrostatic Pressure on Brittleness in Chromium," Phil. Mag. 9:803 (1964).

52. A. Kelly and F. J. P. Clarke, "Tougher Ceramics," Trans. Brit. Ceram. Soc. 62:785-791 (1963).

53. T. S. Liu, R. J. Stokes, and C. H. Li, "Fabrication and Plastic Behavior of Single Crystal MgO-NiO and MgO-MnO Solid-Solution Alloys," J. Am. Ceram. Soc. 47:276-279 (1964).

Chapter 14

The Role of Grain Boundaries in the Plastic Deformation of Calcium Fluoride

P. L. Pratt, C. Roy,* and A. G. Evans

Imperial College of Science and Technology
London, England

Experimental evidence on the plastic deformation of single crystals of calcium fluoride is used to predict the behavior of polycrystalline calcium fluoride. The predictions are compared with experiment, and the role of grain boundaries in determining the brittleness at low temperatures is emphasized.

INTRODUCTION

The study of the mechanisms of plastic deformation of CaF_2 is of general interest in the field of ionic crystals, but it has a special interest in view of the similarity in crystal structure to UO_2, which is used as a nuclear material. The early investigations on the plasticity of ionic crystals were summarized by Schmid and Boas [1], who reported that CaF_2 deforms by slip on a (100) plane in a [1$\bar{1}$0] direction (Fig. 1). By using a decoration technique, Bontinck and Dekeyser [2] studied the configuration of dislocations in natural CaF_2 and reported that dislocations with Burgers vectors $a/2 \langle 100 \rangle$ (a is the side of the unit cell) are the most probable and the expected slip plane should be of the {110} type. Recently, Phillips [3] has studied the deformation and fracture processes in CaF_2 single crystals tested in compression. He found no ductility in as-cleaved or in as-cleaved and vacuum-annealed CaF_2 tested below 400°C. Above 400°C, the critical resolved shear stress decreased exponentially with increase of temperature, and the glide elements were (100) [1$\bar{1}$0], in agreement with Schmid and Boas.

In our work, the plastic properties of CaF_2 single crystals have been studied in compression in the temperature range 20 − 600°C [4]. The chief aim of this paper is to emphasize certain important features of the glide elements which were not observed in the work

*Present address: Atomic Energy of Canada, Chalk River, Ontario, Canada.

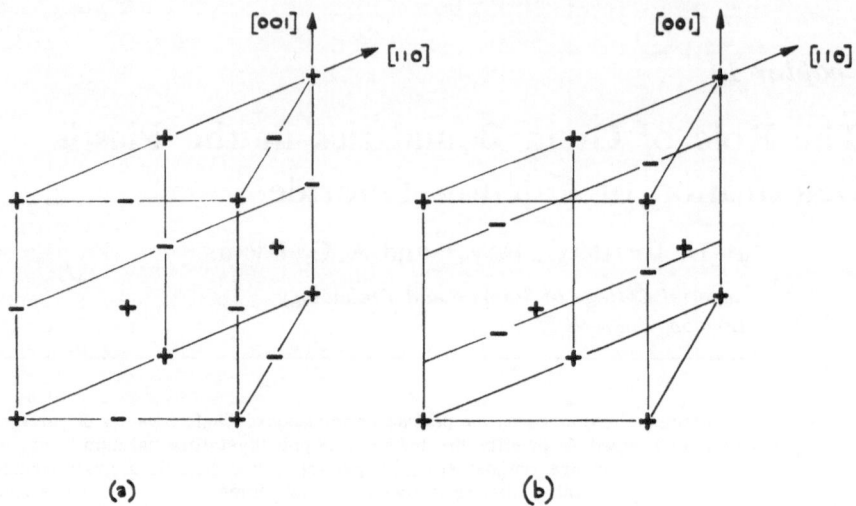

Fig. 1. Slip planes and directions in the CaF_2 structure compared with those in the NaCl
structure — (a) NaCl and (b) CaF_2.

of Phillips. Furthermore, we have studied the plastic properties
of polycrystalline CaF_2 over the same temperature range. Another
purpose of this paper is to emphasize the role of grain boundaries
in determining the onset of brittleness at low temperatures.

EXPERIMENTAL PROCEDURE

The single crystals used in this work were purchased from
Harshaw Chemical Co., and the polycrystalline material was kindly
donated by the Worcester Royal Porcelain Co., Ltd. Although the
nominal composition has not been checked spectrographically, it
is claimed by the manufacturer that impurities amount to a few
tenths of a ppm in the single crystals, and to some 400 ppm in
the polycrystalline material, which also had approximately 4%
porosity. The grain size was in the range $10 - 20\,\mu$ with most of
the porosity in the form of closed pores along the grain boundaries.

Preparation of the Single-Crystal Specimens

Rectangular-shaped specimens with a cross-sectional area of
about 3 mm^2 and a length of 8 mm were used. As CaF_2 cleaves to
an octahedral shape, a wire saw was used for cutting the crystals.

Single crystals with at least one cleaved surface were oriented by the X-ray back-reflection Laue method and then sawed parallel to the desired planes. The accuracy of cutting was kept within 3° of the plane.

A technique has been developed for polishing CaF_2 by chemical means. The {110} and {211} surfaces are reasonably well-polished when agitated for 1 min in a boiling solution of perchloric acid saturated with aluminum chloride and then washed in water and successively rinsed in absolute alcohol and anhydrous ether. Using

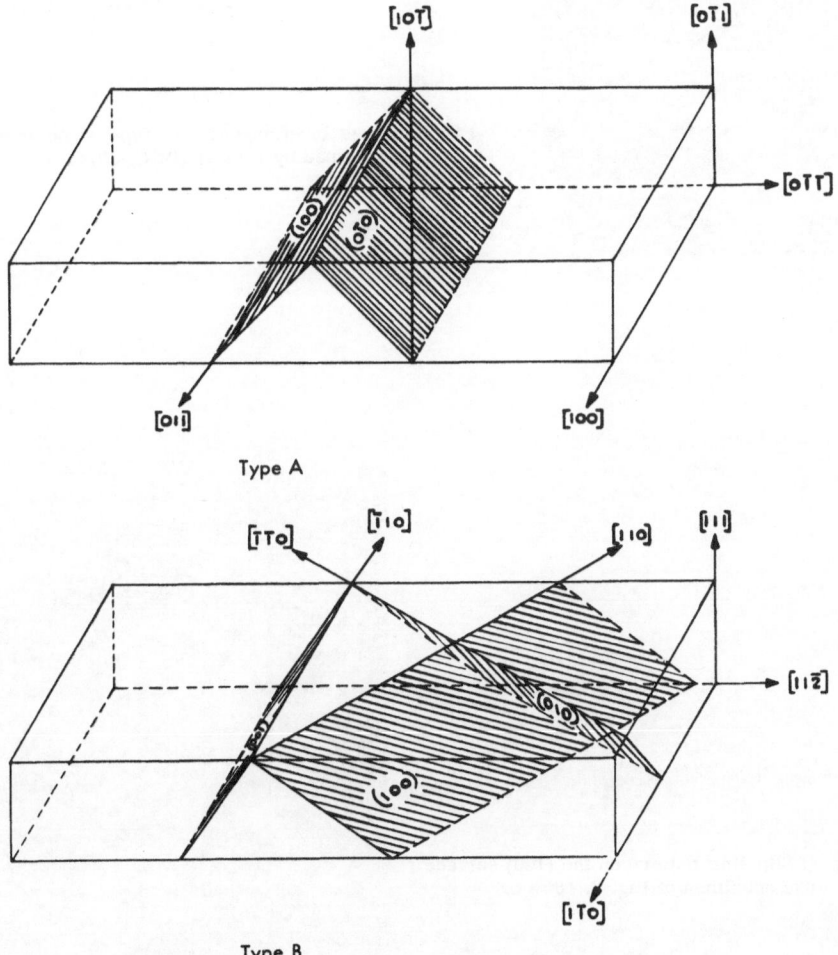

Fig. 2. Crystals oriented for slip on {110} $\langle 1\bar{1}0 \rangle$ systems.

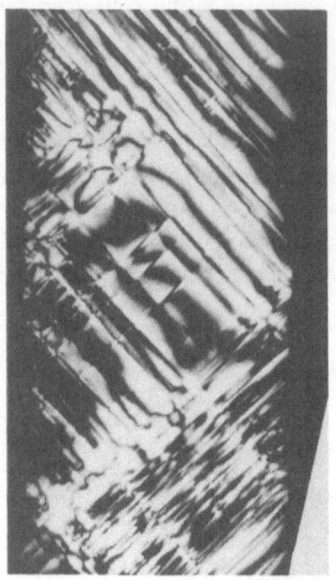

Fig. 3. Stress birefringence on Type A specimens
strained by 1.2% at 104°C, 20×.

Fig. 4. Slip line pattern on the (100) surface
of specimen of Fig. 3, 186×.

the same solution, the {100} and {111} surfaces became deeply etched, making difficult the interpretation of any surface markings produced by deformation; therefore, the {100} faces were mechanically polished and the {111} were cleaved.

Preparation of Polycrystalline Specimens

Polycrystalline specimens were cut from the lid of a slip-cast crucible using a diamond wheel. They were ground and polished to a finished size of 3.50 by 1.50 by 1.50 mm.

Examination

The mechanisms of plastic deformation were studied by using photoelastic techniques to examine stress birefringence, by microscopic examination of slip band traces on the free surfaces, and by dislocation etching techniques. All the mechanical tests were carried out on Instron testing machines and similar strain rates were used for the single-crystal and polycrystalline specimens.

EXPERIMENTAL RESULTS

Determination of Glide Elements in Single Crystals

In order to determine the glide elements and to study the effect of orientation on the stress–strain behavior of CaF_2, crystals of four different orientations were deformed in compression. This enabled different combinations of glide elements to become operative and microscopic studies to be made on different crystallographic surfaces. Figure 2 gives a schematic illustration of the two different crystal orientations which would activate slip on the {100} planes. Compressed in the [110] direction, the Type A crystal has two systems equally favored using the {100} planes, while the ⟨110⟩ directions are inclined at 45° and 60°, respectively, to the stressing axis. Compressed in the [$\bar{1}\bar{1}2$] direction, the Type B crystal has three available {001} ⟨1$\bar{1}$0⟩ systems, but, since the resolved shear stress is greater on one of these systems, single slip should be predominant.

All crystals oriented for slip on {100} were brittle at room temperature, but, in contrast to the results of Phillips, could be deformed plastically at 60°C and above.

Type A Crystal. Stressed in the [110] direction, Type A crystals were found to deform by slip on the two {100} planes inclined at 45° to the compression axis. The birefringence pattern observed normal to the (100) face of a specimen strained by 1.2% at 104°C is shown in Fig. 3. The bands intersecting at 90° and making a 45° angle with the [110] compression axis represent residual stresses after glide on two orthogonal (100) planes. This was accompanied by the formation of corresponding slip lines on both pairs of lateral faces (Fig. 4). The slip bands on the (0$\bar{1}$1) face were running normal to the [110] compression axis. From these observations it can be deduced that slip occurred on the two available {100} planes with a strong indication of the $\langle 1\bar{1}0 \rangle$ slip direction.

Type B Crystal. Stressed in the [$\bar{1}\bar{1}$2] direction, Type B crystals strained by 1% at 80°C deformed initially by single slip on the (001) plane inclined at about 55° to the compression axis. The birefringence bands observed normal to the {110} face, and shown in Fig. 5, make an angle of 55° to the long axis of this specimen and to the {111} pair of side faces. The micrograph of Fig. 6 shows the sharp slip bands developed on the {111} faces; no slip markings

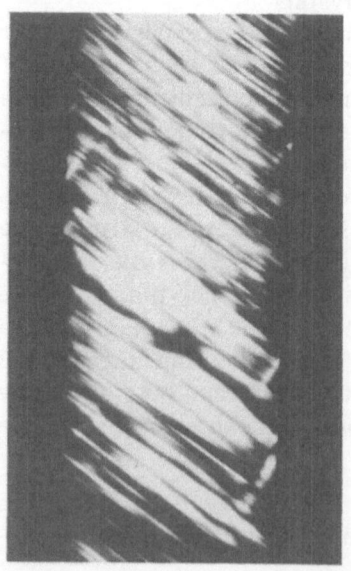

Fig. 5. Birefringence pattern on Type B crystal
strained by 1% at 80°C, 18×.

Fig. 6. Slip line pattern on (111) surface of the specimen of Fig. 5, 141×.

were revealed on {110} crystal surfaces. Figure 7 shows slip bands on a {111} surface after etching.

These observations confirm that the primary slip system in CaF_2 is {001} $\langle 1\bar{1}0 \rangle$.

The two different crystal orientations chosen to activate slip on the {110} $\langle 1\bar{1}0 \rangle$ system are illustrated in Fig. 8. These are defined as Type C and D crystals. The compression axis for both types was the [100] direction, which gives maximum shear stresses on four {110} $\langle 1\bar{1}0 \rangle$ systems, but no resolved shear stress on the {001} $\langle 1\bar{1}0 \rangle$ system. Type C and D crystals are essentially equivalent, except for the crystallography of their side faces. Type C crystals have cubic faces and Type D crystals have side faces of the {110} type. The chemically polished surfaces are smoother, which facilitates microscopic examination.

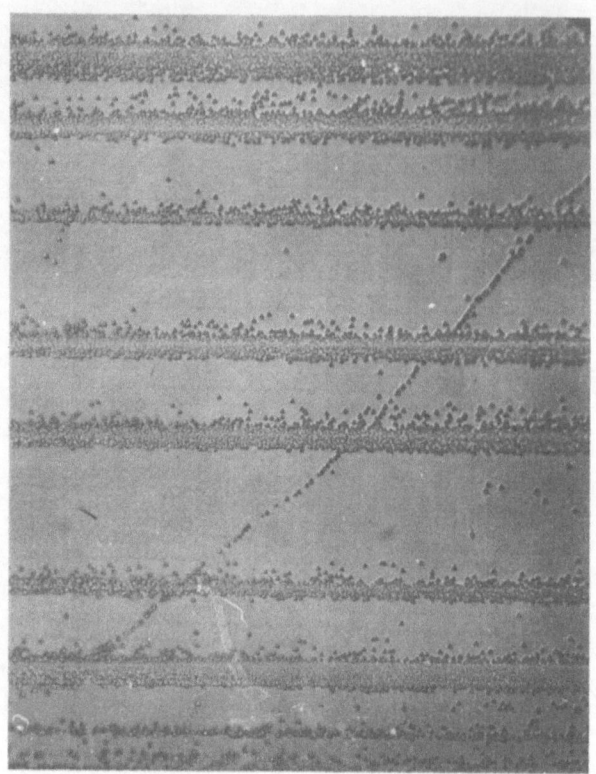

Fig. 7. Etch pits formed at slip bands on (111) surface of Type B crystal strained by 0.6%, 141×.

All the crystals oriented for slip on {110} were brittle at room temperature, but they could be deformed plastically at temperatures above 200°C.

Type C Crystal. Slip on these systems occurred with difficulty at 200°C, but became progressively easier and more homogeneous with increasing temperature. Block slip occurred extensively around 200°C, but was absent at about 400°C. Two orthogonal systems operated preferentially at high temperature (Fig. 9). These were generally situated in such a way that the slip planes intersected the narrow faces in a ⟨110⟩ direction and the wide faces in a ⟨100⟩ direction. This indicates that the slip systems that have the shorter slip distance to reach the surface were more active, as in sodium chloride.

Type D Crystal.   Compressed in the [100] direction, Type D crystals deformed by slip on the {110} planes. Because the lateral faces are all of the {110} type, the specimens had to be rotated by 45° to see birefringent patterns equivalent to those of a Type C crystal.   The stress birefringence of a specimen strained by 6% at 433°C is shown in Fig. 10. As shown in Fig. 11, sharply defined wavy slip bands have developed during deformation.   The slip markings were observed on the four side faces of the specimen

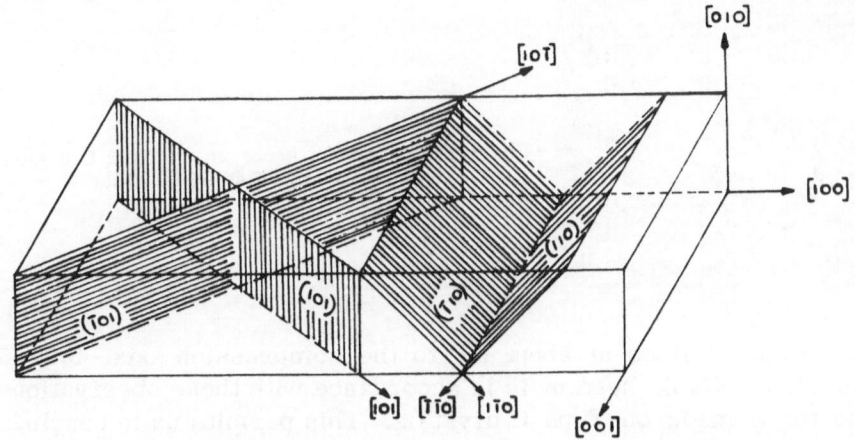

Type C

Type D

Fig. 8. Crystal oriented for slip on {110} ⟨1̄10⟩ systems.

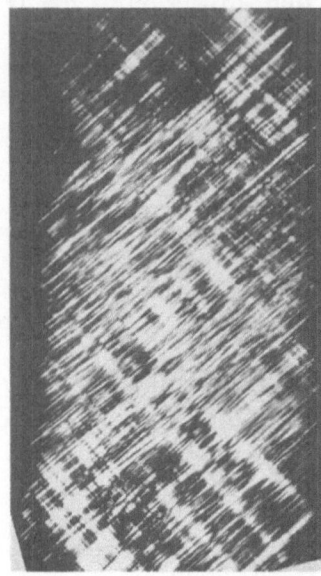

Fig. 9. Stress birefringence on the Type C crystal compressed by 8% at 393°C, 20×.

and were inclined at about 60° to the compression axis. Slip on the {110} ⟨1$\bar{1}$0⟩ system is in accordance with these observations and those made on Type C crystals. This permits us to conclude that, above 200°C, CaF$_2$ can deform by slip on the {110} ⟨1$\bar{1}$0⟩ system.

These slip systems are very reasonable, i.e., ⟨1$\bar{1}$0⟩ is the only glide vector for which the calcium and fluorine sublattices

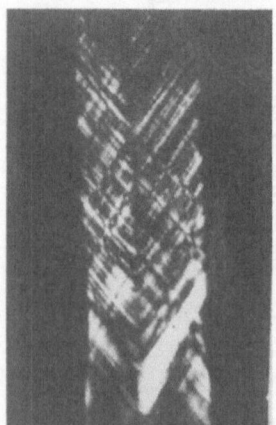

Fig. 10. Stress birefringence on Type D crystal compressed by 6% at 433°C, 18×.

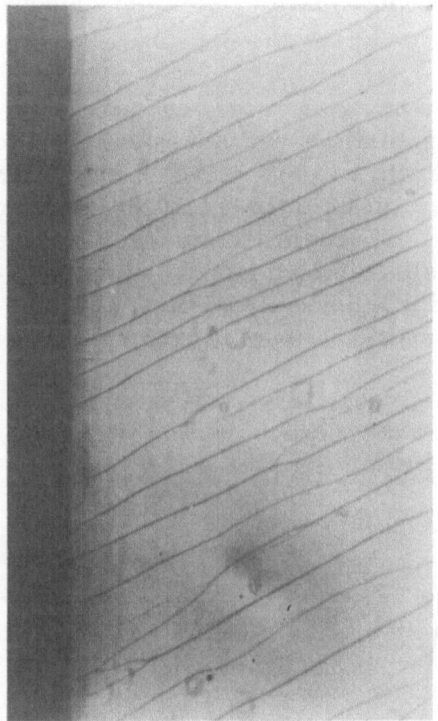

Fig. 11. Wavy slip line pattern on (110)
face of specimen shown in Fig. 10, 141 ×.

are preserved. Furthermore, {100} has a minimum ratio b/h of
1.42, compared to 2.0 for {110}; the elastic energy of a dislocation in {100} is < {110}; and ionic polarizability is low.

Temperature Dependence of Yield Stress for Slip on {100} and {110}

Specimens of the four types of crystal were deformed in compression to study their stress–strain characteristics over a range
of temperatures. The temperature dependence of the critical
resolved shear stress was determined for slip on both systems at
two strain rates. The two lower curves in Fig. 12 are for Type B
specimens slipping on {100} and the two upper curves for Type D
specimens slipping on {110}. The yield stress varies exponentially
with temperature for both specimens at the two strain rates. At
the lower strain rate, the temperature dependence becomes very

large for Type B specimens near 100°C, and for Type D specimens, near 300°C.

Deformation of Polycrystalline Specimens

Groves and Kelly [5] have applied the von Misés criterion to the CaF$_2$ structure, showing that slip on $\{001\}$ $\langle 1\bar{1}0 \rangle$ yields three independent slip systems, while slip on $\{110\}$ $\langle 1\bar{1}0 \rangle$ yields two further independent systems. Thus, a general strain involving five independent slip systems is possible if both families of slip systems operate simultaneously and independently. From the single-crystal results of Fig. 12, the transition temperature for the onset

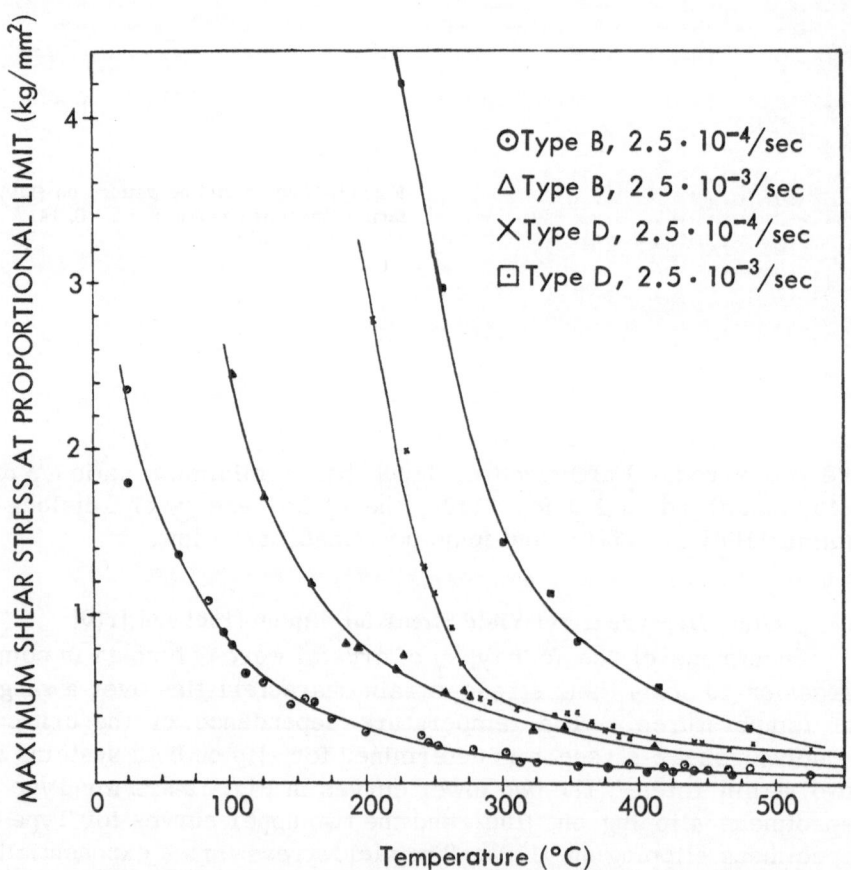

Fig. 12. Temperature dependence of the critical resolved shear stress for slip on $\{100\}$ and $\{1\bar{1}0\}$ planes.

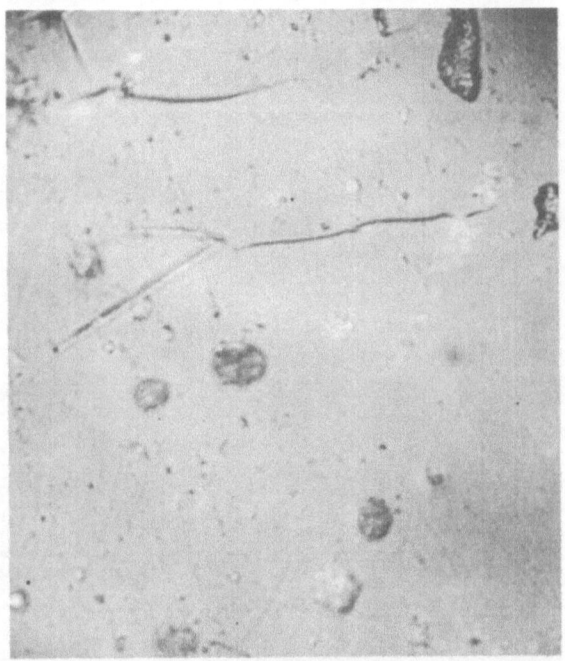

Fig. 13. Microcracks appearing at the yield point of a polycrystalline specimen deformed at 158°C, 980 ×.

of ductility should be in the region of 300°C when slip on {100} becomes possible. At first glance, this seems rather low.

At temperatures less than 300°C, the compression of polycrystalline specimens at the lower strain rate gives rise to microcracks spanning grains at the onset of yielding (Fig. 13). The ductility is very limited in this temperature range, and fracture occurs, for example, at 158°C after only 4% compression. Very fine straight slip lines can be seen in Figs. 13 and 14 after fracture. Wavy slip lines first appear at about 230°C, and all the straight slip lines disappear by 350°C. This transition is accompanied by the appearance of multiple slip in the surface grains (Fig. 15) at 398°C, and by the onset of ductility. Figure 16 shows wavy slip lines on a specimen which fractured after 15% compression at 398°C. Cracks crossing many grains can also be seen.

This onset of ductility is revealed in the stress–strain curves shown in Fig. 17. At low temperatures, the appearance of microcracks is accompanied by a sharp yield drop; this is followed by

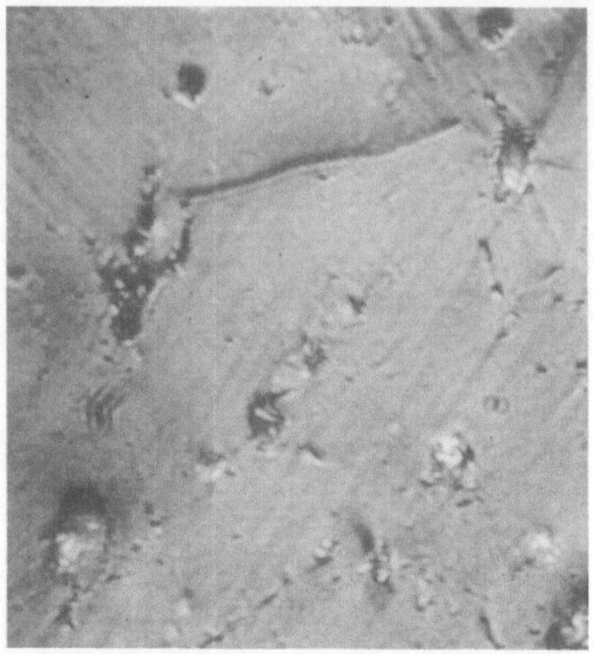

Fig. 14. Straight slip lines on specimen deformed to fracture at 158°C, 1140 ×.

Fig. 15. Multiple slip lines and wavy slip after 7% compression at 398°C, 1140 ×.

Fig. 16. Wavy slip lines and cracks on specimen deformed to fracture at 15% compression, 980×.

rapid work-hardening, leading to fracture. The sharp yield drop disappears by 350°C. At higher temperatures, the rate of work-hardening falls as the ductility increases. It is interesting to compare the results shown in this figure with the temperature dependence of the flow stress for single crystals in Fig. 12. At low temperatures, the onset of cracking at yield shows the same temperature dependence as the yield stress for slip on {100} planes. The onset of ductility and the reduced rate of work-hardening appear around 300°C, when slip on {110} allows a general strain to be achieved.

The agreement between the experimental results of the polycrystalline specimens and the predictions from the single-crystal work is remarkable, especially considering the differing purity of the single crystals and the polycrystalline material. It is not obvious that the impurities in the polycrystalline material should influence the yield stress for slip on {001} and {110} planes in the same manner, and, certainly, the values of the yield stress of the

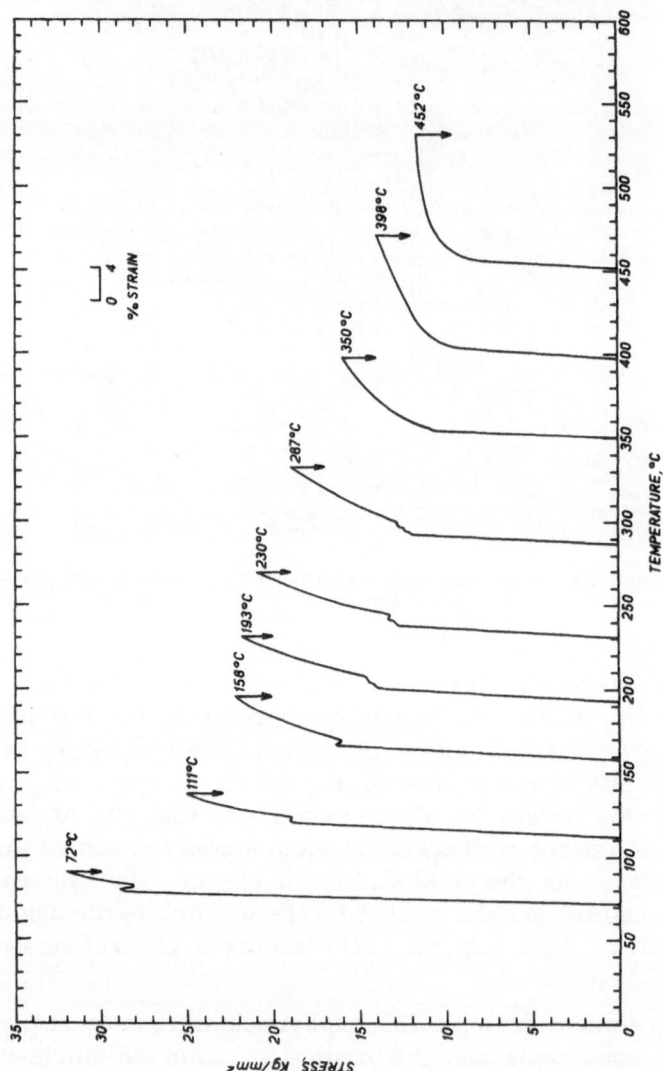

Fig. 17. Effect of temperature on the compression stress—strain curves of polycrystalline CaF$_2$.

polycrystalline material are very much higher than those of the single crystals. However, in view of the success of the application of the von Misés criterion, it seems important to repeat this work using single crystals and polycrystals made from the same starting material.

ACKNOWLEDGMENTS

The authors would like to thank Professor J. G. Ball for providing research facilities in the Department of Metallurgy, the Atomic Energy Research Establishment (Harwell) for continuing financial support and encouragement throughout the whole period of this research, and to the Worcester Royal Porcelain Company for providing the polycrystalline material.

REFERENCES

1. E. Schmid and W. Boas, Plasticity of Crystals, F. A. Hughes (London), 1950, p. 236.
2. W. Bontinck and W. Dekeyser, Physica 22:595 (1956).
3. W. L. Phillips, Jr., J. Am. Ceram. Soc. 44:499 (1961).
4. C. Roy, Canadian Conference of Metallurgists, 1964, to be published.
5. G. W. Groves and A. Kelly, Phil. Mag. 8:877 (1963).

Chapter 15

Grain Boundaries in Sodium Chloride

David M. Martin, Gerald K. Fehr,
and Thomas D. McGee

Iowa State University
Ames, Iowa

Single crystals of optical-grade sodium chloride were joined to produce bicrystals and tricrystals of controlled orientation. The dislocation and small-angle grain boundary configurations of the melt zone were determined. The relative energy of different grain boundary orientations was estimated. The local strain was estimated by etch pit and optical birefringence techniques. Etch pit and electron microscopic determinations were used to study dislocation interaction with grain boundaries.

INTRODUCTION

Many ceramic refractory oxides have the NaCl crystal structure. Study of the properties of refractory oxides, such as MgO, is hampered by their high melting points, the limited availability of single crystals, relatively high impurity levels, and a lack of fundamental knowledge. There is a need for analysis of the effect of grain boundaries on sintering processes, on other diffusional processes, and on the mechanical properties of cubic ionic solids. The halides have low melting temperatures, are available in excellent purity, can be obtained in single-crystal form at low cost, and have received much fundamental study. Research on the alkali halides will provide valuable information which should also apply, within limits, to the cubic ionic refractory oxides. This paper presents some results of research on grain boundaries in NaCl which was conducted to provide a better understanding of the basic properties of grain boundaries in cubic ionic solids.

PREPARATION OF BICRYSTALS

Bicrystals were prepared by joining single crystals with a hot-wire technique [1]. Bulk ingot sections of NaCl were purchased

TABLE I

Semiquantitative Spectrographic Analysis of Typical Ingots of Harshaw Optical–Grade NaCl

	Impurities			
	Si (ppm)	Cu (ppm)	Al (ppm)	Mg (ppm)
A	~10	< 50	~100	< 20
B	~10	< 50	< 50	< 20
C	< 10	< 50	< 50	< 20

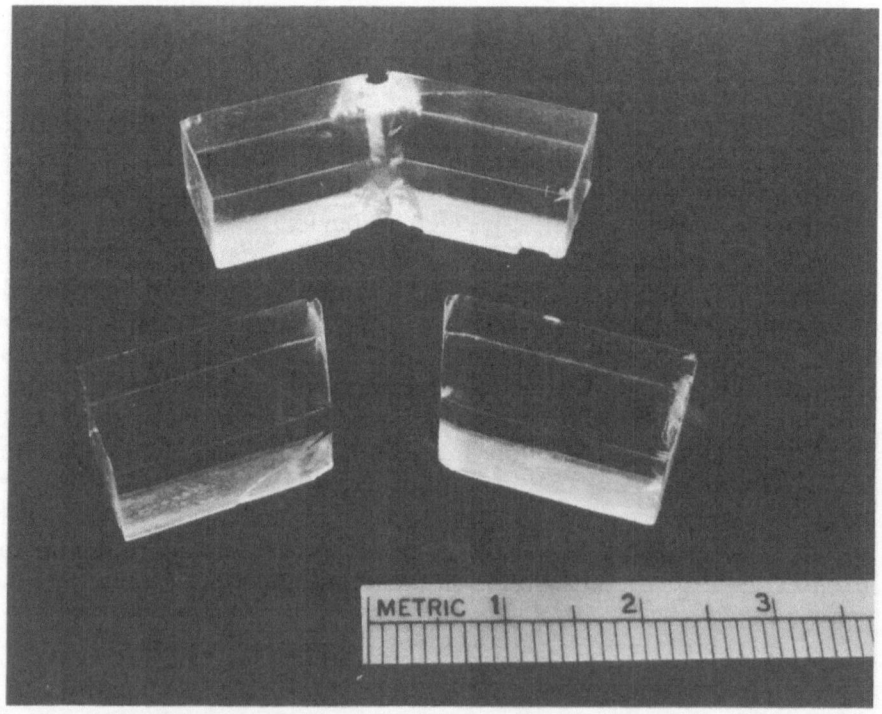

Fig. 1. Bicrystal halves before and after joining.

Fig. 2. Bicrystal after cooling.

from the Harshaw Chemical Company. Spectroscopic analysis usually revealed about 150 ppm total impurities (Table I).

The NaCl ingots were subdivided with a string saw into rectangular prisms with {001} faces. The desired angle of join was produced by controlled solution with a water film on the surface of a nylon lap. In all operations, extreme care was used to prevent mechanical damage to the surfaces. The two halves of each bicrystal were chemically polished with water, methanol, and acetone rinses just prior to joining.

The bicrystal halves were placed on a flat surface in a furnace with the faces to be joined in contact (Fig. 1). The furnace was heated to 600°C and a hot Pt–10%Rh wire was passed down through the interface at a rate of 0.05 cm/min. The wire was heated by resistance to about 810°C, about 10°C above the melting point of the NaCl. This produced a melt zone about 2.5 mm wide. The single crystals re-formed by crystallization as the melt zone cooled. A grain boundary formed where the single crystals met (Fig. 2).

With this technique, grain boundaries of controlled orientation were produced. The method served to standardize the thermal history of the bicrystals. Polycrystals were produced in a similar manner.

STRUCTURE AND PROPERTIES OF THE GRAIN BOUNDARIES

Optical Properties

The grain boundaries were transparent and strain-free in polarized light. Reflected and transmitted light did not reveal defects or inclusions if the wire temperature was not excessive. Bubbles with crystalline symmetry were produced if the joining wire was too hot.

Recently, Dr. Wiederhorn of the National Bureau of Standards examined some of our bicrystals with an ultramicroscope. After irradiation and heat treatment, he found scattering centers (Fig. 3). These were present in the region that had been melted during the joining process, but they were not present elsewhere. We believe these to be tiny voids and highly distributed platinum nuclei. Spectroscopic analysis of the melt zone did not reveal an appreciably greater impurity content although platinum would be suppressed by

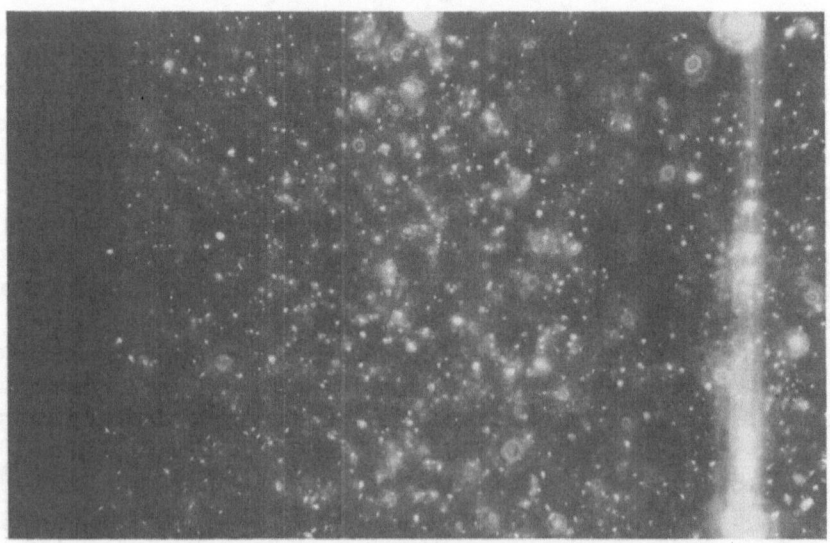

Fig. 3. Scattering centers in the melt zone after irradiation and annealing at 400°C for 6 hr. Dark field. 67×.

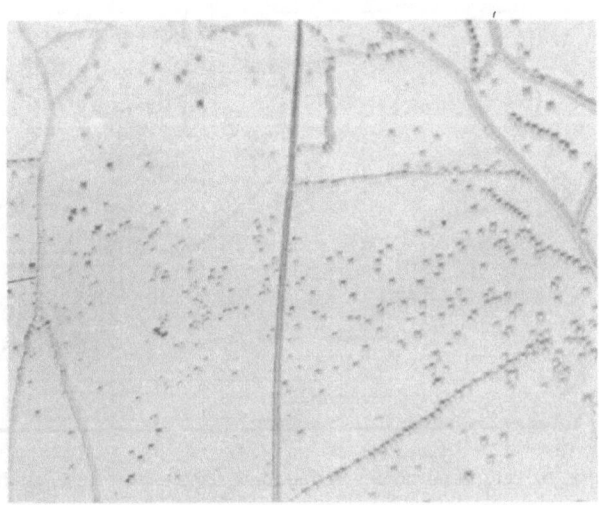

Fig. 4. Typical grain boundary showing the dislocation configuration at the join. 100×.

sodium in the analysis and a trace of rhodium did appear. The room-temperature mechanical properties are not believed to have been affected appreciably. The presence of the scattering centers is presumed to be undesirable. Dislocation distributions in NaCl, decorated by precipitation of metal particles, have been extensively studied with the ultramicroscope by Amelinckx [2]. The scattering centers, in this case, do not have the configurations observed by Amelinckx. They are randomly distributed except at the boundary where they appear as discontinuous lines which are randomly distributed in the plane of the boundary. However, heat treatment does produce precipitation with decoration of {100} planes and small-angle boundaries. The precipitates are tabular or flat hexagonal plates.

Dislocation Distribution

The melt zone produced by the hot-wire technique had a smaller number of individual dislocations, but a larger number of very small-angle grain boundaries than the unmelted region (Fig. 4). The dislocation density, not including the small-angle grain boundary dislocations, was usually $2-4 \cdot 10^5/cm^2$ in the grain boundary region.

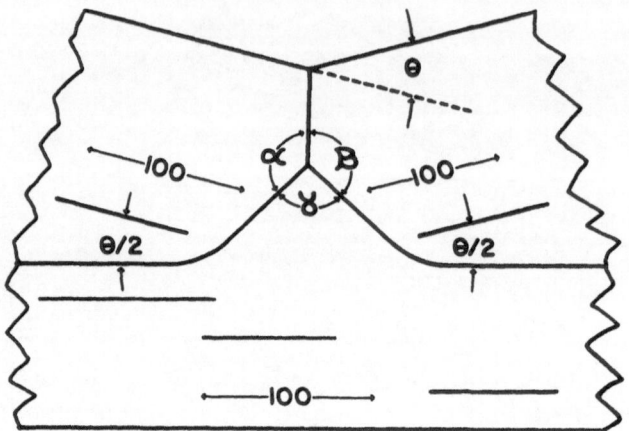

Fig. 5. Tricrystal configuration for a simple tilt energy analysis.

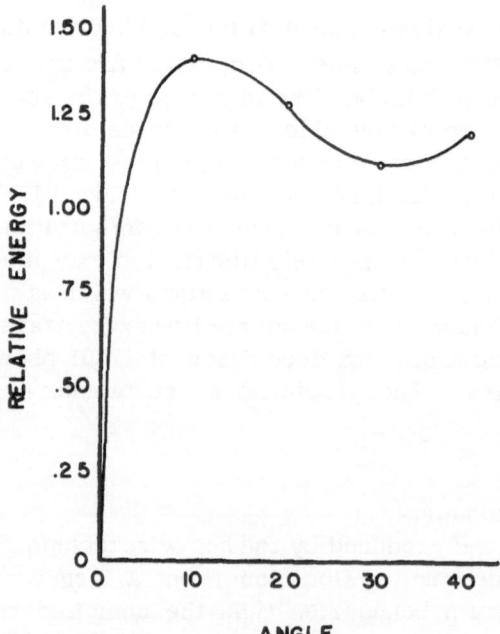

Fig. 6. Relative energy of symmetric simple tilt boundaries comparing the boundary at
θ with a boundary at $\theta/2$.

Grain Boundary Energy

The energy of a grain boundary varies with orientation. The dislocation theory of small-angle grain boundaries predicts that the energy of a boundary is a function of the angle of boundary.

$$E = k\theta (A - \ln\theta) \tag{1}$$

where $k = Gb^2/4\pi(1 - \nu)$ for edge dislocations. Here, G is shear modulus, b is Burgers vector, ν is Poisson's ratio, and A is a constant. Equation (1) has been found to fit measured energies of metals at angles even greater than those to which the dislocation theory is believed to apply [3]. Relative grain boundary energy is often obtained by determining the equilibrium dihedral angle for the grain boundary at the edge of a crystal [4]. This measurement is dependent on a constant surface energy regardless of the orientation of the crystal–vapor interface. In contrast to most metals, the surface energy of ionic crystals is usually strongly dependent upon orientation. Therefore, grain boundary energies must be measured with reference to known interfaces. This can best be done with tricrystals.

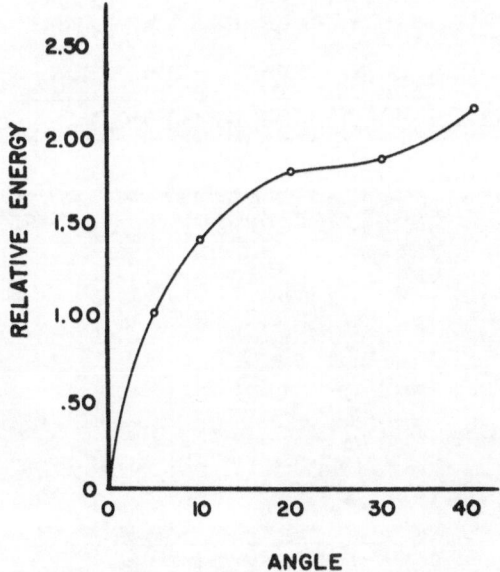

Fig. 7. Relative energy of symmetric simple tilt boundaries as a function of angle.

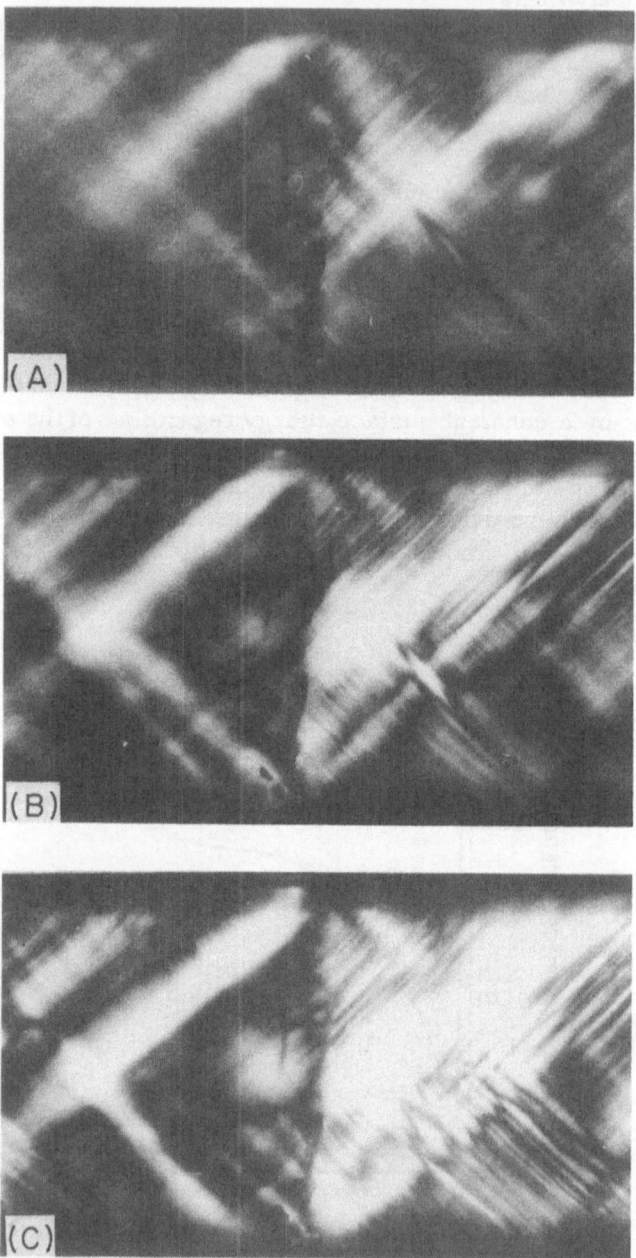

Fig. 8. (A)–(E)—Development of birefringence during loading. (F)—Fracture surface.

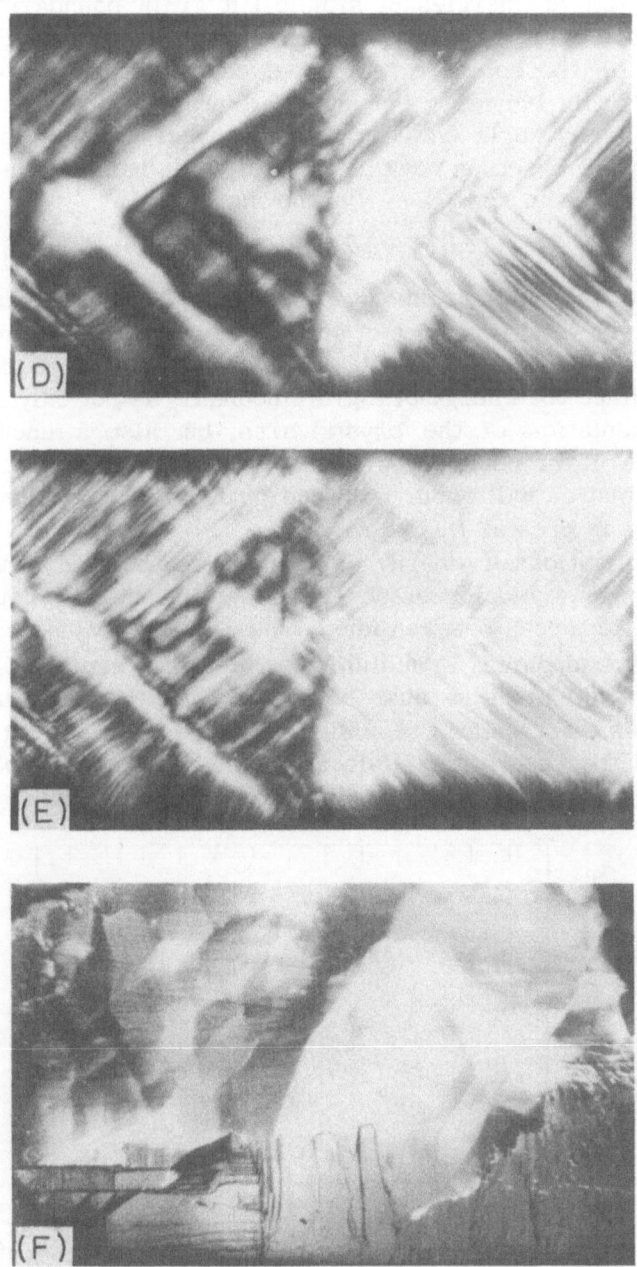

Fig. 8 (cont.).

The relative energies of simple tilt grain boundaries can be easily obtained. Tricrystals of orientation shown in Fig. 5 are produced by the hot-wire technique; each crystal thus produced gives the grain boundary energy of a boundary of angle θ relative to a boundary of angle $\theta/2$ (Fig. 6). A sequence of crystals in which the angles are based on powers of two of some basis angle θ_0 yields a system of n equations in $n + 1$ unknowns.

$$[2 \cos (\gamma_n/2)] \; E(2^n \theta_0/2) = E(2^n \theta_0) \tag{2}$$

where γ_n is the angle subtended by the $2^n (\theta_0/2)$ grain boundaries. Figure 7 shows the results of such a sequence relative to a 5° tilt boundary. Results of such an estimating procedure will be inexact, since the energy of a grain boundary is not only a function of misorientation of the crystal axes, but also a function of the orientation of the boundary with respect to its symmetric position. Underestimation will occur if the energy dependence on the boundary orientation is greater than zero.

The variation of energy in ionic compounds is a function of orientation to a much greater extent than in metals because of the electrostatic forces between ions. Thus, the energy of simple symmetric tilt boundaries in sodium chloride rises rapidly with angle, but, at the 30° angle, a cusp occurs because the {012} symmetry causes less electrostatic repulsion. By careful study of the deviation from the theoretical dislocation model, it may be possible to learn more about the structure of high-angle boundaries.

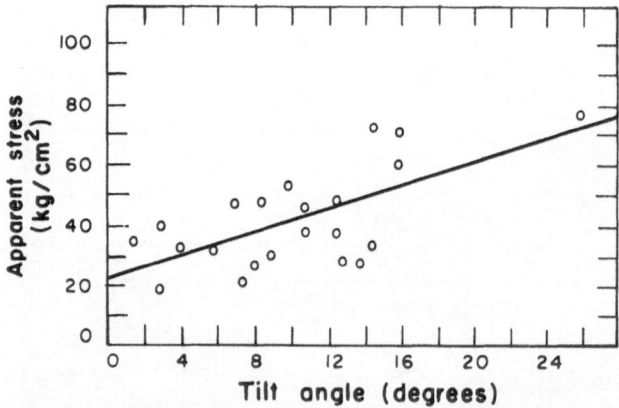

Fig. 9. The apparent stress (calculated from the elastic formula) at which birefringence crossed the grain boundary of symmetric tilt bicrystals as a function of the angle of tilt.

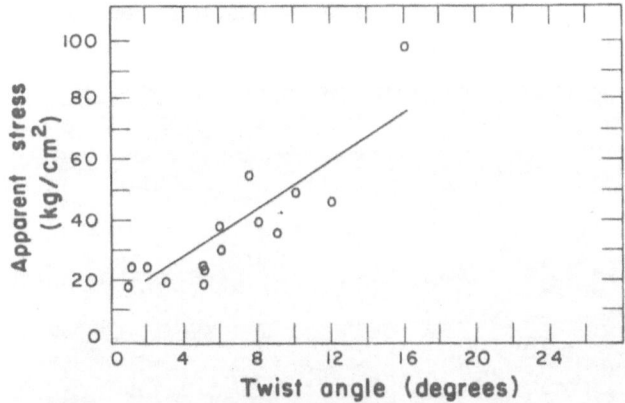

Fig. 10. The apparent stress at which birefringence crossed grain boundaries as a function of the angle of twist.

Fig. 11. Photomicrograph of a bent etched NaCl bicrystal. 365 ×.

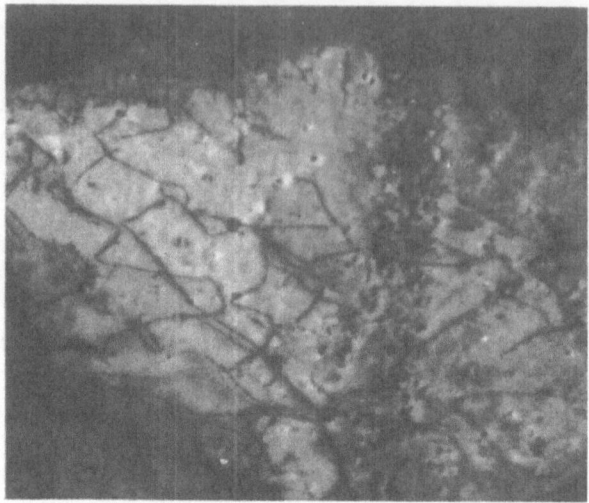

Fig. 12. Dislocations in NaCl. 30,000 ×.

The Grain Boundary and Plastic Deformation

Although single-crystal NaCl is ductile under controlled conditions at room temperature, polycrystalline NaCl is not. Bicrystals were tested as beams with four-point loading to study the interaction of plastic slip systems with grain boundaries. The birefringence was photographed with polarized light when loaded at 0.008 mm/min.

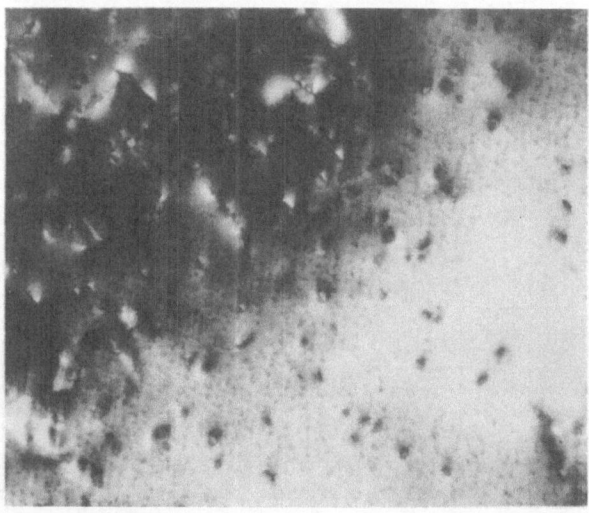

Fig. 13. Dislocations in NaCl showing radiation damage. 60,000×.

Load and deflection measurements were synchronized with time-lapse photographs of the birefringence (Fig. 8). By testing bi-crystals at different orientations, the effect of increasing angles of tilt or twist on grain boundary resistance was observed (Figs. 9 and 10). Of course, only slip bands were seen, and it was impossible to determine whether dislocations crossed the boundary or whether new dislocations were generated on the other side. However, from these experiments, the following observations were made:

1. Resistance to glide by the boundary increases with angle. Twist is more resistant than tilt.
2. Grain boundaries are inherently strong. Failure does not occur at the boundary until stress concentration occurs. Often failure was somewhere else in the crystal.
3. Grain boundary fracture was most frequent in twist crystals.
4. Grain boundaries differ from free surfaces. The slip bands readily approach free surfaces, but not grain boundaries. The presence of a grain boundary, therefore, effectively raises the yield point of the crystals.
5. When a grain boundary blocks a slip plane, propagation of the entire slip plane is inactive, i.e., cross glide requires dislocation motion.
6. The boundaries did not act as a source for slip to start.

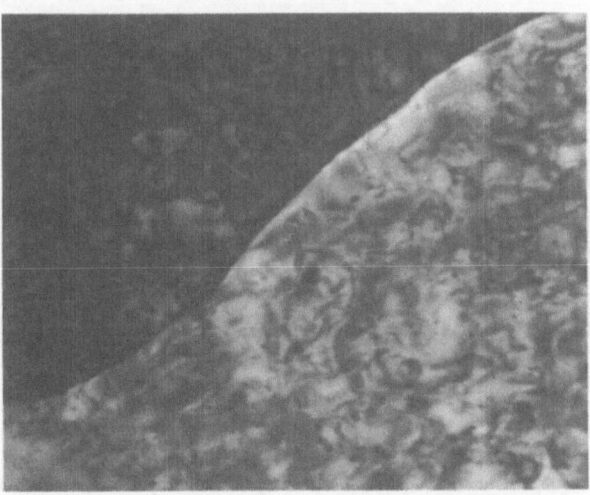

Fig. 14. Grain boundary area in NaCl. 40,000 ×.

Fig. 15. Selected area diffraction pattern of grain boundary in NaCl.

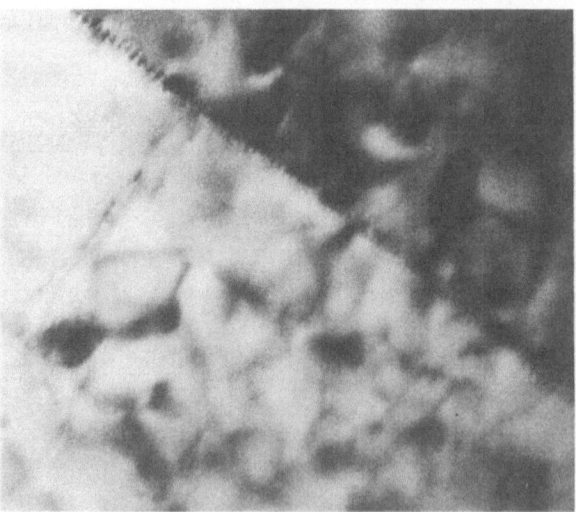

Fig. 16. Symmetric 6° tilt boundary in NaCl looking down the line of the edge dislocations making up the boundary. The spacing between the dislocations is 40 A.

Some of these conclusions are confirmed by etch pit configurations on deformed bicrystals (Fig. 11).

Transmission Electron Microscopy of NaCl Grain Boundaries

The transmission electron microscopy of NaCl is difficult because radiation damage is always present. Sodium chloride is discolored by F-centers after electron-microscopic examination. If excessive current density is employed during examination, the heating may cause volatilization. The degree of damage is a function of the time and intensity of irradiation. The salt rapidly becomes opaque to electrons as the damage progresses. Therefore, interpretation of dislocation structures must always be tempered by allowance for radiation damage.

Dislocations in NaCl are similar to those in MgO and in metals (Figs. 12 and 13). Observations of dislocations often reveal differences in contrast on opposite sides of the boundary (Fig. 14). The dislocation theory of small-angle grain boundaries predicts spacing of dislocations: $d = b/\sin\theta$. This spacing can be observed by transmission electron microscopy (Figs. 15 and 16). Individual dislocations can be discerned. This confirms the simple tilt dislocation model. Angles as high as 12° have been found to conform to this model.

SUMMARY

High-purity sodium chloride polycrystals of controlled orientation were produced by a hot-wire technique and used for experimental determination of relative grain boundary energies and the effect of grain boundaries on mechanical properties. The energy of simple tilt grain boundaries shows a cusp at approximately 30°, which is in accord with crystallographic reasoning. Propagation of slip bands is hampered by the presence of grain boundaries; the degree of interference is higher for twist-type grain boundaries and, in general, increases with angle. Transmission electron microscopy reveals dislocation configurations similar to those expected from studies of MgO and metals and confirms the dislocation model of tilt grain boundaries of angles less than 12°.

ACKNOWLEDGMENTS

The support of the U.S. Army Research Office (Durham) is gratefully acknowledged. We thank Dr. Sheldon Wiederhorn for providing information about scattering centers in joined crystals and for providing specimens that had been irradiated and annealed.

REFERENCES

1. S. A. Long and T. D. McGee, "Effect of Grain Boundaries on Plastic Deformation of Sodium Chloride," J. Am. Ceram. Soc. 46(1): 583–587 (1963).
2. S. Amelinckx, "Dislocations in Ionic Crystals: I. Geometry of Dislocations," Chapt. 2, in: W. W. Kriegel and H. Palmour III (eds.), Mechanical Properties of Engineering Ceramics, Interscience (New York), 1961, pp. 9-34.
3. W. T. Read, Jr., Dislocation in Crystals, McGraw-Hill Book Co. (New York), 1953, p. 168.
4. R. A. Swalin, Thermodynamics of Solids, John Wiley & Sons (New York), 1962.

Chapter 16

The Structure of Twist Grain Boundaries in NaCl

M. L. Gimpl, A. D. McMaster, and N. Fuschillo

Research Division, Melpar, Inc.
Falls Church, Virginia

Grain boundaries in twist bicrystals grown by the Kyropoulos technique were examined by: (1) pulling the crystals apart at the grain boundary, and (2) polishing away one side of the bicrystal to within 0.5 mm of the grain boundary and examining the grain boundary surface by focusing through the layer of NaCl. Both of these techniques showed that the bicrystals were held together by small regions or pips of continuity across the grain boundary. The small bonded regions are separated by large unbonded areas. The pip density ranged from 500 to 20,000 cm^{-2} for 45° and 10° twist bicrystals, respectively.

INTRODUCTION

The effect of grain boundaries on the mechanical properties of ceramics has been reviewed by Stokes [1]. Murray and Burn [2] have recently investigated the mechanical properties of KCl bicrystals. They have shown that twist bicrystals doped with 100 ppm calcium were significantly less ductile than similar tilt bicrystals. Class [3] has shown that KCl twist bicrystals are significantly weaker than tilt bicrystals.

The various models for high-angle grain boundaries that have been proposed are reviewed by McLean [4]. Mott [5] considers a boundary as consisting of islands where the atomic matching is good, separated by regions where it is poor. Our observations show that, in twist bicrystals of NaCl, this is indeed the case, but on a macroscopic scale, as compared with Mott's atomic–dimension model.

Levine et al. [6] have investigated the mechanical behavior of polycrystalline AgCl in various aqueous solutions. They find that certain solutions induce brittle behavior caused by cracks nucleating

TABLE I

Spectrographic Analysis
of NaCl Bicrystal

Element	ppm
Al	15
Bi	< 50
Ca	13
Cr	< 10
Cu	45
Fe	< 6
Pb	< 50
Mg	10
Mn	< 6
Ni	< 50

at grain boundaries of certain orientations. A similar phenomenon is observed in copper tested in liquid-metal environments [7]. Cracks are seen to nucleate at copper grain boundaries with certain critical orientations.

It appears that the structure of the high-angle grain boundaries varies critically with the orientation of the two crystals with respect to each other. Certain grain boundary orientations may well show very little continuity and bonding across the boundary. This is shown to be the case for high-angle twist bicrystals of NaCl.

EXPERIMENTAL METHOD

Bicrystals of NaCl approximately 8 cm long by 2 cm diameter were grown by the Kyropoulos [1] technique in an argon atmosphere. Two seed crystals were held in chucks controlling their relative orientation. Pure tilt and pure twist bicrystals were grown with total mismatches varying from 5° to 45°. The bicrystals were pulled at a rate of 1–2 cm/hr from the melt contained in a platinum crucible. A typical spectrographic analysis of the bicrystals is shown in Table I.

Sections approximately 1 cm long perpendicular to the growth axis were cut from the bicrystals for microscopic examination. In one technique for examining the grain boundary, one side of the bicrystal was polished down to within less than 1 mm of the grain

Fig. 1. Grain boundary surface of 10° twist bicrystal, 65×.

boundary; the boundary then could be examined by focusing directly through the remaining sodium chloride. A crack was first initiated at the grain boundary; by viewing the region immediately in front of the crack in dark-field illumination, a characteristic structure could be observed in the twist bicrystal boundary.

Since the twist bicrystals could be readily cleaved at the grain boundary, the second technique involved direct examination of the

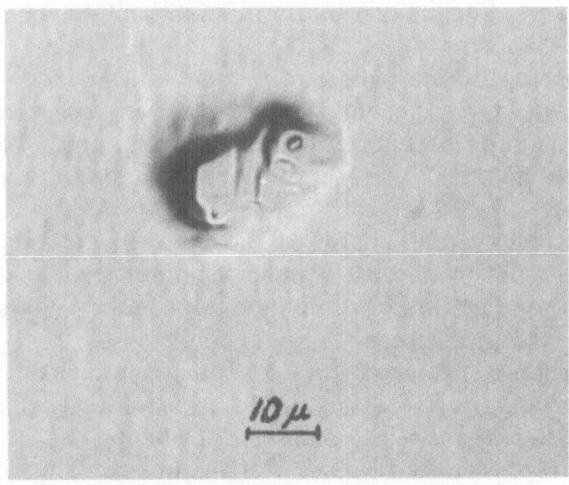

Fig. 2. Grain boundary surface of 10° twist bicrystal, 1000 ×.

Fig. 3. Grain boundary surface of a 30° twist bicrystal, as seen by focusing through one
side of bicrystal, 100×.

exposed surfaces, i.e., fractography. The tilt bicrystals were all
but impossible to cleave at the grain boundaries and had to be
examined by the first technique.

RESULTS AND DISCUSSION

Twist bicrystals with a total angular mismatch of 10°, 30°, and
45° were examined by the two techniques mentioned above.

Fractographic examination of the grain boundary surfaces after
cleaving showed characteristic pip-like regions covering the entire
grain boundary surface. These pips on a 10° twist boundary are
shown in Figs. 1 and 2, and appear to be analogous to spot welds
in metal sheets. The twist grain boundaries are held together by
these small regions of continuity or pips each having a cross section of less than 5 μ^2.

Examination of the twist grain boundary surfaces by the alternate polishing technique showed structures, as shown in Fig. 3,
in which the pips of continuity are visible, too. The pip density
ranged from 500 to 20,000 cm^{-2} for 45° and 10° twist bicrystals,
respectively.

The tilt bicrystals could not be cleaved directly at the grain
boundaries. When the tilt grain boundaries were examined by
polishing away one side of the bicrystal, no pips or other anomalies
were seen at the grain boundary surface.

The structure observations on the twist boundaries are corroborated by qualitative diffusion and mechanical property observations. High-angle twist boundaries when placed in water show
extremely rapid grain boundary diffusion. A 30° twist bicrystal,
8 cm long, was placed in water. The water moved up the boundary

at the rate of about 1 cm/min. In five minutes, the bicrystal was completely separated. Small thermal or mechanical stresses will readily cause the higher-angle twist bicrystals to part at the grain boundary. Cooling the as-grown twist bicrystals in air usually will cause them to fracture and separate at the grain boundary.

Tilt bicrystals did not display the high rates of diffusion of water along the grain boundary. Tilt bicrystals could not be fractured at the grain boundaries by thermal stressing.

DISCUSSION

The twist bicrystals with a total angular mismatch of 10°–45° may be thought of as two single crystals in intimate mechanical contact, bonded only at small regions with an area of $\approx 5\mu^2$ each. The total number of these bonded areas varies from $20 \cdot 10^3$ to $0.5 \cdot 10^3$ for 10° and 45° twist bicrystals, respectively. The cause of the lack of bonding in large areas may be due to impurity segregation and to lattice mismatch.

It is obvious that grain boundaries with structures as seen in the pure twist NaCl bicrystals could well account for lack of ductility in polycrystalline materials. The structure of mixed twist and tilt boundaries has not been studied, but structures with somewhat more extensive regions of continuous bonding than in pure twist crystals should be observed.

The causes of lack of ductility in polycrystalline AgCl [6] when tested in certain aqueous solutions could be explained by the preferential dissolving of pips at twist grain boundaries due to rapid diffusion up these boundaries. The dissolved grain boundary then may act as a crack, causing brittle behavior. The same situation could be true for the liquid-metal embrittlement of copper [7].

REFERENCES

1. R. J. Stokes, "Correlation of Mechanical Properties with Microstructure," in: Microstructure of Ceramic Materials (Proc. Am. Ceram. Soc. Symp., April 27-28, 1963), NBS Misc. Publ. 257:41-72 (1964).
2. G. T. Murray and R. A. Burn, The Role of the Grain Boundary in the Deformation of Ceramic Materials, U. S. Air Force Rept. ASD-TDR-62-225, Part III, March 1964.
3. W. H. Class, Thesis, Columbia University, 1964.
4. D. McLean, "Grain Boundaries in Metals," Oxford (London), 1957, pp. 15-43.
5. N. F. Mott, Proc. Phys. Soc. 60:391-394 (1948).
6. E. Levine, H. Solomon, and L. Cadoff, Acta Met. 12:1119-1124 (1964).
7. S. Rosenberg and L. Cadoff, Fracture of Solids, Interscience (New York), 1962.

Chapter 17

Grain Boundary Sliding in Alumina Bicrystals

M. P. Davis and Hayne Palmour III

North Carolina State University
Raleigh, North Carolina

A technique has been developed for forming bicrystals of alumina with controlled grain boundary orientation by pressure-sintering in vacuo. Bicrystals with tilt misorientations prepared by this technique have been tested in compressive shear and the stress–strain relationship determined. Bicrystal fracture strengths at several temperatures are reported. The microstructural features of the fractured crystal boundaries were examined and related to the observed mechanical behavior.

INTRODUCTION

As more precise measurement techniques are employed for studying crystalline materials, it has become increasingly apparent that boundaries in polycrystalline structures exert a significant effect on the measured properties. Experimental evidence indicates that, at elevated temperatures, the mechanical properties of ceramic and metallic materials are limited by the properties of the grain boundaries. These grain boundaries become the "weak links"; deformation and failure occur by the separation of individual grains. A fundamental knowledge of the structure and mechanical behavior of impurity-free grain boundaries may lead to improvement in the mechanical performance of polycrystalline materials.

There are basically two approaches for studying the effect of grain boundaries on the mechanical behavior of polycrystalline solids, i.e., use of polycrystalline solids with different grain sizes, and use of multicrystals, e.g., bicrystals and tricrystals. The earliest uses of bicrystals for studying the properties of grain boundaries employed metal crystals which could be formed by a strain-anneal process [1, 2]. More recently, bicrystals of ionic compounds have been obtained from blocks containing only a few grains [3]. The resulting bicrystals are of apparently good quality; however, there is little control over orientation or impurity con-

centration in the grain boundary. In an effort to exercise control
over grain boundary orientation, more direct approaches to bicrystal formation have been tried. McGee [4] has succeeded in
forming bicrystals of NaCl by passing a hot wire along the boundary.
Another technique which has shown promise is the direct-sintering
method.

PREPARATION OF ALUMINA BICRYSTALS

The alumina bicrystals used in this study were formed by a
pressure-sintering technique [5] which is illustrated schematically
in Fig. 1. After properly orienting the crystal halves using Laue
back-reflection, the surfaces normal to the load direction are ground
flat and parallel. The two mating surfaces are then polished with
diamond paste on hard wooden laps followed by an ultrasonic
cleaning. Crystals to be sintered are placed directly between the
tungsten loading rods of a high-temperature vacuum furnace and

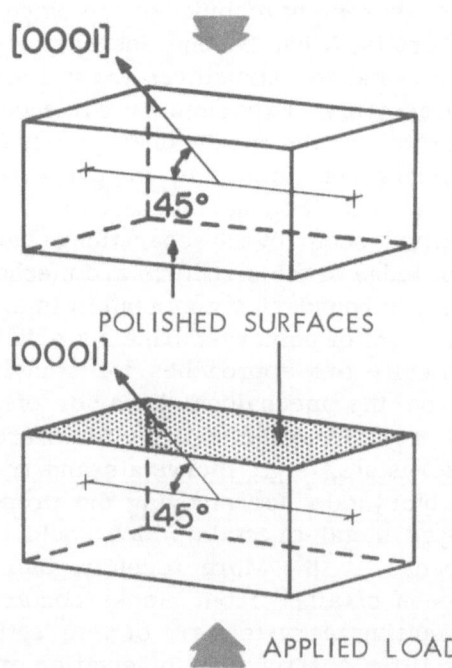

Fig. 1. Schematic illustration of crystal orientation during bicrystal formation.

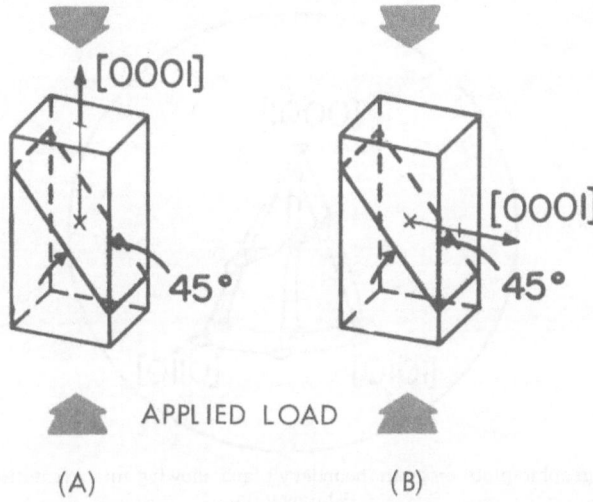

Fig. 2. Crystallographic orientation for type A and B bicrystal test specimens.

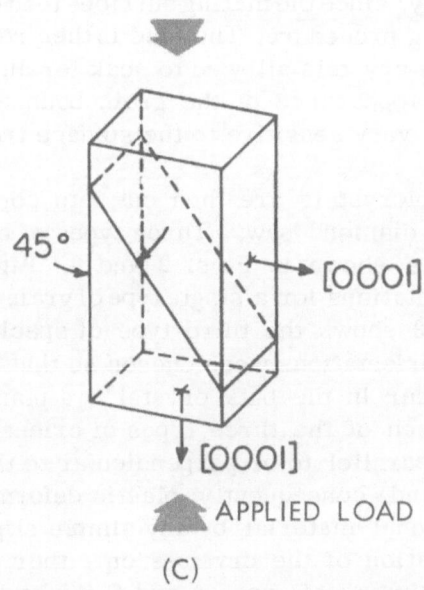

Fig. 3. Crystallographic orientation for type C bicrystal test specimen.

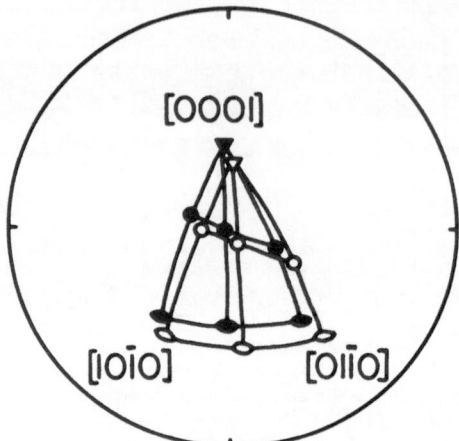

Fig. 4. Stereographic plots on grain boundary plane showing misorientation of type A and
B bicrystals.

heated to a temperature of approximately 2000°C under a small
load, just enough to maintain contact. The crystals are allowed to
soak for 30 min at temperature before increasing the load. A small
amount of plastic deformation is induced in order to smooth and
flatten the boundary, since the mating surfaces tend to round slightly
during the polishing procedure. The load is then reduced below the
yield point and the crystals allowed to soak for an additional hour.
The macroscopic appearance of the grain boundary when formed
in this manner is very sensitive to the surface treatment prior to
sintering.

The sintered bicrystals are then cut into compressive shear
specimens with a diamond saw. Three types of orientations have
been examined, as shown in Figs. 2 and 3. Figure 2 shows two
different test orientations for a single type of grain boundary orientation, and Fig. 3 shows the third type of specimen orientation.
These particular orientations were chosen so that plastic deformation would not occur in the bulk crystal in a plane parallel to the
boundary. For each of the three types of orientation, the c-axis
ideally is either parallel to or perpendicular to the load direction
in each crystal, and, consequently, plastic deformation should not
occur in this trigonal material by any simple slip mechanism [6].
The actual orientation of the crystals on either side of the grain
boundary for specimens of types A and B is shown in Fig. 4. The
misorientation amounts to a 10° tilt boundary with essentially no

component of twist.   Figure 5 shows the stereographic plots for
type C test specimens.   In order to obtain this orientation, the top
crystal was rotated 180° before sintering, thus giving the grain
boundary a "twin" character while retaining the 10° tilt misorientation.

The general macroscopic quality of the sintered alumina bicrystals prior to testing is shown in Figs. 6 and 7. The two photographs
of the same bicrystal were taken using oblique reflected light, so
that well-healed, transparent regions show as dark areas. Figure 6
shows a side view of a bicrystal which has the type A orientation;
Fig. 7 shows a view taken through the grain boundary. The highly
reflective streaks near the sides of the crystal are unhealed regions.
The center of the grain boundary contains hazy areas which are
not completely healed, but show up only when viewed in an intense
light, as was used to take this photograph. The hazy appearance
is caused by very small pores or discontinuities which have not
been completely eliminated.   As the surfaces of the presintered
crystals are improved, the amount of haze observed in the resulting
grain boundary is reduced.

STRESS–STRAIN RELATIONSHIPS FOR ALUMINA BICRYSTALS

The first two bicrystals to be tested in the compressive shear
configuration were of the same approximate optical quality, as

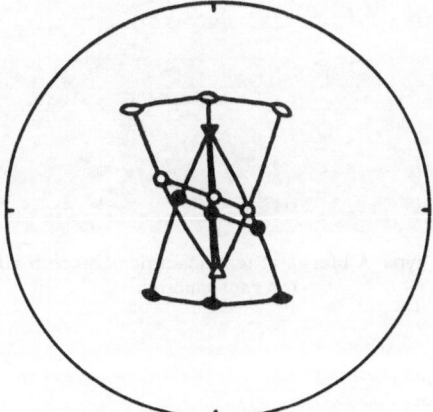

Fig. 5. Stereographic plots on grain boundary plane showing misorientation of type C
bicrystal.

Fig. 6. Side view of a type A bicrystal test specimen before testing. 22 ×; reduced 10%
for reproduction.

Fig. 7. Bicrystal test specimen before testing. Photograph was taken looking through grain boundary. 22 × ; reduced 10% for reproduction.

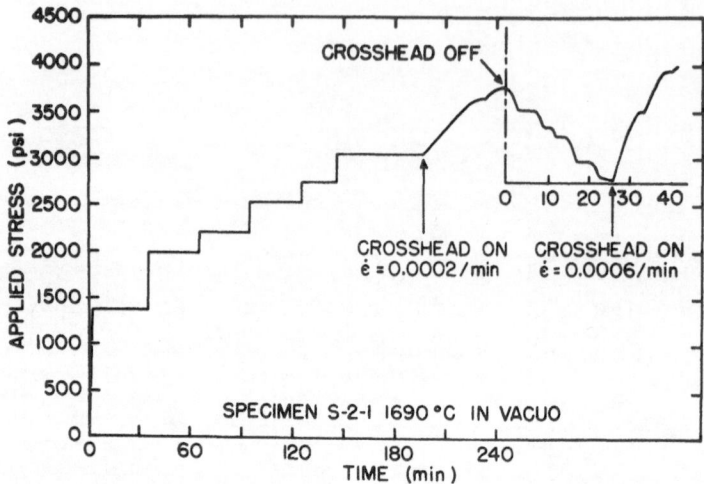

Fig. 8. Stress—time curve for bicrystal specimen with type C orientation.

shown in Figs. 6 and 7.   These two bicrystals were heated to temperature, allowed to come to equilibrium, and then loaded at a constant strain rate. The first bicrystal was tested at 1630°C with a strain rate of 0.01 min^{-1}. There was no evidence of a yield point, and the bicrystal failed at a stress of 3500 psi. The second of these two bicrystals was tested at 1720°C with a strain rate of 0.002 min^{-1}. Again, there was no evidence of yielding, and the bicrystal failed at 2300 psi.

After these two attempts at forcing the grain boundary to slide, it was decided to employ a static load-and-hold approach (stationary crosshead) to allow an incubation period for grain boundary sliding of the type observed by Adams and Murray [3] in MgO. The bicrystals used in this series of tests were of better quality, i.e., there was less haze in the boundary than in the preceding tests. The stress—time curve for bicrystal S-2-1, which has the type C orientation [180° twist plus 10° tilt (Fig. 3)], is shown in Fig. 8. The bicrystal was heated to 1690°C in vacuo and the applied stress increased by increments with waiting periods of 20 to 30 min at each of the various stress levels.   The load remained constant during these periods, and the bicrystal did not deform or slide at stresses up to 3200 psi.   The stress was increased at a constant slow rate ($\dot{\epsilon}$ = 0.0002 min^{-1}) up to 3700 psi, at which point the motion of the crosshead was stopped.   The bicrystal then re-

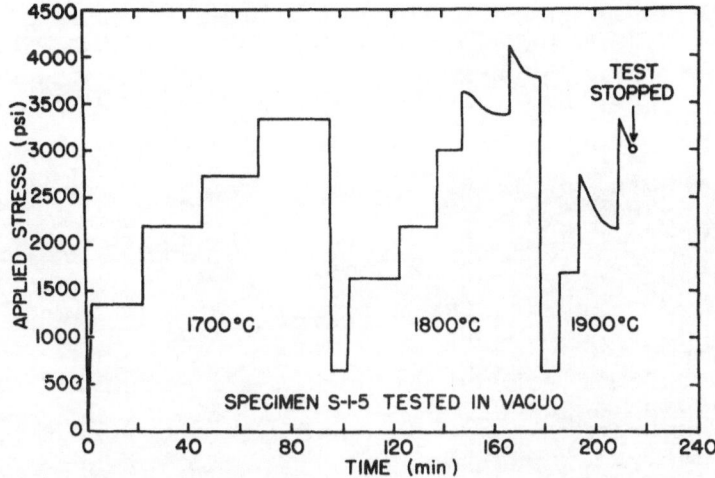

Fig. 9. Stress—time curve for bicrystal specimen with type B orientation.

lieved the applied stress in separate, discrete intervals. After a 25-min waiting period, the crosshead motion was resumed ($\dot{\epsilon} = 0.0006$ min^{-1}). The bicrystal was stressed until it separated at the grain boundary. Microscopic examination of the fractured halves showed that plastic deformation had occurred in the adjacent crystals, but there was no visible evidence of grain boundary sliding.

Another bicrystal (S-1-5) of comparable quality but with type B orientation (Fig. 2) was tested under similar conditions at three temperatures. The stress—time curve for this bicrystal is shown in Fig. 9. At 1700°C, no deformation or relaxation was observed up to 3300 psi. When the temperature was raised to 1800°C, there was evidence of stress relaxation above 3500 psi. At 1900°C, the first evidence of stress relaxation occurred at 2700 psi. The test on this specimen was stopped prior to failure so that the nature of the deformation could be determined. Figures 10a–11b show two views of bicrystal S-1-5 before and after the compressive shear test. The before and after views of Figs. 10a and 10b show a visible change in the shape of the bulk crystals. The grain boundary also has a bent region near its top. Figure 11b shows the development of a white streak across the grain boundary just below the point where the boundary exits from the bicrystal. This streak is a separation or crack which was generated during the deformation of the bicrystals. Figures 12a and 12b show greater detail of the

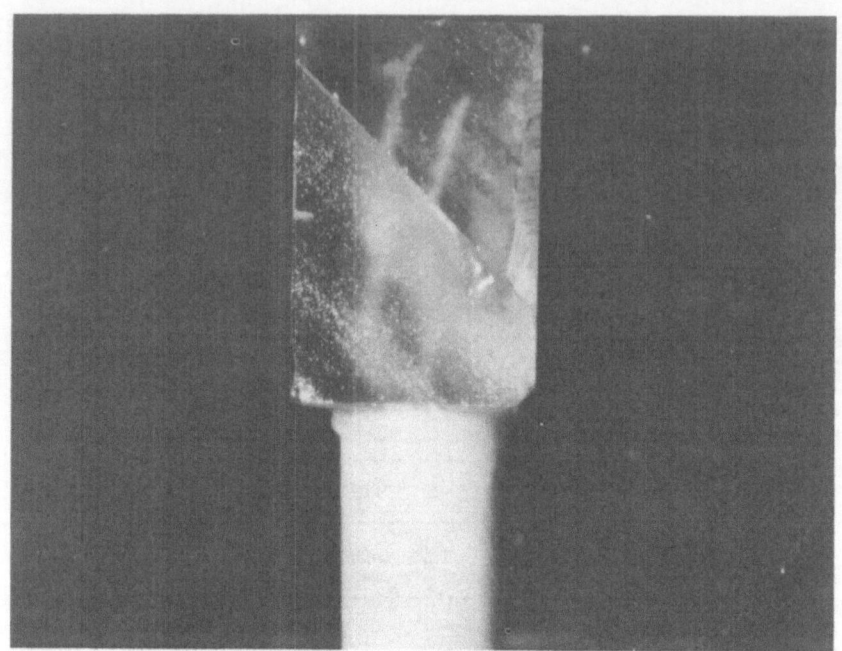

Fig. 10a. Bicrystal S-1-5 before high-temperature compressive testing. 13 ×.

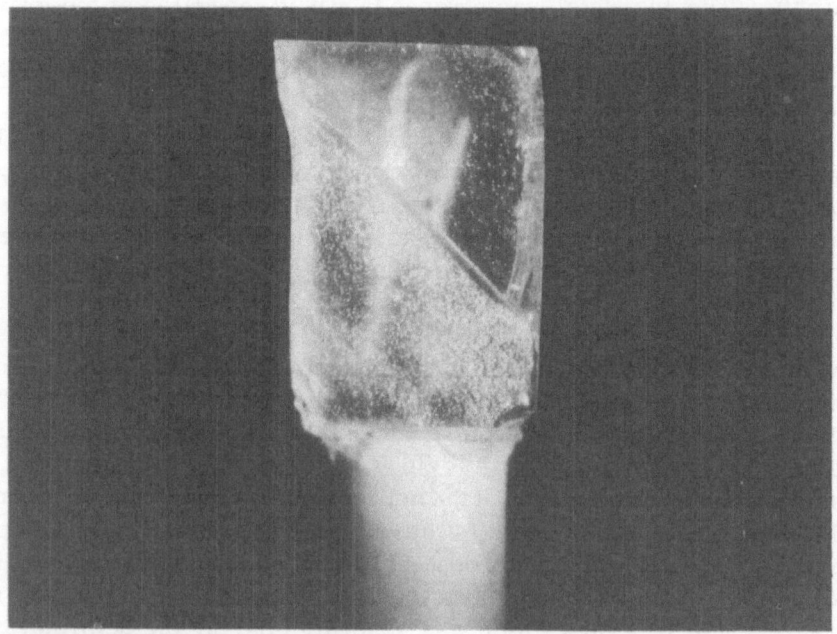

Fig. 10b. Bicrystal S-1-5 after high-temperature compressive testing. 13 ×.

Fig. 11a. Orthogonal view of S-1-5 (through boundary) before testing. 13 ×.

Fig. 11b. Orthogonal view of S-1-5 (through boundary) after high-temperature testing. 13×.

Fig. 12a. Bicrystal S-1-5 viewed through the grain boundary, showing shape and extent of crack formed during compressive shear test. 50 ×; reduced 10% for reproduction.

Fig. 12b. Bicrystal of Fig. 12a viewed from a slightly different angle. 50 ×; reduced 10% for reproduction.

shape and extent of the crack when viewed from two different angles.   Figure 12b also reveals some of the detail in the unparted
grain boundary.   The scratch marks from the original surface
polishing are visible and show no sign of grain boundary sliding.

The crack developed in the bent region of the grain boundary.
If the deformation process had been allowed to continue, the bicrystal would have fractured as a result of stress concentrations
caused by the crack, or by its extension.   It is also quite possible
that bicrystal S-2-1 failed in such a manner, since it was observed
that the adjacent crystals had deformed, and a similar bent portion
of the boundary was noted on the fractured surfaces.

Fig. 13. Fracture surface of bicrystal of type A orientation tested at 1630°C. 215×;
reduced 55% for reproduction.

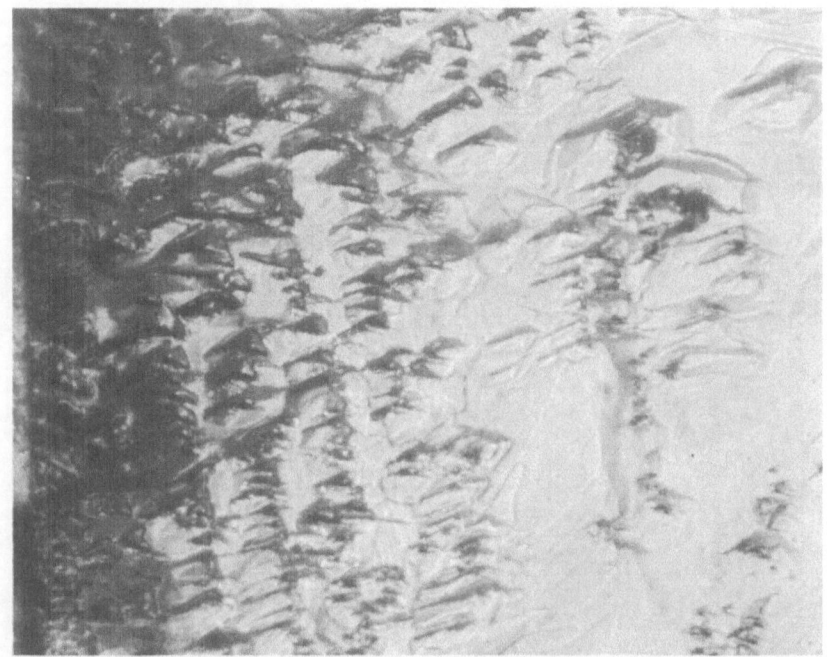

Fig. 14. Fracture surface of bicrystal of type A orientation tested at 1630°C. 540×; reduced 55% for reproduction.

MICROSTRUCTURE OF FRACTURED GRAIN BOUNDARIES

The apparent absence of grain boundary sliding in an alumina bicrystal at 1900°C under a resolved shear stress of 1500 psi (a value three times higher than the resolved shear stress accounting for plastic flow in the single crystals) was quite surprising. Microscopic examination of the fractured grain boundary surfaces shed some light on this subject. Figure 13 shows a portion of the fractured surface of the bicrystal tested at 1630°C. The white flat area on the left is an unhealed region of the bicrystal which was visible before the test was made. The area on the right is in the healed region which was transparent except for small voids. The unhealed region is still flat and smooth, while the healed and fractured region is quite rough. The healed and subsequently fractured region has a very crystallographic character. The pits in the surface have straight edges and are all in the same direction. Figure 14 was taken from the same crystal, but at a higher magnification; the crystallographic nature of the grain boundary is evi-

Fig. 15. Fracture surface of bicrystal of type A orientation which fractured at elevated temperature during bicrystal fabrication. 215×; reduced 55% for reproduction.

denced by the shape and arrangement of the pits which appear to represent the intersection of dislocations with the grain boundary. It appears that the dislocations were pushed into the grain boundary area during bicrystal formation, since the orientation at that time was very favorable for dislocation movement and since some measureable plastic deformation did occur. However, it is interesting that, even after one hour at a temperature just 50 – 75°C below the melting point, the worked structure of the grain boundary did not change appreciably, i.e., the very complex crystallographic boundary did not convert to a microscopically planar one. A more dramatic illustration of this point is shown in Fig. 15. In this instance, the association of dislocations with surface markings is less apparent; however, the crystallographic nature of the boundary is certainly emphasized.

Figures 16a and 16b show the fractured boundary of bicrystal

S-2-1 (180° twist plus 10° tilt boundary). The two photomicrographs were taken from the parted crystal halves at approximately the same position; Fig. 16a was made with dark-field lighting, and Fig. 16b was made with oblique bright-field illumination. A close examination of the two photographs shows that protrusions on one crystal correspond to depressions or pits on the other crystal. The very small white spots in the crystal on the right are considered to be dislocation termini. It is not clear, in the absence of detailed etch pit studies, whether or not the dislocations have traversed the boundary; however, the presence of the dislocations has certainly had an effect on the structure of the grain boundary.

After examining these photomicrographs of fractured grain boundary surfaces, it is possible to account for the absence of grain boundary sliding. The actual area of the boundary which is being loaded is much greater than one would calculate by simple geometry. It is apparent also that the boundary as such is not subjected to a shear loading; instead, the bulk crystals are carrying the load. In order for the two crystal halves to move relative to one another, it would be necessary for the crystals to plastically deform adjacent to the boundary, but this is not easily achieved in the given orientations because of the structure of alumina which limits plastic deformation to basal slip. Any other mechanism by which this type of grain boundary could slide would necessarily involve some diffusion. No evidence of diffusion has been observed within the time periods that these bicrystals were stressed at temperature.

From the work to date on alumina bicrystals, it is apparent that microscopically straight and flat grain boundaries are not possible in high-purity alumina with completely random orientations. There are other factors besides surface preparation and furnacing technique which determine the physical structure of the grain boundary and, thus, dominate its mechanical behavior. These additional factors are related to the crystal structure of the material and the dislocation structure. Interaction is clearly present between dislocations and medium-angle tilt boundaries in clean alumina, a condition which favors the formation of a crinkled, entirely crystallographic surface which very effectively prohibits grain boundary sliding. It is less likely that the same type of interlocking structure will result for pure twist boundaries in alumina; however, this remains to be determined as our research continues.

Fig. 16. Mirror-image surfaces from adjacent crystals of specimen S-2-1 after fracture. 175×; reduced 25% for reproduction. (a) Left surface; dark-field illumination. (b) Right surface; oblique bright-field illumination.

ACKNOWLEDGMENTS

The authors are indebted to the Department of Engineering Research, North Carolina State University, and its director, N. W. Conner, for providing space, facilities, and support for this research.

REFERENCES

1. H. C. Chang and N. J. Grant, "Grain Boundary Sliding, Migration, and Intercrystalline Fracture Under Creep Conditions," Trans. AIME 197:304–312 (1953) and J. Metals (Feb. 1953).
2. J. Intrader and E. S. Machlin, "Grain Boundary Sliding in Copper Bicrystals," J. Inst. Metals 88:305–310 (1960).
3. M. A. Adams and G. T. Murray, "Direct Observation of Grain Boundary Sliding in Bicrystals of NaCl and MgO," J. Appl. Phys. 33(6):2126–2131 (1962).
4. D. M. Martin, G. K. Fehr, and T. D. McGee, "Grain Boundaries in Sodium Chloride," this volume, Chapter 15.
5. M. P. Davis, "Pressure-Sintered Alumina Bicrystals," J. Am. Ceram. Soc. 47(9):463–464 (1964).
6. M. L. Kronberg, "Plastic Deformation of Single Crystal Sapphire, Basal Slip, and Twinning," Acta Met. 5(9):507–523 (1957).

Chapter 18

The Deformation of Single Crystals and Polycrystals of High-Melting Materials

Leo M. Fitzgerald*

Explosive Research and Development Establishment
Ministry of Aviation
Waltham Abbey, Essex, England

A comparison has been made between the modes of deformation of single crystals and polycrystals of some high-melting carbides and oxides. In particular, titanium carbide and aluminum oxide are considered. The contribution of grain boundaries to the low strength properties of the polycrystalline materials at high temperatures is discussed. Detailed examination of surface cracks, caused by the impact of a hard carbide sphere, has been made using both interference microscopy and dislocation etching techniques. Some observations have been made on the effect of temperature on brittle fracture.

INTRODUCTION

A recent paper [1] presents the results of an investigation into the effect of high temperatures on the dynamic hardness of a number of carbides and borides having high melting points. The bulk of the work reported dealt with polycrystalline, hot-pressed, self-bonded specimens. With the exception of boron carbide, all the materials tested showed considerable loss of hardness as the temperature was raised above room temperature. At temperatures around 1000°C, the softening became marked. Boron carbide, on the other hand, retained its room-temperature hardness at temperatures in the region of 1400°C. Density measurements showed that many of the hot-pressed specimens had quite a high degree of porosity — as high as 20% in some cases. However, Brookes and Atkins [2] have shown that hardness (determined by mutual indentation) is virtually independent of porosity at high temperatures, although, for temperatures up to about 500°C, there was a fall in hardness with increasing porosity. These workers have also found that the hardness of these materials is increasingly dependent

*Present address: Morganite Research and Development Ltd., London, England.

on the strain rate at higher temperatures. The dynamic hardness tests described elsewhere [1] involved a very high rate of strain. Atkins [3] has found that, at lower strain rates, the fall in hardness with increasing temperature is considerably more marked and that even boron carbide softens appreciably at temperatures well below 1000°C. Kaufman and Clougherty [4] have reported a similar temperature effect on the hardness of a number of refractory borides. Leipold and Nielsen [5] have shown in a recent paper that the mechanical properties of hot-pressed carbides are strongly influenced by impurities at the grain boundaries.

In an effort to extend our knowledge of the strength properties of the refractory carbides, borides, and oxides and to gain a better idea of the fundamental processes involved as these materials are deformed at very high temperatures, the impact tests described elsewhere [1] were applied, where possible, to single crystals of these materials. The only carbide for which single crystals of dimensions suitable for these tests could be obtained was titanium carbide. A comparison between its impact strength at high temperatures and that of polycrystalline TiC has been given elsewhere [1]. It was seen that, for the high rate of strain applied, the decrease in hardness for the single-crystal material was considerably less than for the corresponding polycrystalline specimen. Permanent indentations did not appear until an impacting temperature of about 1600°C was reached compared with about 800°C for the polycrystals.

The difference between the behavior of the single-crystal and polycrystalline forms, for the same strain rate, may be explained by considering the behavior of the grain boundaries in the polycrystalline specimens. It is possible that some form of grain boundary sliding of the type proposed by Zener [6] is the main mechanism involved in the lowering of the strength properties as the temperature of the polycrystalline material is increased. Wachtman and Lam [7] also have attributed the rapid decrease in strength of polycrystalline materials at high temperatures (above about 1000°C) to grain boundary slip. In the light of the experiments on single crystals, it seems unlikely that the permanent indentations produced on the polycrystalline specimens can be regarded as true plastic deformation of the individual grains of the material, but that they are due, rather, to viscous sliding of the grain boundaries. Recent work by Brookes [8] has confirmed this grain boundary weakness in polycrystalline carbides. However,

Atkins [3] has found, for low strain rates, little difference between the effect of temperature on the hardness of single crystals and polycrystals of TiC. This observation would seem to suggest that grain boundary sliding does not play such an important part in the deformation process, particularly at temperatures up to 1000°C.

Brookes has obtained bulk plastic deformation of single crystals of TiC at temperatures in the region of 1000°C by slip on the {111} planes, in agreement with the observations of Williams and Schaal [9]. In the latter experiments, dead-weight loading was used, and it is likely that the greatly differing strain rates between their experiments and the dynamic tests of Fitzgerald [1] may account for the different temperatures at which plastic flow occurs. Kronberg [10] has in fact shown for single crystals of aluminum oxide that the transition from brittle to plastic flow is a sharp one and that it is sensitively related to the strain rate. In the case of dead-weight loading, we have, effectively, an applied strain rate of zero, and the 800°C reported by Williams and Schaal would correspond to the minimum flow temperature.

The most freely available single crystals of any of the refractory materials are those of synthetic sapphire (Al_2O_3). It is possible to obtain very large crystals of sapphire of known orientation, which affords the possibility of obtaining valuable information concerning the mechanical behavior of single crystals for a range of temperatures. A number of workers [7, 10-14] have studied the behavior of both polycrystalline and single-crystal aluminum oxide. As a result of this work, much is now known about the brittle–ductile transition in alumina and its dependence on temperature, rate of strain, and the presence of grain boundaries. In the course of the dynamic hardness measurements mentioned above, single crystals of Al_2O_3 were impacted over a range of temperatures. A number of interesting observations have been made on the mode of deformation of sapphire under impact conditions. The significance of these observations, in the light of the current knowledge of sapphire, will be discussed in this paper.

DEFORMATION OF SINGLE-CRYSTAL SAPPHIRE DUE TO IMPACT LOADING

A single crystal of synthetic sapphire of cylindrical shape and having a diameter of $5/8$ in. and a length of 1 in. was used in these experiments. The polished planes of the crystal were (22$\bar{4}$3), and

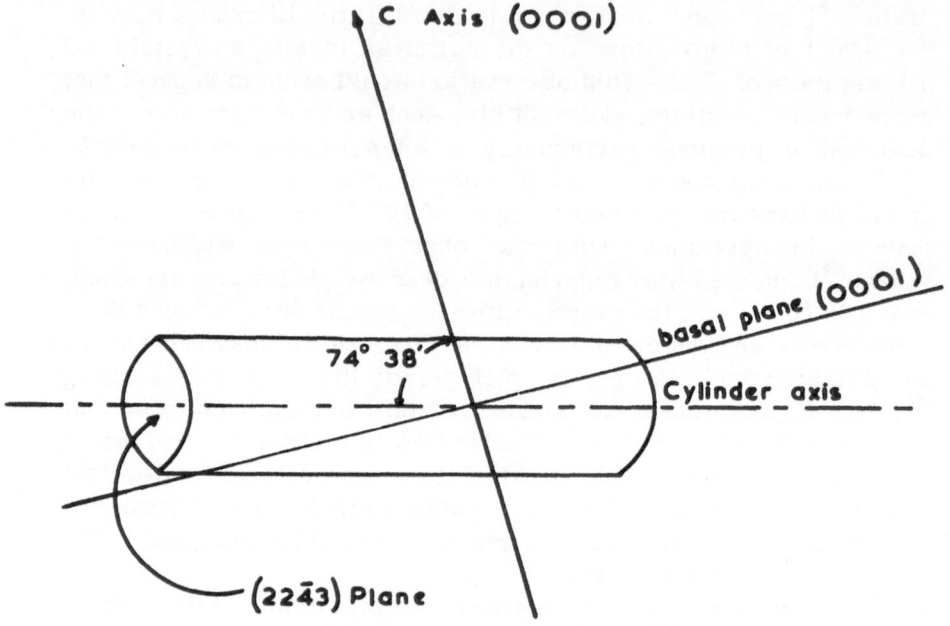

Fig. 1. Orientation of sapphire crystal.

the basal (0001) plane made an angle of 15° 22' with respect to the cylinder axis. The included angle between the $(22\bar{4}3)$ plane and the (0001) plane is 73° 38'. Figure 1 shows the orientation of the crystal. The principle of the dynamic hardness measurements referred to above has been described elsewhere [15]. Briefly, it consisted of measuring the height of rebound of a spherical tungsten carbide indenter at temperatures ranging from room temperature to about 2000°C.

For temperatures up to about 1600°C, the height of rebound of the indenter showed no decrease from the room-temperature value and, in each case, the impact was accompanied by brittle fracture. Other workers have reported plastic flow in sapphire at temperatures as low as 900°C. It was mentioned earlier that Kronberg [10] has shown that the transition from brittle to plastic behavior is a very sharp one in sapphire and is strongly dependent on the strain rate. He found that, for the strain rate obtained with a head velocity of 0.2 in./min, the transition temperature was 1520°C. It is not surprising, therefore, that, for our exceedingly high strain rate, no appreciable reduction in hardness was observed at temperatures up to 1600°C.   On an homologous scale, 1600°C represents $0.8\ T_m$ for sapphire compared to about $0.5\ T_m$ for titanium carbide.

The deformation of the crystal surface due to the impact of the indenter on close examination showed a number of interesting features.   At room temperature, a particularly interesting type of ring crack could be observed.   Figures 2a and 2b show such a crack using transmitted light and reflected light, respectively, for illumination.   The deformation shows a tendency to form the perfect ring crack, such as is obtained in glass.   However, a pair of parallel lines cut across a diameter of the circular pattern and, since they always have a specific crystallographic orientation, they must be either cleavage cracks, mechanical twins, or slip lines. On the side of the parallel marks opposite to that on which the ring-type crack occurred, fracture of a conchoidal nature took place. In some instances, a series of smaller lines parallel to the two principal lines could also be observed.   Figure 3 shows such a crack.

The literature furnishes very little information on the cleavage of sapphire crystals and, in some instances, it has been suggested that sapphire does not exhibit the property of cleavage. It seems more likely that these parallel markings are due either to slip or to lamellar twinning.   Wachtman and Maxwell [16] found that slip occurs in sapphire in the <$11\bar{2}0$> direction on the (0001) plane (basal slip); this was confirmed by Kronberg [17].   Slip on another system {$1\bar{2}10$} <$\bar{1}010$> (prismatic slip) was observed by Klassen-Neklyudova [18].   However, Kronberg [19] has suggested that the observation of slip lines around indentations on sapphire faces is extremely difficult and that the more likely explanation is mechanical twinning.   Stofel and Conrad [13] have recently produced mechanical twins, both (0001) and ($0\bar{1}11$) types, by compression. They found that cracks propagated more readily along the twin boundaries than in nontwinned crystal.

To ascertain the true cause of these line markings, the deformation process was examined more closely.   A spherical indenter, such as was used in the impact hardness measurements, was mounted in the Vickers projection microscope and was slowly pressed into the surface of the sapphire crystal.   It was possible to observe three stages of deformation.   The first stage was the production of the parallel line markings, which was followed by the partial ring crack, and then the conchoidal fracture.   Figure 4 shows the initial stage of deformation on the ($22\bar{4}3$) face of the crystal.

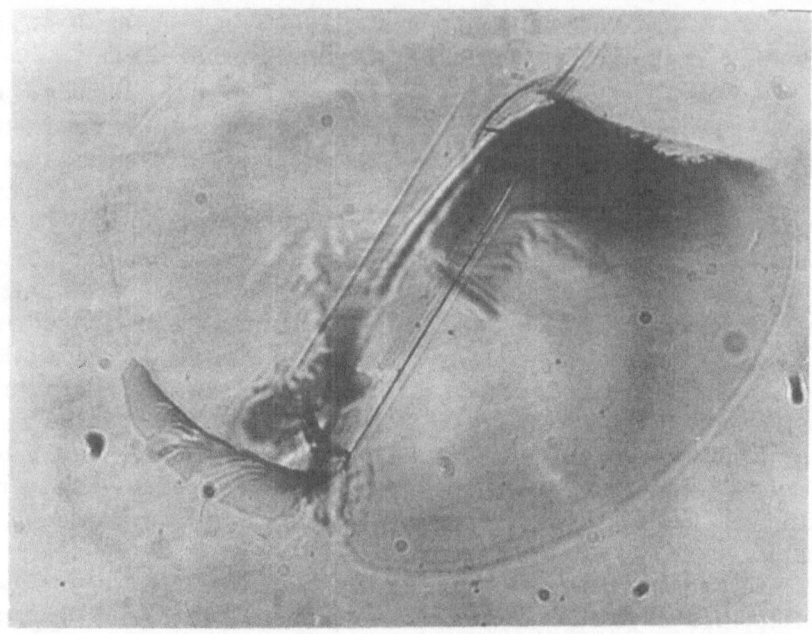

Fig. 2a. Ring crack on sapphire at room temperature seen with transmitted light.

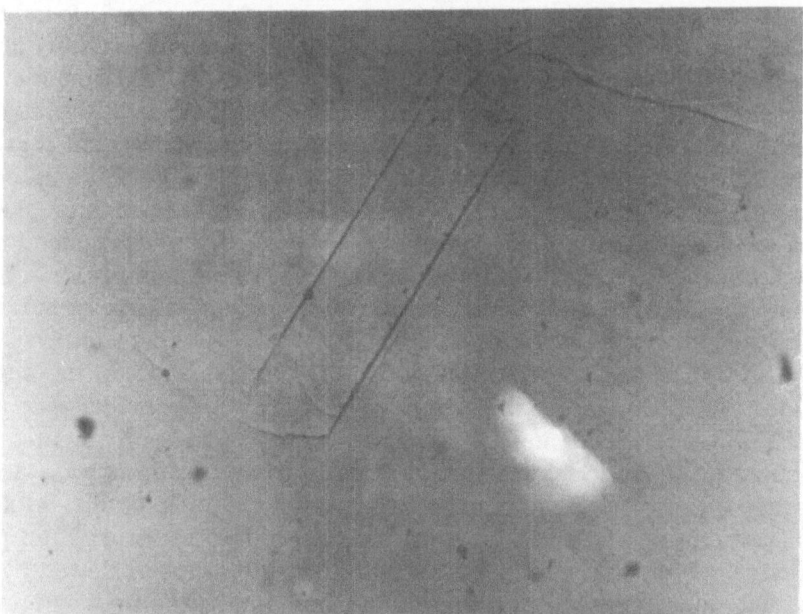

Fig. 2b. Ring crack on sapphire at room temperature seen with reflected light.

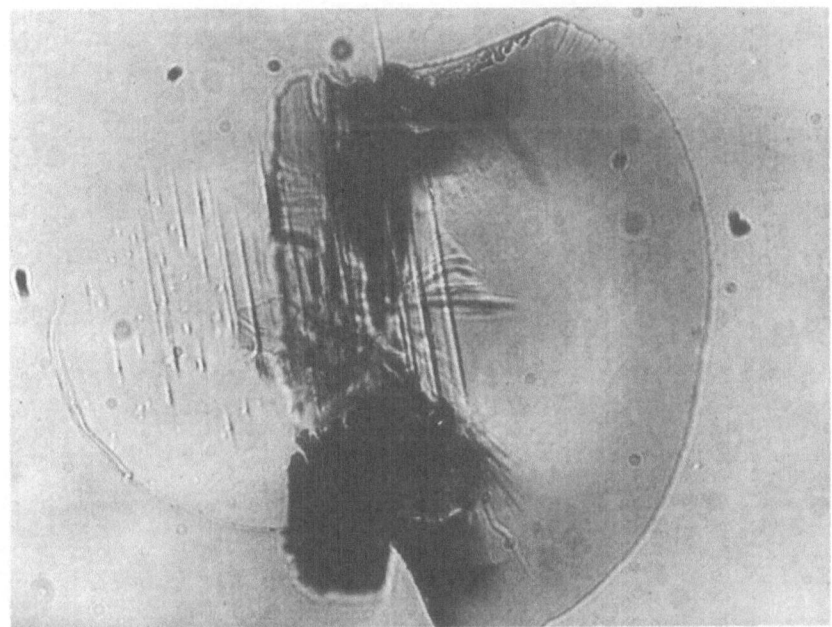

Fig. 3. Room-temperature impact markings on sapphire.

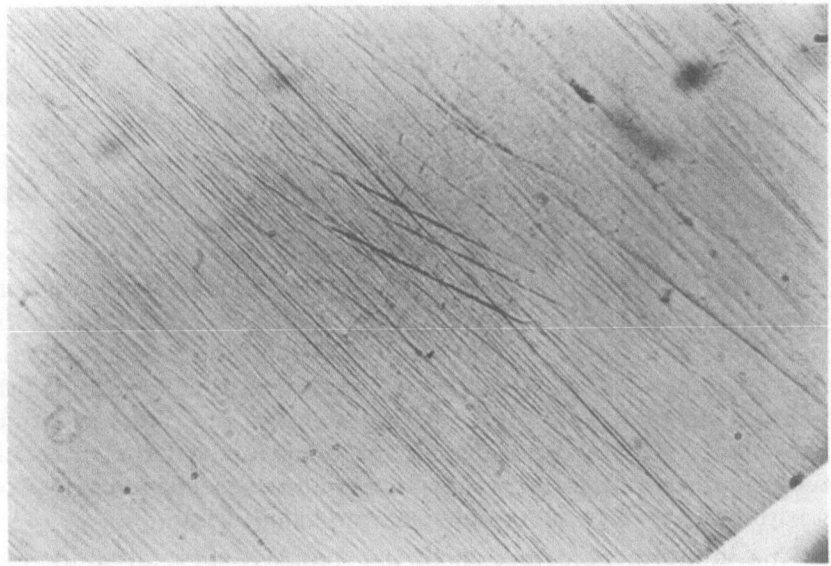

Fig. 4. Initial stage of deformation by impact on sapphire crystal.

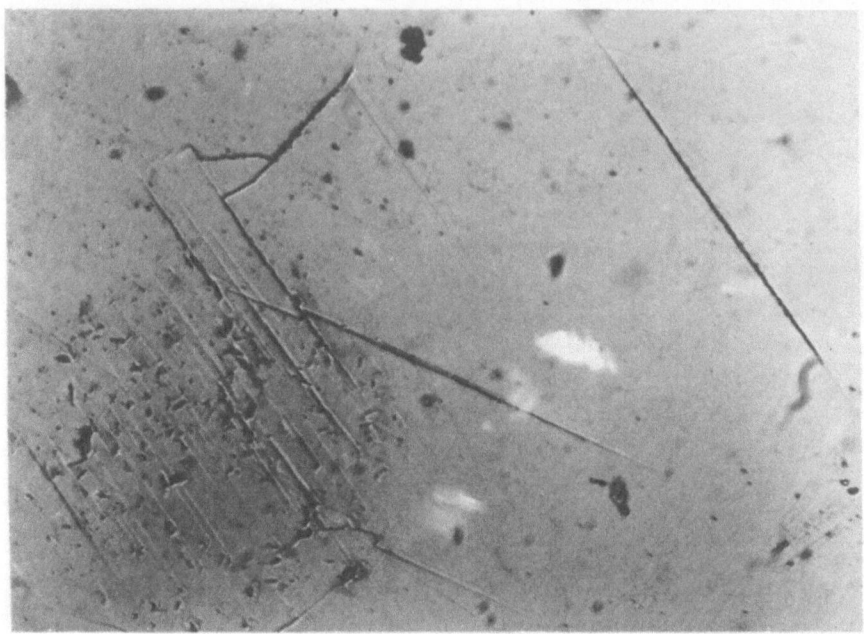

Fig. 5. Deformation of sapphire by impact at 1230°C.

X-ray investigations showed that the direction of these line markings was consistent with the intersection of the basal (0001) plane with the (22$\bar{4}$3) plane. However, a number of other planes have the same line of intersection as (0001) on (22$\bar{4}$3). These are the planes which have the same first two index values (i.e., $h\,k$ in the Bravais–Miller notation $h\,k\,i\,l$; for example, 00il, 11il, 22il, etc.). The values for i and l may vary independently. A new surface was cut and polished parallel to the crystal axis and perpendicular to the direction of the parallel line markings. When this surface was impacted in the same way, a similar deformation process took place. The parallel markings on this new surface made an angle of 15° 25' with the crystal axis, thus confirming that the lines on the (22$\bar{4}$3) plane do represent its intersection with the basal plane. This additional information does not, however, exclusively indicate either slip or twinning as the mechanism involved, as both have been found to occur on the basal plane.

The mode of deformation of the sapphire crystal as the temperature of impact was raised gave additional weight to the hypothesis that mechanical twinning was the cause of the parallel marks observed at room temperature.

Although impacts at temperatures up to about 1600°C showed little change from the nearly completely elastic rebound observed at room temperature, the fracture induced by these impacts remained largely brittle.   However, the microscopic appearance of the deformation marks did change with temperature. The most significant change was that the parallel lines became less accentuated (disappearing altogether for impacts at temperatures above about 1400°C) while the glass-like ring crack became more developed, with minor, concentric ring cracks also appearing. Figures 5 and 6 show the result of impacts at 1230°C and 1470°C, respectively.   In the latter instance, there is no longer any sign of the supposed twin boundaries. The diagonal crack in Fig. 6 is about 80 μ below the surface and can be seen only with transmitted light. By using an interference objective, it can be shown (Figs. 7a and 7b) that a slight permanent indentation has been produced in this case, whereas there is no evidence of plastic flow in the deformations illustrated in Figs. 2a, 2b, 3, and 4. However, it is clear

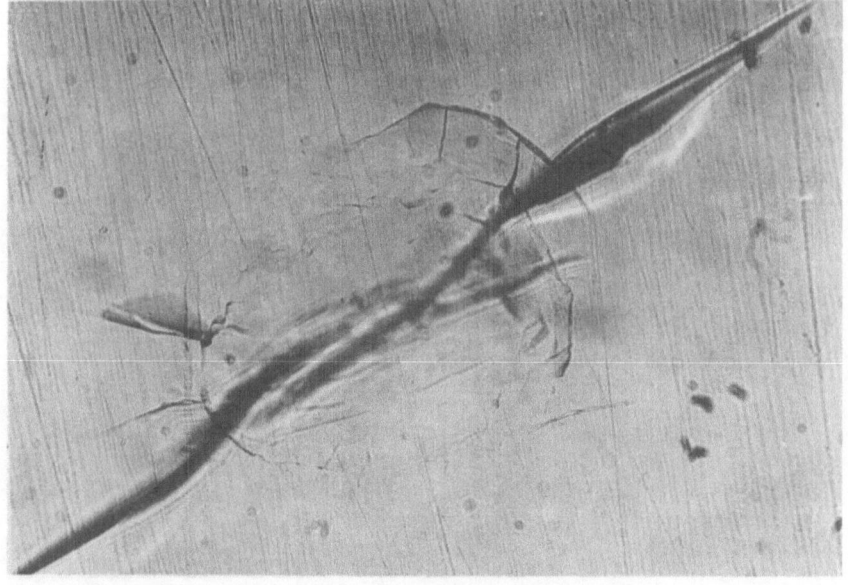

Fig. 6. Deformation of sapphire by impact at 1470°C.

Fig. 7a. Interferogram of surface cracks at room temperature.

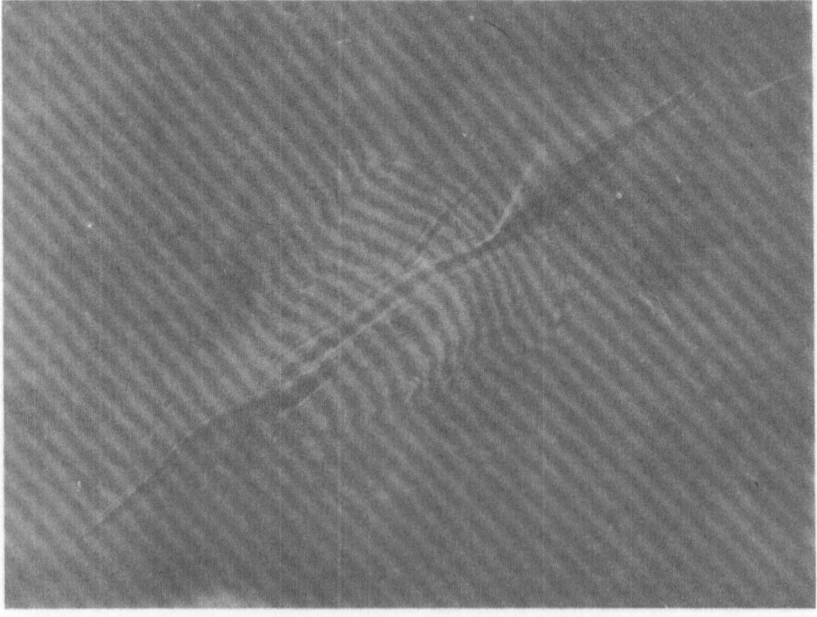

Fig. 7b. Interferogram of surface cracks at 1470°C.

that the cracking induced at 1470°C is still essentially brittle in character.

Alford and Stephens [20] have described an etching technique for revealing dislocations on both the basal and prism planes. They also noted [21] that the etch pits showed a change of orientation across a twin boundary. Using the same etching conditions, viz., fused potassium bisulphate at 675°C, dislocation etch pits were produced on the (22$\bar{4}$3) plane. Figure 8 shows these pits in the region of the room-temperature deformation mark. There appears to be some, although perhaps not conclusive, evidence here of a change in orientation of the pitting across the parallel marks, suggesting that their origin is indeed mechanical twinning.

When the high-temperature cracks are etched in the manner described above, there is no evidence of pitting in any special crystallographic direction and the etch-pit configuration which is observed (Fig. 9) seems to be associated with the onset of plastic flow in the crystal as indicated in Figs. 7a and 7b. These observations are consistent with the theory that it is mechanical twinning rather than slip which gave rise to the line markings observed at room temperature.

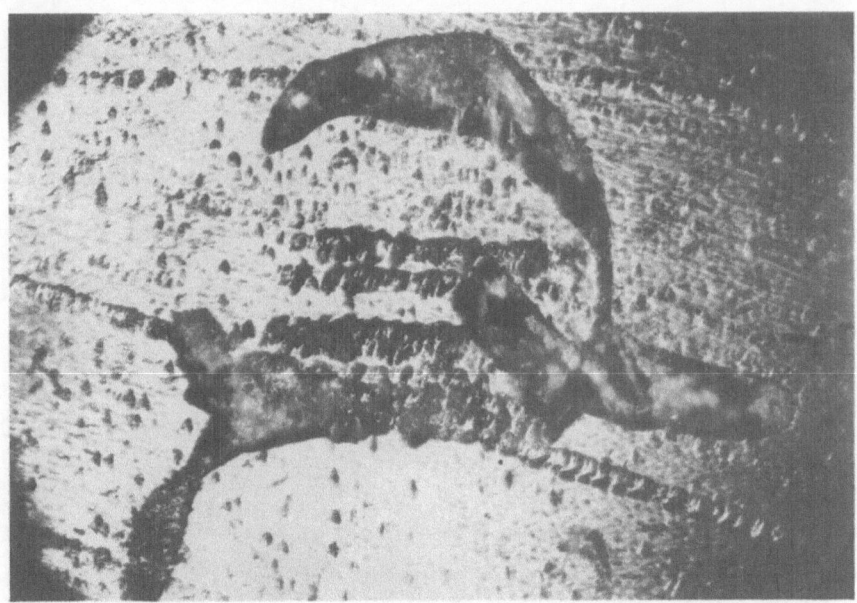

Fig. 8. Etch pits on (2243) surface around room-temperature deformation crack.

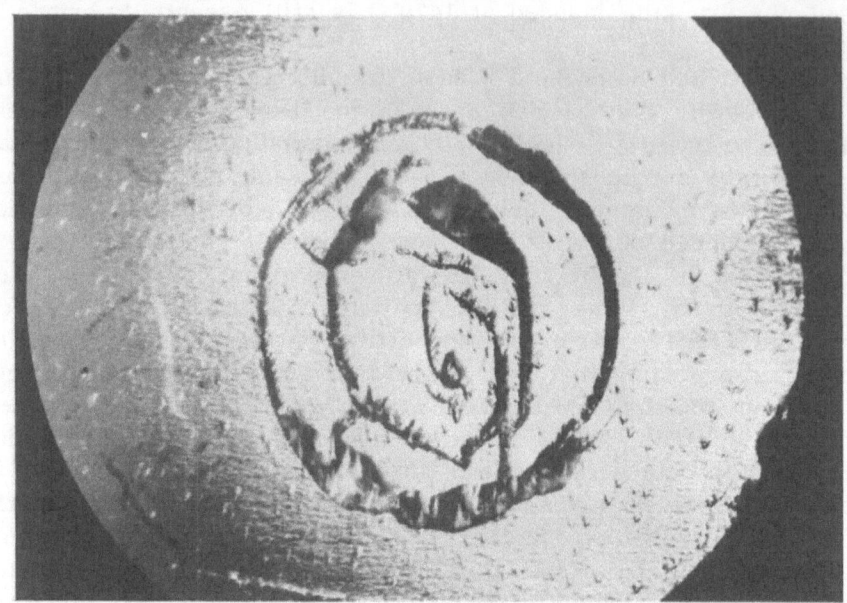

Fig. 9. Deformation of sapphire at 1470°C (etched in KHSO$_4$ at 675°C).

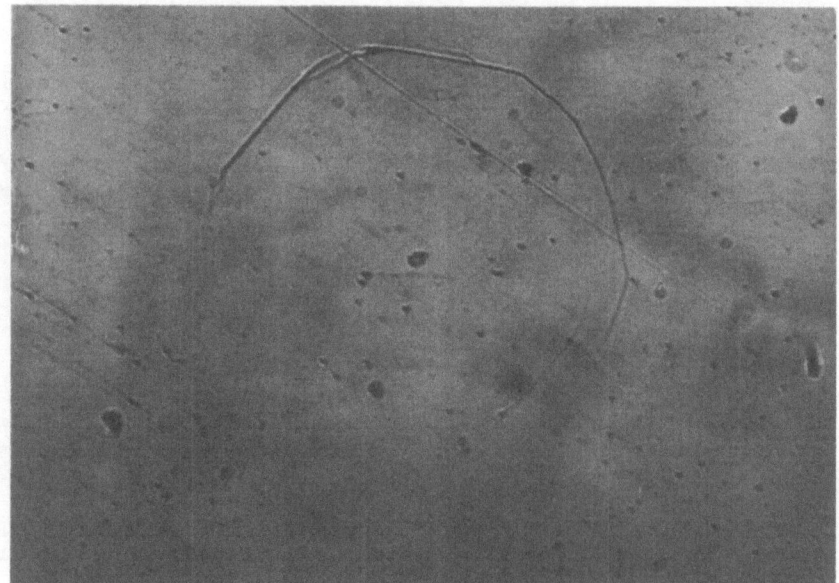

Fig. 10. Hexagonal ring crack on silicon carbide crystal after impact at 850°C.

BRITTLE FRACTURE AT ELEVATED TEMPERATURES

The facilitation of brittle fracture with increase in temperature, as noted in Figs. 5 and 6, suggests a lowering of the free surface energy with temperature. A similar effect has been noted with silicon carbide. Figure 10 shows a single crystal of a-SiC which has been impacted at 850°C on the (0001) basal plane. A hexagonal ring crack appeared at this temperature which was not observed after similar impacts at lower temperatures. The author [22] has also observed this phenomenon with diamond crystals on {111} faces.

ACKNOWLEDGMENTS

The major part of this work was done in the Laboratory of Physics and Chemistry of Solids, Cavendish Laboratory, University of Cambridge. I wish to thank Dr. F. P. Bowden, C.B.E., F.R.S., for his encouragement and advice and also Dr. C. A. Brookes for much valuable discussion.

REFERENCES

1. L. M. Fitzgerald, "The Hardness at High Temperatures of Some Refractory Carbides and Borides," J. Less-Common Metals 5:356-364 (1963).
2. C. A. Brookes and A. G. Atkins, "The Friction and Hardness of Refractory Compounds," Paper No. 44, 5th Plansee Seminar, Reutte/Tyrol, June, 1964.
3. A. G. Atkins, private communication.
4. L. Kaufman and E. V. Clougherty, "Investigation of Boride Compounds for High Temperature Applications," Paper No. 43, 5th Plansee Seminar, Reutte/Tyrol, June, 1964.
5. M. H. Leipold and T. H. Nielsen, "Mechanical Properties of Hot-Pressed Zirconium Carbide Tested to 2600°C," J. Am. Ceram. Soc. 47(9):419-424 (1964).
6. C. Zener, Elasticity and Anelasticity of Metals, University of Chicago Press (Chicago), 1948, p. 150.
7. J. B. Wachtman, Jr., and D. G. Lam, Jr., "Young's Modulus of Various Refractory Materials as a Function of Temperature," J. Am. Ceram. Soc. 42:(5):254-260 (1959).
8. C. A. Brookes, "Friction and Deformation at High Temperature of Titanium Carbide," in: P. Popper (ed.), Special Ceramics 1962, Academic Press (London), 1963, pp. 221-236.
9. W. S. Williams and R. D. Schaal, "Elastic Deformation, Plastic Flow and Dislocations in Single Crystals of Titanium Carbide," J. Appl. Phys. 33(3):955-962 (1962).
10. M. L. Kronberg, "Dynamical Flow Properties of Single Crystals of Sapphire (I)," J. Am. Ceram. Soc. 45(6):274-279 (1962).
11. E. A. Jackman and J. P. Roberts, "On the Strength of Polycrystalline and Single Crystal Corundum," Trans. Brit. Ceram. Soc. 54(7):389-398 (1955).
12. J. B. Wachtman, Jr., and L. H. Maxwell, "Plastic Deformation of Ceramic-Oxide Single Crystals (II)," J. Am. Ceram. Soc. 40(11):377-385 (1957).
13. E. Stofel and H. Conrad, "Fracture and Twinning in Sapphire (Alpha-Al_2O_3 Crystals)," Trans. Met. Soc. AIME 227:1053-1060 (1963).

14. R. M. Spriggs, J. B. Mitchell, and T. Vasilos, "Mechanical Properties of Pure Dense Aluminium Oxide as a Function of Temperature and Grain Size," J. Am. Ceram. Soc. 47:(7):323-327 (1964).
15. L. M. Fitzgerald, "The Measurement of Hardness at Very High Temperatures," Brit. J. Appl. Phys. 11:551-554 (1960).
16. J. B. Wachtman, Jr., and L. H. Maxwell, "Plastic Deformation of Ceramic Oxide Single Crystals," J. Am. Ceram. Soc. 37(7): 291-299 (1954).
17. M. L. Kronberg, "Plastic Deformation of Single Crystals of Sapphire—Basal Slip and Twinning," Acta Met. 5:507-524 (1957).
18. M. V. Klassen-Neklyudova, "Plastic Deformation of Crystals of Synthetic Corundum," J. Tech. Phys. (USSR) 12:519-534, 535-551 (1942).
19. M. L. Kronberg, discussion of paper by H. Palmour, III, W. W. Kriegel, and J. J. Duplessis, in: W. W. Kriegel and H. Palmour, III (eds.), Mechanical Properties of Engineering Ceramics, Interscience (New York and London), 1961, pp. 329-330.
20. W. J. Alford and D. L. Stephens, "Chemical Polishing and Etching Techniques for Al_2O_3 Single Crystals," J. Am. Ceram. Soc. 46(4):193-194 (1963).
21. D. L. Stephens and W. J. Alford, "Dislocation Structures in Single-Crystal Al_2O_3," J. Am. Ceram. Soc. 47(2):81-86 (1964).
22. F. P. Bowden and D. Tabor, "The Friction and Lubrication of Solids, Part II," Oxford University Press (London), 1964, p. 163.

PART IV. Grain Boundary Contributions to Strength and Thermomechanical Behavior

J. A. PASK, Presiding

University of California
Berkeley, California

Chapter 19

Electron Microbeam Probe Techniques for Studying Grain Boundaries in Polycrystalline Ceramics

T. J. Gray and J. K. Zope*

State University of New York College of Ceramics
Alfred University, Alfred, New York

The techniques of electron microbeam probe surface analysis are reviewed with particular emphasis on the special problems encountered with ceramics and minerals. Some limitations are discussed and methods of overcoming these proposed. The unique applicability of this technique to the detailed investigation of grain boundary phenomena is emphasized. Preliminary data on the interaction between molten metals and refractories are presented.

INTRODUCTION

Since the focusing of X-rays by lenses is impracticable, alternate techniques are necessary for the development of satisfactory X-ray microscopy. Of the various methods currently employed, the technique employing a flying spot scanning system similar to familiar television techniques has evoked the most attention. The technique is simply X-ray spectrometry on a microscale using a very small focused electron beam to excite X-ray emission from the specimen. The specimen current, back-scattered electrons, or X-ray emission may all be used to modulate a synchronized cathode-ray display tube or to operate suitable counter and recorder systems. These instruments, generally classified as electron microbeam probes, can therefore be used to provide both a visual display of the specimen and a detailed quantitative microanalysis of its composition.

The static probe instruments were developed almost simultaneously by Castaing and Guinier [1] in France and by Browskii [2] in Russia. Subsequently, however, the important advantages of

*Dr. Zope is Visiting Professor of Physics from Saugor University, Saugor, India.

scanning techniques were rapidly appreciated in the work of Cosslett and Duncumb [3]. Many excellent reviews have been published on both the instrumentation and application of this technique, e.g., those by Wittry [4], Duncumb [5], Castaing [6], and Cosslett [7]. Although electron microbeam probe analysis has been widely used in metallurgy, its application to ceramics has been considerably restricted.

PRINCIPLES OF MICROBEAM PROBE ANALYSIS

In a typical system an electron beam of 10^{-5} to 10^{-8} A accelerated to 5 to 50 kV is focused by a system of two or more electron lenses to a spot size of 0.1 to $1\,\mu$ at the specimen surface. The volume element in the immediate vicinity of the spot emits a complex X-ray spectrum characteristic of the immediate environment which may then be investigated by typical X-ray spectroscopic techniques. Generally, the electron beam is deflected in a raster pattern, and the specimen current or back-scattered electrons are employed to modulate a synchronized cathode-ray tube display system. In a typical case, the characteristic X-ray intensity for a particular element is simultaneously monitored by a relatively conventional bent crystal monochromator and counter system.

Penetration of the specimen by the electron beam is a function of the energy of the incident electron beam and the material parameter (atomic number/atomic weight). Some incident electrons will be back-scattered while the remainder will produce continuous and characteristic line emission and some heating. The critical range within the specimen for the production of X-rays varies considerably over a range of $5\,\mu$ or more for aluminum and lighter elements at 30 kV to less than $1\,\mu$ for heavy elements under the same conditions.

Castaing [6] has shown that for first approximation the line intensity of a particular element in the specimen relative to the corresponding intensity of a pure-metal reference is proportional to the mass concentration in the specimen. This assumes that the stopping power of equal masses of different elements is roughly equal, which is far from accurate, and also ignores X-ray absorption in the sample itself and the secondary fluorescent emission excited both by characteristic and continuous emission generated within the sample. The first assumption is susceptible to a relatively simple correction based on the introduction of a stopping power coefficient which decreases for increasing atomic number. The stopping power may be calculated after the work of Poole and Thomas [8] and may vary by a factor of two between light and heavy elements. A further improvement can also be achieved by making allowance for that portion of the electron beam back-scattered with energy greater than the critical excitation energy. This error increases with atomic number and can be compensated by introducing a back-scattering coefficient which represents the ratio between the intensity which would be emitted with no back-scattering to that which is actually emitted. This leads to a second approximation expression (after Castaing [6])

$$I_A/I_{(A)} = a_A \cdot C_A/\Sigma a_i C_i \tag{1}$$

where I_A and $I_{(A)}$ are the intensities for the same characteristic radiation in the sample and pure element, respectively; C_A is the mass concentration of the element in the sample; and $a_i = S_i \cdot \lambda_i$, i.e., the product of a stopping power coefficient S_i and a back-scattering coefficient λ_i.

These preliminary observations relate to the intensities of characteristic radiation generated within the specimen, and it is necessary to consider how these relate to the measured intensities at the X-ray spectrometer detector. In addition to the characteristic line radiation, the typical bent crystal spectrometer receives some continuous radiation which is small. Primary radiation will, however, have been lost by absorption in the specimen, while fluorescent radiation may have been occasioned both by the continuous radiation or by characteristic radiation from some other element which may be present. The absorption correction may be very large, particularly if the characteristic line is of shorter wavelength than that of the absorption edges of the matrix material. The enhancement can also be very significant, as high as 20 to 30%, in typical cases. Corrections for target absorption have been developed by Castaing [9], Birks [10], and Philibert [11] and studied experimentally by Green [12].

The corrections for fluorescent emission are elegantly reviewed by Duncumb [13], after the work of Castaing [14], Wittry [15], and Birks [16]. Application of these various corrections permits the fundamental development of a scheme of accurate microanalysis by means of the electron microprobe technique. However, there are still very profound difficulties involved in the application of these techniques when phase boundaries are encountered.

THE GRAIN BOUNDARY PROBLEM

When measurements by electron microbeam probe are undertaken in the vicinity of a grain boundary, many complicating factors are introduced. Consideration of the theoretical and experimental papers to which reference has been made will emphasize the difficulties involved in the derivation of corrections for homogeneous phases. Grain boundary intersections are often regions of heavy enrichment. At or near a grain boundary, electron penetration occurs in a region where there is usually a significant variation in concentration of the elements to be analyzed so the apparent concentration will, in any event, be averaged. In addition, the absorption and fluorescent corrections can be determined only to a reasonable degree of accuracy for a homogeneous body. An error may be introduced when the region traversed by the emerging X-rays is of different composition than the area analyzed. The absorption errors can be minimized by high take-off angle and by choosing the orientation of the specimen properly.

The fluorescent errors allow a serious degree of uncertainty. A fraction of the measured X-ray intensity consists of fluorescence excited by continuous radiation and sometimes by characteristic radiation from other elements. The volume from which the fluorescence radiation originates is much larger than that penetrated by the electrons producing the primary radiation. Inhomogeneities in this volume give rise to errors in concentration measurements. The fluorescent uncertainty occurs when measurements are made close enough to a phase boundary. This effect is most serious when the analyzed element is in small concentration in one phase and in large concentration in the adjacent phase. Only when the composition of the phases and the angle between the interface and the surface are known is it possible to calculate the necessary correction, and even then significant uncertainty remains.

While the fluorescent uncertainty is very slightly increased at high take-off angle, this is insignificant compared with the advantage realized in reduction in absorption correction and the effect of surface irregularities.

The electron probe microanalyzer measures the characteristic X-rays from a layer a few microns thick on the specimen surface. This layer must be representative of the volume. Several factors about the surface are very critical and they include flatness of surface, electrical conductivity, and X-ray take-off angle. Etching and repolishing are often performed in metallurgical practice, but this is harmful at grain boundaries where it may not only remove some constituents, but may also redeposit other elements. Residual unevenness at grain boundaries, even after most careful polishing, introduces a significant error. This is essentially avoided by the use of a normal incidence beam.

For electrical conductivity, the common method is to evaporate a thin layer of metal onto the ceramic specimen surface. The metal used should not be one of the elements present in the sample. The layer should be very thin, so that the electrons will have no difficulty in reaching the surface. Layers of the order of 100 A are normally used. They are prepared by the technique used for shadowing electron microscope specimens. Carbon also can be used for the deposition. One of the advantages of carbon coating in place of metal film lies in the fact that it obscures less of the surface details as seen in reflected light.

It will be apparent from these studies that, for a study of phase and grain boundaries, a high take-off angle is most advantageous. Fixed-angle instruments (usually with 25–35° range) are at a serious disadvantage in this respect. Suitable selection of angle and orientation can minimize errors of the type experienced when investigating grain boundary phenomena, particularly when operating at a high angle. One form of probe modification to overcome limitations of this type has been developed by Shirai and Onoguchi [17]; a magnetic deflecting system is employed to bend the electron beam through approximately 90° and is illustrated in Fig. 1. It has been shown that for a copper–aluminum alloy with less than 5% aluminum the AlK_{α} line is hardly detectable at a take-off angle of 10°, whereas at normal take-off angle it is clearly detectable.

A further advantage from this approach is the minimizing of the white-radiation background with the corresponding improvement in sound/noise ratio. It has been shown by Shirai that the intensity of white radiation in the 90° direction is almost one-third that in the 10° direction for copper–aluminum alloy at an electron energy of 12 kV. The absorption of characteristic X-rays in the sample itself is reduced. This increases not only the accuracy of quantitative analysis, but also the sensitivity of light-element detection in a heavy matrix. Unfortunately, these improvements are not achieved without some disadvantages. The working distance of the second electron lens is materially increased. Reduction of the

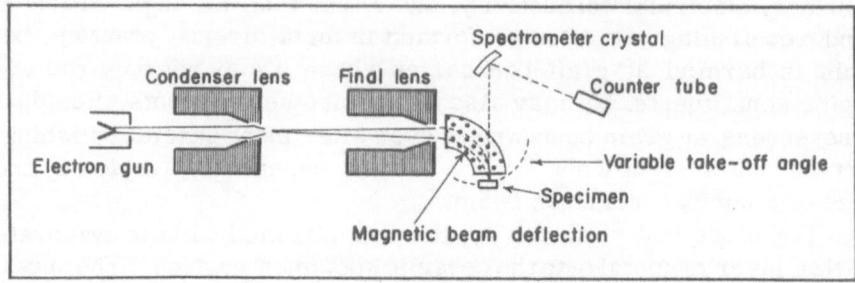

Fig. 1. High take-off angle microprobe for grain boundary investigations (after Shirai and Onoguchi [17]).

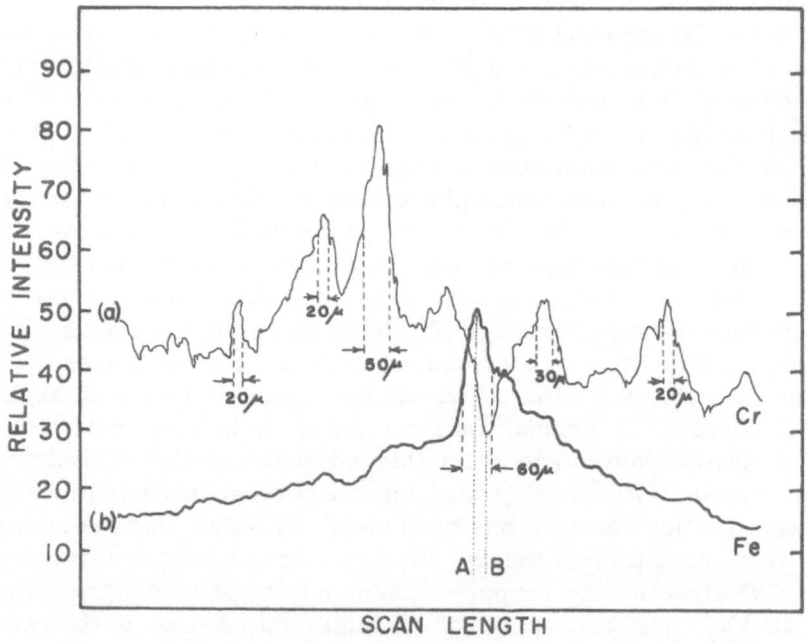

Fig. 2. Microprobe X-ray analysis for chromium and iron on chrome–magnesite ceramic with iron penetration of grain boundary: (a) Cr K_α (69.5°); peak intensity, 5000 counts/sec; (b) Fe K_α (57.7°); peak intensity, 20,000 counts/sec. LiF monochromator crystal, $R = 16.90$ in.; Geiger–Mueller detector, 30-kV accelerating voltage, background intensity, 25 counts/sec.

electron beam current is caused by the spherical aberration of the second condenser lens and astigmatism is introduced by the deflecting field. These disadvantages may eventually be overcome by increasing the beam current which is at present usually reduced by a factor of ten in this type of instrument and by correcting for astigmatism and spherical aberration, after Archard [18].

The difficulty of extending X-ray microanalysis techniques to the light elements is well known. The problem is one of detecting and discriminating between the wavelength K lines with an energy resolution good enough to distinguish between the elements with an overall quantum efficiency high enough to produce scanning images at a useful rate. Dolby [19] has used, for detection and discrimination, proportional counters followed by a pulse analysis method. It was used successfully to produce scanning images of synthesized specimens containing the elements magnesium, aluminum, and silicon. For light elements, a thin (1000–2000 A) collodion window was used. The film is sufficiently strong to withstand the counter gas pressure and gives 90% transmission for carbon, nitrogen, and oxygen wavelengths. Because of absorption, the detection of light elements in a heavy-element matrix is much more difficult than in

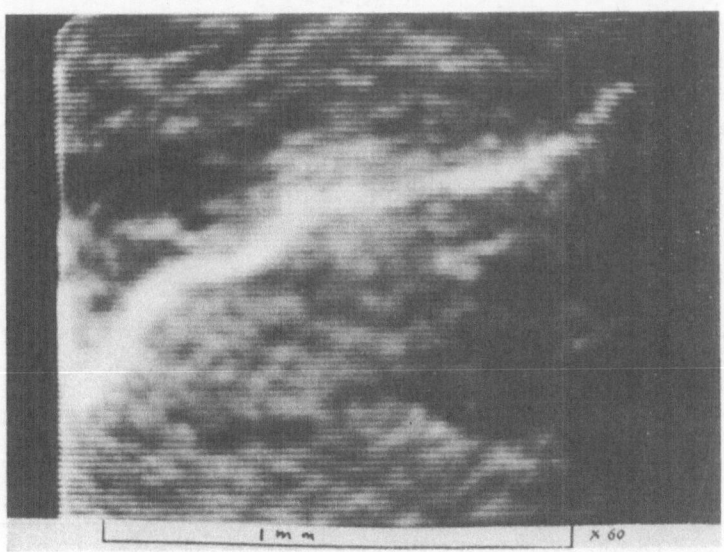

Fig. 3. Visual display of chrome–magnesite specimen showing iron penetration of grain boundary.

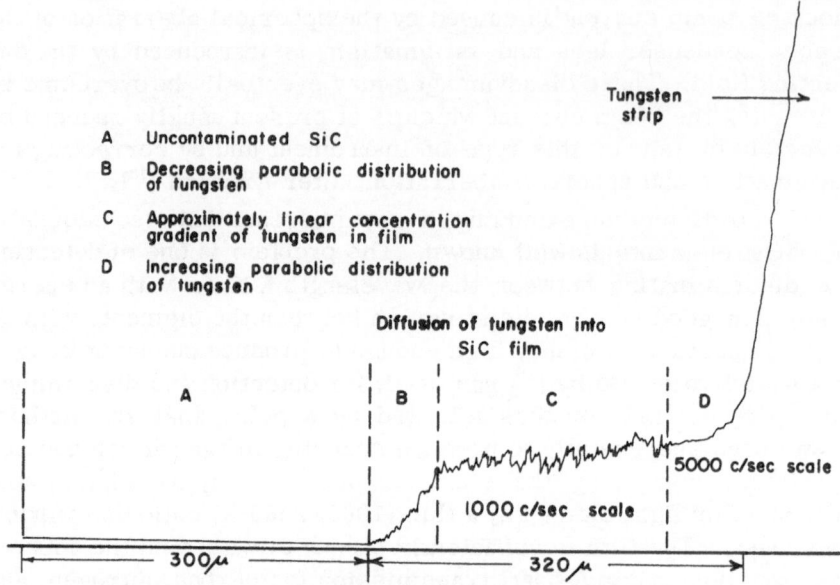

A   Uncontaminated SiC

B   Decreasing parabolic distribution
     of tungsten

C   Approximately linear concentration
     gradient of tungsten in film

D   Increasing parabolic distribution
     of tungsten

Fig. 4. Microprobe analysis for tungsten in pyrolytic silicon carbide coating, on tungsten strip.   Sectioned specimen, W K_α(36.6°); LiF monochromator crystal, 30 kV, 0.2 μA.

a light matrix. Moreover, there are no mass absorption coefficients available for the very long wavelengths in heavy elements. For the detection of carbon, the low efficiency of the counter is usually the most troublesome problem.   To improve detection efficiency, Nagatani and Sakaki [20] made their study on a proportional counter having a hot wire anode. By means of the temperature rise of the

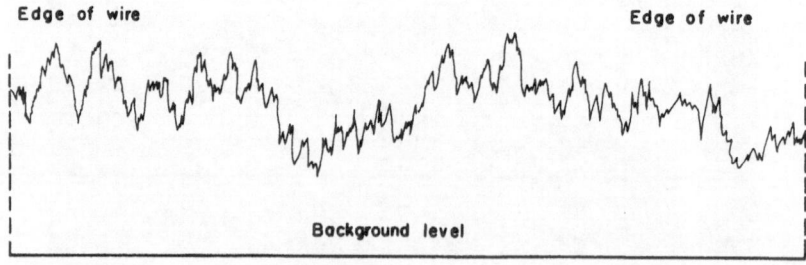

Fig. 5. Microprobe analysis for tungsten in pyrolytic silicon carbide coating on tungsten wire. Emission observed is from subsurface contamination layer, and is not as informative as that from sectional analysis. Crystallinity of the film is readily observed.

portion of the enclosed gas around the hot anode, efficiency of ion multiplication by gaseous molecules may be increased without deteriorating the sound/noise ratio. The windows of the counter consist of $6\,\mu$ mylar and collodion film of thickness less than 5000 A.

EXAMPLES OF MICROBEAM PROBE ANALYSIS

Iron Penetration into Chrome-Magnesite Ceramics

Some indication of the difficulties experienced in the investigation of a typical refractory ceramic body is given in the X-ray scan shown in Fig. 2, corresponding to the visual display in Fig. 3. These data were obtained on a modified Elion microbeam probe with a take-off angle of 37.5°. In this case, the sample is a section from a chrome–magnesia thermocouple protection tube that has been exposed to molten iron. The entry of the iron into the major grain boundary at A is clearly observed as is the corresponding reduction of chromium concentration in this region. Displacement of the points A and B indicates that the iron-filled grain boundary is at an angle to the surface. Variations in chromium concentration are indicative of the individual crystallites present in the composite body.

Interface Diffusion of Tungsten into Pyrolytic Silicon Carbide

Whereas the principal object of this presentation has been to emphasize the importance of microbeam probe techniques in studying grain boundary phenomena, it is expedient to consider its application to interface characterization. The technique possesses unique advantages in investigations of this nature, as exemplified in Fig. 4, which illustrates the distribution of tungsten contamination in a pyrolytic silicon carbide film on a tungsten strip substrate. The silicon carbide film is applied to the tungsten strip by the thermal decompositions of a mixture of silane and hydrocarbon at temperatures in the range 1200–1500°C for varying periods of time. Examination of the analysis of a typical section establishes conclusively that there is a region of carbide contamination by tungsten stretching outwards from the tungsten–carbide interface for a considerable distance. At the metal interface, the concentration of tungsten varies parabolically, indicating a typical diffusion mechanism from the bulk metal into the carbide. There is essentially a uniform concentration gradient in the contaminated region at the edge of which the tungsten concentration falls parabolically to a very low value (below 10 ppm) in the outer uncontaminated silicon carbide layer.

It will be evident from these data that the rate of diffusion of tungsten into the contamination region and the attenuation at the contaminated—uncontaminated carbide interface can be studied in detail by this technique.   Fundamental data can, therefore, be obtained under conditions not readily amenable to alternate methods of investigation.   The extension of this technique into the many different areas of diffusion and corrosion processes to derive detailed fundamental kinetics can readily be accomplished.

Figure 5 represents the type of analysis obtained from the surface of the identical pyrolytic silicon carbide on tungsten where it will be observed that penetration of the electron beam is exciting the subsurface contaminated layer.   However, such a display is inadequate for an understanding of the nature of the deposit.   Nevertheless, this analysis does suggest the microcrystallinity of the contamination layer which is not so clearly evident from the sectional analyses, perhaps due to the growth of platelets parallel to the surface of the tungsten substrate.

These observations indicate an important area of application for the electron microbeam probe technique. The considerations already discussed relative to quantitative investigations apply equally in these cases although they may have lesser significance when deriving kinetic relationships based on rate of change under specific conditions.   There is no doubt that these techniques will see major developments in the years ahead.

REFERENCES

1. R. Castaing and A. Guinier, Proceedings of the Conference on Electron Microscopy, Delft, 1950.
2. I. B. Borovskii, Problems in Metallurgy, Acad. Sci. USSR (1953), p. 135.
3. V. E. Cosslett and P. Duncumb, Proceedings of the European Conference on Electron Microscopy, Stockholm, 1956; P. Duncumb, Ph. D. Thesis, Cambridge University, England (1957).
4. D. B. Wittry, J. Appl. Phys. 29: 1543 (1958).
5. P. Duncumb, Brit. J. Appl. Phys. 10: 920 (1959); 11: 169 (1960).
6. R. Castaing, Advan. Electron. and Electron Phys. 13: 317 (1960).
7. V. E. Cosslett, Met. Rev. 5: 225 (1960).
8. D. M. Poole and P. M. Thomas, J. Inst. Metals 90: 228 (1961).
9. R. Castaing and J. Descamps, J. Phys. Radium 16: 304 (1955).

10. L. S. Birks, J. Appl. Phys. 33: 233 (1962).
11. J. Philibert, in: H. H. Pattee, V. E. Cosslett, A. Engstrom et al. (eds.), X-Ray Optics and X-Ray Microanalysis, Academic Press, Inc. (New York), 1963.
12. M. L. Green and V. E. Cosslett, Proc. Phys. Soc. (London) 78: 1206 (1961).
13. P. Duncumb and P. K. Shields, in: H. H. Pattee, V. E. Cosslett, A. Engstrom, et al. (eds.), X-Ray Optics and X-Ray Microanalysis, Academic Press, Inc. (New York), 1963.
14. R. Castaing, Ph.D. Thesis, University of Paris, 1951.
15. D. B. Wittry, Ph. D. Thesis, California Institute of Technology, 1957.
16. L. S. Birks, J. Appl. Phys. 32: 387 (1961).
17. S. Shirai and A. Onoguchi, in: H. H. Pattee, V. E. Cosslett, A. Engstrom, et al. (eds.), X-Ray Optics and X-Ray Microanalysis, Academic Press, Inc. (New York), 1963.
18. G. D. Archard, J. Appl. Phys. 32: 1505 (1961).
19. R. M. Dolby, Brit. J. Appl. Phys. 11: 64 (1960).
20. T. Nagatani and Y. Sakaki, Rev. Sci. Instr. 33 (5): 556 (1962).

Chapter 20

Grain Size Effects in Polycrystalline Ceramics

R. M. Spriggs,* T. Vasilos, and L. A. Brissette

Avco Corporation
Wilmington, Massachusetts

The interrelationships of pressure-sintering processing conditions in the graphite-die, ceramic-die, and ultrahigh pressure regimes, resulting microstructures, and mechanical properties are described, particularly for dense, pure, submicron and near-submicron grain size ceramic oxides. Strength increases with decreasing grain size appear to be limited by structural subtleties which result from the processing techniques employed. Very fine-grained, highly dense ceramics, as initially prepared, have strengths which are less than would be predicted on the basis of decreasing grain size. However, they respond to additional processing variations and post-fabrication treatments with changes in surface and interior structure and considerable relative strengthening.

INTRODUCTION

It is currently recognized that an increased final density and fine grain size are of particular importance with respect to many of the properties of single-phase, polycrystalline ceramics. In particular, mechanical properties, such as tensile, bend and compressive strengths, and elastic moduli, have been shown to have a strong dependence on relative density, and, in some instances, increase exponentially with it. Further, while the precise effect of grain size on mechanical properties has not yet been completely established for ceramic systems, available data for alumina [1-6], magnesia [2,5,7,8], beryllia [9-12], spinel [13], and perhaps thoria [14] and chromium carbide [14] do indicate increasing strength with decreasing grain size.

Complicating factors, such as surface condition [7], stress corrosion or static fatigue [3], residual strains [12], an apparently beneficial role played by relatively large pores [6], and even difficulties in measuring grain size, tend to obscure the intrinsic relationship between strength and grain size. Nonetheless, for

*Present address: Lehigh University, Bethlehem, Pennsylvania.

313

TABLE I

Uniaxial Pressure-Sintering Die Materials

Die material	Maximum use temperature (°C)	Maximum pressure (psi)	Remarks
Graphite	2500	10,000	Inert atmosphere usually required
Al_2O_3	1200	30,000	Difficult to machine;
ZrO_2	1180	unknown	careful alignment and
BeO	1000	15,000	loading needed; limited
MgO	unknown	unknown	thermal shock resistance; creep limited
SiC	1500	40,000	Difficult to machine;
TaC	1700	8,000	reactive with some
WC, TiC	1400	10,000	materials; expensive
TiB_2	1200	15,000	Difficult to machine; expensive
W	1500	3,500	Easily oxidized;
Mo	1100	3,000	creep limited
Inconel X, hastelloy, stainless steels	1100	varies	Used primarily for halides; creep limited

essentially purely brittle ceramics, such as alumina, where little or no evidence of plastic deformation during fracture exists, it is now believed that brittle fracture is controlled by a Griffith–Orowan mechanism in which pre-existing surface defects, produced during material preparation, are propagated to fracture at a critical level of stress given by the Griffith–Orowan equation, with the crack length being equal to the grain size, within a factor of one to three [6].

It is thus tentatively established that a major route toward improved mechanical properties is the development of fabrication techniques which result in pore-free polycrystalline ceramics with ever-decreasing grain sizes. In this regard, the pressure-sintering (or hot-pressing) process, with its demonstrated capability for grain growth control coupled with very high relative density, is known to be a very useful technique for furnishing high-strength, polycrystalline ceramic materials. Indeed, many of the highest strengths and elastic moduli for polycrystalline ceramics (e.g., oxides, borides, carbides, and nitrides) have been obtained with pressure-sintered materials.

Most pressure-sintering has been performed with graphite as the die and plunger material, though not without some disadvantages. Contamination due to corrosion and wear is a problem, particularly with high-purity materials. The low strength of graphite limits applied pressures usually to less than 10,000 psi.

To overcome these limitations, those working with pressure-sintering have used dies and plungers constructed from ceramic materials, such as oxides, carbides, borides, and refractory and oxidation-resistant metals and alloys. Table I is a representative listing of the materials being used or considered for use as die materials [15,16], and includes their present upper limits of pressure and temperature. With such dies, many ceramic materials have been pressure-sintered (at lower temperatures, but higher pressures) to higher relative densities and finer grain sizes than previously possible with graphite. Mechanical property data for such materials have not yet been reported.

Table I indicates that the useful strengths of nongraphite die materials, while higher, are still less than an order of magnitude greater than that of graphite in typical uniaxial pressure-sintering configurations. Although experience with nongraphite dies has been limited, the early results of pressure-sintering, particularly with ceramic dies (see, for example, Spriggs et al. [17]), were encouraging enough to foster interest in methods of attaining even higher pressures. Thus, an extension of pressure-sintering into the very high and ultrahigh pressure regimes represented a logical augmentation of the investigations and experience at lower pressures.

The purpose of the present work, therefore, is to describe some of the interrelationships among (1) pressure-sintering processing conditions in the graphite-die, ceramic-die, and ultrahigh pressure regimes; (2) resulting microstructures; and (3) mechanical properties, particularly for dense, pure, submicron, and near-submicron grain size ceramic oxides.

EXPERIMENTAL PROCEDURE

All samples were prepared by pressure-sintering by techniques which, in general, have been recently described [4,6,8,16−22]. Specific variations in processing, however, were employed to

Fig. 1. Ultrahigh pressure-sintering apparatus.

produce the ultrafine grain sizes and very high relative densities. These variations included pressure-sintering in graphite dies at higher pressures (up to 12,000 psi) and lower temperatures than normally employed [4,8,18,19], pressure-sintering in nongraphite ceramic dies (up to 40,000 psi) [17,20,24], controlled-atmosphere pressure-sintering [16], and very high pressure-sintering (up to 750,000 psi) [22].

The ultrahigh-pressure unit (Fig. 1) is a modified piston and cylinder device, based on the concept of lateral support and consisting of a floating die assembly and two punch assemblies. Each of the three assemblies is constructed of a series of interference, press fit, concentric rings of varying hardness. The outermost ring in each case is mild steel, while the innermost item (die body or punch) is either hardened tool steel or cemented carbide. The unit is capable of reaching 1,000,000 psi (≈ 70 kb). The sample, which is 0.5 in. in diameter by 1.0 in. long, is contained within pyrophyllite and can be resistance-heated (with electrical contact made through the punches) to temperatures exceeding 1500°C. Provisions have also been made to conduct the high-pressure investigations in controlled atmospheres.

Most of this report is concerned with Al_2O_3 and, to a lesser extent, MgO. The starting materials were Linde A aluminum oxide or Fisher electronic-grade magnesium oxide. Many other refractory ceramics have also been studied, e.g., NiO, CoO, Cr_2O_3, Fe_2O_3, Y_2O_3, TiO_2, and CeO_2; none were less than 99.9% pure.

Strengths were obtained from diametral-compression (tensile) [23,24] and transverse-bend tests. Small right circular cylinders ($\approx 0.3-0.5$ in. in diameter and $0.05-0.10$ in. thick) were used for the former and rectangular parallelepipeds for the latter. Bend-test sample sizes for those pressed in graphite were 0.25 by 0.15 by at least 1.50 in. Smaller ones, 0.15 by 0.10 by 0.50 in., were employed in bend-testing ceramic die and ultrahigh-pressure specimens. Precision diamond machining and grinding were employed throughout. Extreme care was taken to ensure that the specimens were flat and parallel and free from nicks and scratches. To reduce possible scatter in test results attributable to stress concentrations at sharp edges, a radius of $\frac{1}{64}$ in. was ground onto the tensile surfaces.

The larger bend-test bars were tested by four-point loading at the third point over a 1.5-in. span. The smaller samples were tested in three-point loading over a 0.375-in. span. A detailed description of the apparatus and testing technique has been given previously [18].

Detailed electron microscopic studies, especially of fractured surfaces, were made of the samples utilizing negative carbon replicas shadowed with palladium.

RESULTS AND DISCUSSION

Graphite Die Pressure-Sintering

As-Machined. As mentioned, prior work had indicated that higher relative densities and finer grain sizes could be obtained in graphite dies through the use of higher pressures and lower temperatures than those normally employed. Thus, a large number of pressure-sintering experiments were conducted with alumina under these conditions. At 1400°C, variations in pressure from 4000 to 12,000 psi resulted in specimens which were all in excess of 99% relative density and had grain sizes of less than 2μ. The finest grain sizes were associated with the highest pressures. At a constant 6000 psi, variations in temperature from 1275 to 1400°C yielded specimens which were nearly equally dense, and, in addition, had grain sizes equal to or less than 1μ. A summary of these

TABLE II

Effect of Fabrication Pressure on Microstructure and Transverse
Bend Strength of Pure Alumina (4 hr at 1400°C and indicated
pressure)

Specimen number	Pressure (psi)	Average relative density (%)	Average grain size (μ)	Room-temperature transverse bend strength (psi)	Average strength (psi)
21-A				78,000	
21-B	4000	99.5	1−2	88,800	83,400
21-C				83,500	
21-1M*				78,900	
21-2M*	4000	99.5	1.40	115,400	83,700
21-3M*				56,700	
22-1	5000	99.6	1−2	84,300	84,300[†]
23-1				83,800	
23-2	6000	99.8	1−2	82,000	85,400
23-3				90,500	
24-1				91,900	
24-2	6000	99.5	1−2	95,000	93,500
25-1				94,300	
25-2	10,000	99.4	<1−2	63,300	77,900
25-3				76,200	
26-1				53,300	
26-2	12,000	99.3	<1−2	54,400	54,600
26-3				56,000	
26-1M*				45,800	
26-2M*	12,000	99.3	<1−2	63,000	54,800
26-3M*				55,600	

*Microbend test bars (0.10 by 0.15 by 0.50 in.), tested in three-point loading over 0.375-in. span; machined from larger bend-test bars.

†Value obtained for one measurement only.

results is given in Tables II and III, along with room-temperature bend strengths.

Table II indicates that fabrication pressures of 4000–10,000 psi resulted in average bend strengths in the range 78,000–93,000 psi. Individual maximum strength values indicated a general trend toward higher strengths with increasing pressures, but, at the

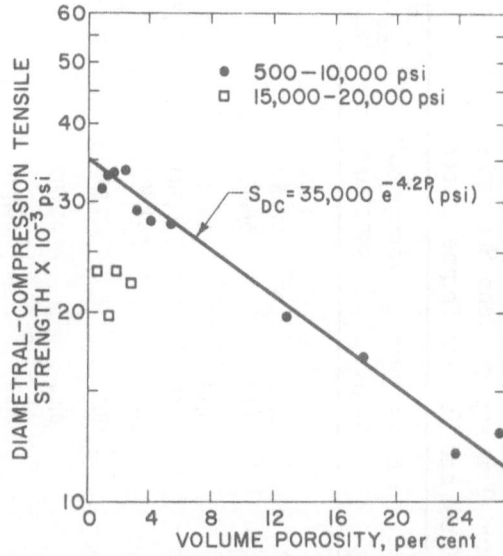

Fig. 2. Diametral-compression tensile strength versus porosity for pressure-sintered NiO.

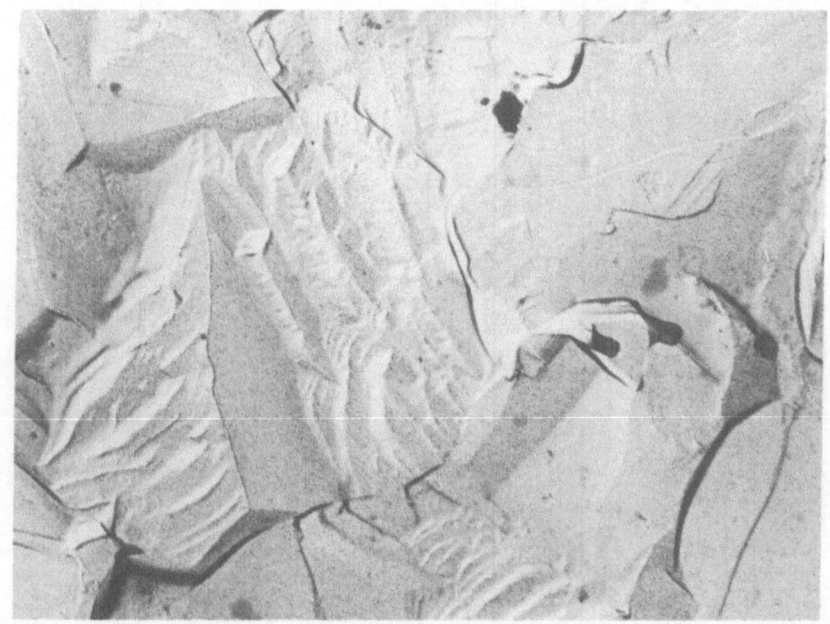

Fig. 3a. Electron fractograph of NiO. Lower pressure. Strength, 31,500 psi. Carbon-Pd replica. 20,000×.

TABLE III

Processing Conditions, Microstructures, and Transverse Bend Strengths for
Pure Alumina Pressure–Sintered at 6000 psi and Various Temperatures

Specimen number	Fabrication time (hr)	Temperature (°C)	Average relative density (%)	Average grain size (μ)	Average bend strength (psi)	Average strength (psi)
31-1	4	1275—1300	99.4	0.53	75,200	
31-2	4	1275—1300	99.4	0.53	75,600	65,800
31-3	4	1275—1300	99.4	0.53	56,600	
32-1	4	1300	98.7	0.65	74,700	
32-2	4	1300	98.7	0.65	49,700	65,300
32-3	4	1300	98.7	0.65	81,400	
33-1	4.5	1380	99.5	0.77	37,600	
34-1	4.5	1400	99.6	0.97	55,500	
35-1	4	1350	99.1	0.98	65,900	

Fig. 3b. Electron fractograph of NiO. Higher pressure. Strength, 22,300 psi. Carbon-Pd
replica. 20,000 ×.

highest pressures (10,000–12,000 psi), there were definite indications of variations and decreases in average strengths. Optimum conditions appeared to exist at 6000 psi. Table II also shows the results of strength measurements on smaller test bars which had been machined from broken larger bars. These results were quite comparable for the two testing spans, span-to-height ratios, and methods of loading (three-point versus four-point).

Very recent work with nickel oxide [21] had revealed that the substitution of pressure for temperature to achieve maximum density and minimum grain size resulted in structures which appeared to be densely packed, but with incompletely bonded grains. This phenomenon usually manifested itself as follows: Specimens prepared at lower pressures, despite larger grain sizes (or with equal grain sizes) show higher strengths and predominantly transgranular fracture characteristics. Specimens prepared at higher pressures, on the other hand, with smaller grain sizes (particularly in the submicron region), show lower strengths and predominantly intergranular fracture. This was shown graphically for nickel oxide in a plot of strength versus porosity (Fig. 2), and in the

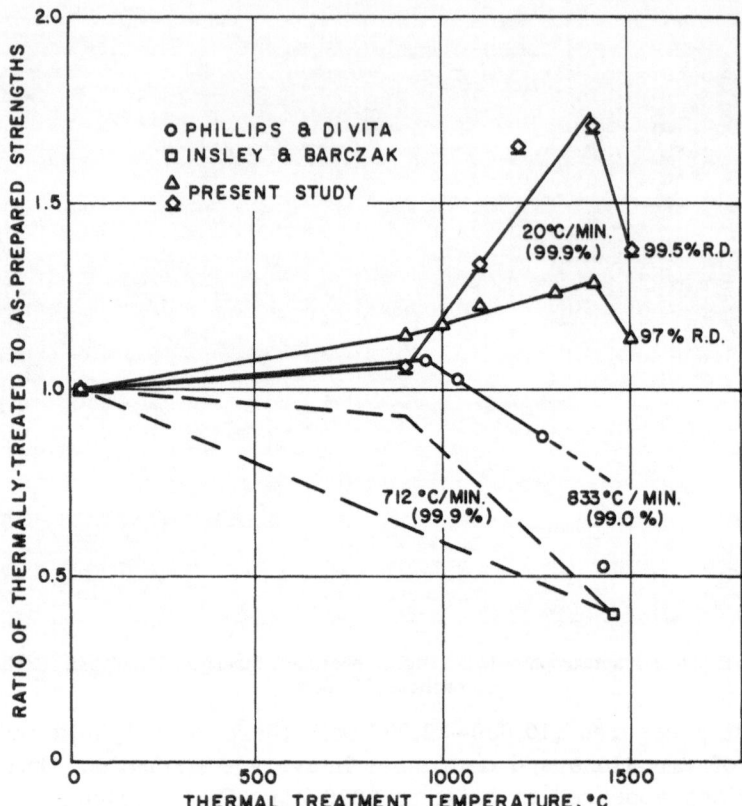

Fig. 4. Relative strength changes of dense, pure alumina as a function of post-fabrication thermal treatment and cooling rate.

electron fractographs (Figs. 3a and 3b) [21]. The present results with alumina showed the same behavior.

A change in densification mechanism, such as suggested for nickel oxide, is again hypothesized, with intergranular diffusion playing a smaller role at higher pressures or lower temperatures, or both. A necessary criterion appears to be that sufficient diffusion, perhaps in the form of grain boundary migration, be allowed to occur to cause strengthening of the grain boundaries, even without major grain size or porosity changes.

In Table III, concurrent experiments are summarized (for alumina pressed at 1275–1400°C at 6000 psi), which also indicate the "incompletely bonded" phenomenon. Specimens prepared under these conditions were quite dense (98.7–99.6%) and had grain sizes of less than $1\,\mu$ $(0.53-0.98\mu)$. These specimens, however, showed as-machined strengths of only 38,000–81,000 psi despite their high density and submicron grain size. Such strengths are much lower than would be predicted on the basis of their grain size [2,4].

It was found that subtle structural modifications were possible, and could bring about strengthening of these densely packed, but apparently incompletely bonded materials. One such technique employed thermal treatments after fabrication. Variations in the fabrication process, e.g., changing temperatures and pressures, and cycling them, also showed considerable promise. Combinations also appeared possible, in which materials prepared by variations in fabrication parameters could be strengthened through post-fabrication treatments.

Post-Fabrication Thermal Treatments. In a detailed series of experiments with almost dense alumina (99.5% of theoretical) of $1-1.5$-μ grain size, very careful post-fabrication thermal treatments of the machined specimens in vacuo ($\approx 0.5\,\mu$) resulted in progressive strength increases with increasing treatment temperature from 900 up to 1400°C (comparable to original fabrication temperature). Tensile strengths of specimens treated at 1400°C were 75–94% higher than those of untreated (64,400–71,300 psi vice 36,800 psi in diametral compression, which is usually about one-half of the transverse bend strength value). Figure 4 shows this response to thermal treatment in a plot of relative strength as a function of thermal treatment temperature. Also shown are data for a similar, but less dense (97%) alumina, where the response to thermal treatment appeared to be dampened by the additional porosity. In addition, recent results of Phillips and DiVita [25] and Insley and Barczak [26] are shown for so-called "thermally conditioned" high-alumina samples. The present thermal treatments employ a rather slow rate of cooling (20°C/min), whereas thermal conditioning, as employed by others, utilizes a very rapid rate of cooling (e.g., 712–833°C/min). Thermal conditioning with such rapid cooling or quenching rates has been found to be effective for alumina ceramics containing a glassy phase, perhaps because it prevents the exsolution of components soluble in the crystalline phase (or phases), thus relieving internal stresses. As shown in Fig. 4, it is not effective for essentially pure alumina; data of this type have usually been included in such studies [25,26] only for purposes of comparison or completeness.

TABLE IV

Effect of Post-Fabrication Thermal Treatments on the Transverse Bend Strength of Pure Alumina Pressure-Sintered in Graphite at Various Temperatures

Specimen number	Fabrication temperature (°C)	Post-fabrication treatment			Average relative density (%)	Average grain size (μ)	Transverse bend strength (psi)	Average strength (psi)
		Time (hr)	Temperature (°C)	Atmosphere				
21-1M—3M	1400	—	—	—	99.5	1.40		83,700
21-4M		0.1	1400	vacuum	99.5		64,800	87,700
21-5M		0.1	1400	vacuum		1.75	95,500	
21-6M		0.1	1400	vacuum			102,900	
31-1-3	1275—1300	—	—	—	99.4	0.53		65,800
32-1-3	1300	—	—	—	98.7	0.65		65,300
32-4	1300	16	1200	vacuum	98.7	0.65	79,200	(79,200)
33-1	1380	—	—	—	99.5	0.77	37,600	(37,600)
33-2	1380	0.1	1200	vacuum	99.5	0.77	75,700	(75,700)
34-1	1400	—	—	—	99.6	0.97	55,500	(55,500)
34-2	1400	0.1	1200	vacuum	99.6	0.97	63,400	(63,400)
35-1	1350	—	—	—	99.1	0.98	65,900	(65,900)
35-2	1350	1	900	vacuum	99.1	0.98	84,900	(84,900)
35-3	1350	48	900	air	99.1	0.98	71,000	74,100
35-4	1350	48	900	air	99.1	0.98	77,200	
35-5	1350	0.1	1400	vacuum	99.1	0.98	80,200	67,200
35-6	1350	0.1	1400	vacuum	99.1	0.98	54,200	

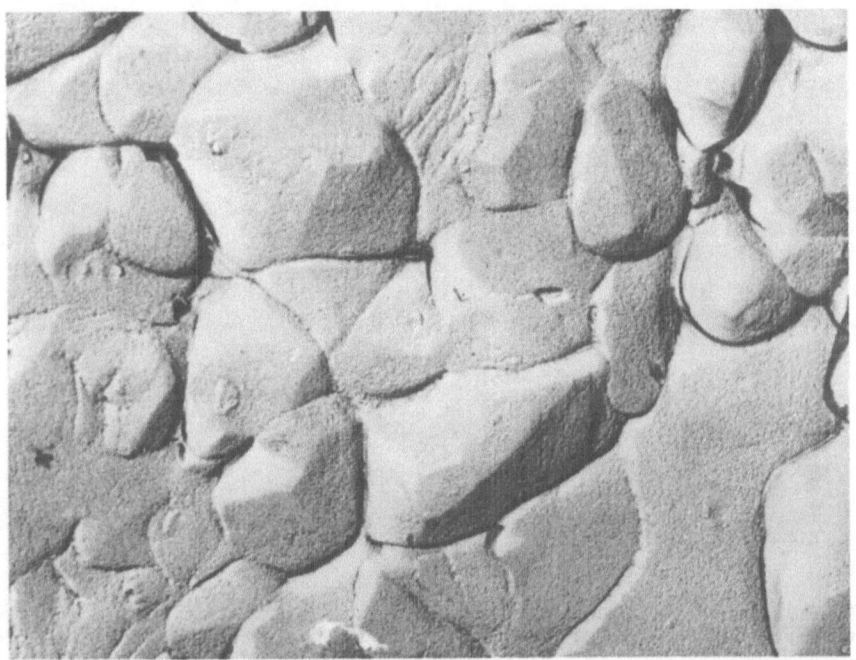

Fig. 5a. Fracture characteristics of submicron (0.77μ) alumina as-machined. Strength, 37,600 psi. Carbon-Pd replica. 20,000 ×; reduced 5% for reproduction.

Similar thermal treatments were also given to selected samples from previously pressure-sintered aluminas in graphite dies (Table III). The results are shown in Table IV and in the electron fractographs of Figs. 5a, 5b, 6a, and 6b. As mentioned, the as-machined strengths were low for those specimens with grain sizes of less than 1μ. Definite responses to the post-fabrication thermal treatments are evident in Table IV, ranging from a few percent increase in strength to over 100% for the Number 33 series. From fractographs of these specimens, it was apparent that those exhibiting large strength increases also showed greater proportions of transgranular fracture as illustrated in Figs. 5a and 5b for the Number 33 series, and their grain boundaries demonstrated a greater degree of angularity. On the other hand, the relatively larger grain size specimens, which had exhibited higher as-machined strengths, showed smaller changes in angularity and in proportionate fractions of transgranular fracture, as illustrated in Figs. 6a and 6b for the Number 34 series.

Fig. 5b. Fracture characteristics of submicron (0.77 μ) alumina after thermal treatment in vacuo at 1200°C. Strength, 75,700 psi. Carbon-Pd replica. 20,000×; reduced 5% for reproduction.

Fig. 6a. Fracture characteristics of submicron (0.97 μ) alumina as-machined. Strength, 55,000 psi. Carbon-Pd replica. 20,000×; reduced 5% for reproduction.

Fig. 6b. Fracture characteristics of submicron (0.97 μ) alumina after thermal treatment in vacuo at 1200°C. Strength, 63,400 psi. Carbon-Pd replica. 20,000×; reduced 5% for reproduction.

Additional tests were conducted on samples from the 33, 34, and 35 series, which first had been given a thermal treatment and then machined. For the particular treatment used (50 hr at 1200°C in air), the strengths were not too different from the original, as-machined values. This suggested that a surface response to the thermal treatments, e.g., healing of surface defects caused by machining or removal of adsorbed layers or both, as well as a microstructural response, may be involved.

Fabrication Process Variations and Additional Treatments. Tables V and VI present a summary of preliminary information indicating some major trends of response to (1) variations in the fabrication process, (2) some additional post-fabrication treatments, and (3) attempts to observe strengthening through a combination of them. The 53 series was heat-treated at temperatures up to 1450°C after primary densification had occurred at 1300°C. These treatments produced initial grain sizes of about 1.5 μ, apparently too large for the structure to respond to the subsequent treatments with any great increase in strength. The 51 series was fabricated at temperatures

TABLE V

Processing Conditions, Microstructures, and Transverse Bend Strengths for Pure Alumina Pressure-Sintered Under Various Conditions

Specimen number	Fabrication conditions*			Post-fabrication treatment			Average relative density (%)	Average grain size (μ)	Transverse bend strength (psi)
	Time (hr)	Temperature (°C)	Pressure (psi)	Time (hr)	Temperature (°C)	Atmosphere			
51B-1	3.5	1250	6,000		As-machined		93.0	<1	37,400
51B-2	0.5	1250	10,000	20	1200	vacuum	93.0	<1	49,900
51T-1†	0.1	1250-1350	10,000		As-machined		94.4	<1	48,900
51T-2	0.2	1350	10,000	20	1200	vacuum	94.4	<1	61,500
52-1					As-machined		~100	1.2-2.4†	56,100
52-2					As-machined		~100	1.2-2.4	40,600
52-3	3	1380	5,000	16	900	air	~100	1.2-2.4	50,700
52-4				16	1240	air	~100	1.2-2.4	60,600
52-5				16	1220	vacuum	~100	1.2-2.4	57,000
52-6				16	1225	hydrogen	~100	1.2-2.4	51,400
53T-1	4	1300	6,000		As-machined		99.2	~1.5	65,600
53T-2	0	1300	10,000		As-machined		99.4	~1.5	56,700
53T-3	0.1	1300-1450	10,000		Polished		99.3	~1.5	51,600
53T-4	0.25	1450-1400	10,000	16	1200	vacuum	99.3	~1.5	72,000
53T-5	0	1400	0	16	1200	vacuum	99.3	~1.5	62,700
53B-1					As-machined		99.4	~1.5	70,200
53B-2	Same as 53T above			2	1200	vacuum	99.4	~1.5	66,700
53B-3				2	1200	vacuum	99.4	~1.5	69,100
61‡							99.8	~1.0	48,900-82,600

*Entire set of fabrication conditions apply to all specimens in a given series.
†Smaller measurement is by optical microscopy while larger one is by electron microscopy.
‡See Table VI for full information.

TABLE VI

Effect of Post-Fabrication Thermal Treatments on the Transverse Bend Strength of Dense, Pure, Fine-Grained Alumina (99.8% relative density)

Specimen number	Fabrication conditions†			Post-fabrication treatment			Transverse bend strength (psi)
	Time (hr)	Temperature (°C)	Pressure (psi)	Time (hr)	Temperature (°C)	Atmosphere	
61-1					As-machined		48,900
61-2	2	1250	6,000	16	1100	vacuum	62,400
61-3	0	1250	10,000	17	1200	vacuum	80,100
61-4	0.1	1250-1300	10,000	20	1200	air*	74,100
61-5	2	1300	10,000	15	1250	vacuum	82,600
61-6	0.2	1300-1375	10,000	1 min	1400	air*	68,500

*Cooling rates in air were more rapid than in vacuum.
†Entire set of fabrication conditions applies to all six specimens.

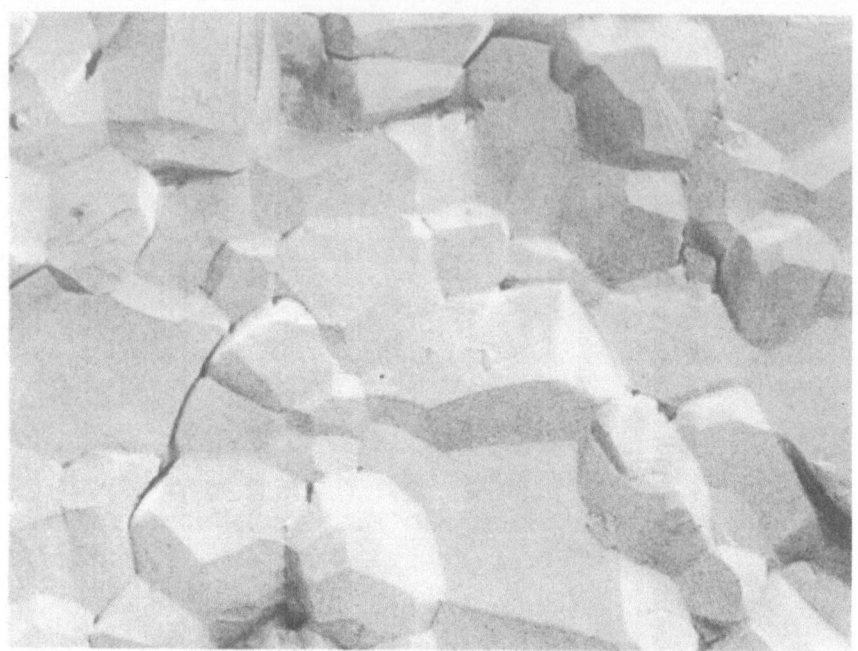

Fig. 7a. Fracture characteristics of submicron (0.96 μ) alumina as-machined. Strength, 48,900 psi. Process variations used to achieve original structure. Carbon-Pd replica. 20,000×; reduced 5% for reproduction.

up to 1350°C, with the bulk of the densification taking place at 1250°C, yielding grain sizes from 0.62 μ (51T) to 0.79 μ (51B). These samples also responded to post-fabrication thermal treatments; the actual strengths, however, were lower due to the presence of 5.6–7.0% porosity.

The 61 series (Table VI), fabricated at temperatures up to 1375°C (with steps, or "holds," in the densification cycle at 1250°C, 6000 psi; and 1300°C, 10,000 psi) produced almost dense (99.8%) specimens with grain sizes of ≈0.96 μ, as-machined. They revealed substantial strength increases after post-fabrication treatments both in vacuo and in air at temperatures up to 1250°C (Table VI). Representative electron fractographs of this material as-machined, and after a 1250°C treatment in vacuo, are shown in Figs. 7a and 7b. Once again, the higher-strength material showed a greater proportion of transgranular fracture.

Examples of electron microscopic replicas of the tensile surfaces of such specimens are shown in Figs. 8a–9c. Distinct differences are apparent between as-machined and thermally treated materials. The latter showed decided roughening and evidences of substructural features as a result of the thermal treatments. The implications of these observations are not as yet fully known and are the subject of current research.

On the basis of the work conducted to date, it appears that a "pivotal" point occurs at about $1\,\mu$; structures of less than $1\,\mu$ respond more to post-fabrication thermal treatments than those of greater than $1\,\mu$. (As an additional example, the 52 series shown in Table V with a grain size of 1.2 to 2.4 μ appears to be relatively insensitive to a wide variety of treatments.) This finding, of course, applies specifically to the particular pure alumina (Linde A) under study and to the particular pressure-sintering techniques employed. Still to be explored in greater detail are other starting materials, other pressure-sintering techniques, and other post-fabrication treatments.

Fig. 7b. Fracture characteristics of submicron $(0.96\,\mu)$ alumina after thermal treatment in vacuo at 1250°C. Strength, 82,600 psi. Process variations used to achieve original strucutre. Carbon-Pd replica. 20,000×; reduced 5% for reproduction.

Fig. 8a. Tension surface replica of submicron (0.96 μ) alumina as-machined. Strength, 48,900 psi. Carbon-Pd replica. 7500×.

Ceramic Die Pressure-Sintering

Several hundred pressure-sintering experiments (using ceramic dies) have been conducted with Al_2O_3, MgO, NiO, CoO, Cr_2O_3, Fe_2O_3, Y_2O_3, TiO_2, CeO_2, and other refractory ceramic materials. Present understanding of the sintering process at higher pressures has been enhanced by these experiments, and they have led to many of the processing variations and post-fabrication treatments described here. Processing conditions, microstructures, and mechanical properties of these oxides will be reported in detail in a future paper. However, some of the results are included here: (1) For Al_2O_3 and MgO, see discussion below; (2) for NiO, pressure variations resulted in an increase in tensile strength from 22,300 to 31,500 psi (Fig. 2); (3) for Cr_2O_3, increase of the temperature upon completion of densification yielded increases in tensile strength from 18,000 to greater than 25,000 psi; and (4) for TiO_2, similar treatment increased strength from 18,000 to 22,500 psi.

In a similar manner, increasing the pressure upon completion of densification resulted in dense alumina with grain sizes of less than 0.5 μ and bend strengths of about 90,000 to 110,000 psi. Data for alumina pressure-sintered in ceramic dies in air and in vacuo are listed in Table VII. At these lower temperatures (1150–1300°C), it has not been possible to achieve theoretical density, although the higher pressures and shorter times have resulted in submicron grain sizes. High strengths and increased strengths resulting from post-fabrication treatments are suggested by these limited tests.

Samples of high-density magnesia have also been fabricated by pressure-sintering in ceramic dies in air and in vacuo. Temperatures from 900–1140°C and pressures of 15,000–30,000 psi were

Fig. 8b. Tension surface replica of submicron (0.96 μ) alumina after thermal treatment at 1200°C in air. Strength, 74,100 psi. Carbon-Pd replica. 7500×.

Fig. 8c. Tension surface replica of submicron (0.96 μ) alumina after thermal treatment at 1250°C in vacuo. Strength, 82,600 psi. Carbon-Pd replica. 7500 ×.

employed. Specimens prepared in air were highly translucent, while magnesia pressure-sintered in vacuo was transparent. Grain sizes were equal to or less than 0.5μ. Bend strengths up to 28,600 psi were observed for those prepared in air and up to 48,800 psi for material prepared in vacuo. This latter value represents a significant increase over previous results [2,8] and is the highest yet measured for this material. (By comparison, commercially available, transparent magnesia used for infrared transmitting elements measured ≈35,000 psi.) It may be that the vacuum environment aids in removing adsorbed and otherwise entrapped gases, yielding densities more nearly approaching theoretical, and also improves intergranular bonding. Difficulty in preparing fractographic replicas of such specimens has been experienced, perhaps as a result of the improved nature and perfection of the grain boundaries. More work along this line is necessary and is in progress.

Ultrahigh Pressure-Sintering

Specimens of Al_2O_3, MgO, NiO, and Cr_2O_3 have also been pressure-sintered in the apparatus shown in Fig. 1 at temperatures in the range 800–1200°C and at pressures of 100,000–750,000 psi. Densities in excess of 99% have been achieved for each of these, as well as grain sizes below 1μ.

Unusual specimens of fully dense, highly transparent MgO have been obtained by the ultrahigh pressure-sintering. Knoop micro-hardnesses nearly twice single-crystal values (1170 vice 660 KHN)

Fig. 9a. Tension surface replica of submicron (0.96 μ) alumina as–machined. Strength, 48,900 psi. Carbon-Pd replica. 45,000×.

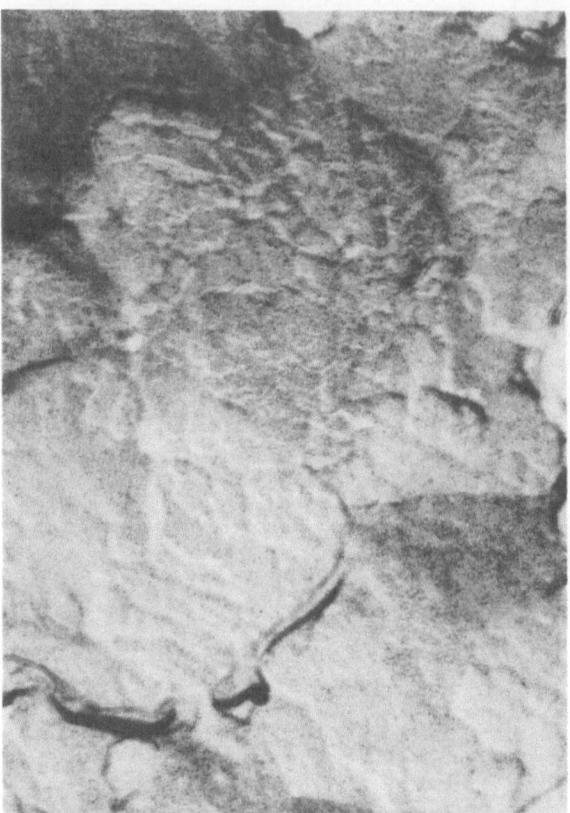

Fig. 9b. Tension surface replica of submicron (0.96 μ) alumina after thermal treatment at 1200°C in air. Strength, 75,100 psi. Carbon-Pd replica. 45,000×.

have been measured. In addition, the grain size could not be determined by standard electron microscopic replicating techniques, but was estimated by X-ray line broadening to be about 0.05 μ. This has since been confirmed by newly developed electron transmission techniques. Transmission micrographs of thin films of MgO, such as those shown in Figs. 10a, 10b, 11a, and 11b, have shown that essentially no grain growth or interdiffusion occurred during pressure-sintering and that the grain size was about 0.04 to 0.05 μ, i.e., comparable to the starting particle size. The achievement of very high density without grain growth suggests that the responsible densification mechanism was not diffusional, but probably one of deformation and plastic flow.

Bend strengths of the specimens of Al_2O_3, MgO, NiO, and Cr_2O_3 prepared at very high pressures were extremely low; when initially measured maximum values were only 5700 psi (Cr_2O_3) and 7500 psi (NiO). Because of the frequent occurrence of cracks and laminations, particularly with the MgO, it was felt that a true estimate of the strength had not been obtained. Additional samples of MgO which appeared to be free of such flaws were tested and gave strengths of $\approx$20,000–25,000 psi. This represented a substantial improvement in strength over the original measurements, but still did not conform to the exceptionally high strength that is expected for ultrafine grain sizes. Again, it is suggested that the lack of higher strengths for such materials was due to incomplete bonding

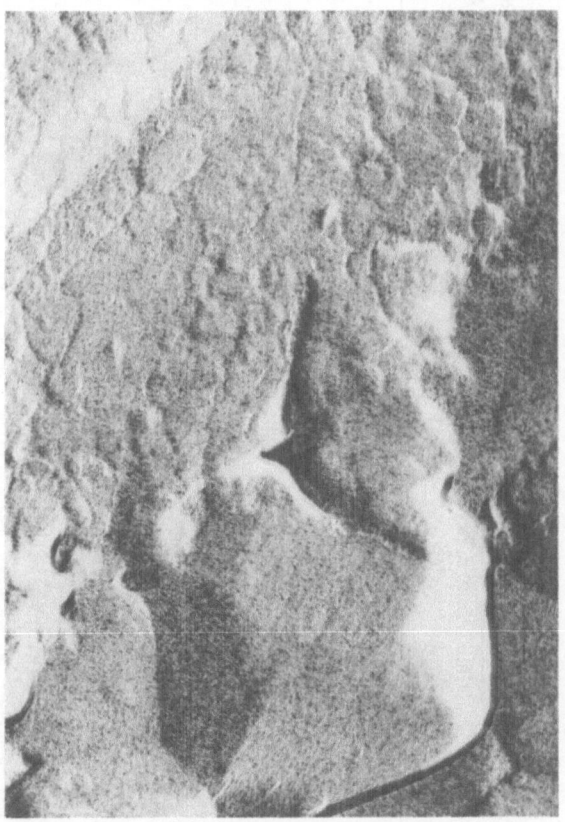

Fig. 9c. Tension surface replica of submicron (0.96 μ) alumina after thermal treatment at 1250°C in vacuo. Strength, 82,600 psi. Carbon–Pd replica. 45,000×.

TABLE VII

Transverse Bend Strengths of Very-Fine-Grained Alumina Prepared by Higher Pressure-Sintering in Ceramic Dies

Specimen number	Fabrication conditions			Post-fabrication treatment			Relative density (%)	Grain size (μ)	Transverse bend strength (psi)
	Time (min)	Temperature (°C)	Pressure (psi)	Time (hr)	Temperature (°C)	Atmosphere			
71-1*	50	1200	16,000		As-machined		96.9	0.39	57,600
71-2*	50	1200	16,000	16‡	1200	vacuum	96.9	0.39	96,700
72-1†	90	1200	20,000		As-machined		98.5	~0.50	77,000
72-2†§	90	1200	20,000						
73-1*	50	1300	20,000—30,000		As-machined		97.3	0.33	91,400
73-2*	50	1300	20,000—30,000	16‡	1200	vacuum	97.3	0.33	112,500
74-1†	480	1150	15,000—20,000	88	1500	hydrogen	98.7	4.22	62,500
74-2†	480	1150	15,000—20,000	88	1500	hydrogen	98.7	4.22	50,800

*Air atmosphere pressure-sintering.
†Vacuum pressure-sintering.
‡Polished.
§Specimen lost in post-fabrication treatment.

Fig. 10a. Preshadowed negative carbon replica of magnesia prepared by high pressure-sintering (10 min, 880°C, 250,000 psi) showing ultrafine grain size (400 A, 0.04 μ). 55,000×.

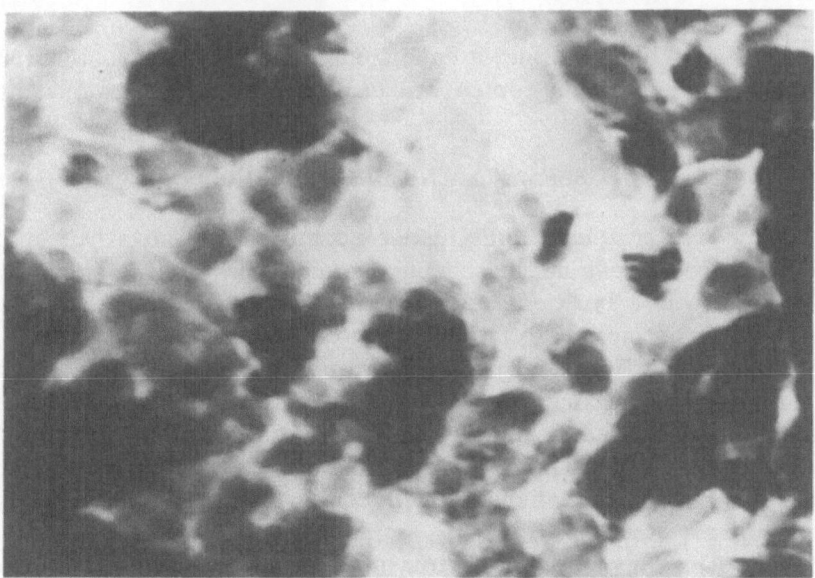

Fig. 10b. Transmission electron micrograph of magnesia prepared as in Fig. 10a. 160,000×.

Fig. 11a. Preshadowed negative carbon replica of magnesia prepared by high pressure-sintering (10 min, 920°C, 250,000 psi) showing ultrafine grain size (500 A, 0.05 μ). 55,000×.

between individual grains, a condition caused by the substitution of very high pressure for temperatures in achieving maximum density and minimum grain size.

SUMMARY AND CONCLUSIONS

Very fine-grained, highly dense ceramics, as initially prepared by high-pressure-sintering, have strengths which, in general, are less than would be predicted on the basis of their decreased grain size. Microstructurally, they appeared to be densely packed but with incompletely bonded grains. Lower strengths and predominantly intergranular fractures were associated with smaller grain sizes in materials prepared at higher pressures or lower temperatures or both. Higher strengths and predominantly transgranular fractures were characteristic of those prepared at lower pressures, despite larger grain sizes.

Depending upon the refractory material in question and the pressure-sintering technique employed, the above phenomenon manifested itself at pressures of only 10,000−40,000 psi, and was greatly accentuated at pressures of 100,000–750,000 psi, where the measured bend strengths were usually less than 10,000 psi (except for MgO, where strengths of 20,000–25,000 psi were measured). A difference in densification mechanisms has been suggested, with intergranular diffusion playing a smaller role at higher pressures, where plastic flow may be the predominant process. A necessary criterion for increased strength appears to be that sufficient diffusion be allowed, perhaps in the form of grain boundary migration, to cause strengthening of the grain boundaries, even though no significant grain growth occurs.

It was possible to bring about structural modifications capable of strengthening densely packed, but weakly bonded materials. Process variations during fabrication, post-fabrication thermal treatments, and combinations of them yielded such strengthening. However, for highly dense, fine-grained alumina pressure-sintered in graphite at higher than normal pressures, a pivotal condition appeared to exist at approximately 1μ. Materials with grain sizes of less than 1μ were much more responsive to strengthening based upon thermal treatments than were coarse-grained ones.

Fig. 11b. Transmission electron micrograph of magnesia prepared as in Fig. 11a. 240,000×.

Large-grained alumina ($>2\mu$) may respond to post-fabrication thermal treatments principally by changes in surface, e.g., healing of surface defects and removal of adsorbed layers. Finer-grained specimens ($<1-2\mu$) also showed an apparent surface response, but in addition exhibited microstructural changes, i.e., very slight increases in grain size, increased angularity of grain boundaries, apparently improved intergranular bonding with increases in transgranular fracture characteristics, and large relative increases in strength. It may be, however, that the very fine-grained specimens, because of their lower actual strengths, were more susceptible to machining damage than the larger-grained materials, which because of their higher actual strengths were not as badly damaged. In this view, the fine-grained materials may have responded more to such thermal treatments because of damage sustained prior to it.

For the particular pure alumina under study and for the graphite-die pressure-sintering technique employed, observed strengths have been limited to about 90,000–120,000 psi. A possible technique for overcoming this apparent limitation may be pressure-sintering in nongraphite dies in vacuum or air atmospheres; initial results have shown strengths in the range 90,000–110,000 psi for pure alumina with about 3% porosity.

Magnesia specimens with submicron grain sizes ($<0.5\mu$) prepared by very high pressure-sintering showed relatively low strengths (20,000–25,000 psi). On the other hand, a sample of submicron magnesia pressed at 15,000–30,000 psi in a ceramic die in vacuo had a strength of nearly 49,000 psi, the highest value yet reported for this material.

ACKNOWLEDGMENTS

This work was sponsored by the U.S. Army Research Office, Durham, North Carolina, under Contract No. DA-31-124-ARO-D-168; the U.S. Navy, Bureau of Naval Weapons, under Contract NOw 64-0217-d; and the Avco Corporation. Their permission to publish this paper is acknowledged. The writers are also pleased to acknowledge the contributions of the following individuals: P. F. Jahn and A. J. DeLai for materials preparation; J. Hogan for mechanical testing; R. E. Gardner and G. W. Robinson, Jr., for ceramographic preparation; and P. L. Burnett and R. H. Duff for electron microscopic studies.

REFERENCES

1. W. B. Crandall, D. H. Chung, and T. J. Gray, "Mechanical Properties of Ultrafine Hot-Pressed Alumina," in: W. W. Kriegel and Hayne Palmour, III (eds.), Mechanical Properties of Engineering Ceramics, Interscience (New York), 1961, pp. 349–379.
2. R. M. Spriggs and T. Vasilos, "Effect of Grain Size on Transverse Bend Strength of Alumina and Magnesia," J. Am. Ceram. Soc. 46(5): 224–228 (1963).
3. R. J. Charles, "Delayed Fracture: Static Fatigue," Task 9 in: Studies of the Brittle Behavior of Ceramic Materials, ASD Technical Documentary Report No. ASD-TR-61-628 (April, 1962).
4. R. M. Spriggs, J. B. Mitchell, and T. Vasilos, "Mechanical Properties of Pure, Dense Aluminum Oxide as a Function of Temperature and Grain Size," J. Am. Ceram. Soc. 47(7): 323–327 (1964).
5. P. R. V. Evans, "Effect of Microstructure on Fracture Strength of Alumina and Magnesia," Task 4 in: Studies of the Brittle Behavior of Ceramic Materials, ASD Technical Documentary Report No. ASD-TDR-61-628 (Part II), (May, 1963).
6. E. M. Passmore, R. M. Spriggs, and T. Vasilos, "Strength–Grain Size–Porosity Relations in Alumina," J. Am. Ceram. Soc. 48(1):1–7 (1965).
7. W. B. Harrison, "Fracture Behavior of Dense Polycrystalline MgO," presented at the Sixty-Third Annual Meeting, The American Ceramic Society, Toronto, Canada, April 24, 1961 (Basic Science Division, No. 2-B-61); for abstract, see Am. Ceram. Soc. Bull. 40(4): 187 (1961).
8. T. Vasilos, J. B. Mitchell, and R. M. Spriggs, "Mechanical Properties of Pure, Dense Magnesium Oxide as a Function of Temperature and Grain Size," J. Am. Ceram. Soc. 47(12):606–610 (1964).
9. J. F. Quirk, N. B. Mosley, and W. H. Duckworth, "Characterization of Sinterable Oxide Powders: I, BeO," J. Am. Ceram. Soc. 40(12): 416–419 (1957).
10. R. E. Fryxell and B. A. Chandler, "Creep, Strength, Expansion and Elastic Moduli of Sintered BeO as a Function of Grain Size, Porosity, and Grain Orientation," J. Am. Ceram. Soc. 47(6): 283 (1964).
11. G. C. Bentle, "Deformation in Single and Polycrystal BeO," presented at the Annual Meeting, The American Ceramic Society, Chicago, Illinois, April 22, 1964.
12. F. J. P. Clarke, "Residual Strain and the Fracture Stress–Grain Size Relationship in Brittle Solids," Acta Met. 12(2): 139–143 (1964).
13. H. Palmour, III, D. M. Choi, and W. W. Kriegel, "Mechanical Properties and Microstructure of Hot-Pressed Magnesium Aluminate," Report of U.S. Army Research Office, Durham, North Carolina, Contract DA-01-009-ORD-903, North Carolina State College, Raleigh, North Carolina (August, 1962).
14. F. P. Knudsen, "Dependence of Mechanical Strength of Brittle Polycrystalline Specimens on Porosity and Grain Size," J. Am. Ceram. Soc. 42(8): 376–387 (1959).
15. R. M. Fulrath, "Hot Forming Processes," in: Critical Compilation of Ceramic Forming Methods, Air Force Materials Laboratory Technical Documentary Report No. RTD-TDR-63-4069 (January, 1964), pp. 33–43.
16. R. M. Spriggs and T. Vasilos, "Processing of Ceramics at Subconventional Temperatures by Pressure-Sintering," presented at the Sixty-Sixth Annual Meeting, The American Ceramic Society, Chicago, Illinois, April 22, 1964.
17. R. M. Spriggs, L. A. Brissette, M. Rosetti, and T. Vasilos, "Hot-Pressing Ceramics in Alumina Dies," Am. Ceram. Soc. Bull. 42(9): 477–479 (1963).
18. J. B. Mitchell, R. M. Spriggs, and T. Vasilos, "Microstructure Studies of Polycrystalline Refractory Oxides," Summary Technical Report on U.S. Navy, Bureau of Weapons Contract NOw 62-0648c, Avco Corporation, Wilmington, Massachusetts, Technical Report No. RAD-TR-63-32, August 7, 1963, U.S. Dept. of Commerce, OTS, No. AD 413,994.
19. T. Vasilos and R. M. Spriggs, "Pressure-Sintering: Mechanisms and Microstructures for Alumina and Magnesia," J. Am. Ceram. Soc. 46(10): 493–496 (1963).

20. R. M. Spriggs, L. A. Brissette, and T. Vasilos, "Preparation of Magnesium Oxide of Submicron Grain Size and Very High Density," J. Am. Ceram. Soc. 46(10):508-509 (1963).

21. R. M. Spriggs, L. A. Brissette, and T. Vasilos, "Pressure-Sintered Nickel Oxide," Am. Ceram. Soc. Bull. 43(8): 572-577 (1964).

22. A. J. DeLai, R. M. Spriggs, R. M. Haag, and T. Vasilos, "Ultrahigh-Density, Submicron-Grain-Size Magnesia Prepared by Very High Pressures," J. Am. Ceram. Soc., to be submitted.

23. A. Rudnick, A. R. Hunter, and F. C. Holden, "Analysis of Diametral-Compression Test," Mater. Res. Std. 3(4): 283-289 (1963). 351

24. R. M. Spriggs, L. A. Brissette, and T. Vasilos, "Tensile Strengths of Dense, Polycrystalline Ceramics by the Diametral-Compression Test," Mater. Res. Std. 4(5): 218-220 (1964).

25. C. J. Phillips and S. DiVita, "Thermal Conditioning of Ceramic Materials," Am. Ceram. Soc. Bull. 43(1): 6-8 (1964).

26. R. H. Insley and V. J. Barczak, "Thermal Conditioning of Polycrystalline Alumina Ceramics," J. Am. Ceram. Soc. 47(1): 1-4 (1964).

Chapter 21

The Activation Energy for Grain Growth in Alumina

P. E. Evans

University of Manchester
Manchester, England

High-purity alumina has been isostatically cold-compacted and sintered. Nickel was used as a grain-growth inhibitor. Grain size distribution in the sintered materials has been measured by electron microscopy. The activation energy for grain growth in the temperature range 1600–1800°C was determined as 138 ± 7 kcal/mole; other workers have found a value of 150 kcal/mole. Electron probe microanalysis has revealed a nickel-rich phase present not only at the original sites of the nickel particles, but also as a fine dispersion whose mean spacing corresponds to the particle diameter of the alumina before compaction. The estimated nickel content of the spinel phase (15 wt.%) accords with that of a nonstoichiometric spinel found by other workers. A grain-growth inhibitor might be expected to increase the activation energy for grain growth, but it is shown that the necessary increase in activation energy is of the same order as the limits of error of the activation energy. However, this is further considered with reference to the Arrhenius rate equation, and it is suggested that the presence of nickel could alter the "constant" term, thus affecting grain growth, while leaving the activation energy for the process essentially unchanged.

INTRODUCTION

The work reported here forms a small part of an investigation into the factors governing the production of strong, fine-grained alumina by cold isostatic pressing and sintering.

EXPERIMENTAL

The alpha alumina powder used in this work came from one batch* manufactured from the chloride hexahydrate. Spectrographic analysis showed 210 ppm silicon and 10 ppm iron to be present. Flame photometry gave 21 ppm sodium.

After the addition of 2.0 wt.% of the grain-growth inhibitor in

*Supplied by Plessey Research Laboratories, Towcester, England.

the form of 6.5-μ metallic nickel powder, the batch was milled for 45 hr in a rubber-lined drum loaded with alumina balls. (Previous experiments on the effect of milling time had shown that a maximum surface area was reached in 45 hr.) The surface area of the powder (measured by a modified B.E.T. method [1]) after milling was 37,000 cm^2/g and the modal particle size was 0.41 μ. The powder was isostatically pressed at 67,200 psi (47 kg/mm^2) by a method described by Penrice [2] to a density of about 2.40 g/cm^3. The compacts were sintered in air at temperatures of 1600, 1665, 1730, and 1800°C. They then measured about 1.5 by 1.5 by 0.7 cm and had densities ranging from 3.60 to 3.90 g/cm^3. They were cut in two, perpendicular to their principal plane, and etched in boiling

Fig. 1. Typical microstructure of pressed and sintered alumina.

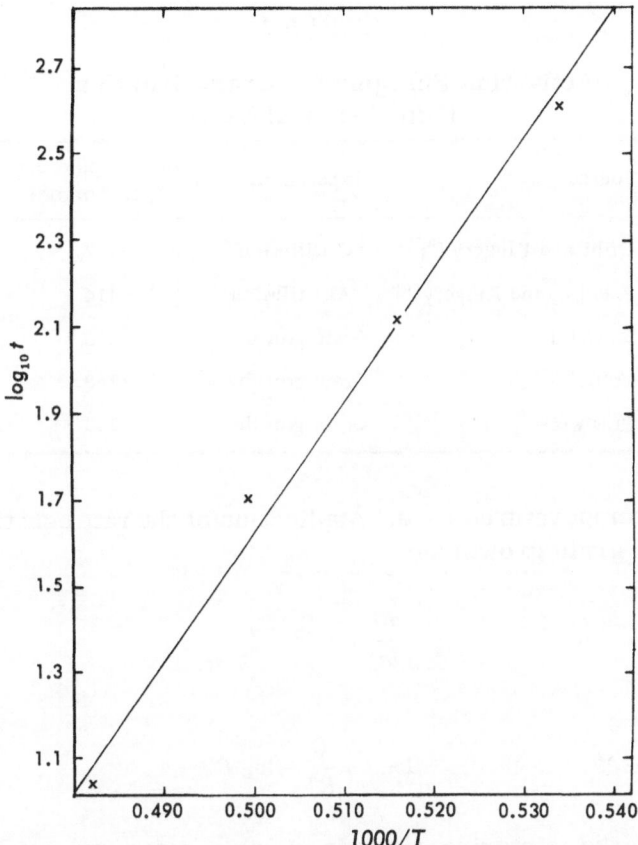

Fig. 2. Least-squares line fitted to plot of log t versus $1000/T$. Activation energy for grain growth is 138 ± 7 kcal/mole (95% confidence limits).

KHF_2. Two–stage preshadowed carbon replicas were made. The replicas were examined in an electron microscope, after examination on an optical microscope had shown that any area of the compact of about 2.5 mm^2 was representative of the whole surface in terms of grain-size variations. No exaggerated grain growth was observed in any specimen. A typical electron micrograph is shown in Fig. 1. An arbitrary series of lines was drawn on photographic prints (standard magnification 10,000 ×) and the Martin's accidental diameter [4] was measured for 500 grains per specimen. The grain size, corrected for three-dimensional distribution, was plotted against frequency and the modal grain size determined.

Modal grain size was plotted against sintering time for the four

TABLE I

Activation Energies for Grain Growth and
Diffusion in Alumina

Source	Experiment	Activation energy (kcal/mole)
Oishi and Kingery [5]	^{18}O Diffusion	152
Paladino and Kingery [6]	^{26}Al Diffusion	114
Bruch [7]	Grain growth	153.7
Coble [8]	Grain growth	153
This work	Grain growth	138

sintering temperatures used. Application of the rate equation gives
the rate of grain growth as:

$$\frac{dG}{dt} = Ae^{-Q/RT} \tag{1}$$

or

$$\log t = \frac{Q}{RT} - \log B \tag{2}$$

where t is the sintering time to reach a given modal grain size
(4μ in this case) at each value of the sintering temperature T (in
degrees Kelvin). The plot of t against $1/T$ is shown in Fig. 2. If
a straight line is fitted to these four points, a value for Q of
138 ± 7 kcal/mole is obtained (95% confidence limits). We may
compare this with the values found by others (Table I). Extrapolation
of Fig. 2 gives $B = 10^{-12}$ [equation (2)].

A preliminary, qualitative examination of one of these specimens
was carried out on an electron probe microanalyzer (Cambridge
Microscan) in an attempt to determine the distribution of nickel.
Since the electron beam is about 1μ across, the specimen with the
biggest grain size (12.6μ) was used.

The electron probe (Fig. 3) showed the original sites of the
nickel particles (the large light areas) and also a finely dispersed
nickel-rich plane. An estimate of the nickel content at the original
sites gives 14 to 15 wt.% Ni (cf., 33 wt.% Ni in $NiAl_2O_4$). No estimate

was possible for the fine dispersion since its size is smaller than the probe diameter. The density of the fine dispersion is about $6 \cdot 10^8$ particles/cm^2. This gives a mean linear spacing of 0.4 μ, which corresponds to the modal particle size of the alumina (0.41 μ) before compaction. Thus, the nickel probably diffused between the alumina particles in the early stages of sintering or else the original interparticle regions provided rapid diffusion paths for some considerable part of the sintering period.

In some separate work [9] in which 1.9 wt.% MgO was added (to the same alumina, sintered for 30 min at 1780°C), we obtained results from the probe analyzer similar to those shown for nickel.

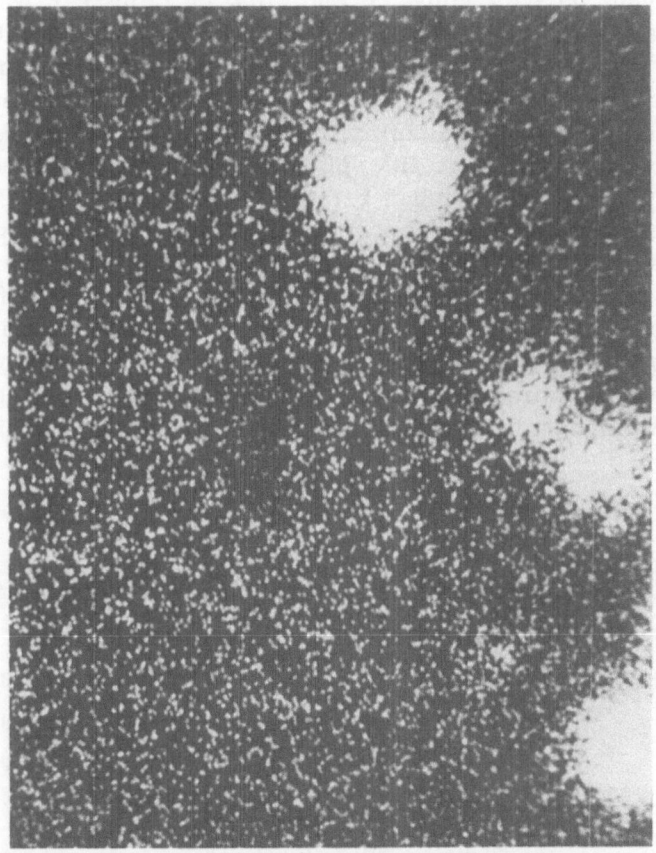

Fig. 3. Electron probe microanalysis; typical X-ray image photograph showing nickel-rich regions (light areas).

The magnesium probe analysis was quantitative and showed that the maximum peak values on the magnesium trace were equivalent to 30 wt.% magnesia. This corresponds, within experimental error (for magnesium, about 5%), to the spinel of composition $MgAl_2O_4$.

The grain boundary regions as distinct from the high-nickel "islands" may still contain appreciably more nickel in solution than the overall 2% average represented by the total addition. In other words, the electron microprobe has insufficient resolution either to prove or refute Jorgensen and Westbrook's work [10], in which it was shown by microhardness measurements, by autoradiography, and by measuring lattice parameters as a function of grain size, that in alumina doped with 0.05–0.1 wt.% NiO or MgO, the solute segregates at the grain boundaries.

Relevant here is some recent work by Lejus [11], who studied the systems $MgO-Al_2O_3$ and $NiO-Al_2O_3$. In brief, the phase diagram determined by Lejus for the nickel oxide–alumina system shows no solubility of NiO in the spinel at any temperature and vanishingly small solubility of Al_2O_3 in the spinel below 1300°C. However, nonstoichiometric spinels, containing between 50 and 83 mol.% Al_2O_3, were found between 1300 and about 1900°C, i.e., an increasing solid solution of Al_2O_3 was found in the spinel with rising temperature. Lejus also concludes from the X-ray evidence that two aluminum ions replace three nickel ions with the formation of holes in the cation lattice and that the distribution differs increasingly from the normal distribution as the aluminum content of the sample increases. Thus, the tetrahedral sites in the normal, stoichiometric spinel are increasingly occupied by aluminum ions

TABLE II

Deduced Spinel Compositions[*]

Temperature (°C)	Molecular % Al_2O_3	Limit of solid solution	Lattice parameter (A)
1000	50	$NiAl_2O_4$	8.043
1300	52	$Ni_{0.94}Al_{2.04}\square_{0.02}O_4$	8.038
1450	56	$Ni_{0.83}Al_{2.11}\square_{0.06}O_4$	8.019
1600	60	$Ni_{0.72}Al_{2.19}\square_{0.09}O_4$	8.004
1900	80	$Ni_{0.30}Al_{2.46}\square_{0.24}O_4$	7.956

*After Lejus [11].

with progressive displacement of the nickel. With reference to the present work, therefore, we might interpret the nickel-rich islands shown by the probe analyzer as probably nonstoichiometric spinels. The spinel compositions deduced by Lejus for different temperatures are shown in Table II. The estimated nickel content, 15 wt.%, from our electron probe analysis is in good agreement with Lejus' findings. Linear interpolation in her results (Table II) for 1800°C, the temperature at which this specimen was sintered, gives a solubility limit of $Ni_{0.45}Al_{2.36}\square_{0.19}O_4$ which corresponds to 17 wt.% Ni. This suggests that equilibrium was reached at least at the sites of the original nickel particles and also that the high-temperature, nonstoichiometric spinel composition was essentially retained in cooling to room temperature.

DISCUSSION

Let us examine, briefly, the concept of grain growth. For uniform grain growth to occur, the growing, as opposed to the consumed, grains must be statistically uniformly distributed in the volume of the specimen. Furthermore, the growing side of the boundary must, in the overall view, have a lower surface energy than the consumed side of the boundary. Hence, the activation energy for grain growth must be that necessary to transfer ions to the more stable configuration of the growing side of the boundary.

It might be expected that the presence of a grain-growth inhibitor, as used in this work, would appreciably increase the activation energy for grain growth. That nickel is an effective grain-growth inhibitor is demonstrated by experiments carried out, under otherwise identical conditions, both with and without this addition; e.g., after 15 min at 1800°C, the modal grain size was $10.0\,\mu$ without the inhibitor compared with $2.9\,\mu$ when the inhibitor was present.

The activation energy with the inhibitor (138 kcal/mole) is lower than that found by Bruch [7] and by Coble [8]. However, from the internal evidence of the present work, we cannot say that the activation energy without the inhibitor would not be lower.

If we take the rate of grain growth in the absence of the inhibitor to be approximately three times greater than when an inhibitor is present, as in the example quoted, then this could be accounted for by an activation energy only 5 kcal/mole less than that found when the inhibitor is present. The limits of error in this work are at best about ± 7 kcal/mole, so that quite obviously any increase in activation energy due to the inhibitor may well be masked by the limits of error of the activation energy.

However, it is of interest to examine the ways in which the grain growth could conceivably be affected without any change in the activation energy for the process. Let us re-examine the rate equation.

According to Zener [12], the rate equation can be written as

$$\text{Rate} = n\nu e^{-Q/RT} \tag{3}$$

where the reaction rate equals the number of atoms n in metastable position multiplied by the frequency ν with which each such atom attempts to scale the energy barrier, multiplied by the probability that, during such an attempt, the atom has the necessary activation energy Q.

To meet the real situation where the potential energy of the migrating atom is dependent on and influences the motion of neighboring atoms, we must replace the potential energy Q by the free energy F. Therefore, equation (3) becomes

$$\text{Rate} = n\nu e^{S/R} e^{-E/RT} \tag{4}$$

where S is the entropy of activation, and equation (2) becomes

$$\log t = E/RT - (\log B + S/R) \tag{5}$$

Thus, an effective grain-growth inhibitor could also lower the rate of grain growth by reducing the term $n\nu$. It should be clear that we are implicitly discounting the disperse-phase mechanism of grain-growth inhibition. Although no continuous nickel was detected in solution at the grain boundaries, Jorgensen and Westbrook's work offers conlcusive evidence that segregation of the nickel solute does in fact occur. Thus, ν would be reduced if the presence of the nickel locally decreased the strength of binding of the alumina lattice. (In comparison with aluminum, the greater mass of nickel would also tend to decrease ν.) Since ν is related to the elastic properties of the material [13], measurements of these for pure alumina and for a nickel—alumina spinel might give a lead here.

Further experiments are desirable to determine the relative part played in the rate equation by the $n\nu$ term and the exponential term in the presence and absence of a grain-growth inhibitor.

ACKNOWLEDGMENTS

The writer wishes to thank his research students, B.P. Hardiman, B. C. Mathur, and W. S. Rimmer, on whose experimental work this contribution is based; B. W. Lambert for carrying out the probe analysis; and Professor K. M. Entwistle for the provision of facilities.

REFERENCES

1. W. V. Loebenstein and V. R. Dietz, "Surface Area Determination by Adsorption of Nitrogen from Nitrogen—Helium Mixtures," J. Res. Nat. Bur. Std. 46: 51 (1951).
2. T. W. Penrice, "Compacting Powders Using Molds Made from Reversible Gels," Powder Met. 1/2:79 (1958).
3. D. E. Bradley, Brit. J. Appl. Phys. 5: 165 (1964); and J. Inst. Metals 83: 35 (1954).
4. G. Martin, C. E. Blythe, and H. Tongue, Trans. Brit. Ceram. Soc. 23 (2): 61 (1923).
5. Y. H. Oishi and W. D. Kingery, "Self-Diffusion of Oxygen in Single-Crystal and Poly-crystalline Aluminium Oxide," J. Chem. Phys. 33 (2): 480 (1960).
6. A. E. Paladino and W. D. Kingery, "Aluminum-Ion Diffusion in Aluminum Oxide," J. Chem. Phys. 37 (5): 957 (1962).
7. C. A. Bruch, "Sintering Kinetics for High-Density Alumina Process," Am. Ceram. Soc. Bull. 41 (12): 799 (1962).
8. R. L. Coble, "Sintering Crystalline Solids: 11, Experimental Test of Diffusion Models in Powder Compacts," J. Appl. Phys. 32 (5): 793 (1961).
9. P. E. Evans and M. Chappell, unpublished work.
10. P. J. Jorgensen and J. H. Westbrook, "Role of Solute Segregation at Grain Boundaries During Final-Stage Sintering of Alumina," J. Am. Ceram. Soc. 47 (7): 332 (1964).
11. A.-M. Lejus, Thesis, Faculté des Sciences, l'Université de Paris, 1964.
12. C. Zener, "The Role of Statistical Mechanics in Physical Metallurgy," in: Thermo-dynamics of Physical Metallurgy, A. S. M., Cleveland, Ohio, 1950, p. 16.
13. N. F. Mott and H. Jones, The Theory of the Properties of Metals and Alloys, Oxford University Press, 1936, p. 4.

Chapter 22

Grain Boundaries and the Mechanical Behavior of Magnesium Oxide

R. B. Day and R. J. Stokes

Honeywell Research Center
Hopkins, Minnesota

A comparison of the mechanical behavior of magnesium oxide single crystals and polycrystals at different temperatures is presented. Single crystals show increasing interpenetrability and multiplicity of slip systems at elevated temperatures. Above 1700°C, slip is very flexible, i.e., slip systems with different Burgers vectors completely interpenetrate each other and slip occurs over a variety of slip planes. Furthermore, above 1600°C, rearrangement of dislocations by polygonization is observed, and, in certain cases, recrystallization can be obtained. Polycrystals prepared by the recrystallization of single crystals show a brittle-to-ductile transition with increasing temperatures. At low temperatures, any structural discontinuity capable of blocking slip dislocations, including subgrain and grain boundaries, induces cleavage crack nucleation and brittle fracture. At intermediate temperatures (i.e., 1400 – 1700°C), more slip occurs, but it is not flexible enough for grains to conform to each other's change in shape, and constraints develop which can be relaxed only by intergranular sliding. This leads to intergranular crack nucleation and brittle fracture. At elevated temperatures (i.e., above 1700°C), slip is very flexible, and the polycrystalline matrix deforms plastically and necks down to a ductile fracture. Polygonization and recrystallization of the deforming grains also contribute to relaxation of internal stresses. These high-temperature deformation processes are remarkably similar to face-centered cubic metals undergoing creep. Slight porosity causes an increase in the brittle-to-ductile transition temperature. While high-density, hot-pressed magnesia shows plasticity above 2000°C, it is limited by intergranular failure.

INTRODUCTION

The past ten years have seen a tremendous research effort into the mechanical properties of ceramics. The reasons for this intense interest include their potential application as structural components capable of supporting loads at high temperatures. While this effort has involved materials such as the intermetallic compounds, carbides, borides, and nitrides, the greatest attention has been given to the mechanical behavior of the oxides. Their most attrac-

355

TABLE I

Various Types of Polycrystalline Magnesium Oxide

Source	Density (%)	Grain size (μ)		Microstructure	
		As-received	After anneal at 2000°C for 1 hr	As-received	After anneal at 2000°C for 1 hr
Illinois Institute of Technology Research Institute (IITRI)	99.5	50—60	100	Porosity in grain boundaries and within grains	Same as for as-received
Eastman Kodak Company	100	3	30	Clear, only slight porosity	Residual porosity chiefly in grain boundaries, some in grains
Avco Corporation	100	20	40—150	Clear, only slight porosity along triple lines	Residual porosity chiefly in grain boundaries, some within grains
The Boeing Company	100	20—25	600—800	Clear, no porosity. Some impurity particles	Clear, slight evidence for pores along triple lines. Some impurity particles
Recrystallized single crystals	100	~1000		Perfect. No particles, no porosity, and no defects to be seen	

tive property is their chemical stability at high temperatures in an oxidizing environment. Research on the mechanical properties of oxide ceramics was initially concentrated on the behavior of magnesium oxide single crystals at low temperatures, but more recently the field has broadened to include single crystals of other oxides, and the emphasis has switched to polycrystalline material and to the mechanical behavior at high temperatures. This is a reasonable trend in view of the ultimate application of ceramics.

General aspects of the strength and fracture of polycrystalline ceramics at different temperatures have been known for some time and their limitations recognized. However, an understanding of the fundamental physical processes involved, a necessary prerequisite to the initiation of improvement, has been lacking. One of the reasons for this has been the difficulty in obtaining high-quality material on which significant and reproducible experiments can be performed. This difficulty has now been largely overcome as will be described later and some significant observations have been made.

This paper reviews recent work in our laboratory, work concerning the extension of knowledge of the mechanical behavior of magnesia, both single and polycrystalline, to include properties at high temperatures. Our attitude is that it is logical first to study single crystals, since they can be tested and analyzed in more detail than polycrystalline material. Then, the additional micro-structural variables are introduced gradually so that the relative contributions of the simple grain boundary interface, the triple lines along which grain boundaries meet, the triple points where triple lines meet, and finally porosity may be assessed. In this respect, we are examining a wide range of materials including single crystals, single crystals containing subgrains, recrystallized single crystals, fused bicrystals, simple polycrystalline material (i.e., a large-grained, hot-pressed, and sintered magnesia of theoretical density), and, finally, hot-pressed magnesia of high density.

MATERIAL AND PROCEDURE

Mechanical tests were performed on single crystals and poly-crystalline material. The polycrystalline material was obtained from a number of sources and was of particularly high quality as indicated in Table I. All of these materials were prepared by hot-pressing techniques and, with the exception of the IITRI (Illinois Institute of Technology Research Institute) material, were initially of theoretical density and quite transparent. As a prelude to tests at very high temperatures, it was considered necessary to anneal all material at 2000°C for one hour to produce a stable microstructure. As Table I indicates, annealing resulted in considerable grain growth and, with the exception of the Boeing Company material, also resulted in opacity due to the evolution of some porosity (approx. 0.1%). This porosity presumably derived from the evaporation of low-melting-point phases present in the original hot-pressed grain boundaries. Although the Boeing Company material remained transparent, and thereby provided a beautiful polycrystalline material to work with, it nevertheless still contained a low density of defects in the form of elongated pores along triple lines, a few pores in grain boundary interfaces, and precipitate particles within the matrix. As will be made clear later, these defects had detrimental effects on the mechanical properties. The best source of polycrystalline material for fundamental studies of deformation mechanisms proved to be recrystallized single crystals whose preparation will be described later.

All of the tests to be described here were performed in tension using techniques which have been described elsewhere [1, 2]. Tension has a number of advantages over compression or bending as a means of testing. Compression suffers disadvantages due to end constraints and buckling of the specimen which may in some instances limit the ductility. Bending leads to load–deflection curves which are difficult to analyze when the specimens become very plastic. Tension tests, on the other hand, give a real and useful measure of the fracture strength and ductility, and, furthermore, the test may be run to failure so that the fracture mode and origin may be determined.

Tension specimens were cut ultrasonically from slices initially prepared either by cleaving single crystals or by cutting the polycrystalline material with a diamond wheel. The tension tests were performed at various temperatures from room temperature up to 2000°C. For room-temperature tests, the tapered grip ends of the tension specimens were glued into grips with epoxy cement [1]. For high-temperature tests (i.e., above 1000°C), they were held loosely in graphite grips and tested in an inert argon environment in a carbon-element furnace [2]. They were all tested in the Instron machine at a strain rate of approximately $5 \cdot 10^{-4}$/sec.

GENERATION OF SUBGRAIN BOUNDARIES AND GRAIN BOUNDARIES IN SINGLE CRYSTALS BY HIGH-TEMPERATURE DEFORMATION

As a background to the discussion of the effect of subgrain and grain boundaries on mechanical behavior of magnesium oxide, some experimental observations on the high-temperature deformation of single crystals must be described. These observations were important for two reasons; first, they told us something about the multiplicity of slip modes operating at high temperatures and the degree to which slip systems interpenetrated, and second, they led us to a new and reproducible procedure for producing perfect grain boundaries by the recrystallization of single crystals.

High-Temperature Deformation of Single Crystals

In a recent paper [2], we described the different modes of deformation observed on certain single crystals as the test temperature was raised from room temperature up to the vicinity of 2000°C. These crystals had a <100> tensile axis so that four <110> {110} slip systems were equally stressed, and it was shown that the mechanical behavior could be subdivided according to the number and variety of slip systems operating concurrently within a given volume of the crystal.

From room temperature up to approximately 1300°C, massive slip in any given volume of the gage length was confined to a single <110> {110} planar system. The slip zones could not interpenetrate and the crystals work-hardened rapidly before fracturing in a brittle manner due to slip interaction at the slip zone interface.

From 1300 to 1700°C (and at a moderate strain rate of 10^{-4}/sec), massive slip on any pair of systems having <110> Burgers vectors at 60° still could not interpenetrate, whereas those making 90° could. In this range, mechanical behavior depended on the initial slip distribution along the gage length. When slip occurred primarily on two 90° systems, a simple one-dimensional reduction in cross section was possible and a knife-edge ductile fracture resulted; but when slip occurred on systems at 60°, "kink" subgrain boundaries were generated at the interface between the two slip zones, and brittle fracture ensued.

Above 1700°C, slip on all systems with Burgers vectors both at 60° and 90° could interpenetrate. Then, after an initial period of yielding with little or no work-hardening (approx. 20% elongation), the crystals started to work-harden and continued to do so for over

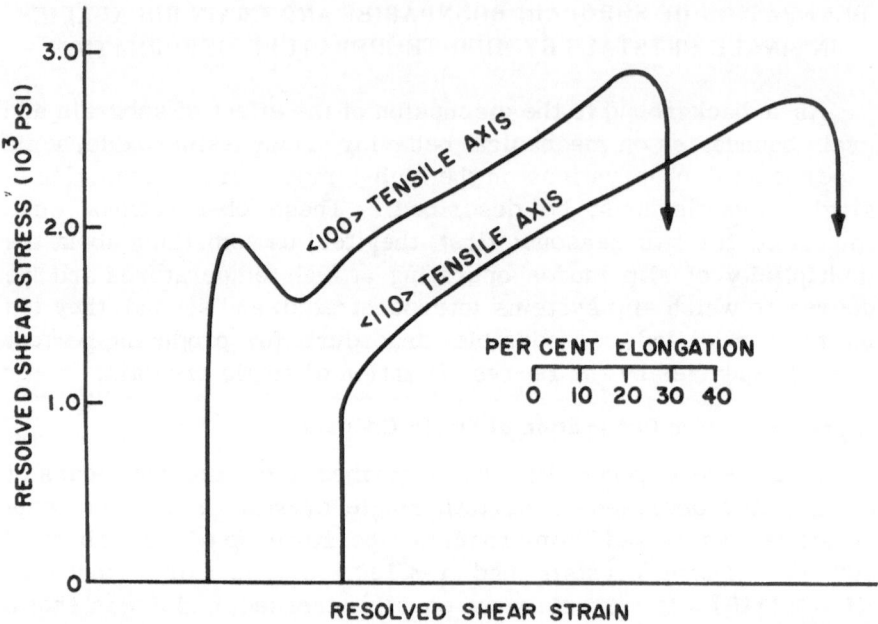

Fig. 1. Tensile stress—strain curves for magnesium oxide single crystals at 1800°C.

100% elongation. (See the stress—strain curve in Fig. 1.) At this stage, they began to neck down in a completely ductile manner with contraction in both specimen dimensions. Reasons for the work-hardening will be discussed later.

From these observations, we learn that as the temperature is raised, the degree of interpenetration of slip on a variety of systems increases.

In these experiments, it was also found that "wavy slip," i.e., slip in a given <110> direction but on a variety of planes (e.g., {110}, {100}, {111}) having this direction as a zone axis, occurred above 1500°C. It was particularly prominent in those regions such as kink boundaries subject to plastic constraint. These observations agree with the measurements of Hulse et al. [3] and Copley and Pask [4, 5] on the relative stresses to initiate slip on different planes. They also agree with recent observations by us on the tensile behavior of single crystals of the <110> tensile axis orientation, an orientation which favors slip on the {100} and {111} planes. From these observations we learn that as the temperature is raised, the multiplicity of slip planes for a given slip vector increases.

These two factors—slip interpenetrability and slip multiplicity—are very important when considering the mechanical properties of polycrystalline material, since together they constitute slip flexibility, a feature which has been demonstrated to be a necessary prerequisite for polycrystalline ductility in a number of materials [6−10]. On the basis of these observations on single crystals, we cannot reasonably anticipate polycrystalline ductility in magnesium oxide until the temperature exceeds 1500 − 1700°C.

Polygonization and Recrystallization of Single Crystals

Above 1600°C, additional features accompanied plastic deformation. Due to the high temperature ($0.6\ T_m$) of the test, dislocations could rearrange by climb and then, under stress, glide to annihilate or interact with each other to form a stable polygonized substructure. An example of a polygonized network revealed by etching the surface of a <100> crystal deformed to fracture at 1900°C is illustrated in Fig. 2. It could be seen that most of the dislocations

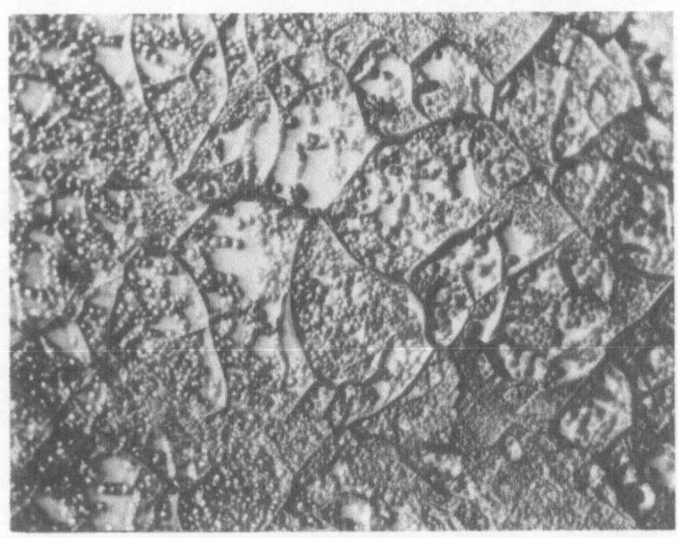

Fig. 2. Substructure in a single crystal deformed to fracture at 1900°C. <100> tensile axis. Etched to reveal dislocation distribution. 250 ×.

were located in sub-boundaries; the regions within the sub-boundaries were relatively clear of dislocations.

Electron transmission studies showed that fully developed sub-grain boundaries consisted of fairly regular dislocation networks of the kind illustrated in Figs. 3a and 3b. These boundaries formed as a result of the interaction of dislocations with Burgers vectors either at 90° or 60° to each other. Dislocations with Burgers vectors at 90° could possibly react according to the equation

$$\frac{a}{2}[1\bar{1}0] + \frac{a}{2}[110] \rightarrow a[100] \tag{1}$$

While this reaction results in no net decrease in elastic energy, the decrease in core energy for certain orientations of the dislocation lines involved might favor it. In general, however, this reaction was not observed; instead, the dislocations formed a

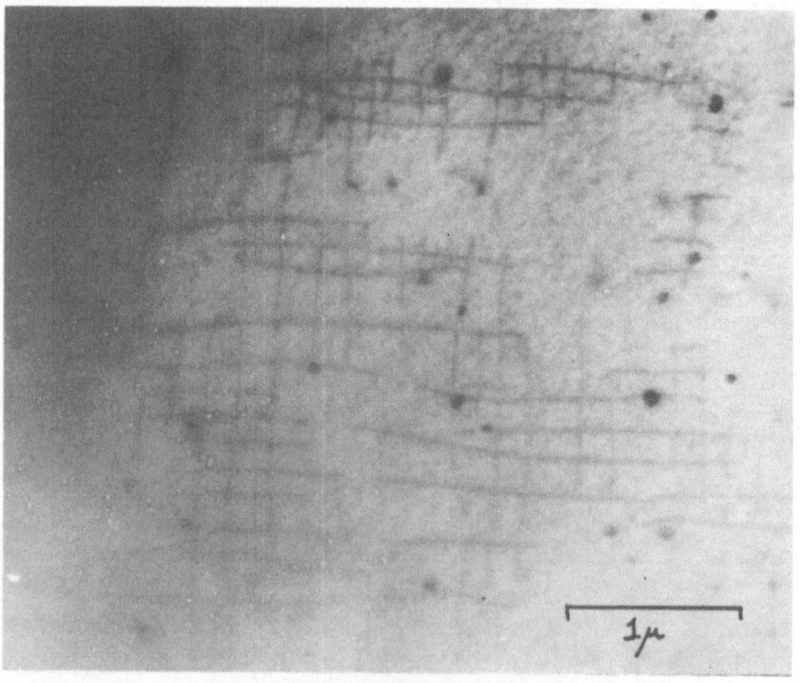

Fig. 3a. Electron transmission micrograph of dislocation networks in a magnesium oxide single crystal deformed at 1800°C. Burgers vectors at 90°.

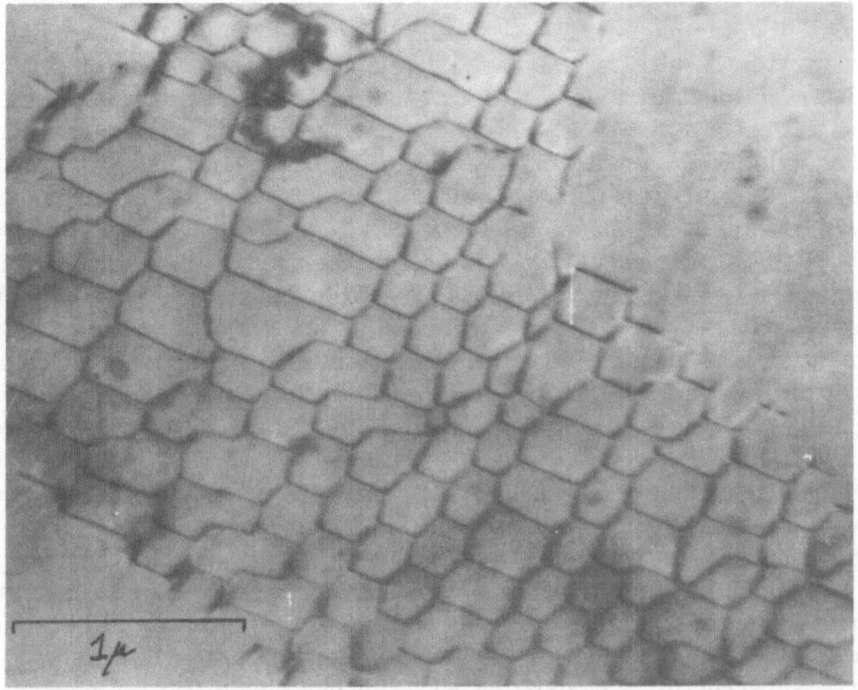

Fig. 3b. Electron transmission micrograph of dislocation networks in a magnesium oxide single crystal deformed at 1800°C. Burgers vectors at 60°.

square net with the lines parallel to [110] and [110], respectively, as shown in Fig. 3a. Dislocations with Burgers vectors at 60° could react according to the equation

$$\frac{a}{2}[1\bar{1}0] + \frac{a}{2}[\bar{1}01] \rightarrow \frac{a}{2}[0\bar{1}1] \qquad (2)$$

with a reduction in elastic energy. This reaction was observed. The dislocations formed a hexagonal net, as shown in Fig. 3b, with the three dislocations in the reaction above constituting the segments of the net. The resultant dislocation segments, $a/2[0\bar{1}1]$, lie parallel to [110]. Irregularities or singularities in the networks arise from the absorption at the network interface of dislocations having other Burgers vectors [11-13].

Networks similar to those described above have been observed

and analyzed in detail by Amelinckx [11-13]. He decorated dislocations in deformed and annealed alkali halide crystals and examined
them with the ultramicroscope. While the dislocation distributions
and reactions were identical to those described here, the greater
magnification available with the electron microscope meant that
the dislocation density in the magnesium oxide networks was much
higher, and, therefore, the angular misorientation between subgrains
was larger, corresponding to 10 − 20°.

Two points should be noted in connection with the hexagonal
networks. First, they were more stable because a reaction took
place at their nodal points. Second, extensive occurrence of this
reaction was possible only when considerable interpenetration of
slip on systems at 60° occurred. The development of stable subgrains, therefore, was most likely in single crystals tested above
1700°C. It was found that, after etching a <100> tensile axis
orientation specimen pulled to fracture, the heavily necked region
was in fact highly polygonized with sharply defined subgrain or even
grain boundaries.

It is our opinion that the work-hardening shown by the <100>
orientation specimens after approx. 20% elongation (see Fig. 1) was
a direct consequence of the development of a stable hexagonal-
type substructure. Consistent with this hypothesis was the behavior
of <110> orientation specimens. The particular feature of this
orientation was that slip was forced to take place on systems with
Burgers vectors at 60° in order to maintain axial alignment during
elongation. Thus, a stable substructure was developed right from
the initiation of yielding throughout the whole gage length. At high
strains, the polygonized substructure was fully developed, as
shown in Fig. 4a. This substructure accounted for the continuous
work-hardening measured on these specimens as indicated in Fig. 1.

Now, it was also noticed that a <100> orientation specimen,
which had been pulled to fracture at a temperature above 1700°C
and then annealed at a higher temperature, say 2000°C, recrystallized to develop a single large grain which extended only as far as
the boundary of the two-dimensionally necked region. This observation suggested that the ability to recrystallize was in some manner
associated with deformation and work-hardening in the presence of
stable hexagonal networks. If this is the case, then it was argued
that a <110> orientation specimen should recrystallize throughout
its whole gage length. This, in fact, proved to be so.

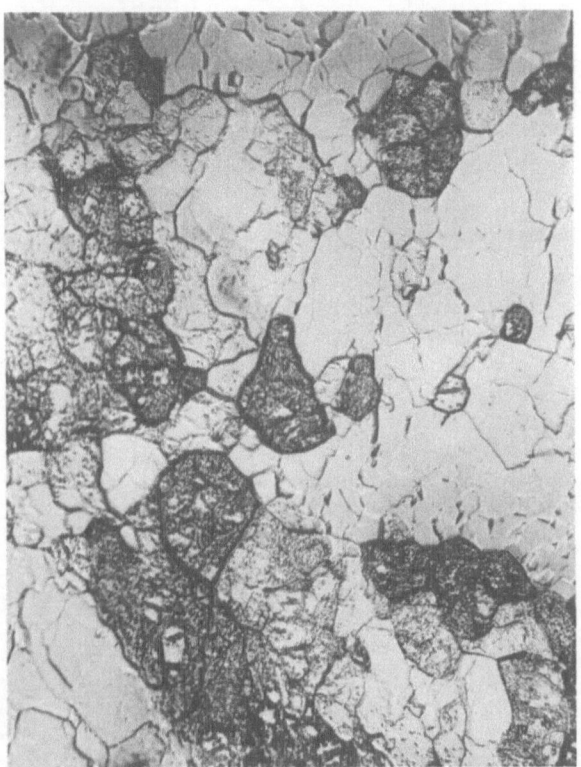

Fig. 4a. The substructure of recrystallized magnésium oxide single crystals after 60% elongation at 1800°C. <110> tensile axis. 150 ×.

The procedure for producing large-grained polycrystalline specimens of magnesium oxide by recrystallization is as follows. Single crystals of <110> tensile axis orientation were deformed by approximately 60% at 1800°C and then annealed for 15 – 30 min at 2000°C. During deformation at 1800°C, they showed continuous work-hardening with a fairly uniform elongation within the gage section. On examination at this stage, they had the highly poly-gonized appearance shown in Fig. 4a. After annealing at 2000°C, they recrystallized to produce the grain structure revealed by etching in Fig. 4b. Laue back-reflection photographs before and after recrystallization (Figs. 5a and 5b) showed that the grains

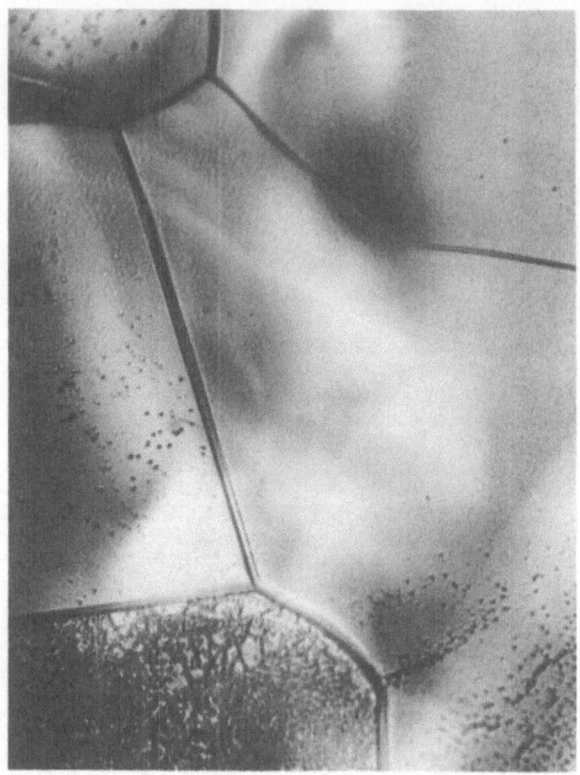

Fig. 4b. Grain structure of the same specimen as in Fig. 4a after 30-min anneal at 2000°C.
150 ×.

had a random orientation in general. Prolonged annealing at 2000°C
produced a more stable polycrystalline configuration consisting
of either a simple bicrystal or even a single crystal of new orientation. There were some instances where recrystallization could
not be induced even after long annealing periods and the grains
retained their original highly polygonized structure.

This process of recrystallization produced grain boundaries
and a form of polycrystalline material which was particularly well-
suited for fundamental mechanical deformation studies. The grain
boundaries were absolutely clean and free from any pores or flaws;
there was no evidence for impurity precipitation at the grain
boundaries or along the triple lines. Finally, because the poly-

crystalline tensile specimens were prepared in situ by high-temperature deformation and recrystallization, there was no chance for flaws to be introduced during preparation. All of these variables were significant, and it was important that they could be eliminated from consideration as a possible limiting factor in the study of mechanical behavior.

The observations made on the deformation, polygonization, and recrystallization of single crystals are summarized in Table II.

EFFECT OF SUBGRAIN BOUNDARIES AND GRAIN BOUNDARIES ON ROOM-TEMPERATURE MECHANICAL PROPERTIES

Results

Although this subject has been discussed in detail elsewhere [1,7], the main conclusions will be restated here for the sake of

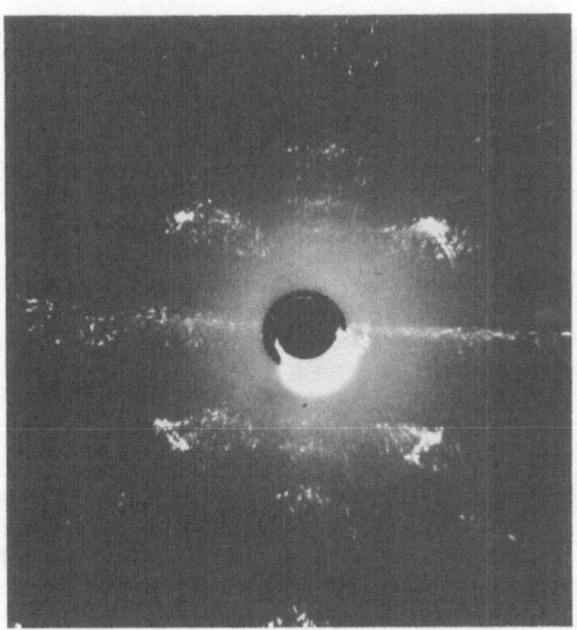

Fig. 5a. Laue X-ray back-reflection photograph showing the breakup of asterism due to polygonization before recrystallization.

Fig. 5b. Laue X-ray back-reflection photograph after recrystallization.

TABLE II

Summary of Relevant Findings on Mechanical Behavior of Magnesium Oxide Single Crystals at High Temperatures

Orientation	Slip interpenetrability	Polygonization and recrystallization	Slip multiplicity
<100> Tensile axis	90° Interpenetration above 1300°C 60° Interpenetration above 1700°C	Polygonization above 1500°C Recrystallization difficult (only in necked region above 1700°C)	Wavy slip above 1500°C
<110> Tensile axis	90° Interpenetration above 1500°C 60° Interpenetration above 1500°C	Polygonization during deformation above 1500°C Recrystallization throughout gage length when deformed 50% at 1800°C and annealed at 2000°C	Slip on {100}, {110}, {111} planes above 1500°C. Complex modes

completeness and to allow the introduction of new supplementary observations.

Research on mechanical properties at room temperature has shown that magnesium oxide single crystals are plastic only in the presence of suitably oriented, mobile slip dislocation sources. In the absence of these fresh dislocation sources, single crystals behave elastically and support stresses far in excess of the normal yield stress [1, 7, 14]. The mechanical behavior of fused bicrystals also depends on the presence of slip dislocation sources, but the consequences are rather different. In the absence of slip sources, bicrystals are also strong and elastic, but, in the presence of mobile dislocations, they are weak and brittle in tension [1, 7]. This weakness is due to the fact that mobile dislocations moving in slip bands become blocked by the grain boundary. This results in the generation of high stress concentrations which eventually leads to the nucleation of a crack and intergranular separation of the two crystals. Figures 6a and 6b illustrate such a crack nucleus and the mechanism considered to be operative.

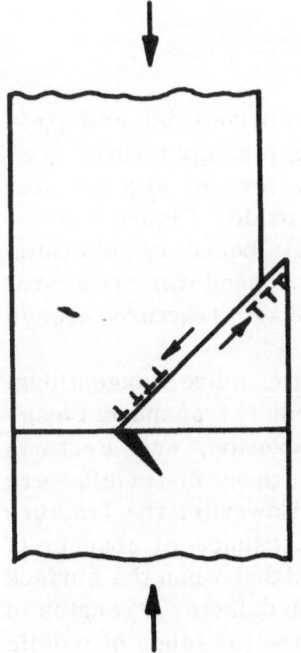

Fig. 6a. Mechanism of crack generation at a grain boundary due to blockage of room-temperature slip.

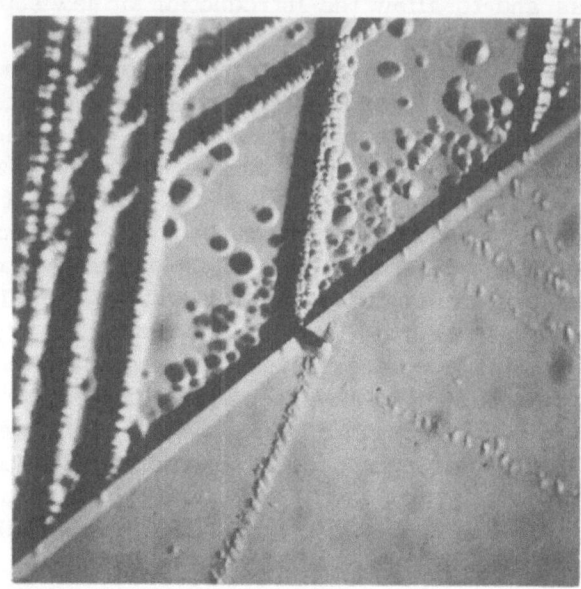

Fig. 6b. Crack generation at a grain boundary due to blockage of room-temperature slip.
500 ×.

Recent work has shown that the subgrain boundaries and grain boundaries generated by deformation at high temperatures (described above) are equally effective as barriers to slip and also result in the embrittlement of magnesium oxide. Figure 7 illustrates the interaction of slip with a subgrain boundary generated at 2000°C. While some of the slip bands crossed this structural discontinuity, others were blocked. Cleavage fracture always originated from such a discontinuity.

Similar tests have been performed on the more conventional polycrystalline material, such as the high-density transparent magnesia from Boeing Company, Avco Corporation, and Eastman Kodak Company (see Table I). In all cases, these materials were completely brittle at room temperature. However, the fracture strength appeared to be sensitive to the presence or absence of mobile slip dislocation sources. It was found that when the surface was very carefully treated to eliminate all defects, strengths of the order 30,000 psi were attained, but, in the presence of mobile dislocations, this dropped by one-third to a value between 15,000 and 20,000 psi [1].

Conclusions

The conclusion from all this work is that any structural discontinuity capable of blocking slip dislocations, i.e., subgrain boundaries or grain boundaries, will result in embrittlement of high-density polycrystalline magnesium oxide at room temperature. The effect is quite fundamental and is a consequence of the extreme stress concentration generated at the tip of a slip band in this material; it does not depend upon the character or quality of the boundary and is not due to the presence of impurity there. As proposed elsewhere [7], the particular sensitivity of grain boundaries in magnesium oxide to blocked slip is related to the dislocation distribution within the slip band. The dislocation density is very high and localized by the restrictions on cross slip; thus, the stress concentration due to a blocked slip band is strongly focused at the

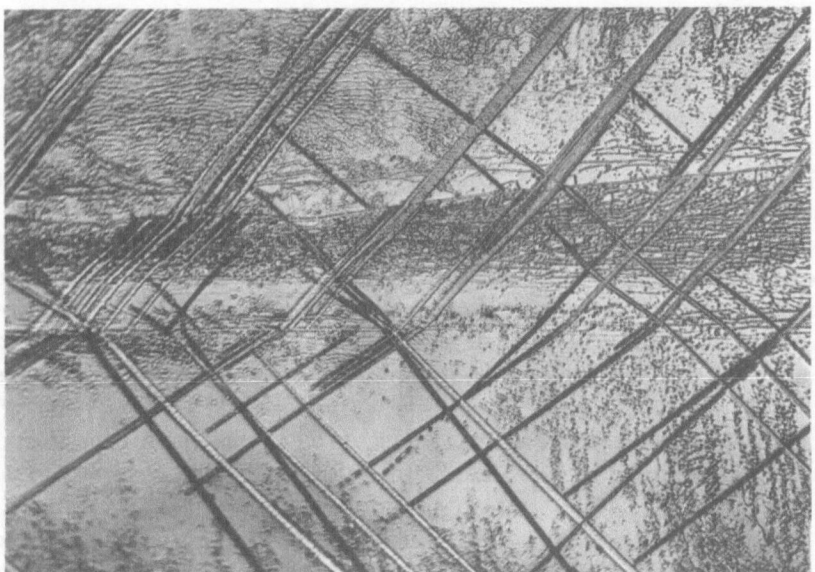

Fig. 7. Interaction of room-temperature slip (straight bands) with subgrain boundary produced in a single crystal by deformation at 2000°C. 150 ×.

grain boundary. Only simple tilt boundaries of a few degrees misorientation can relax the stress concentration by the transfer of slip from one grain to another.

EFFECT OF GRAIN BOUNDARIES ON HIGH-TEMPERATURE MECHANICAL PROPERTIES

The role of microstructural discontinuities, such as subgrain and grain boundaries, as barriers to slip and, consequently, the poor tensile properties of magnesium oxide just described for room temperature persisted up to relatively high temperatures. It was not until temperatures were raised to 1700°C and above (depending on the quality of the polycrystalline material) that recognizable changes in fracture mechanism and deformation behavior appeared. At these temperatures, the increased multiplicity of slip planes, interpenetrability of slip vectors (see Table II), polygonization, and grain boundary sliding all contributed to the change in deformation behavior. These effects were best demonstrated by recrystallized single-crystal "polycrystalline" material and to a lesser extent by the Boeing Company high-density polycrystalline magnesia, and it was on these materials that the most detailed qualitative studies were made.

Results

Observations on Grain Boundaries Formed by Recrystallization. As mentioned earlier, recrystallized single-crystal specimens contained the most perfect kind of grain boundary and provided the best medium for studies of the fundamental mechanisms of poly-crystalline deformation. Our studies showed that below 1700°C these specimens were relatively brittle and behaved in a manner similar to that at room temperature, but above 1700°C they were very ductile and yielded, deformed, and necked down to a ductile fracture just as if the grain boundaries did not exist. Thus, coarse-grained recrystallized magnesia might be said to go through a brittle-to-ductile transition around 1700°C for the present strain rate. The manner in which this transition occurred was very interesting.

In the brittle range, i.e., below 1700°C, the deformation before fracture was so slight that it could not be picked up on the load-elongation curve and was therefore less than 1%. However, there was no doubt that a slight amount of slip was occurring at 1600°C

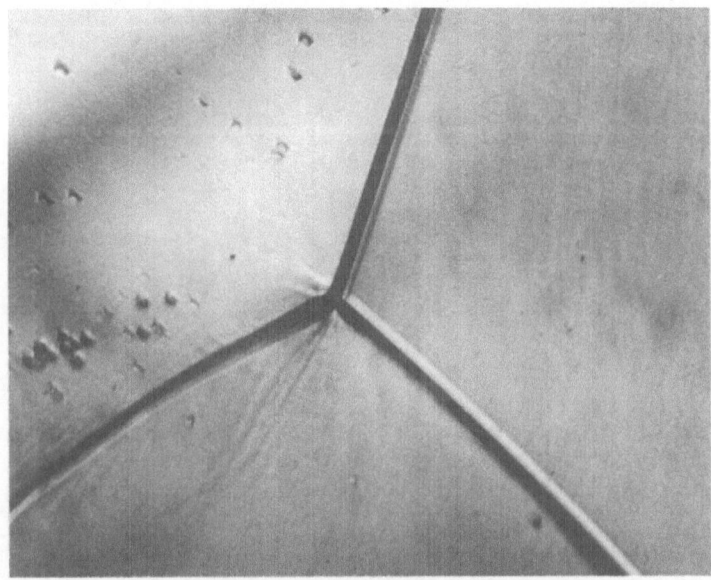

Fig. 8. Slip markings in the vicinity of a triple point at 1600°C. Recrystallized specimen deformed to fracture (i.e., approx. 1% elongation). 290 ×.

as manifested by the few slip traces which appeared on the surface of the specimen in the vicinity of the fracture (see Fig. 8). There was evidence of wavy slip (or surface "folds") in the vicinity of some of the triple points. Of much greater significance, however, were the flaws which developed on interior intergranular surfaces, particularly along the triple lines. Figure 9a, taken with transmitted light, shows an intergranular flaw, and Fig. 9b, taken with transmitted polarized light, shows evidence for localized plastic deformation in the same area. It was considered that these flaws generated as a consequence of stress concentrations set up by the local plastic deformation. Such stress concentrations eventually led to intergranular rupture and brittle failure. The fracture surface was a mixture of cleavage and intergranular, but the source was always clearly intergranular in this temperature range.

In the transition range, i.e., around 1700°C, there was considerable ductility in tension, but the deformation was accompanied by profuse intergranular sliding and intergranular rupture (cavita-

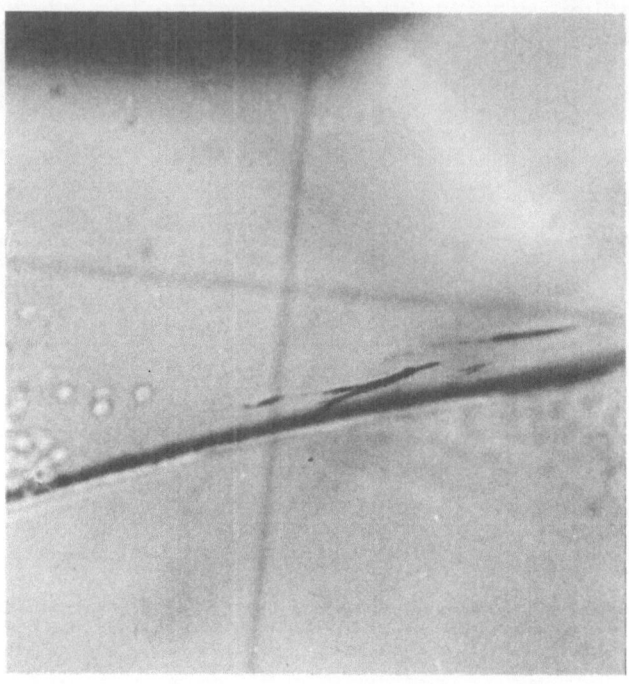

Fig. 9a. Transmitted light showing intergranular flaws which developed in a recrystallized specimen deformed to fracture (i.e., <1% elongation) at 1600°C. 150 ×.

tion).   Figure 10 shows a triple point where sliding had obviously taken place during deformation. The grain boundary (vertical in the picture) had sheared most, whereas shearing of the other grain boundaries had caused a complex folding of the surface in their immediate vicinity. In some areas, this rumpling was accompanied by a corrugation of the grain boundary interface. (Note particularly the grain boundary region in the vicinity of the arrow.)

In the ductile range, i.e., above 1700°C, the recrystallized specimens were extremely plastic and eventually failed by necking down to a ductile fracture. In order to understand the reasons for such ductility, the development of slip in the vicinity of triple points and grain boundaries was followed. At first, the observations of slip were similar to those just described for specimens deformed at 1600°C, i.e., folding due to wavy slip developed near the triple points as in Fig. 8 (see also Fig. 15) and rumpling and corrugations

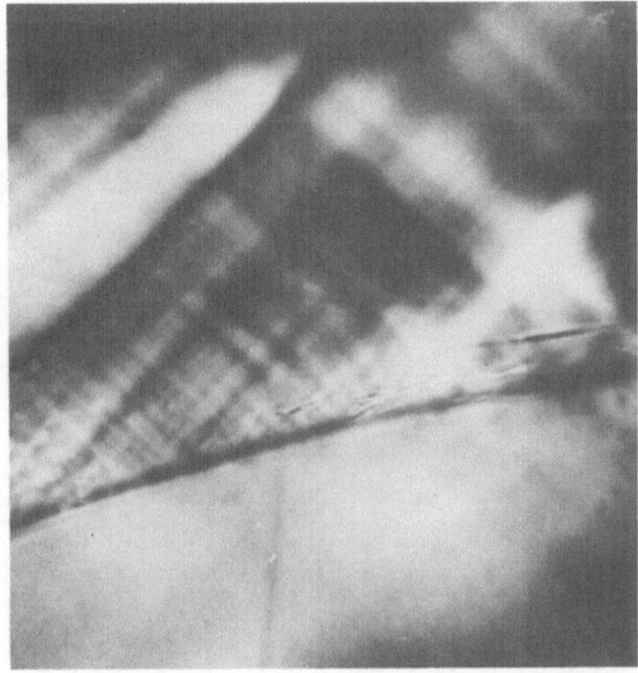

Fig. 9b. Transmitted polarized light showing slip distribution. 150 ×.

developed along the grain boundaries.   However, there was no
evidence of massive intergranular sliding or cavitation. At larger
deformations, the surface markings became more intense and, in
addition to slip traces, the surface in the vicinity of triple points
and along the grain boundaries developed outlines of a cellular
substructure.   An example of this kind of surface may be seen in
the lower part of Fig. 10 (at ×). Etching these specimens revealed
that the dislocations participating in deformation had indeed re-
arranged by polygonization, and the individual grains were themselves breaking up by polygonization.   Figure 11 shows the dislocation arrangement in the vicinity of a triple point after 5%
elongation.   After about 20% elongation, the breakup of the grains
by plastic deformation and polygonization had proceeded to the
point where the original grain boundaries were losing their
identity.   The grain boundaries were no longer sharp and straight,

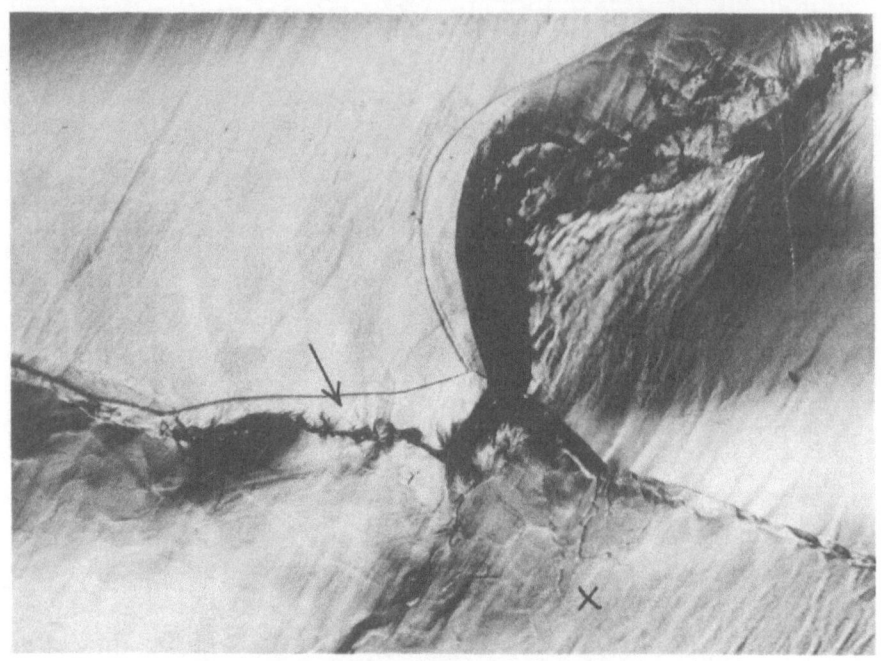

Fig. 10. Grain boundary sliding and corrugation at 1700°C. Recrystallized specimen
deformed to fracture (i.e., approx. 10% elongation). 150 ×.

but distinctly corrugated, as shown in Figs. 12a and 12b, with the
corrugations now defining outlines of the new subgrain (or grain)
boundaries.   The process whereby a grain boundary becomes
corrugated and the material around it broken up is summarized
diagrammatically in Fig. 13.   It was very important to note that
there was never any indication of intergranular rupture in these
specimens until the final stages of ductile rupture.

Our current understanding of the brittle-to-ductile transition
in these recrystallized polycrystalline specimens is, therefore, as
follows.   First, it should be noted that the brittle range (below
1700°C) is below the temperature at which single crystals show
multiple interpenetrating slip (Table II).   Thus, even though wavy
slip is possible, the lack of interpenetrability does not permit the
concurrent operation of five independent slip systems in any given
volume of material and the von Mises criterion is not completely

satisfied.   (For a more detailed discussion of the von Mises criterion, see papers by Copley and Pask [5] and Groves and Kelly [15].)   Thus, slip in the individual grains cannot be completely accommodated across a grain boundary interface and stress concentrations arise.   These stress concentrations can be relaxed by grain boundary sliding at high enough temperatures (i.e., above approx. 1400°C [16, 17]), but this then results in the development of internal flaws (cavitation) along the triple lines. [See Fig. 14 (a).] Eventually these flaws grow so large that they can no longer be supported by the specimen, and brittle intergranular fracture and cleavage fracture ensue.   Thus, fracture always originates from an intergranular defect inside the specimen.

The ductile range (above 1700°C) is in the temperature range where multiple interpenetrating slip can occur, and it is possible for the grains to conform to each other's change in shape. If

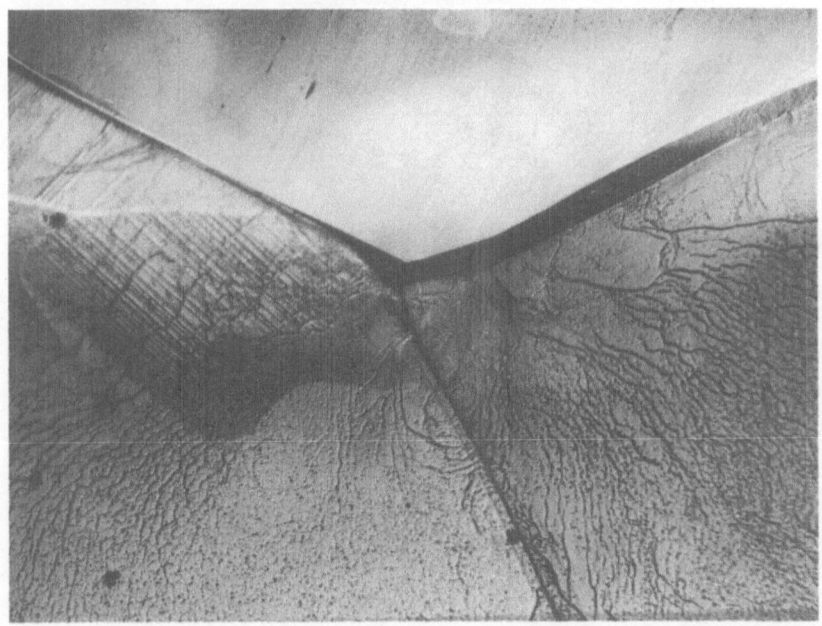

Fig. 11. Polygonization in the vicinity of a triple point at 1800°C. Recrystallized specimen deformed 5% and etched. 125 ×.

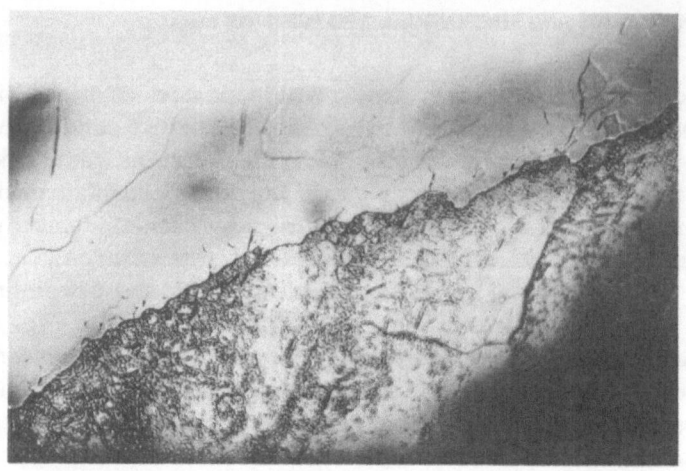

Fig. 12a. Grain boundary corrugation and polygonization at 1800°C in a grain boundary interface of a recrystallized specimen deformed 15%. 125 ×.

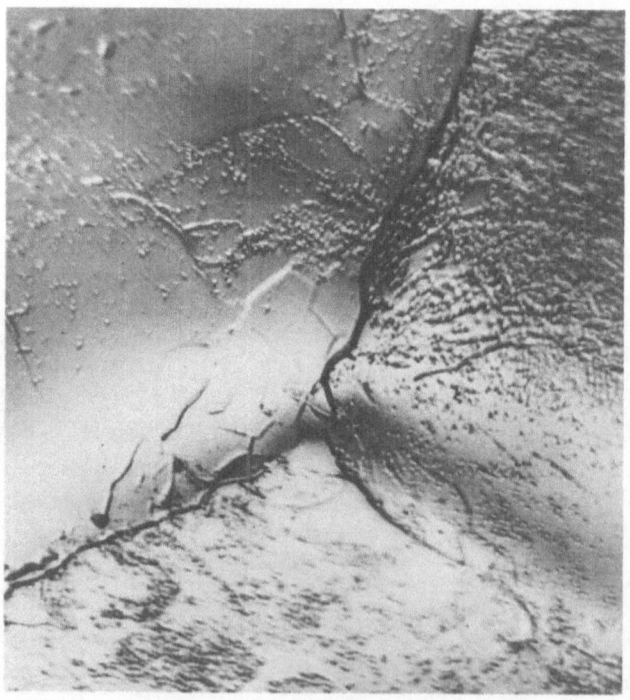

Fig. 12b. Grain boundary corrugation and polygonization at 1800°C at a triple point of a recrystallized specimen deformed 15%. 325 ×.

grain boundary sliding should occur, then it can be accommodated by plastic flow within the grains along the flow lines illustrated in Fig. 14 (b) without the generation of internal flaws. Furthermore, the temperature is high enough for the dislocations participating in plastic flow to polygonize. This permits the material in the vicinity of triple points to become redistributed, as shown in Figs. 12b and 13 (c), and permits grain boundary corrugation to be accommodated by transfer of material from one grain to another through polygonization and grain growth, as shown in Figs. 12a and 13 (c). It is important to note that the breakup of the individual grains and corrugation of the grain boundary interface makes grain boundary sliding more difficult, and this probably accounts for the lack of intergranular cavitation at temperatures above the brittle-to-ductile transition. So the grains continue to deform and polygonize, and the original grain boundaries are eventually obliterated and replaced by a highly polygonized microstructure. Since this microstructure is essentially similar to that of single crystals after 20% elongation, it is not surprising that the mechanical behavior of the recrystallized polycrystalline specimens should be similar. They also show a period of work-hardening prior to necking down to a ductile fracture similar to the single crystals in Fig. 1.

The transition range (1700°C) is one where the grains still cannot conform to each other and intergranular flaws develop, but the material is now sufficiently plastic that the flaws do not propagate immediately for brittle cleavage-type fracture; instead, fracture occurs by intergranular separation.

Grain Boundaries Formed by Hot-Pressing. As mentioned earlier, we were fortunate to obtain from R. W. Rice of the Boeing Company some hot-pressed magnesium oxide of theoretical density which retained its transparency after a high-temperature anneal at 2000°C. There was considerable grain growth (see Table I) during heat treatment, and the grain size became almost as large as in the recrystallized single crystals. While this material looked clear and defect-free and the grain boundaries on the surface were straight and sharp, there was, nevertheless, within the material slight residual porosity which appeared as elongated bubbles along some triple lines, together with some impurity particles within the grains and on the grain boundary interfaces. The difference between recrystallized single crystals and this high-density, hot-

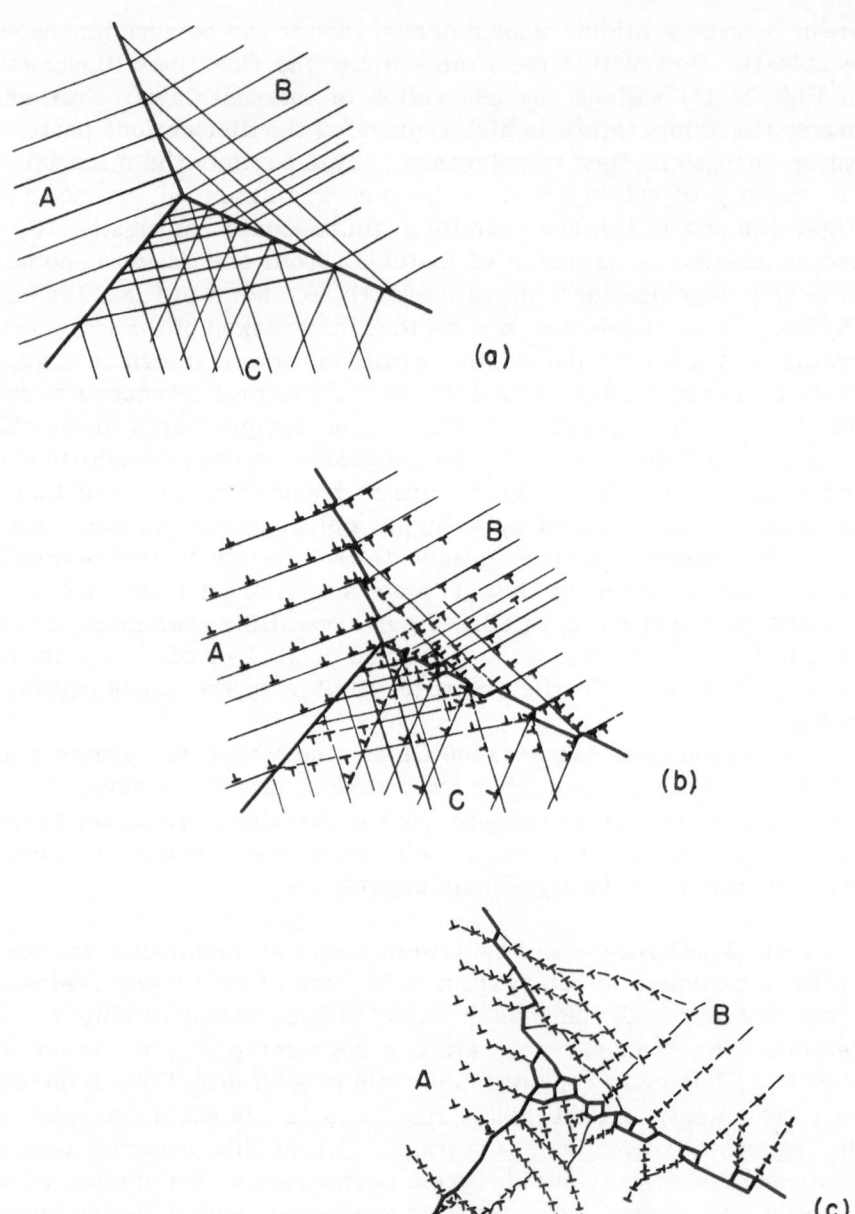

Fig. 13. Diagrammatic illustration of grain boundary corrugation by slip (b) and polygonization (c).   Boundary between B and C in (b) may represent situation above arrow in
Fig. 10, and in (c) may represent situation in Fig. 12a.

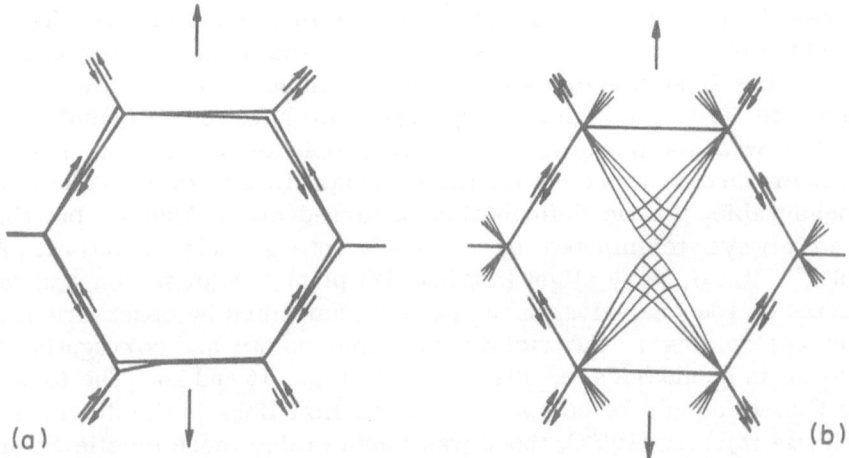

Fig. 14. Grain boundary sliding resulting in: (a) the formation of voids along triple lines at low temperatures, and (b) surface folds along flow lines at high temperatures (after Brunner and Grant [22]).

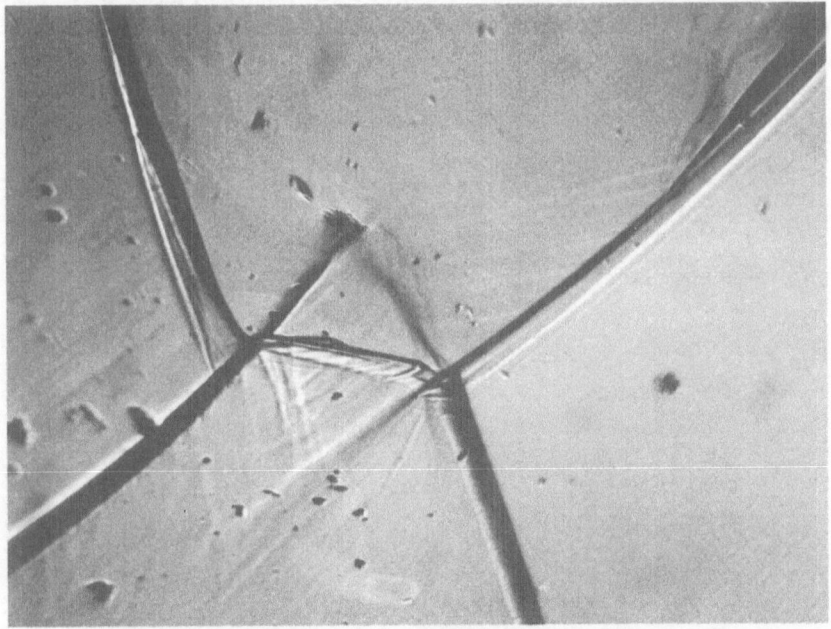

Fig. 15. Slip markings in the vicinity of triple points. Boeing Company high-density, hot-pressed magnesia deformed to fracture (i.e., approx. 1% elongation) at 1900°C. Compare with Fig. 8 and Fig. 14(b). 250 ×.

pressed material was, therefore, one of origin and quality. These slight differences had a marked effect on the mechanical behavior.

In the first place, there was no example of completely ductile fracture up to the highest temperature we have tested (2100°C), so that it was not possible to identify a brittle-to-ductile transition temperature as was done for the recrystallized material. The first measurable plastic deformation occurred above 1900°C, but this was always terminated by a brittle intergranular fracture. At 1900°C itself, very slight (approx. 1%) plastic deformation was detected on the load-elongation curve accompanied by observations of surface folding in the vicinity of triple points and corrugation of the grain boundaries, as illustrated in Figs. 15 and 16. The folding in Fig. 15 should be compared with the flow lines in the diagram of Fig. 14 (b). At 2100°C, there was considerably more tensile deformation (approx. 10%) before intergranular fracture. At this temperature, grain-boundary corrugation was accommodated to a certain extent by polygonization of the dislocations participating in plastic flow, as shown in Fig. 17. This figure should be compared with Fig. 12 and the diagram of Fig. 13 (c) between grains A and C. Thus, similar deformation characteristics were identified in the

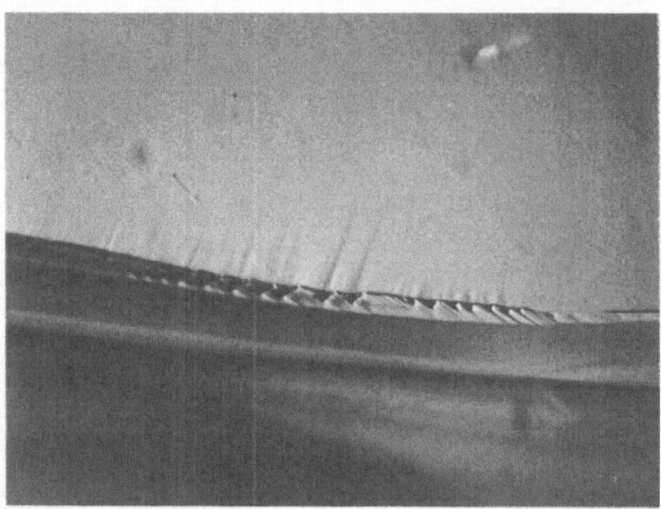

Fig. 16. Slip markings and grain boundary corrugation in Boeing Company magnesia deformed to fracture (i.e., approx. 1% elongation) at 1900°C. 500 ×.

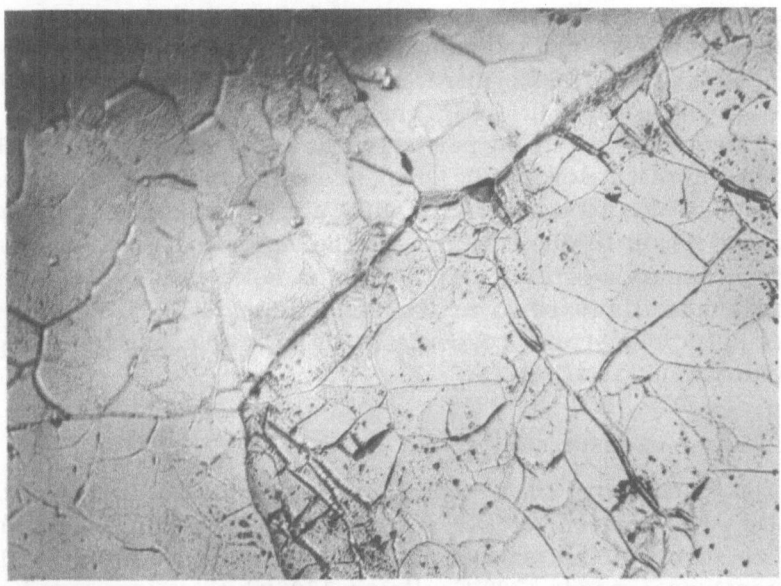

Fig. 17. Grain boundary corrugation and polygonization in Boeing Company magnesia deformed to fracture (i.e., approx. 10% elongation) at 2100°C. Compare with Fig. 12b and Fig. 13 (c). 150 ×.

fully dense hot-pressed material as in the recrystallized single crystal, but at higher temperatures.

High-density, hot-pressed magnesia of much finer grain size, from other sources (Eastman Kodak Company, Avco Corporation, and IITRI), were completely brittle up to the highest test temperature, 2100°C. Load—elongation curves gave only the slightest indication of plastic yielding before fracture. Unfortunately, it was not possible to determine from optical examination of the surface whether any yielding had occurred because the surfaces were so badly thermally eroded as to be quite unsuitable for resolution of slip traces. Thermal erosion was more prevalent in these specimens because of the residual porosity and smaller grain size after the 2000°C pre-anneal (see Table I).

The greater brittleness of the hot-pressed material was considered to be due to the slight intergranular porosity present in the material beforehand. The porosity had two effects. First, it meant that, even at temperatures above 1700°C where the grains could conform to each other's change in shape, intergranular flaws and, therefore, potential sources of intergranular rupture were present at the start of deformation, and the situation illustrated in Fig. 14 (a) always prevailed. Second, the presence of porosity made it more difficult for grain boundary corrugation to be accommodated by localized polygonization and grain growth. It is well-known that macroscopic grain growth is inhibited by porosity, presumably due to the relative difficulty of transferring material from one grain to another through the vapor phase rather than by a simple rearrangement of ions in the solid phase. Thus, these intergranular flaws could neither be eliminated nor avoided, and under sufficient stress concentration they were able to propagate for intergranular failure.

Conclusions and Discussion

The conclusion from these high-temperature studies is that plastic deformation of polycrystalline magnesium oxide under tension, even in the absence of porosity, is not possible until the temperature exceeds 1700°C. Below 1700°C, there are insufficient independent slip systems capable of interpenetrating and operating concurrently within a given volume of material to satisfy the von Mises criterion. Consequently, grain boundary constraints arise which, when relaxed by intergranular sliding, lead to intergranular flaws and eventually intergranular rupture. Above 1700°C, there are sufficient independent slip systems to satisfy the von Mises criterion and complete continuity can be maintained between adjacent grains. The additional ability of the material to polygonize and recrystallize permits a continuous relaxation of internal stresses and contributes to the overall plasticity and change of shape under tension. Deformation proceeds without intergranular separation, and fracture continues in a truly ductile manner.

In the presence of porosity, intergranular flaws are always present and propagate under the influence of stress concentration for intergranular rupture at temperatures as high as 2100°C. Consequently, although plastic deformation is observed above 1900°C in hot-pressed material, there is no indication that completely ductile fracture will occur.

On the basis of these observations, it is quite obvious that the high-temperature deformation and creep of magnesium oxide is a process controlled by the movement, interaction, and climb of dislocations, in agreement with an earlier observation by Cummerow [18]. This is in contrast to the behavior of most other oxide ceramics where high-temperature creep apparently occurs by the Nabarro—Herring mechanism of vacancy diffusion to regions of high dilatation [19, 20]. However, in most other oxide ceramics, dislocation movement occurs only with difficulty at the higher temperatures, and even then there are insufficient independent slip systems for polycrystalline plasticity due to slip [21]. Consequently, grain boundaries play a different, but still very important, role in these materials, by providing an easy path for vacancy diffusion.

It is fascinating to realize how similar the observations reported here on magnesium oxide are to those made on ductile metals undergoing creep at a similar homologous temperature [22—24]. In fact, the descriptive terms we have used here—"folding" of the surface at triple points, "corrugation" of the grain boundaries, "cavitation" in internal surfaces, and "cellular" substructure—are all taken from the metallurgical literature. Also, Fig. 14 (b) has been used previously to describe observations on the role of grain boundary sliding in the creep of metals. Actually, it is not too surprising that the deformation of magnesium oxide should become "metallic" at high enough temperatures, since, of all the refractory oxides, it has the simplest crystal structure; when multiple interpenetrating slip and polygonization occurs, it possesses all the attributes of a ductile face-centered cubic metal undergoing creep.

ACKNOWLEDGMENTS

It is a pleasure to acknowledge the experimental assistance of P. N. Johnson. The authors are extremely grateful to Dr. R. M. Spriggs and P. F. John of Avco Corporation, R. W. Rice of the Boeing Company, and S. Firestone of IITRI for supplying high-density magnesium oxide. This work was supported by the U. S. Air Force, Systems Engineering Group (Research and Technology Division) under Contract No. AF 33 (615)-1282.

REFERENCES

1. R. J. Stokes and C. H. Li, "Dislocations and the Tensile Strength of Magnesium Oxide," J. Am. Ceram. Soc. 46:423 (1963).
2. R. B. Day and R. J. Stokes, "Mechanical Behavior of Magnesium Oxide at High Temperatures," J. Am. Ceram. Soc. 47:493 (1964).

3. C. O. Hulse, S. M. Copley, and J. A. Pask, "Effect of Crystal Orientation on Plastic Deformation of Magnesium Oxide," J. Am. Ceram. Soc. 46:317 (1963).

4. S. M. Copley and J. A. Pask, "Plastic Deformation of MgO Single Crystals Up to 1600°C," J. Am. Ceram. Soc. 48:139 (1965).

5. S. M. Copley and J. A. Pask, "Deformation of Polycrystalline Ceramics," this volume, Chapter 13.

6. R. D. Carnahan, T. L. Johnston, R. J. Stokes, and C. H. Li, "Effect of Grain Size on the Deformation of Silver Chloride at Various Temperatures," Trans AIME 221:45 (1961).

7. R. J. Stokes and C. H. Li, "Dislocations and the Strength of Polycrystalline Ceramics," in: H. H. Stadelmaier and W. W. Austin (eds.), Materials Science Research, Vol. 1, Plenum Press (New York), 1963, p. 133.

8. D. W. Budworth and J. A. Pask, "Effect of Temperature on the Plasticity of Polycrystalline Lithium Fluoride," Trans. Brit. Ceram. Soc. 62:763 (1963).

9. D. W. Budworth and J. A. Pask, "Flow Stress on the {100} and {110} Planes in LiF, and the Plasticity of Polycrystals," J. Am. Ceram. Soc. 46:560 (1963).

10. P. L. Pratt, C. Roy, and A. G. Evans, "The Role of Grain Boundaries in the Plastic Deformation of Calcium Fluoride," this volume, Chapter 14.

11. S. Amelinckx, "The Direct Observation of Dislocation Patterns in Transparent Crystals," in: J. C. Fisher, W. G. Johnston, R. Thompson, and T. Vreeland, Jr. (eds.), John Wiley & Sons, Inc. (New York), 1957, p. 3.

12. S. Amelinckx, "Dislocations in Ionic Crystals," Nuovo Cimento, Supplement 2, 7:569 (1958).

13. S. Amelinckx, "Dislocations in Ionic Crystals: I. Geometry of Dislocations," in: W. W. Kriegel and H. Palmour, III (eds.), Mechanical Properties of Engineering Ceramics, Interscience (New York), 1961, p. 9.

14. R. J. Stokes, "Dislocation Source and the Strength of Magnesium Oxide Single Crystals," Trans. AIME 224:1227 (1962).

15. G. W. Groves and A. Kelly, "Independent Slip Systems in Crystals," Phil. Mag. 8:837 (1963).

16. M. A. Adams and G. T. Murray, "Direct Observations of Grain Boundary Sliding in Bicrystals of Sodium Chloride and Magnesia," J. Appl. Phys. 33:2126 (1962).

17. G. T. Murray, J. Silgalis, and A. J. Mountvala, "Creep Rupture Behavior of MgO Bicrystals," J. Am. Ceram. Soc. 47:531 (1964).

18. R. L. Cummerow, "High-Temperature Steady-State Creep Rate in Single-Crystal MgO," J. Appl. Phys. 34:1724 (1963).

19. R. C. Folweiler, "Creep Behavior of Pore-Free Polycrystalline Aluminum Oxide," J. Appl. Phys. 32:773 (1961).

20. J. E. Burke, "Grain Boundary Effects in Ceramics," in: H. H. Stadelmaier and W. W. Austin (eds.), Material Science Research, Vol. 1, Plenum Press (New York), 1963, p. 69.

21. R. J. Stokes, "Correlation of Mechanical Properties with Microstructure," in: Microstructure of Ceramic Materials, National Bureau of Standards Miscellaneous Publication No. 257, 1964, p. 41.

22. H. Brunner and N. J. Grant, "Deformation Resulting from Grain Boundary Sliding," Trans. AIME 215:48 (1959).

23. H. Brunner and N. J. Grant, "Measurement of Deformation Resulting from Grain Boundary Sliding in Aluminum and Aluminum–Magnesium from 410°F to 940°F," Trans. AIME 218:122 (1960).

24. D. McLean, "Creep Processes in Coarse-Grained Aluminum," J. Inst. Metals 80:507 (1951-2).

Chapter 23

Internal Surfaces of MgO

Roy W. Rice

The Boeing Company
Seattle, Washington

Grain boundaries, phase boundaries, and void surfaces are discussed. The need for better grain boundary models for ionic materials is demonstrated by noting some charge problems of such boundaries. Second-phase accumulations at grain boundaries in bicrystalline and polycrystalline specimens are shown and the relationship of such accumulations to sintering, grain growth, and mechanical properties is discussed. Methods to improve and extend the study of grain boundaries are described. Boundaries between different phases are also briefly discussed, with particular reference to ceramic alloying. Interrelationships between voids, impurities, and dislocation phenomena are considered, and evidence of impurity accumulations at or near voids in single and polycrystalline bodies is presented. Methods of reducing mechanical problems resulting from these internal surfaces are discussed.

INTRODUCTION

The internal surfaces discussed in this paper are grain boundaries, boundaries between two phases, and void surfaces. The structure of these surfaces and their interrelationships with one another and with dislocations and impurities are considered. Particular emphasis is given to grain boundaries, with some discussion of their relationship to sintering, grain growth, and mechanical properties. Methods of improving study techniques and mechanical properties for ceramics are also discussed.

GRAIN BOUNDARIES

Structure

Grain boundaries are transition regions and are necessarily disordered to accommodate the lattice misorientations between adjacent grains. Misorientations of a few degrees or less can be accommodated by various combinations of (1) tilt boundaries, composed of edge dislocations resulting from rotations of two adjoining grains about an axis within the boundary (Fig. 1), and (2) twist

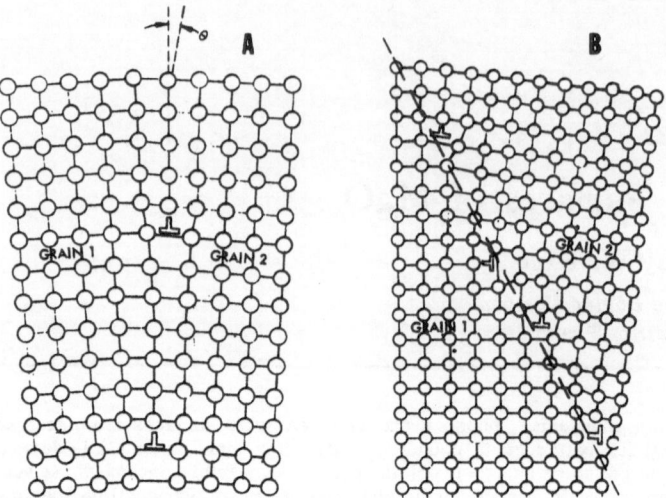

Fig. 1. Tilt grain boundary dislocation model. (A) Symmetrical tilt boundary formed by equal rotation of both grains. Note that only one set of dislocations between the grains is required. (B) Asymmetrical tilt boundary formed by unequal rotation of grains. Note that this requires two sets of dislocations, neither along the boundary.

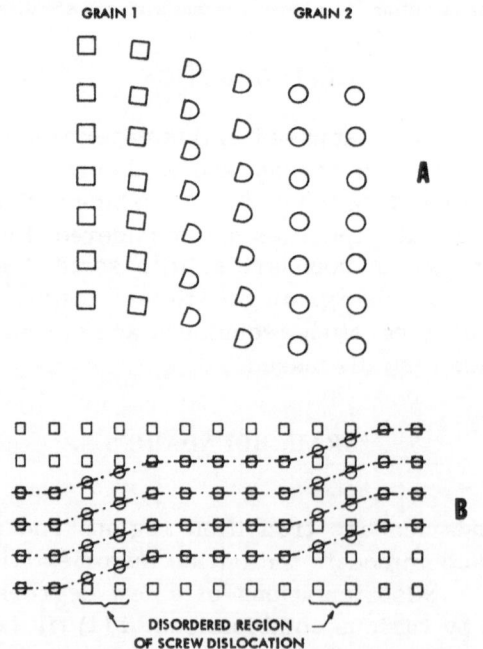

Fig. 2. Twist grain boundary dislocation model. (A) Cross section of boundary. (B) Overlap of two planes in the boundary.

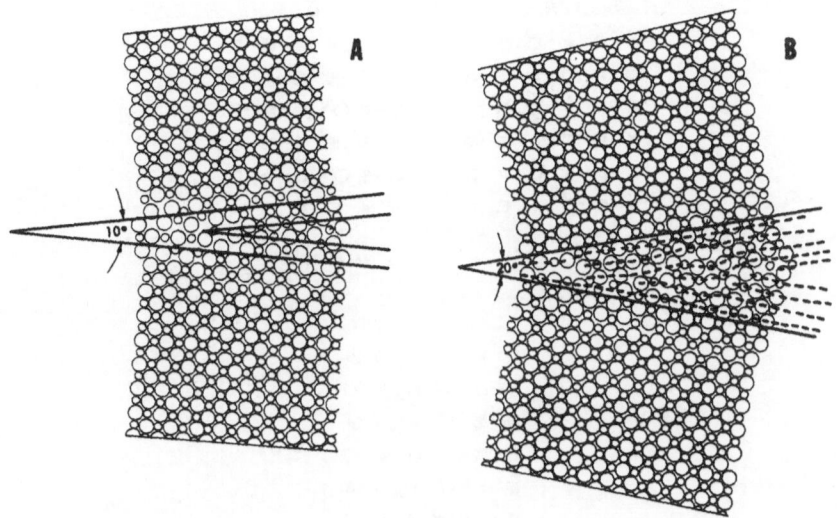

Fig. 3. Diatomic tilt boundaries.   (A) 10° and (B) 20° boundary between {100} surfaces.

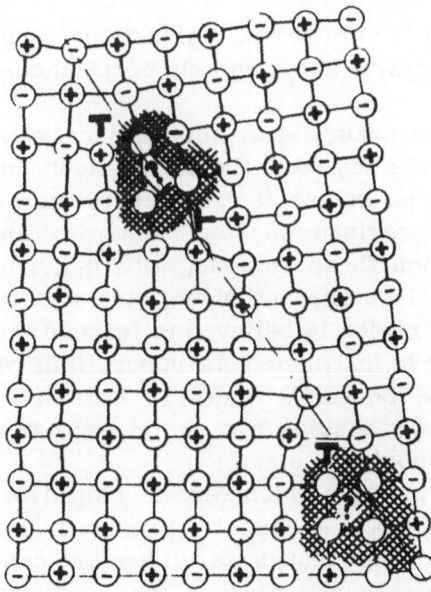

Fig. 4. Charge mismatch in a 10° asymmetrical tilt boundary.

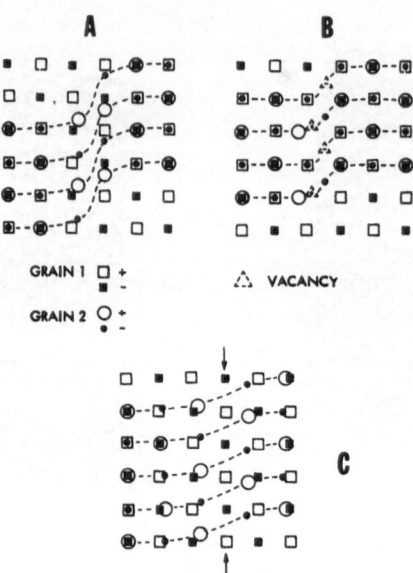

Fig. 5. Diatomic twist boundaries. Like charges cannot occupy adjacent lattice positions in an ionic material, therefore, complicating twist boundaries. Three possibilities are: (A) Bond stretching with a jump to the second plane of atoms, (B) bond compression and vacancy creation, and (C) bond stretching and no bonding to one row of atoms (arrow).

boundaries, composed of screw dislocations resulting from rotation of two adjoining grains about an axis perpendicular to the boundary (Fig. 2).

Some disorder occurs at or near dislocations with near-normal order between and away from the dislocations. Increasing the misorientation between grains is believed to increase the amount of disorder in the regions of some disorder and to decrease the separation between these regions, with individual dislocations no longer being distinguished at higher angles. However, the width of the boundary region is believed to be relatively constant. Seitz and Read [1] have estimated an upper limit of about five lattice spacings for the boundary width in materials characterized by strong short-range bonding forces. A width of about three lattice spacings is commonly accepted [2].

Models and concepts which were developed for essentially monatomic materials, such as metals, are generally applied to crystalline ceramics even though ceramics are complicated by unlike atomic components and different atomic forces. This general applicability is justified because of basically similar dislocation phenomena in such ceramics and metals and detailed experimental observations on a number of materials, such as NaCl-structure materials [3], Al_2O_3[4], and CaF_2[5]. Figure 3 shows sketches of symmetrical pure tilt boundaries of 10 and 20° in a diatomic system with radii ratios of about 2:1 as for MgO, NiO, or FeO. These indicate that the symmetrical tilt boundary model is applicable to such systems, at least in the chosen orientation, and give a general feeling for the increase in disorder with increasing misorientation. However, Fig. 4 shows that a problem of matching charges at the boundary arises with an asymmetrical tilt boundary in a material with ionic bonding, such as MgO, a condition that can be resolved only by the creation of boundary vacancies.

Similarly, greater disorder may occur in ionic twist boundaries as sketched in Fig. 5, since like charges are repulsed from adjacent lattice positions. These charge problems have been generally recognized, but they have not yet been thoroughly studied. It is felt, therefore, that some caution is required in directly applying monatomic boundary models to many ceramics and that it would be quite useful to determine if preferred grain orientations, increased boundary jogging, impurity accumulation, or changes in mechanical properties could be correlated with charge problems revealed in more detailed models.

Impurities

Boundary models and estimates must be used with caution for impure materials, due to the tendency of impurities to accumulate at the grain boundaries because of the greater accommodation the boundary disorder offers. As Kingery [2] points out, this problem of impurities may be more acute in ceramics, particularly the oxides, which can form metastable, noncrystalline, viscous substances in a variety of mixtures. It must also be noted that the inhomogeneous nature of the grain boundary could in itself extend the range of possible glass compositions over those found in bulk quantities. Hence, the concept of a submicroscopic glassy phase in grain boundaries, although now disregarded in metals, may be a potential factor in some ceramics. The importance of grain boundary impurities can be gauged to some extent by simple calculations showing that a boundary film 10A thick can be formed on all grains of a body having a 30-μ grain size by only about 0.01 vol.% impurities.

Several grain boundary surfaces exposed by intergranular

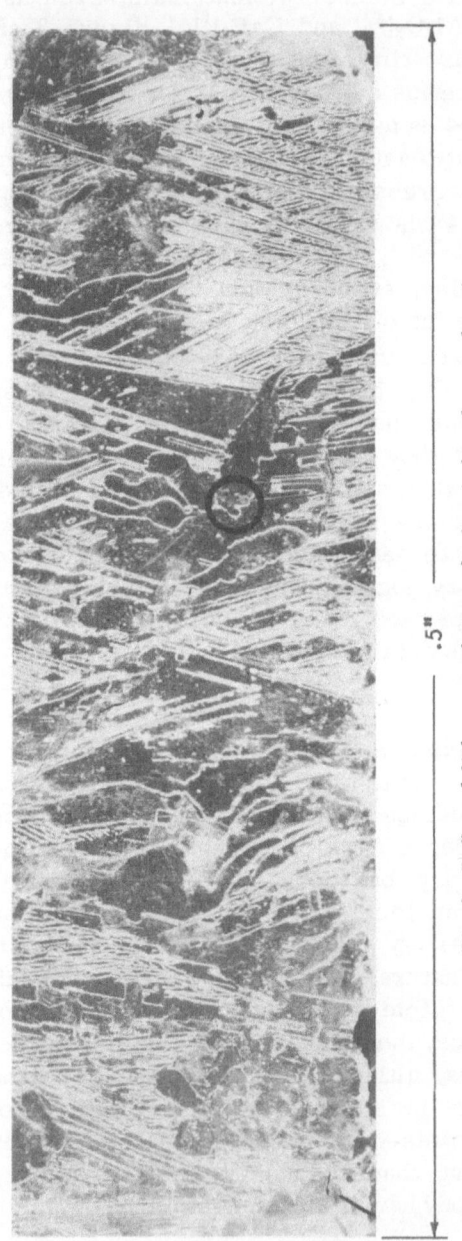

.5"

Fig. 6a. Reflected-light macrophotograph of grain boundary surface A.

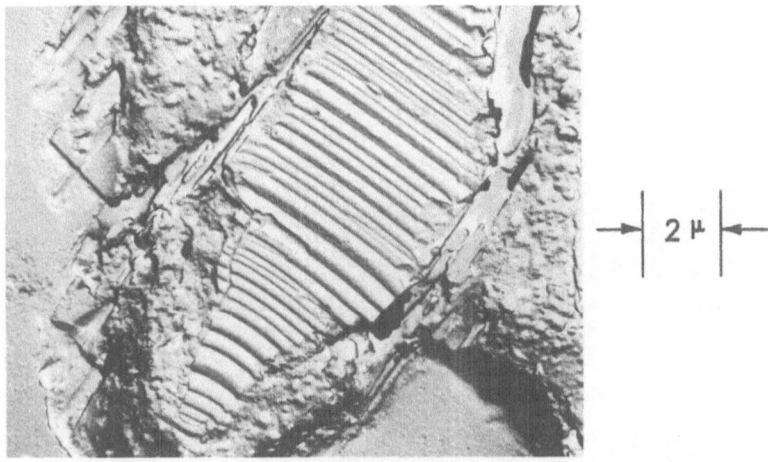

Fig. 6b. Sample electron micrograph of grain boundary surface A.

fracture of MgO bicrystals were examined for possible impurity accumulations. Figures 6a–6c and 7 show boundaries designated A and B, respectively, from Norton Company (Worcester, Mass.) fused MgO. The varying topography of both of these boundaries suggests extensive and varied impurity distribution and possible formation by solidification of a liquid film between previously crystallized grains. The possible existence of such a film is corroborated by the recent conclusion of Davis and Tallan [6] that

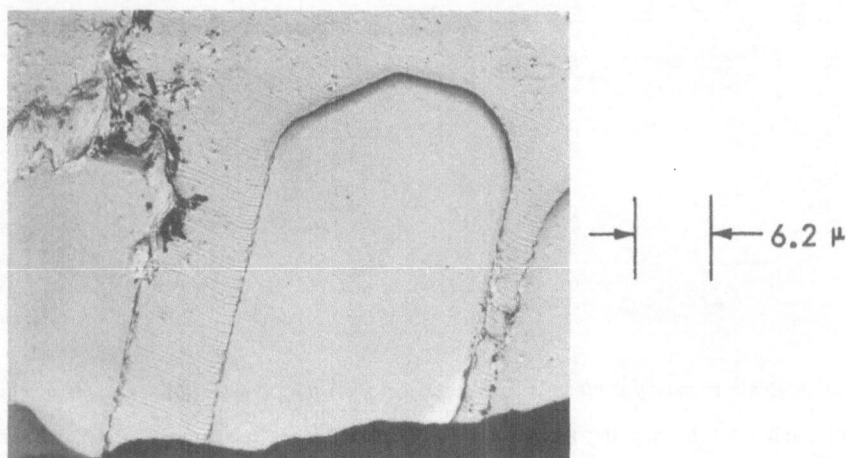

Fig. 6c. Another electron micrograph of grain boundary surface A.

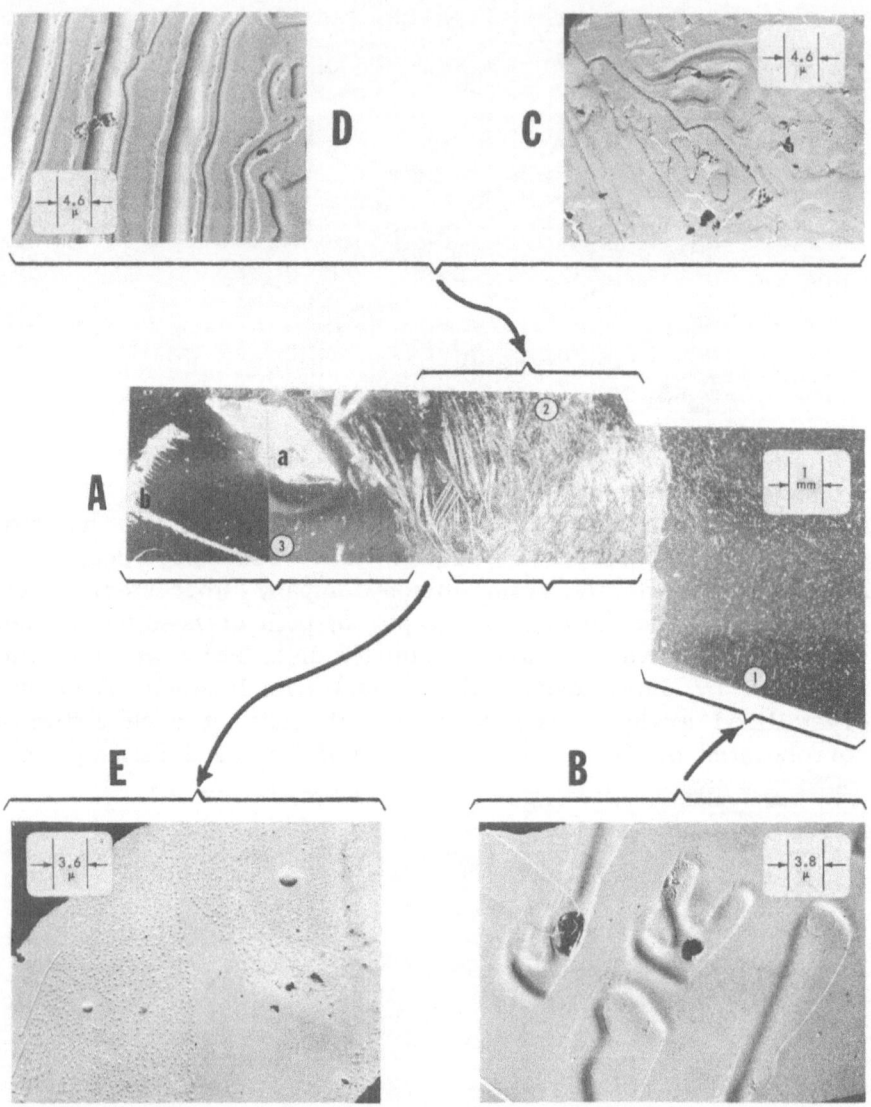

Fig. 7. Grain boundary surface B. (A) Reflected light microphoto—(a) chip from adjacent grain and (b) scratch. Sample electron micrographs (two-stage carbon replica) are: (B) Region 1. (C) and (D) Region 2—"Feather" or "grass" structure. (E) Transition—Region 2 to 3.

a liquid grain boundary film is formed between two LiF crystals being sintered 20–30°C below their melting point. The types of topography shown in Figs. 6a–6c and 7 were typical of several other boundary surfaces of crystals from the same source. Electron probe examination of surfaces A and B, shown in Figs. 8 and 9, identifies zirconium and calcium (probably present as oxides) as the most frequent impurities. Some iron and copper were also found. Silicon was identified as a significant constituent of much of the "feather" structure of boundary B.

It was noted that in areas of irregular topography, as in boundary A, regions or particles of impurities occurred predominantly along the edges of distinct changes in surface contour. Whether this represents a larger impurity region protruding from under a region of changing fracture character or precipitates formed at the edge of these features, possibly due to exolution from a liquid, has not been positively determined. Boundary C (Fig. 10), from Muscle Shoals Electrochemical Corporation, Tuscumbia, Alabama, showed a different form of precipitate over extensive areas, and some areas of the "feather"-type structure of boundary B (Fig. 9). Again, zirconium and calcium were the predominant cation impurities.

All of the fracture surfaces examined had been obtained by breaking away the adjacent grain. Since these specimens exhibited failure predominantly along the boundary rather than partial or complete transgranular fracture, they may be more irregular or impure than the average boundary. In order to check this, cross sections of a 45° twist, 20° tilt boundary and a 5° twist, 12° tilt boundary were examined with an electron probe. No impurities were found. Whether this is due to a lack of impurities at the surface or due to impurities too thin to be detected when viewed end-on cannot yet be positively determined.

Thin sections parallel to and including the boundary have been made of the above two boundary specimens, and of a third one not examined by the probe. The boundary, which was about 10° tilt on one axis and 5° on the other axis, showed sporadic patches of objects, such as those shown in Fig. 11a. The 5° twist 12° tilt boundary showed some sections of objects as those of Fig. 11a though at a lower concentration, but more tightly grouped. This specimen also showed a rectangular object about 5 by 15 μ at or near one section of the boundary. Figure 11b shows a typical bean-shaped object in the 45° twist 20° tilt boundary. These beans were fairly uniformly

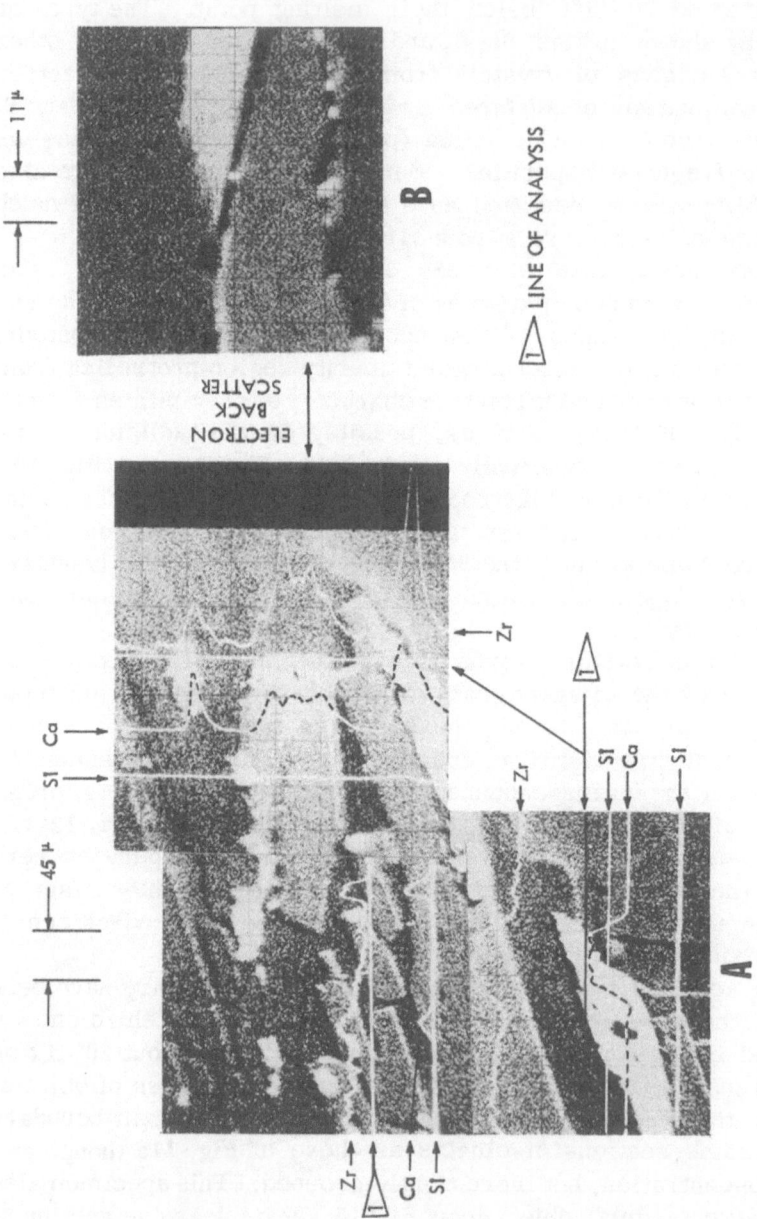

Fig. 8. Electron-probe examination of boundary surface A, electron back-scatter photos. (A) Examination of circled region of Fig. 6 with superimposed relative impurity–concentration curves. (B) Typical higher magnification showing impurities (light) at edge of topographic feathers.

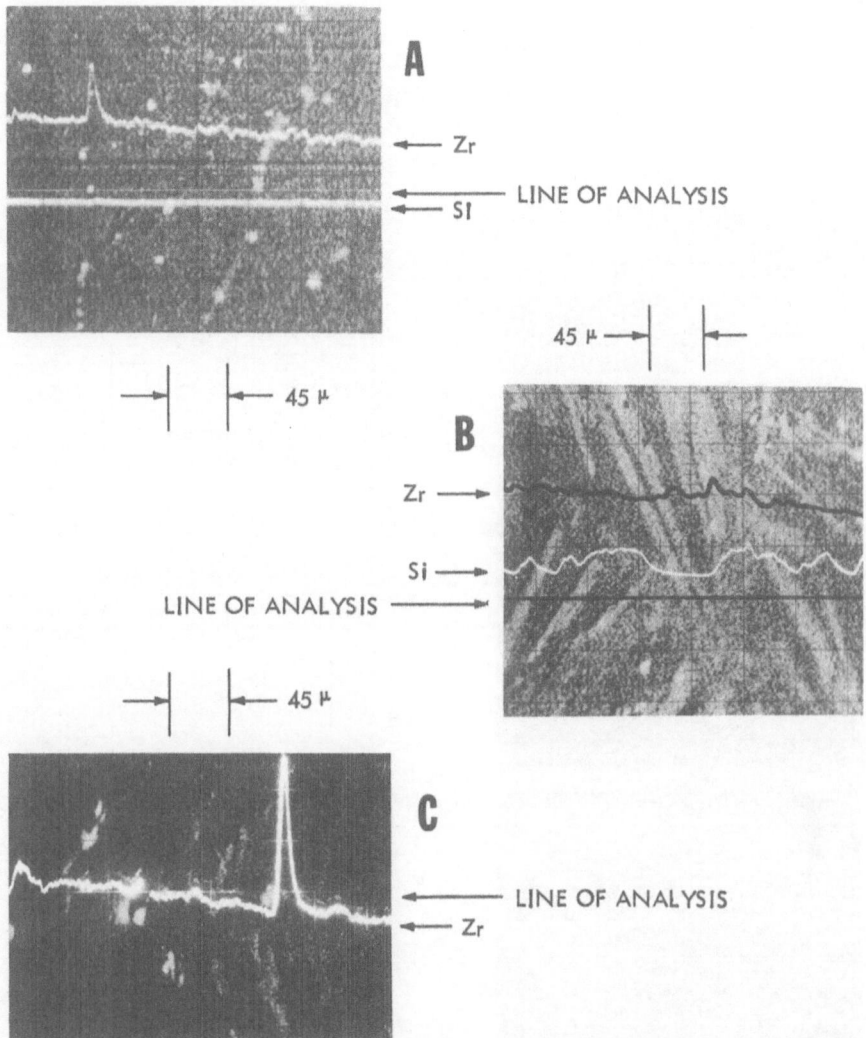

Fig. 9. Electron-probe examination of boundary surface B, electron back-scatter photos of (A) region 1 of Fig. 7, (B) region 2 of Fig. 7, (C) region 3 of Fig. 7.

distributed over a 0.1 in.^{2} area or more, a few hundred microns apart, and had a common orientation. Probe examinations are being planned to determine if the objects of Figs. 11a and 11b are precipitates and, if so, what their constituents are. Further correlation also is needed to see if thin sections reveal as much as electron probe or other surface examinations. These preliminary investi-

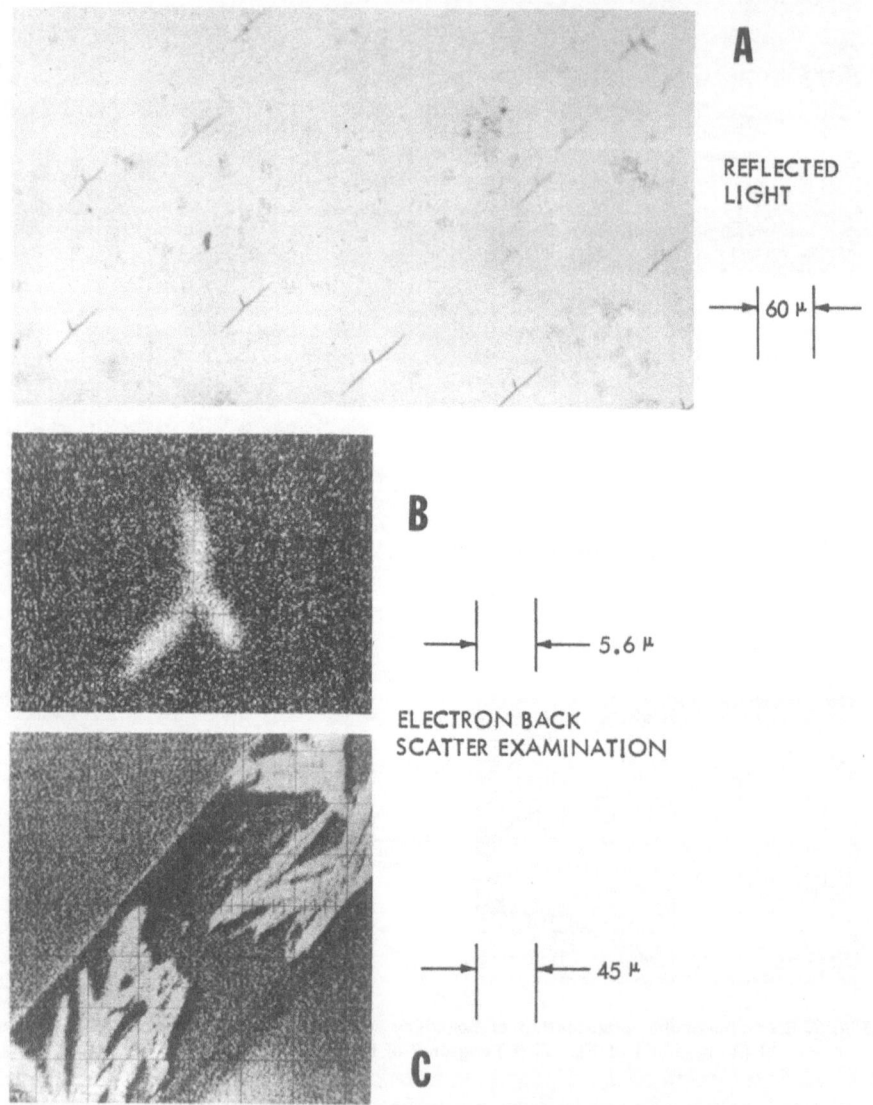

A

REFLECTED
LIGHT

→| 60 μ |←

B

→| |← 5.6 μ

ELECTRON BACK
SCATTER EXAMINATION

→| |← 45 μ

C

Fig. 10. Grain boundary surface C. (A) Reflected light microphoto of markings typical of about 4 cm². (B) and (C) Probe electron back-scatter photo of (B) typical markings. (C) Some "feather" structure; bright areas contained Zr and some Ca.

Fig. 11a. Thin-section examination of a grain boundary. Sample of objects in tilt boundary of 10° and 5° about tilt axes.

gations conclusively show that fused MgO crystal boundaries contain extensive areas of varied impurities and surface irregularities.

The boundary surface of an MgO bicrystal obtained by hot-pressing two single crystals together and then breaking them at the boundary is shown in Fig. 12. Again, substantial amounts of impurities are indicated at the boundary and sub-boundaries intersecting the boundary.

Fig. 11b. Thin-section examination of a grain boundary. Typical oriented object in 45° twist 20° tilt boundary.

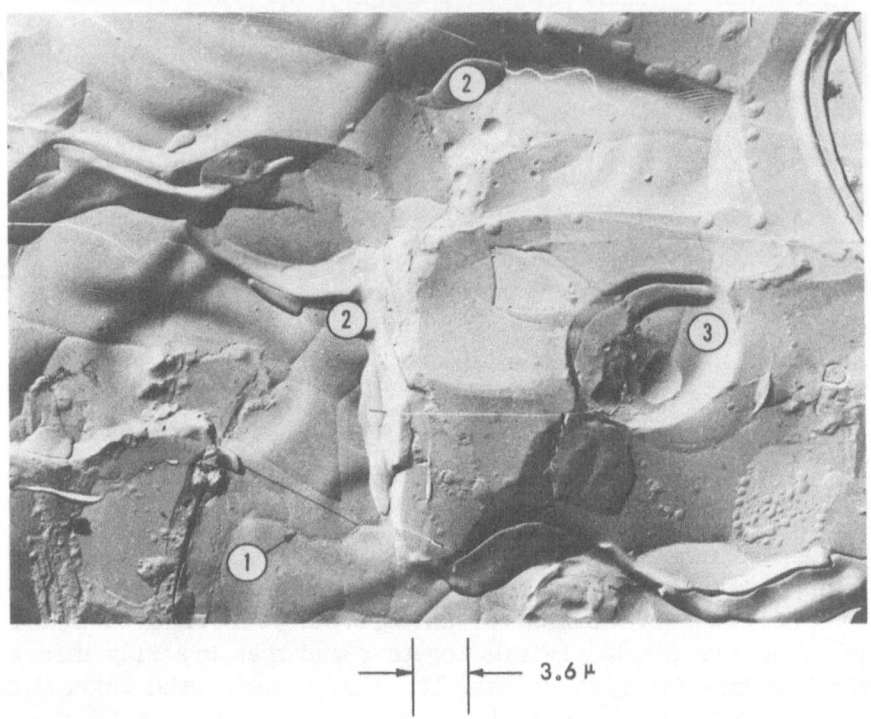

— 3.6 μ

Fig. 12. Grain boundary surface of a hot-pressed MgO bicrystal. Note features indicating
(1) sub-boundaries with precipitates, (2) second phases, and (3) large pore with included
second phase and porosity.

Preferential accumulation of second phases at the grain boundary
in MgO–CaO alloys has recently been reported by Rice [7]. This
occurred in as-received fused alloy ingots, and in bodies fired to
about 2000°C after hot-pressing of mechanically mixed powders.
It was also observed that the boundary phase could be taken into
solution by firing to temperatures of 2200°C or higher, with solution
occurring last from the triple points in both types of bodies; similar
accumulation of a zirconia-rich phase is observed in portions of
MgO – 2 wt.% ZrO_2 fusion,* even after heating to about 2300°C, as
shown in Fig. 13. Hafnium, calcium, and silicon cations have also
been found preferentially associated with these ZrO_2–rich areas.

Accumulation of impurities at grain boundaries in polycrystalline
MgO is suggested by examination of the fracture surfaces shown
in Figs. 14a and 14b. Here, electron micrographs show a structure

*Supplied by the Tennessee Electro-Mineral Corporation, Greenville, Tennessee.

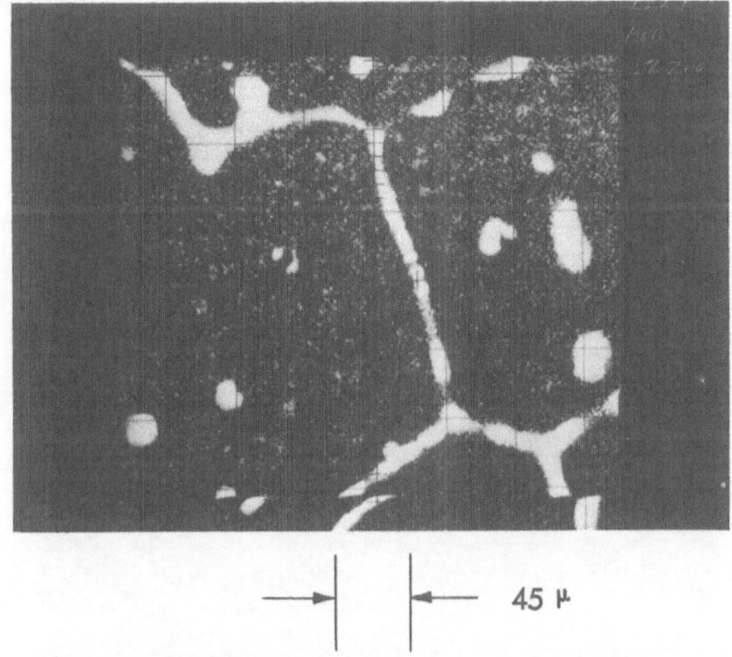

Fig. 13. Grain boundary accumulation of alloy phase in fused MgO-2 wt.% ZrO$_2$ body. Electron back-scatter: Bright (white) is ZrO$_2$-rich.

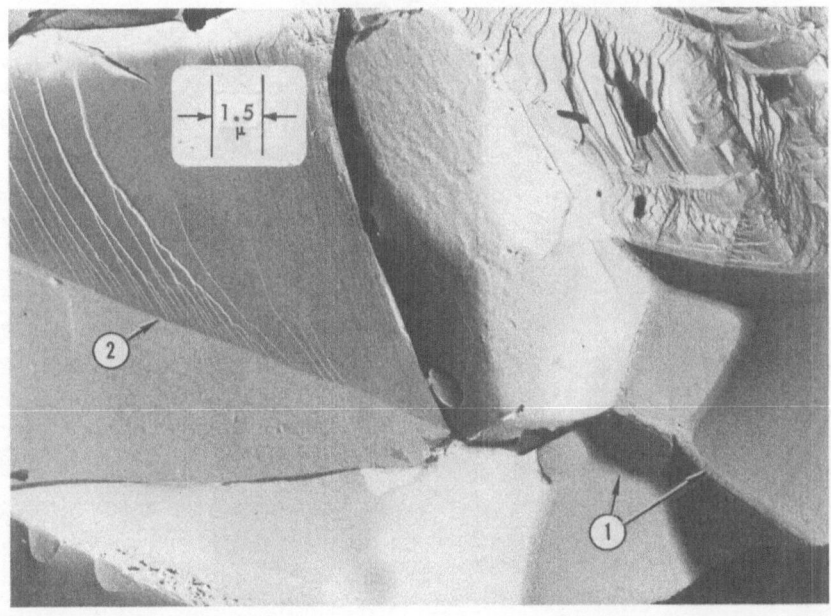

Fig. 14a. Polycrystalline MgO fracture surface. Note indicated second phase at triple lines from intergranular fracture (1), but not at boundary between cleaved grains (2).

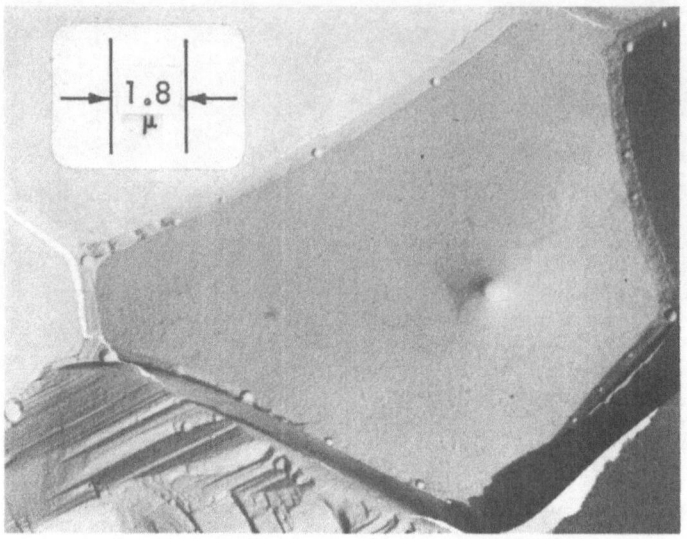

Fig. 14b. Polycrystalline MgO fracture surface. Probable second phase with included porosity at triple line around grain.

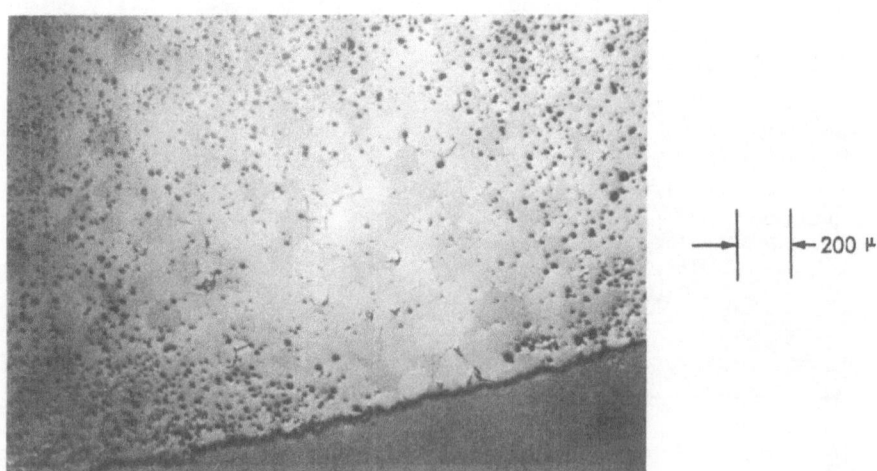

Fig. 15a. Boundary phase in dense section of CaO fabricated with LiF. Dense section in more porous body.

Fig. 15b. Boundary phase in dense section of CaO fabricated with LiF. Portion of the dense section.

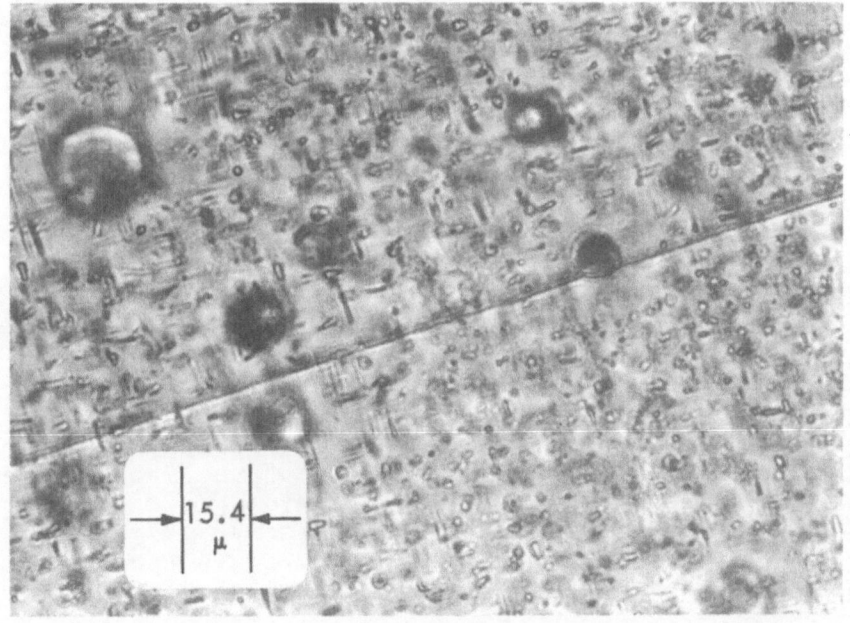

Fig. 16a. Rod precipitates in fused MgO–2 wt.%ZrO$_2$. Thin section of piece fired to 2300°F in transmitted light. Note cleavage crack showing orientation of rods.

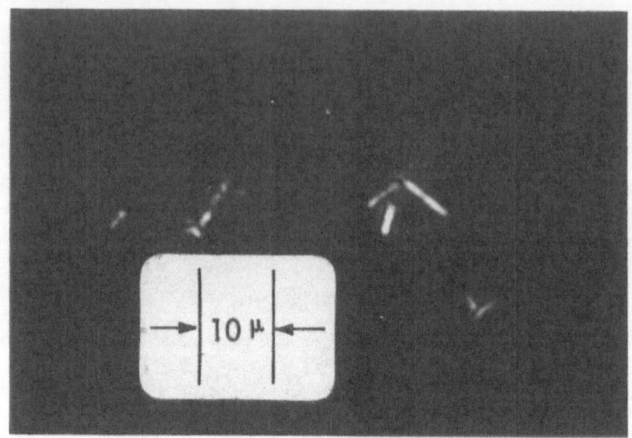

Fig. 16b. Rod precipitates in fused MgO–2 wt.%ZrO$_2$. Rods left after etching away MgO.

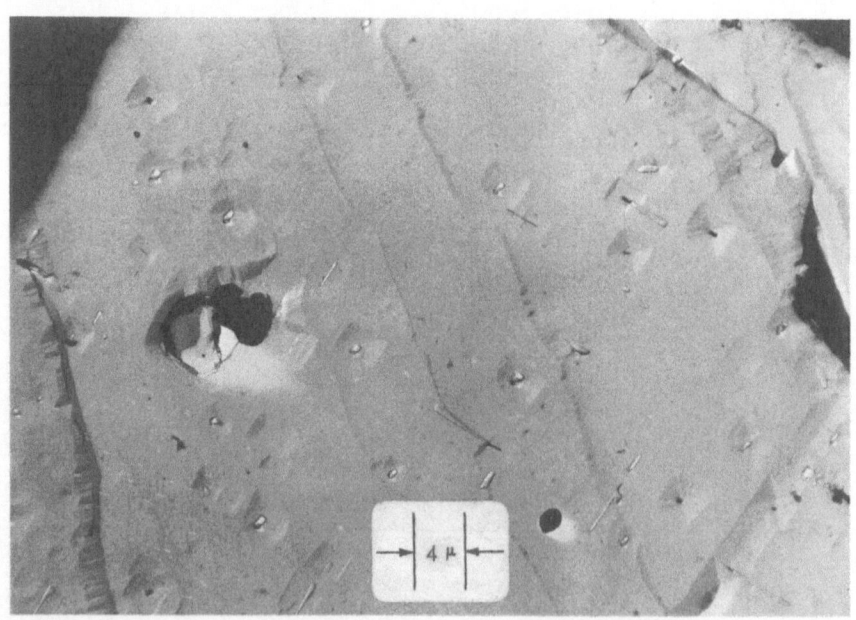

Fig. 16c. Rod precipitates in fused MgO–2 wt.%ZrO$_2$. Electron micrography of cleavage fracture surface of as-received ingot etched 10 sec in boiling chromic acid. Two-stage carbon replica.

suggestive of a second phase at the triple line. This is in agreement with the observations of Stokes and Li [8] in microstructural
examination of polycrystalline MgO fabricated in a fashion similar
to the specimens of Figs. 14a and 14b. LiF has been reported as
a very successful hot-pressing aid for both MgO [9] and CaO [10]. All
lithium is lost on subsequent firing of these specimens; however,
several hundred to several thousand parts per million fluoride ion
(probably as MgF_2 and CaF_2) remain depending on specimen size,
porosity, and processing. Figures 15a and 15b show a second phase,
probably CaF_2, at the boundaries and especially at the triple points
in a dense section of CaO fabricated with LiF.

PHASE BOUNDARIES

Phase boundaries begin to form when alloy constituents or impurities become concentrated in regions with dimensions larger
than several lattice spacings. If this accumulation occurs at a
grain boundary, as in Figs. 13, 14a, 14b, 15a, and 15b, then the
effected portions of that boundary are replaced by two phase
boundaries. The size of such accumulations may substantially effect
the nature of the resulting phase boundaries. For example, the
electron probe analysis of MgO – 2 wt.% ZrO_2 (Fig. 13) shows
particles of a ZrO_2-rich phase about 20μ in diameter which appear
to have no special shape or orientation. However, examination of
thin sections and fracture surfaces of this same ingot (Figs. 16a–
16c) shows rods of an approximately rectangular parallelepipedal
shape about 0.3 by 0.3 by 3 μ in size. Further, these rods appear
to be on {100} surfaces oriented in<100>directions, as is indicated
by the fact that they are either parallel or perpendicular to MgO
cleavage surfaces. X-ray diffraction analysis of the ZrO_2-rich
phase obtained by dissolving the MgO matrix with HCl showed that
it was in the cubic state with a lattice parameter of 5.125 A,
indicating calcia stabilization, in agreement with electron probe
detection of calcium.

Such phase boundaries are often dependent on processing. For
example, powder ground from the above MgO – ZrO_2 fused ingot,
when hot-pressed, results in a body of much coarser, generally
spherical precipitates with essentially no rods remaining (Fig. 17).
Similarly, hot-pressing of MgO and ZrO_2 obtained by calcining co-
precipitated salts shows no special precipitate orientation or structure in preliminary analysis. However, examination shows that

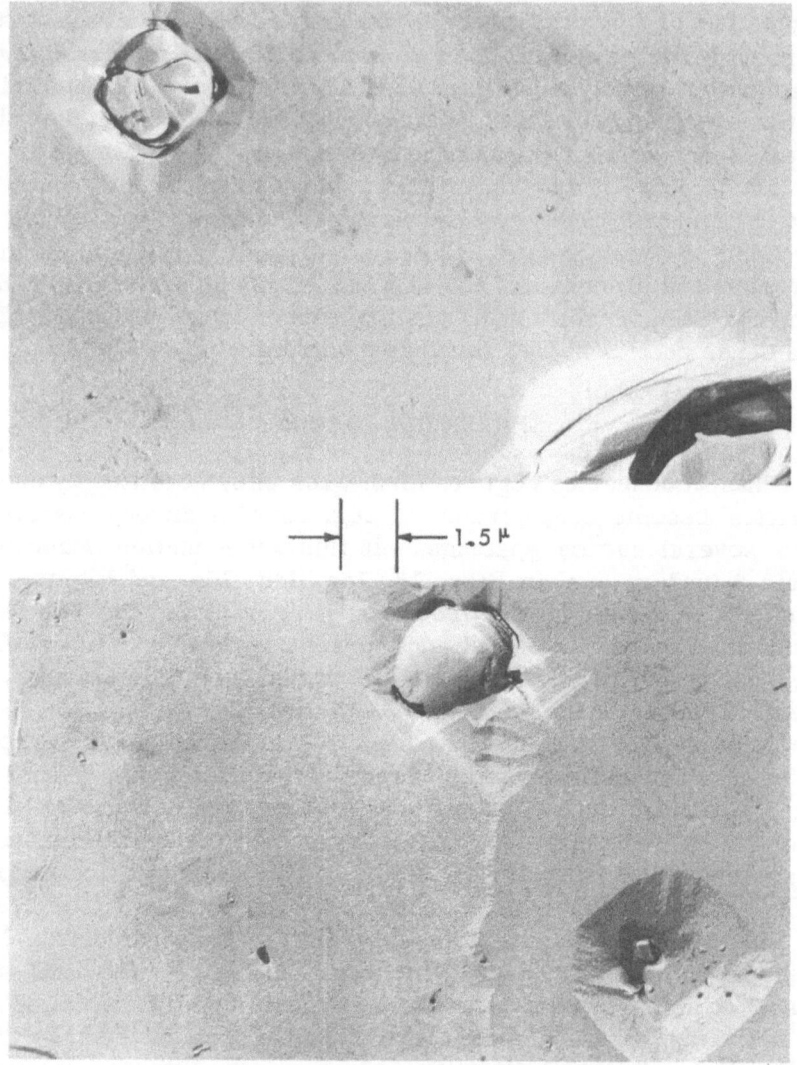

Fig. 17. ZrO$_2$ precipitates in hot-pressed MgO–ZrO$_2$. Fracture surface etched 10 sec
in boiling chromic acid. Electron micrographs (two-stage carbon replica).

some rods may start to form in this latter body after subsequent
firings to about 1500°C. The rods in the original fused ingot show
only a possible shortening and rounding trend after heating to 2300°C
or more.

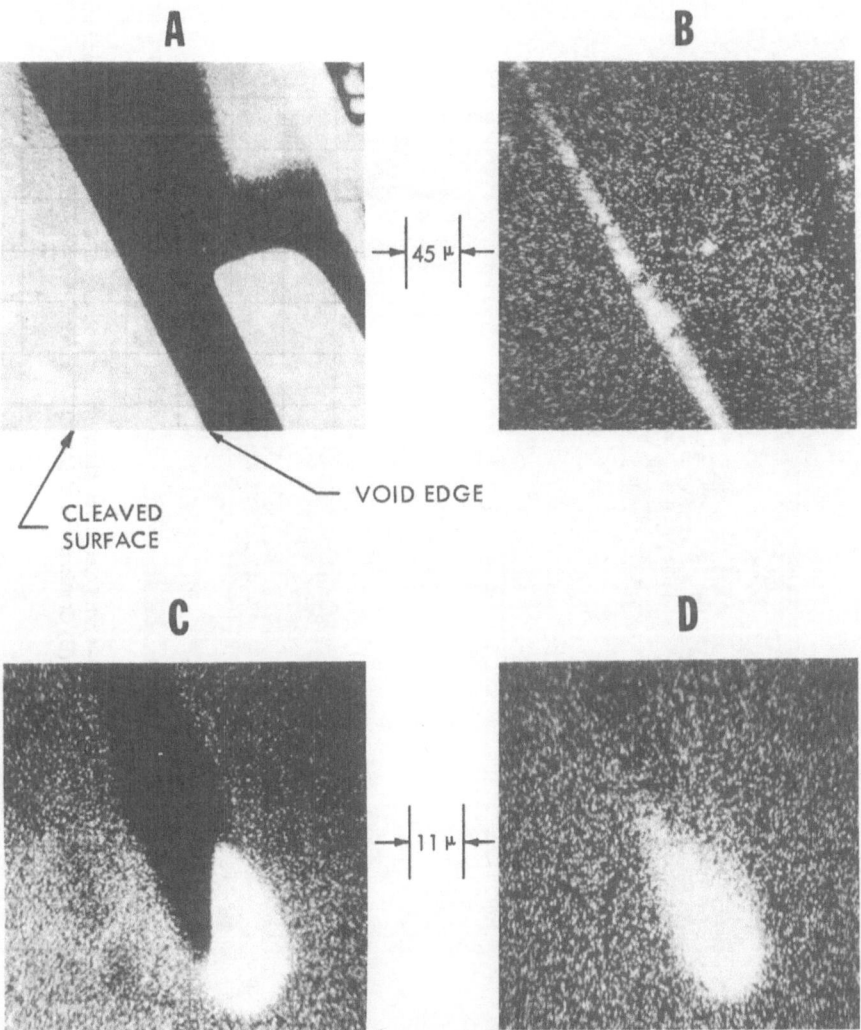

Fig. 18. Impurities in MgO crystal voids. (A) Electron back-scatter; dark: void walls; light: cleaved surface or void bottom. (B) Ca X-ray fluorescence of same area showing Ca along void wall. (C) Electron back-scatter of a second void. (D) Ca X-ray fluorescence of same area as (C) showing accumulation of Ca at one end of void.

Models of such phase boundaries, though more difficult to develop, would also be useful. Development of phase boundary models can be aided by our improved knowledge of modes of formation, thermal stability, crystal structure, and orientation of the

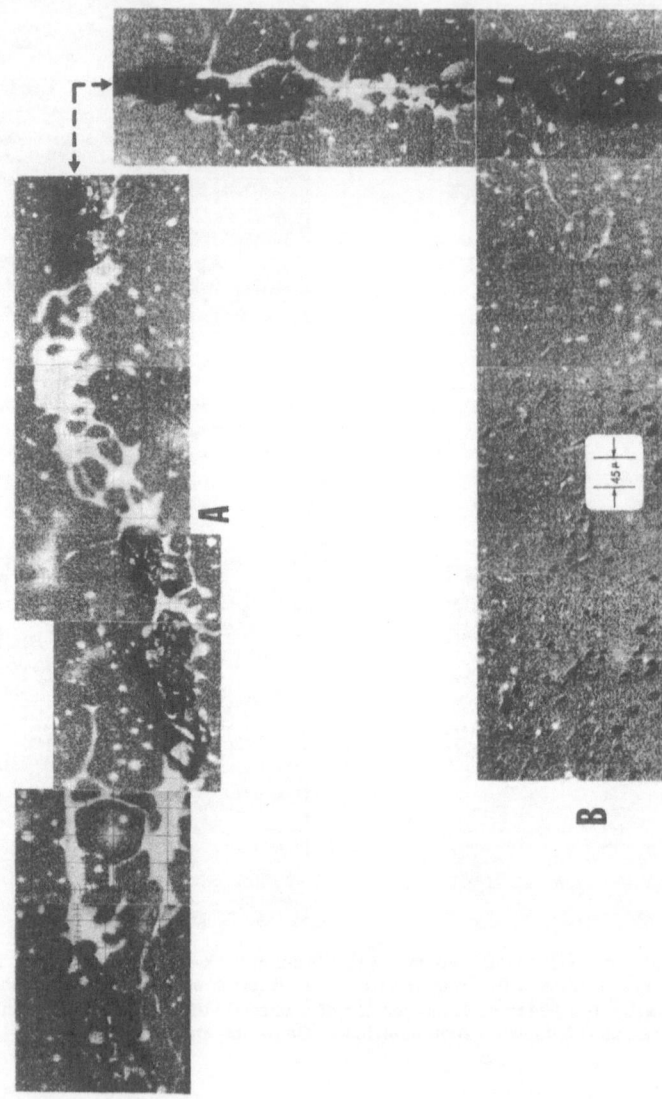

Fig. 19. Electron-probe examination of a hot-pressed and fired MgO—4 wt.%CaO alloy showing accumulation of CaO near a crack or void. (A) Electron back section; bright white; CaO; black; crack. (B) Continuation of (A) with a portion of the surrounding body.

phases. Such models could also assist in identifying other characteristics of the second phase. For example, the general rectangular shape of the above ZrO_2-rich rods would suggest that the boundaries must start from orthogonally intersecting planes, such as {100}, or else be a series of many steps of other planes. The latter would mean a highly jogged boundary. If the surfaces are near-perfect {100} planes, then greater boundary disorder would be expected, since the boundary would have to accommodate the checkerboard alternate-charge pattern of {100} MgO with the repeated like-charge pattern of the {100} cubic ZrO_2. Neither of these initial considerations indicates why a rod, rather than a cube, is formed.

VOIDS

Voids have long been recognized as a source of mechanical failure, and considerable effort has been and continues to be expended to reduce porosity. Porosities well below 1% [9, 11] have been achieved; however, the scope of the task of eliminating porosity can be indicated by the fact that a 1-μ-diameter pore has a volume of the order of $5 \cdot 10^{-13}$ cm^3; therefore, 0.01 vol.% porosity represents about 10^8 pores 1 μ in diameter per cubic centimeter, or about 10^{11} pores 0.1 μ in diameter per cubic centimeter.

There is, however, another aspect of porosity that seems to be neglected. Impurities and alloying agents should find greater accommodation and, hence, should tend to accumulate at pores. To investigate this, Norton MgO crystals with porosity in cloudy areas (indicating impurities) were selected and cleaved to expose the voids (which often are partly cubical in shape). Electron probe examination frequently revealed significant impurity accumulations (primarily calcium, probably as CaO) at the pore surfaces (Fig. 18). An association between voids and impurities was also shown in a recent study of TiB_2 [12], where some void areas were found rich in copper, silicon, aluminum, and iron. The low-porosity region with a second boundary phase in the CaO specimen of Figs. 15a and 15b was also found to be surrounding a macroscopic pore. Whether the large pore growth was aided by the second phase, or the second phase preferentially migrated to the pore cannot be ascertained. Figure 19 indicates an association between a series of voids or internal cracks and the alloy agent, CaO. Again, it cannot be deter-

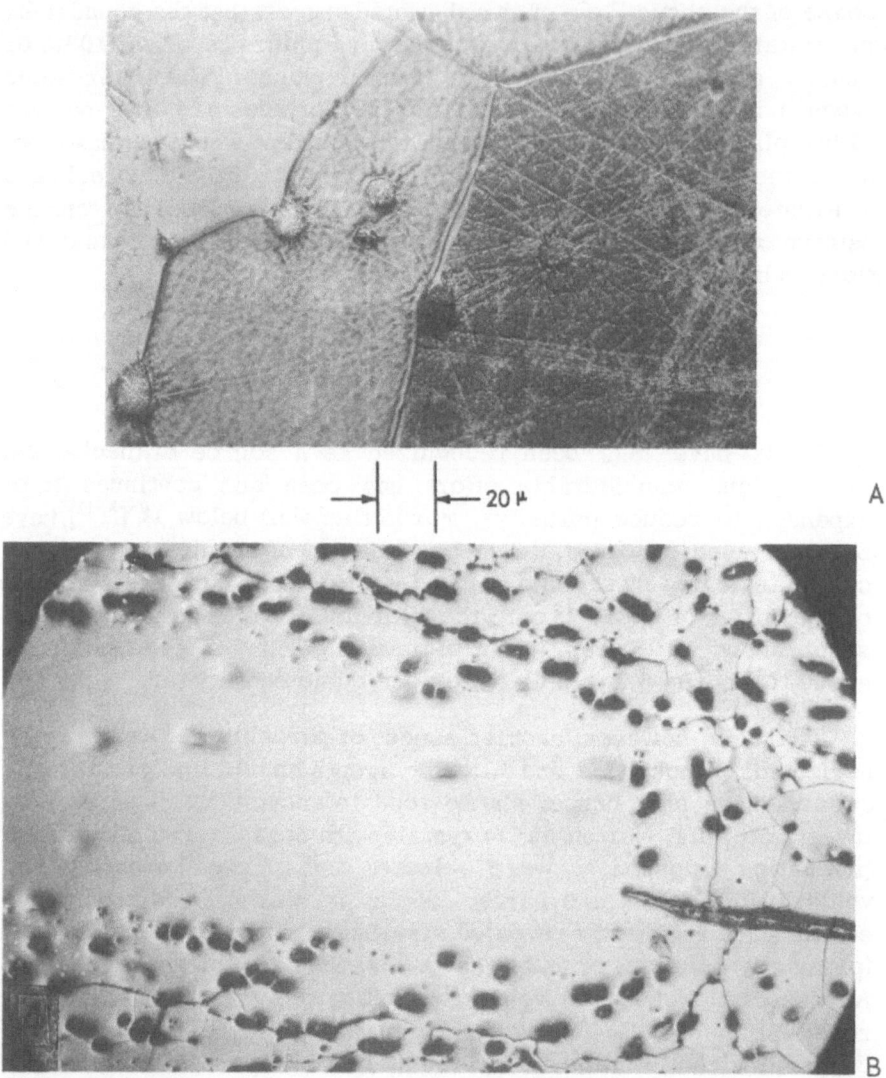

Fig. 20. Voids and grain growth.   (A) CaO specimen hot-pressed with LiF and fired to
1510°C (reflected light).   (B) MgO hot-pressed with LiF and embedded MgO crystal
(original crystal boundary outlined by porosity). Note etching emphasizes porosity; the
body has only about 1% porosity.

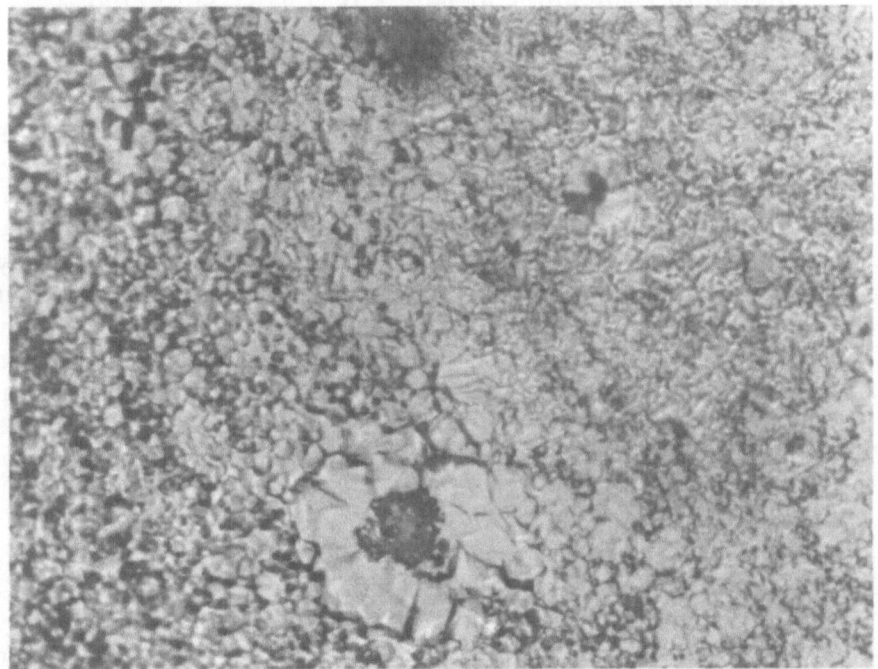

Fig. 21. Excessive grain growth around voids in MgO fired after hot-pressing with LiF.

mined whether the void or crack preferentially occurred in a CaO-rich region, or the CaO preferentially migrated to the void or crack. However, the fact that this accumulation was observed in hot-pressed specimens that were subsequently fired to 2200°C, and that the ends are blunted and filled by CaO while the sides are not completely lined, would suggest that the CaO migrated to the crack. This might have occurred as the crack was propagating, due to thermal stresses during heating. Accumulation of the ZrO_2-alloy phase similar to the accumulation of CaO along the sides of cracks or voids of Fig. 19 has also been observed in the fused MgO – 2 wt.%ZrO_2 body discussed earlier.

Only relatively large voids have been examined in this study, and detailed relationships between the occurrence of impurity accumulations at or near voids, and the character of the voids themselves, have not been clearly established. However, it is clearly demonstrated that impurities and alloying agents can and do preferentially associate with voids.

INTERRELATIONSHIPS BETWEEN INTERNAL SURFACES
AND PROPERTIES OF MgO

Interrelationships between Internal Surfaces

All three of the internal surfaces discussed are commonly interrelated. Voids start at grain boundaries in the initial compaction of a powder. Second phases preferentially migrate to grain boundaries, thus replacing portions of those boundaries with phase boundaries. Similarly, the migration of second phases to voids may result in the formation of phase boundaries. One relationship that does not appear to be commonly recognized is that impurities may accumulate, then volatilize or decompose, producing gases which in turn produce voids. These interrelationships are indicated in Figs. 12, 14a, and 14b, where apparent second phases are in voids at a boundary or these phases are at a triple line with voids in them.

Internal Surfaces and Sintering and Grain Growth

Sintering involves the reduction of void content, and it is generally accepted that only voids at grain boundaries are readily removed. That voids themselves inhibit grain growth is shown in Fig. 20(A). Nearly pore-free areas of a specimen have been observed having almost three times the grain size as portions of the same sample with about 2% porosity. However, as shown in Fig. 20, pores are commonly entrapped in grains by the boundaries sweeping around them. Conversely, voids, at least isolated ones, may assist grain growth as shown in Fig. 21 and to a lesser degree in the CaO specimen of Figs. 15a and 15b. This void enhancement of the growth of surrounding grains may possibly occur by localization of impurities (which may on occasion enhance, rather than inhibit, grain growth), or it may be that porosity in boundaries which would otherwise be inhibiting grain growth is migrating to the void instead.

Impurities that produce volatile or gaseous products have become an important problem with the achievement of high densities which inhibit the diffusion of these products out of the body. This problem can be especially acute with hot-pressed bodies due to the rapid achievement of low porosities at low temperatures in short times. For example, the author commonly finds significant infrared absorption identified as carbonate bands in hot-pressed MgO. Such gas-producing impurities would explain the explosions

[10], blistering [9], bloating [13], or clouding [14] observed either
during hot-pressing or subsequent refiring of low-porosity specimens.   Such processes, of course, entail the creation or enlarge-
ment of voids, and, hence, counteract sintering.

Recently it has been demonstrated that alloying agents, including those forming solid solutions, inhibit grain growth by about
one order of magnitude [7].   The accumulation of impurities and
alloy agents at grain boundaries may substantially affect grain
boundary diffusion, and may be an important factor in the distance
(as much as 200 A) [15] over which boundaries affect diffusion.
Condit and Holt [16] have recently reported greater grain boundary

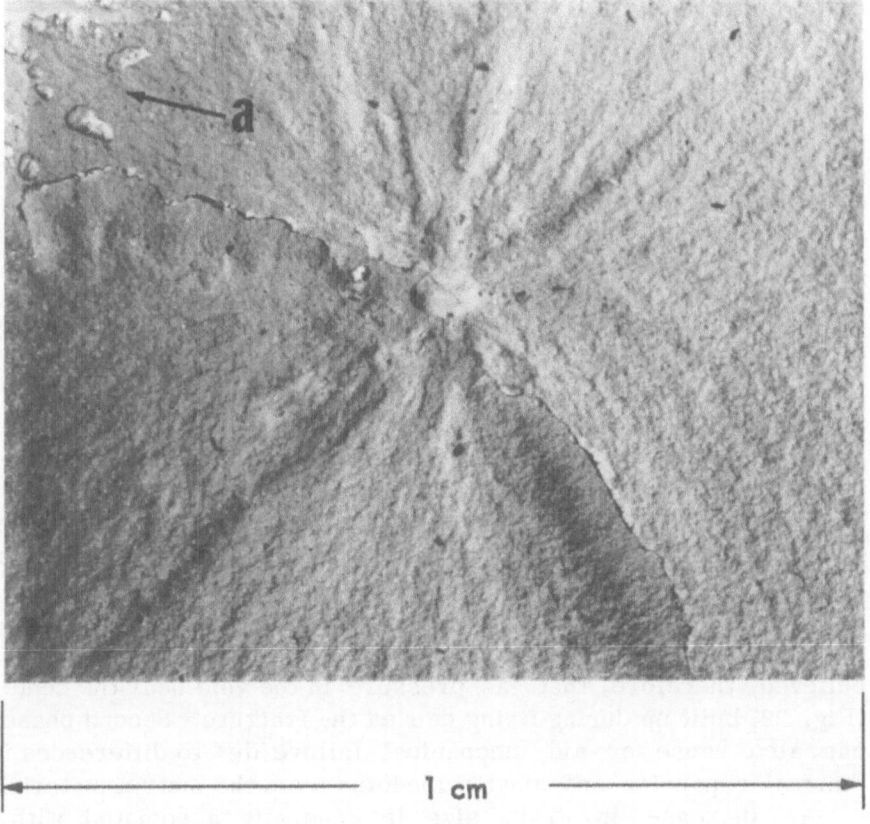

1 cm

Fig. 22. Fracture origin at void. Shadowed replica used for contrast [(a) replica blisters],
but features checked on original specimen. Note ridges and branch cracks radiating from
pore.

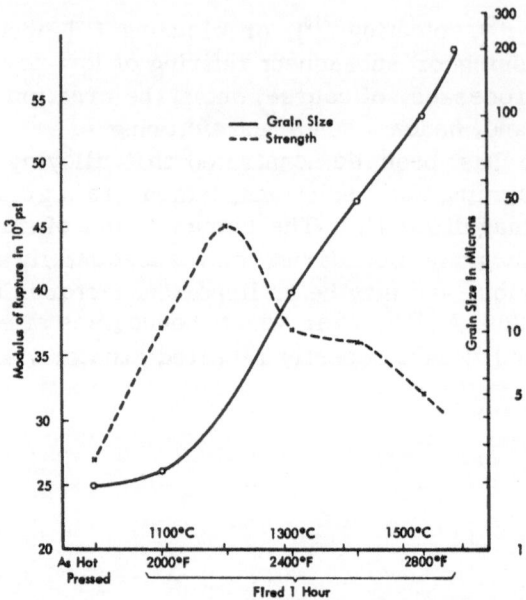

Fig. 23. MgO strength and grain size versus firing temperature. All strength points represent at least five specimens tested in three-point flexure on a 0.78-in. span. Firing to 1050°C was over a period of three days. All specimens were fabricated by hot-pressing at 1205°C.

diffusion in MgO bicrystals which apparently had a higher impurity content.

Internal Surfaces and Mechanical Properties

It has long been recognized that internal surfaces have important effects on mechanical properties. For example, voids are recognized as stress concentrators and, hence, reduce strength. Gas pressure in voids can compound this problem. For example, an MgO cylinder was found with a layer about 0.2 in. thick completely fractured off the top after firing. No sign of voids or flaws was observed in the hot-pressed body prior to firing. It is believed, therefore, that gas pressure in the void near the center (Fig. 22) built up during firing caused the fracture. Second phases can also cause or aid mechanical failure due to differences in thermal expansion and elastic modulus from the matrix material.

An increase in grain size is generally associated with a decrease in strength; however, other factors can reverse this trend, at least temporarily, as shown in Fig. 23. These specimens show a rapid loss of the carbonate infrared absorption band at about 8 μ upon firing, suggesting that the increase in strength is due to the slow diffusion of volatile products out of the body. The author believes that most of these gas-producing impurities are located at the grain boundaries and their removal, therefore, allows better intergranular bonding. A similar initial increase in strength upon heat treatment of MgO specimens hot-pressed with LiF was previously reported [9].

The study of dislocation phenomena in conjunction with the mechanical properties of ceramics and isomorphous materials of related bonding has been most revealing. It has been shown that grain boundaries act as barriers to dislocations [17-19], resulting in an increasing probability of crack nucleation from resulting dislocation pileups with increasing grain misorientation. Ku and Johnston [20] have demonstrated that a Petch-type relation holds for fracture stress σ_F in MgO.

$$\sigma_F = \sigma_0 + KD^{-\frac{1}{2}} \tag{1}$$

In equation (1), σ_0 is the stress to multiply dislocation, K is a constant related to the cohesive strength at the crack tip, and D is the distance between dislocation source and barrier. However, these results are from studies of bicrystals from fused MgO similar to those discussed earlier. Therefore, it may be that impurities at the boundaries, such as those shown earlier, may substantially increase the effectiveness of these boundaries in producing dislocation pileups. This would be in agreement with both the observation of Clarke et al. [19] that some regions of the boundaries in their study appeared to allow passage of slip bands, and with the results of Long and McGee [21], which show that welded bicrystals of NaCl tolerated much more slip in the boundary region than did fused bicrystals which probably contain more impurities than the welded specimens. Such increased tolerance to slip by purer grain boundaries would be similar to results in metals where evidence indicates that boundaries per se may offer no resistance to plastic deformation [22]. However, it cannot necessarily be expected that impurity-free ceramic boundaries offer no resistance to plastic deformation because of their generally fewer slip systems, the charged nature of their dislocations, the possible greater disorder, and charge problems of boundaries mentioned earlier. The importance of the number of slip systems on the effects of

Fig. 24a. Reflected-light microphotograph of a recrystallized MgO crystal.

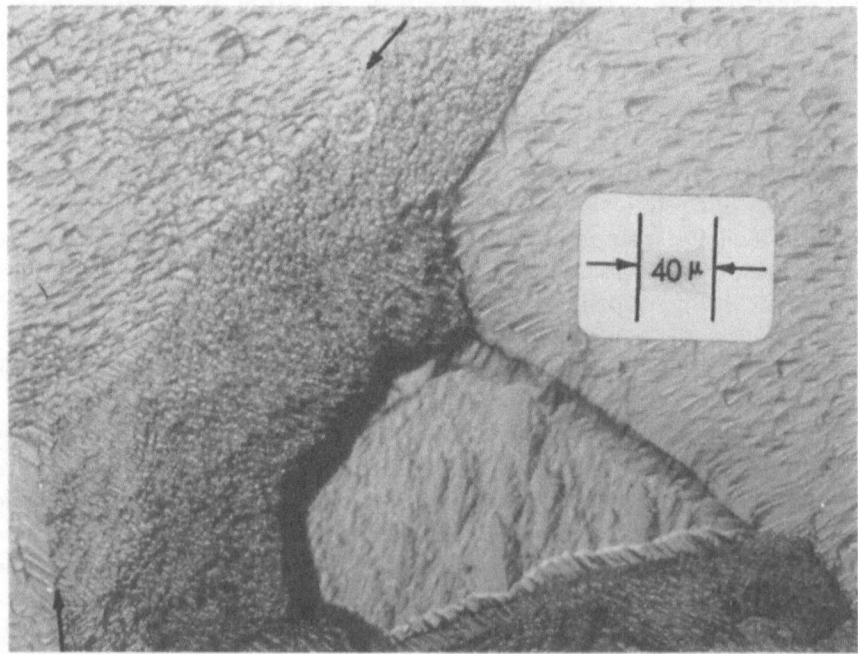

Fig. 24b. Heavy etch in boiling chromic acid to reveal low-angle boundaries, as indicated by the arrow, in the recrystallized MgO crystal.

grain boundaries is illustrated by the almost total loss of the effect of grain size on the mechanical properties of AgCl when its secondary slip system becomes operative to produce wavy glide and five independent slip systems [18].

Phase boundaries should have similar (and probably greater) effectiveness than grain boundaries as barriers to dislocations due to the expected greater disorder and charge effects mentioned earlier. This effectiveness is also suggested by the impurity effects at grain boundaries.

Voids might interact with dislocations either as subcritical cracks which can grow due to insufficient slip for general ductility [23] or as barriers to dislocation glide. This latter case would arise if the surface charges of an ionically bonded material, such as MgO, provide a repulsive force for charged dislocations, of if the void is at least partially lined with impurities. Voids acting as barriers to dislocations could result in some strengthening if they were located within the grain and not at the boundary and if dislocation failure mechanisms were operative at lower stresses than those at which the void itself leads to failure.

Grain boundaries and impurities would appear to be the most likely sources of dislocations in ceramics (as they are in metals) [24]. The etch pits around the ZrO_2 precipitates of Figs. 16a–16c, 17a, and 17b would suggest that they may be sources of dislocations. This would agree with the observation by Henderson [25] of dislocations around precipitates in MgO crystals. It is believed that voids are also possible sources of dislocations.

Grain boundary sliding, which becomes an important mechanism of failure above one-half the absolute melting temperature, should be substantially enhanced by grain boundary porosity, especially porosity on triple lines. Impurities, especially SiO_2 as observed in the grain boundaries (discussed earlier), and low-melting phases should also substantially increase such sliding and are probably an important reason why Adams and Murray observed sliding in MgO to commence at about $0.5\,T_m$, but not until about $0.8\,T_m$ in purer NaCl [26]. Such impurities would also offer an explanation for the catastrophic nature of some of the sliding they observed in MgO. Adams and Murray's study also showed that grain boundary sliding increased with increasing grain misorientation, as might be expected from the structure of boundaries. This is in agreement with results in metals which also suggest that boundary sliding is related to dislocation motion near the boundary [27–29].

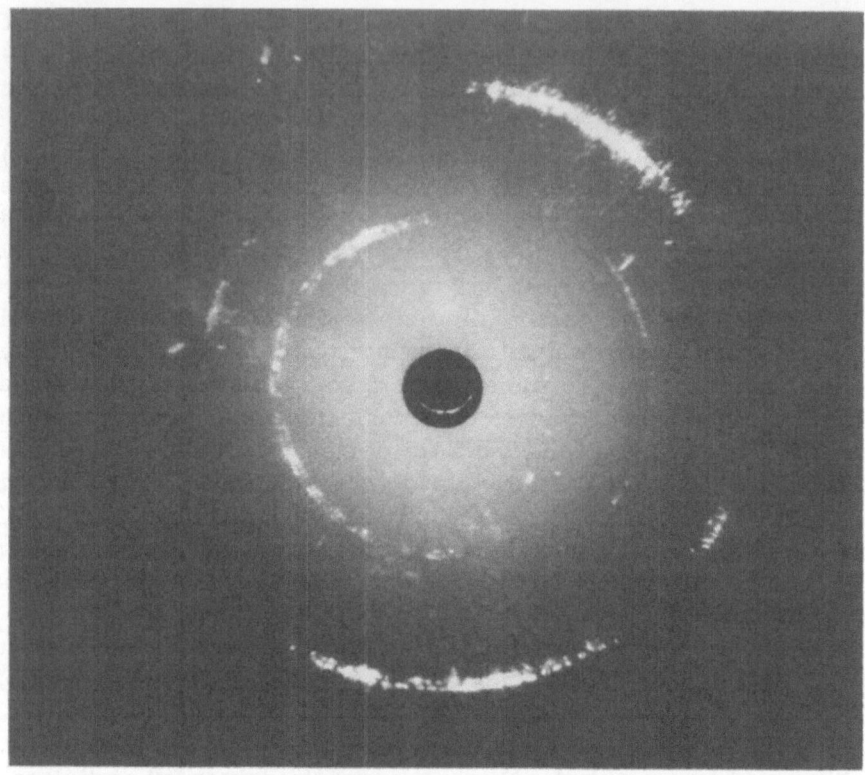

Fig. 24c. Sample Laue back-reflection pattern of the recrystallized MgO crystal.

STUDY TECHNIQUE

The above results show that more detailed and better controlled experiments are still needed, especially on purer systems. Some experimental techniques for such studies are outlined in this section.

Rice [30] has previously reported the successful fabrication of MgO bicrystals of predetermined orientations by hot-pressing two single crystals together at temperatures of 1750–1900°C in graphite dies with a pressure of about 4000 psi. Modulus of rupture up to $17 \cdot 10^3$ psi and fractures along only part or none of the boundary surface show that substantial bonding was achieved. The bicrystal fracture surface of Fig. 12 was one of these specimens. Wuensch and Vasilos [31] and Scott and Roberts [32] have independently reported similar fabrication of MgO and Al_2O_3 bicrystals, respec-

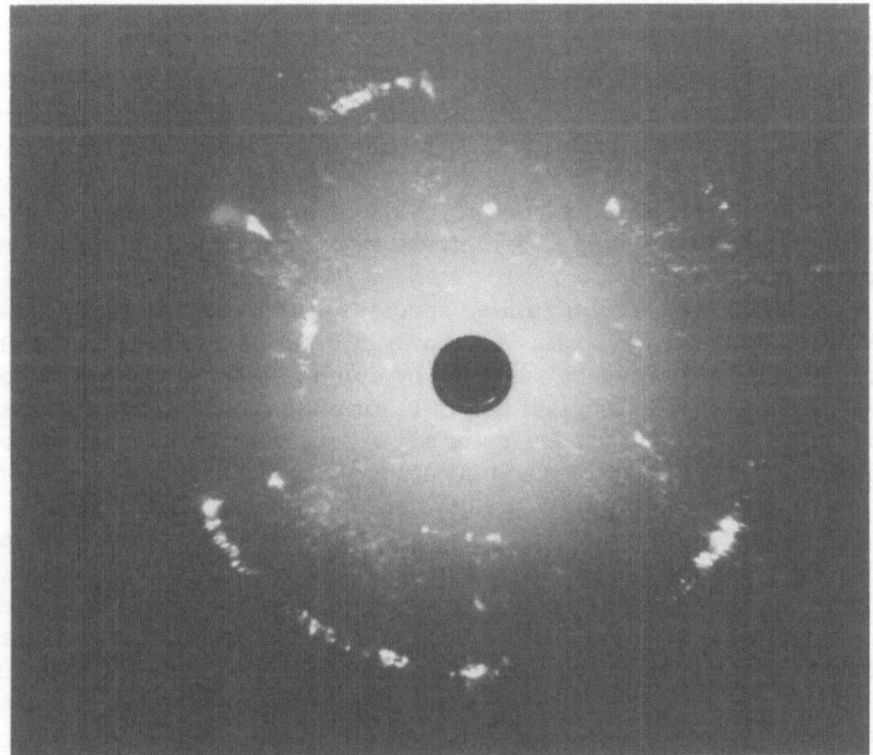

Fig. 24d. Another Laue back-reflection pattern of the recrystallized MgO crystal.

tively. * This method has the advantage of more closely approximating the conditions of hot-pressing, i.e., temperature to allow diffusion of impurities to the boundary and possible atmospheric effects
on the boundary-forming surfaces.   One possible disadvantage at
least with MgO is the formation of substructure near the boundary,
as in Fig. 12.   Recent experiments by the author have shown that
tricrystals may also be formed by this method.

A second method of forming bicrystals of predetermined orientation is by welding, as Long and McGee [21] † have demonstrated.
From recent trials with MgO crystals and successful butt welding
of polycrystalline Al_2O_3 using modified electron-beam welding
techniques, the author believes that MgO and Al_2O_3 bicrystals also
can be made by welding.

*See also Davis and Palmour, this volume, Chapter 17.
†See also Martin, Fehr, and McGee, this volume, Chapter 15.

Special grain boundary studies can be aided by hot-pressing small single crystals in polycrystalline bodies. This allows pretreatment and identification of specific grains. In the original report of this technique [30], it was also noted that substantial growth of these imbedded grains could be achieved. The growth shown by such a grain at only 1750°C even with about 1% porosity in the body (Fig. 20A and B) confirms this and shows that valuable multicrystal specimens might be achieved by growth of closely imbedded crystals.

Another method of obtaining special specimens is by recrystallization of single crystals. As shown in Figs. 24a–24d, substantial recrystallization can be obtained by compressing MgO crystals to a fraction of their original height in graphite dies at temperatures of about 1900°C and pressures of a few thousand psi. Many of these techniques can also be used to study phase boundaries by use of coated or alloyed crystals, or crystals of a second phase.*

Cutting of specimens, especially bi- and tricrystals, with a diamond saw has been the greatest source of specimen breakage. This has been substantially reduced by ultrasonic cutting. An acid saw has shown promise for cutting where ultrasonic methods are not desired. With the crystal heated to about 150°C and phosphoric acid at about 90°C dropped at the rate of 20 ml/hr, cutting rates of about $\frac{1}{4}$ in.^{2}/hr were indicated in initial tests. For such sawing, as with diamond sawing, cutting into the boundary from opposite sides may be advisable.

Grinding of bicrystal specimens is also often required. The technique of grinding along the boundary line (normal to the boundary plane), then grinding the adjoining grains (usually in a direction away from the boundary to minimize cleavage) has been found useful. The common technique for getting various degrees of porosity is to vary fabrication conditions. However, the effects of impurities and grain size on this technique are not clearly established.

SUMMARY AND CONCLUSION

In summary, it has been suggested that development of grain and phase boundary models and their correlation with other properties would be quite useful, especially as purer boundaries are

---

*See Sockel and Schmalzried, this volume, Chapter 5.

obtained.   Both grain boundaries and voids have been shown to be sites for impurities and the type and extent of the common impurities at these sites have been identified in MgO. It was suggested that these impurities may have been responsible for substantially increased crack nucleation and sliding observed with MgO grain boundaries as compared with those in which the boundaries were purer.   Submicroscopic ZrO_2-rich precipitates were shown to have preferred orientation of approximately $<100>$ and to be probable sources of dislocations.   Grain boundaries and voids were also suggested as dislocation sources, and it was noted that voids might act as barriers to dislocations and, thus, have a strengthening effect under certain conditions.   The interrelationship between grain and phase boundaries and voids was discussed with particular note of volatile and gaseous products of second phases creating porosity, leading to fracture, and hindering adequate intergranular bonding.

These observations and discussions reaffirm that the mechanical properties of ceramics should be improved by reducing porosity, grain size, and impurities.   However, it is concluded that control of dislocation phenomena to reduce crack nucleation and probably grain boundary sliding is the most important step to improve ceramics.   Following this are reducing impurities (especially volatile, gas producing, and low-melting impurities) and reducing grain misorientations to decrease grain boundary sliding and crack nucleation.   Control of dislocation phenomena and grain misorientations should substantially reduce the importance of grain size. These factors, plus low levels of porosity now being achieved, and the possible greater tolerance if some porosity remains, reduce the relative importance of decreasing porosity.

Ceramic alloying, as recently discussed in detail by Rice [7], should provide the desired control of dislocation phenomena, while also limiting grain size as noted earlier. The ZrO_2 precipitation shown in MgO indicates another promising ceramic alloy system. Extensive grain orientation, as well as limiting grain size, has recently been demonstrated as a result of successful hot extrusion of MgO [33].   Work in support of these two types of processing to minimize problems due to internal surfaces should receive increasing emphasis, as our knowledge of the detailed nature and effects of such surfaces continues to advance.

ACKNOWLEDGMENT

The following assistance in the above studies is gratefully acknowledged: A. Jenkins, optical microscopy; C. Smith, electron microscopy; R. Racus, electron microprobe; R. Baggerly, X-ray analysis; F. Simpson, MgO – colloidal ZrO_2 specimen preparation; and W. Butterfield, collaboration in welding experiments.

REFERENCES

1. F. Seitz and T. A. Read, "The Theory of the Plastic Properties of Solids, IV," J. Appl. Phys. 12:538 (1941).
2. W. D. Kingery, Introduction to Ceramics, John Wiley and Sons (New York), 1960, pp. 119-201.
3. S. Amelinckx and W. Dekeyser, "The Structure and Properties of Grain Boundaries," in: F. Seitz and D. Turnbull (eds.), Solid State Physics, Vol. 8, Academic Press (New York), 1959, pp. 325-499.
4. R. Scheuplein and P. Gibbs, "Surface Structures of Corundum: 1, Etching of Disloca-tions," J. Am. Ceram. Soc. 43(9):458-472 (1960).
5. W. L. Phillips, "Deformation and Fracture in CaF_2 Single Crystals," J. Am. Ceram. Soc. 44(10):499-506 (1961).
6. M. P. Davis and N. M. Tallan, "Grain Boundary Migration in LiF Bicrystals," J. Am. Ceram. Soc. 47(4):175-178 (1964).
7. R. W. Rice, "Ceramic Alloying," presented at the Sixty-Sixth Annual American Ceramic Society Meeting, Chicago, Illinois, April, 1964.
8. R. J. Stokes and C. H. Li, "Dislocations and the Tensile Strength of Magnesium Oxide," J. Am. Ceram. Soc. 46(9):423-434 (1963).
9. R. W. Rice, "Production of Transparent MgO at Moderate Temperatures and Pres-sures," presented at the Sixty-Fourth Annual American Ceramic Society Meeting, New York City, May, 1962.
10. R. W. Rice, "Fabrication of Dense CaO," presented at the Sixty-Fifth Annual Ameri-can Ceramic Society Meeting, Pittsburgh, Pennsylvania, May, 1963.
11. R. M. Spriggs and T. Vasilos, "Processing Ceramics at Subconventional Temperatures by Pressure Sintering," presented at the Sixty-Sixth Annual American Ceramic Society Meeting, Chicago, Illinois, April, 1964.
12. S. A. Mersol, C. T. Lynch, and F. W. Vahldick, "Investigation of Single-Crystal and Polycrystalline Titanium Diboride: Microstructural Features and Microhardness," U. S. Air Force Technical Documentary Report No. ML-TDR-64-32, April, 1964.
13. T. H. Nielsen and M. H. Leipold, "Thermal Expansion of Ceramic Oxides to 2200°C," J. Am. Ceram. Soc. 46(8):381-387 (1963).
14. R. B. Day and R. J. Stokes, "Research Investigation of Mechanical Properties of Selected High-Purity Magnesium Oxide," Quarterly Progress Report No. 1, Contract AF 33(615)-1282, March, 1964.
15. A. E. Paladino and R. L. Coble, "Effect of Grain Boundaries on Diffusion-Controlled Processes in Aluminum Oxide," J. Am. Ceram. Soc. 46(3):133-136 (1963).
16. R. H. Condit and J. B. Holt, "Oxygen Diffusion in the Grain Boundary of MgO Bi-Crystals," presented at the Sixty-Sixth Annual American Ceramic Society Meeting, Chicago, Illinois, April, 1964.
17. A. R. C. Westwood, "On the Fracture Behavior of MgO Bi-Crystals," Phil. Mag. 6:195-200 (1964).
18. R. J. Stokes and C. H. Li, "Dislocation and the Strength of Polycrystalline Ceramics," in: H. H. Stadelmaier and W. W. Austin (eds.), Materials Science Research, Vol. 1, Plenum Press (New York), 1963, pp. 133-157.

19. F. J. P. Clarke, R. A. J. Sambell, and H. G. Tattersall, "Cracking at Grain Boundaries Due to Dislocation Pile-Ups," Trans. Brit. Ceram. Soc. 61:61-66 (1962).
20. R. C. Ku and T. L. Johnston, "Fracture Strength of MgO Bi-Crystals," Phil. Mag. 9(98): 231-247 (1964).
21. S. A. Long and T. D. McGee, "Effects of Grain Boundaries on Plastic Deformation of NaCl," J. Am. Ceram. Soc. 46(12):583-587 (1963).
22. J. E. Dorn and J. D. Mote, "On the Plastic Behavior of Polycrystalline Aggregates," in: H. H. Stadelmaier and W. W. Austin (eds.), Materials Science Research, Vol. 1, Plenum Press (New York), 1963, pp. 12-56.
23. F. J. P. Clarke and R. A. J. Sambell, "Microcracks and Their Relation to Flow and Fracture in Single Crystals of MgO," Phil. Mag. 5(55):697-707 (1960).
24. G. Thomas, "Dislocations — Their Origin and Multiplication," J. Metals 16(4):365-369 (1964).
25. B. Henderson, "Incoherent Impurity Precipitates in Magnesium Oxide," Phil. Mag. 9:153 (1964).
26. M. A. Adams and G. T. Murray, "Direct Observations of Grain-Boundary Sliding in Bi-Crystals of Sodium Chloride and Magnesia," J. Appl. Phys. 33(6):2126-2131 (1962).
27. F. Rhines, W. Bond, and M. Kissel, "Grain Boundary Creep in Aluminum Bi-Crystals," Trans. Am. Soc. Metals 48:919-945 (1956).
28. R. G. Gifkin, "Some Studies of Grain-Boundary Sliding," in: D. Littler (ed.), Properties of Reactor Materials and the Effect of Radiation Damage, Butterworth (Washington, D. C.), 1962, pp. 335-342.
29. A. Millendore and N. Grant, "Grain Boundary Sliding During Creep of an Al—2%Mg Alloy," Trans. Met. Soc. AIME 227:325-330 (1963).
30. R. W. Rice, "Some Considerations on the Study of Mechanical Effects of Grain Boundaries in Ceramics," presented at the Pacific Coast Regional American Ceramic Society Meeting, Seattle, Washington, October, 1962.
31. B. J. Wuensch and T. Vasilos, "Grain Boundary Diffusion in MgO," J. Am. Ceram. Soc. 47(2):63-68 (1964).
32. W. D. Scott and J. P. Roberts, "Mechanical Properties of Ionic Crystals," Progress Report No. 4, UKAEA Agreement No. 13/5/165/1371, prepared at the Houldsworth School of Applied Science, University of Leeds, February, 1964.
33. R. W. Rice and J. G. Hunt, "Hot Extrusion of MgO," presented at the Sixty-Sixth Annual Ceramic Society Meeting, Chicago, Illinois, April, 1964.

Chapter 24

Grain Boundary and Surface Influence on Mechanical Behavior of Refractory Oxides — Experimental and Deductive Evidence

S. C. Carniglia

Atomics International
Division of North American Aviation, Inc.
Canoga Park, California

A model for grain boundaries of MgO, Al_2O_3, and BeO is postulated and tested against empirical elastic, plastic flow, and fracture data. Descriptions of the various atomistic processes are given and are found to be consistent with experience. New conclusions are reached concerning the strengthening of ceramics.

INTRODUCTION

This treatise on the role of grain boundaries in the mechanical behavior of refractory oxide ceramics deals specifically with the "single-phase" systems MgO, Al_2O_3, and BeO. For convenience in handling a very large subject, it is divided into four sections as follows:

The first section, entitled "A Postulated Structure of Oxide Grain Boundaries," presents a plausible and self-consistent set of structural and behavioral characteristics of grain boundaries in refractory oxides.

In the second section, entitled "Elastic Behavior," those postulates of the preceding section dealing with elasticity are scrutinized in the light of available information concerning single crystals and polycrystals of MgO, Al_2O_3, and BeO. The role of grain boundaries and of impurities in the temperature dependence of the Young's modulus is emphasized.

The third section, entitled "Flow Behavior," presents a discussion of three discrete topics — examination of a postulate developed previously that grain boundary microshear processes governed by impurities should exhibit an amplitude-dependent activation energy; brief reflections on the phenomenology of creep; and a compilation and examination of the plastic flow behavior of single-crystal MgO, Al_2O_3, and BeO as it affects the behavior of polycrystals. Conclusions regarding potential ductility in these oxide ceramics are formulated.

The last section, entitled "Fracture Behavior and Short-Time Strength," is the section in which fracture mechanisms in tensile loading at moderately high strain rates are developed, utilizing postulates from the first section, current brittle-fracture theory, plastic flow behavior as discussed in the third section, and a detailed re-examination of data on strength and grain size. A new depth of insight into the processes leading to catastrophic failure is achieved, thereby laying a firm foundation for conclusions regarding potential strength in refractory oxide ceramics.

A POSTULATED STRUCTURE OF OXIDE GRAIN BOUNDARIES

Grain Boundary Structure

This discussion will be introduced by inferring a model for the structure and characteristics of the grain boundaries in oxide polycrystals, not from the mechanical behavior of refractory oxides, but from their chemical constitution.

MgO, Al_2O_3, and BeO are partly ionic and partly covalent in their chemical bonding. It is beyond the scope of the present work to debate the fraction of each bond type in each bond represented; suffice it to say that both types result in short-range ordering and high rigidity. Restrictions on bond bending, relevant to dislocation generation and mobility, generally increase with increasing covalency. This simplified statement is in accord with the plastic behavior of crystalline MgO vis-à-vis Al_2O_3 and BeO, with that of MgO vis-à-vis NaCl, and with that of the refractory compounds in general, as compared to metallic-bonded elemental crystals.

Rice [1], Gilman [2], and others have considered the influence of ionic and covalent bonding on the arrangement of atoms at grain boundaries. The bubble raft [3] or dislocation array [4] model suitable for metals is not quite satisfactory for refractory oxide grain boundaries owing to disparate atom sizes, charge separation, and electron orbital overlap. The consequent occurrence of large local lattice strain, charge faulting, and the necessary frequency of holes in the structure should produce a substantially lower atom density and a substantially higher average energy than is the case for purely metallic-bonded structures.

If a model similar in outline to that proposed by Mott [5] for metals is adopted, then an oxide grain boundary consists of "island" regions of reasonable lattice fit (probably of the large-atom sub-lattice), bordered by dense dislocations and separated by inter-vening "seaways" or ligaments in which local disorder and charge imbalances predominate, and average energy is high. Because strain is transmitted into the conjugate lattices, the high-energy regions should extend for some distance normal to the boundary. Finally, grain boundary segments formed at high temperatures should adopt a minimum energy configuration which, in general, would not be expected to be planar on an atomic scale, but would undulate about their midplanes.

Figure 1 illustrates schematically the suggested appearance of a grain boundary in a refractory oxide. It deduces a periodic variation in atom density in a sheet-like volume element which contains the nominal plane of the boundary. Figure 2 (lower curve)

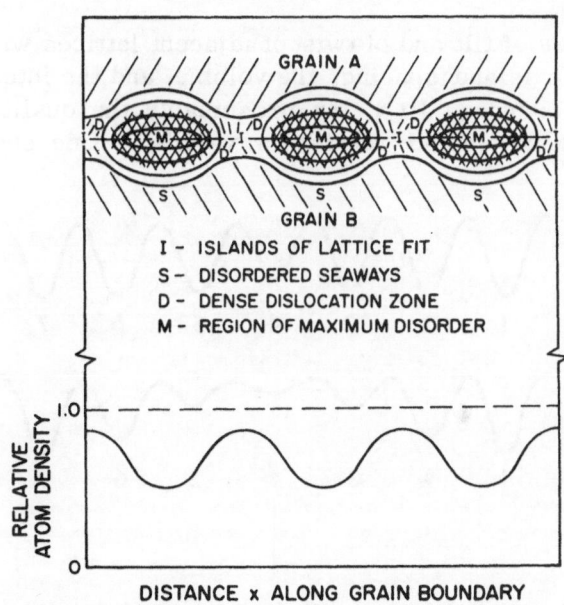

Fig. 1. Schematic section of oxide grain boundary on atomic scale.

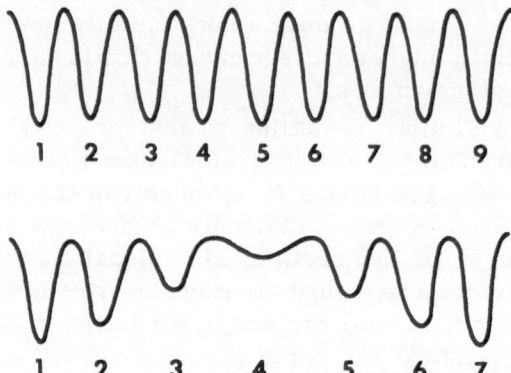

Fig. 2. Schematic energy traces through undisturbed crystal (upper curve) and along center line of grain boundary (lower curve).

illustrates both the increased average atom spacing in the seaways relative to islands and the increase in energy, due to strain, relative to the undisturbed crystal. Figure 3 illustrates the relatively small increase in energy at an island center (position 4, upper curve) above that of the crystal and the transmission of strain and of energy increase out into the conjugate grains at a seaway (lower curve).

The degree of tilt and of twist of adjacent lattices will affect the characteristic island spacing, the volume, and the intensity of defects in the boundary [1,2], but presumably the qualitative model described above should retain its validity. One should expect,

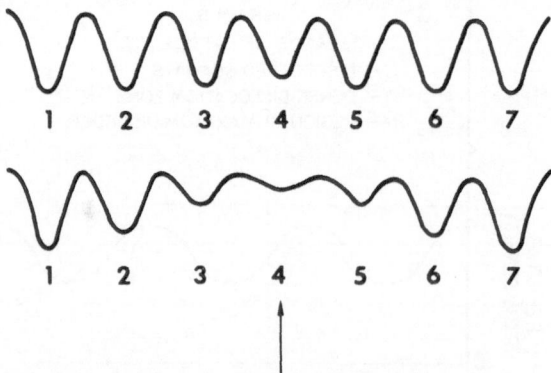

Fig. 3. Schematic energy traces normal to grain boundary plane (arrow): through island (upper curve) and through seaway (lower curve).

therefore, to identify both a periodic variation and an average value with a number of important properties of a given grain boundary segment, such as thickness, atom density, energy, and dislocation and vacancy concentration, and, hence, with a number of related mechanical properties, such as elastic moduli, cleavage energy, and yield and flow properties. For a greater understanding of these related mechanical properties, it is necessary to emphasize that a grain boundary segment is not only anisometric (viz., greatly extended in x and y directions, but short in z), but also anisotropic in its structure and properties owing to coupling with widely differing structures in the $x-y$ and z directions, respectively.

Real polycrystals contain a greater or lesser amount of porosity, some fraction of which is ordinarily located in grain boundaries. The treatment of interaction of pores with stress fields is excessively difficult, and only qualitative recognition will be given here to the existence of pores. However, real polycrystals also contain impurities, and because impurities may greatly influence grain boundary behavior [1, 6, 7], an attempt will be made here to characterize their effect.

Highly stable lattices as represented by MgO, Al_2O_3, and BeO are selective in the formation of substitutional solid solutions. Further, the faulting, disorder, and strain at the grain boundaries of these oxides provide an effective sink for foreign atoms, wherein departures in atom size and charge from those of the lattice atoms may actually lower the energy [6]. A number of studies, e.g., the single-crystal studies of Austerman [8], have suggested that the solubilities of nearly all compounds in BeO are very small. Segregation of a number of oxides to the grain boundaries of Al_2O_3 in the course of sintering was observed by Jorgensen and Westbrook [9] and segregation of MgF_2 impurity in BeO in the course of hot-pressing was described by Carniglia [10]. Impurity precipitates on subgrain boundaries of MgO were seen by Bowen and Clarke [7]. Many other such observations are on record.

The arguments in support of any particular estimate of the thickness of an oxide grain boundary are still vague. If this is taken to average about 10 A (i.e., about five atom spacings), the grain boundary volume in a pure polycrystal is a fraction of the total volume which is determined by the mean grain size $\bar{D}$ and can be estimated. Even small volumes of an impurity phase, distributed over the grain boundaries, will greatly increase their average thickness and volume, as well as affect their properties. Some refer-

TABLE I

Grain Boundary Thickness and Volume Fraction

| $\overline{D}$ | Grain boundary impurity (vol. %) | | | | | |
| | 0 | | 0.1 | | 1.0 | |
(μ)	z (A)	V_b/V_t (%)	z (A)	V_b/V_t (%)	z (A)	V_b/V_t (%)
100	10	0.002	500	0.1	5000	1
10	10	0.02	50	0.1	500	1
1	10	0.2	>10	>0.2	50	1

ence numbers are given in Table I. Whether such magnitudes are actually achieved, and whether the impurity appears as a separate phase or as a gradually enriched solution, will depend upon phase relations, temperature history, and segregation kinetics. The numbers indicate, however, that impurities measured in tenths of a percent may be extremely significant. For MgO, Al_2O_3, and BeO, most segregated impurity compositions will lower the melting point at the grain boundaries, so that high-temperature mechanical behavior will be the most strongly affected. At temperatures near $(T_m)_{gb}$, diffusion of the impurities into the islands may occur, particularly accompanying shear, thus tending to produce a continuous film rather than a periodic net of low-melting material.

Grain Boundary Properties and Behavior

It may be instructive to estimate some ranges of the undulating values of key grain boundary characteristics relative to those of bulk crystals. These are inferential, based on the model described, and may best be regarded as postulates essentially unsupported by direct evidence.

Atom density will vary with lattice mismatch down to approximately 70% of the crystal density along the boundary midplane, or an average of approximately 10% below the crystal density over the whole grain boundary volume (cf. a decrease of 35% in the bond density at the midplane cited by Gilman [2] for a large-angle tilt boundary in metals).

Melting points will usually be lower in grain boundaries than in grains. In segregated impure systems, $(T_m)_{gb}$ may be estimated if the composition and phase relations are known [10].

Energy of a pure grain boundary may be estimated for these oxides (from the heats of fusion) as varying with the lattice mismatch, up to 2000–3000 cal/mole (averaged over the grain boundary volume) above the level in the crystal. Individual atomic-scale domains may be 10,000–15,000 cal/mole above the crystal energy.

Assuming that defects near the boundary midplane diffuse to the surface in the course of cleavage, so that the surface energy is about the same along a cleaved grain boundary as across a crystal, the cleavage energy of a grain boundary may be expected to range down to 50–70% of that of a crystal. Differences in roughness factor have not been considered.

Elastic moduli will be lower in the grain boundary volume than in grains, owing to the existence of broader, shallower energy wells in the strained atom network. Displacement and wave propagation vectors which have a z–component (normal to the boundary) will be strongly influenced by coupling to the adjacent grains; but in particular, shear displacements in the $x-y$ (boundary) plane should disclose the greater compliance of the grain boundary. Values of E and G for the grain boundary volume could be as much as 30–50% below those of crystals. At temperatures near $(T_m)_{gb}$, at which grain boundaries become viscoelastic and then viscous, their behavior, rather than properties of grains, should dominate the elastic behavior of a polycrystal.

Yield and flow properties are difficult to estimate even roughly. As a working hypothesis, it is suggested that the high energy levels and high defect population in the grain boundary volume provide ample generators for shear processes. Shear vectors with a substantial z–component should be more influenced by the lattices and orientations of the conjugate crystals than by the intervening boundary.

However, it is postulated that shear in the $x-y$ (boundary) plane can consist of a relatively short-range process of transition of atoms from the edges of seaways onto the edges of islands, and the reverse, on opposite sides of each island. The critical resolved shear stress for initiation of this process should be much lower than for crystal shear because of the relatively lower density and higher energy states of atoms in the seaways.

Nevertheless, to transmit the shear displacement across islands requires dislocation motion, which passes through the regions of best lattice fit and hence of near-normal energy wells. Since the propagation of shear alternately through islands and sea-ways connects the processes in series, that through islands should control the rate; thus, in a chemically pure system, the activation energy for grain boundary shear will be nearly the same as for crystal shear.

Grain boundary deformation will also interact with jogs and angles in the boundary, as well as with junctions of grain boundary segments (e.g., triple points), so that the elements of pileup, work-hardening, and fracture initiation are present. The plastic work done in shear of grain boundaries will always be small, on account of their small volume fraction in a polycrystal.

It was previously suggested that segregated low-melting im-purities might, under the influence of high temperature and shear deformation under load, diffuse into the islands of the grain boundary and substitute a continuous film of lower-melting composition for the normal island network. This should occur only fairly near to $(T_m)_{gb}$, where viscoelastic or viscous behavior may be expected. When this combination of events occurs, both the critical shear stress and its activation energy should be lowered, the work-hardening rate in the grain boundary should decrease, and the shear rate and internal friction should increase, relative to bulk grain properties.

An interesting consequence of this model of high-temperature, impurity-induced grain boundary shear may be predicted. The model for grain boundary structure included the suggestion that boundary segments are not flat but jogged, based on thermo-dynamic reasoning. Further, triple points, angles, and foreign phase particles present other obstacles to indefinitely extended shear. If the temperature, strain, and stress are high enough, viscous parting and failure can occur; but if not, these obstacles can be by-passed by grain deformation, viz., by extending vacancy diffusion or dislocation motion or both out of the midplane of the boundary and into the surrounding grains. Thus again, a process of low activation energy would become series-connected to a slower one of normal activation energy, and the latter will dominate. It is concluded that impurity-induced grain boundary slide processes of low activation energy will actually show an activation energy which is amplitude-dependent, increasing with increasing displace-ment until the bulk crystal-diffusion characteristics are reached or until viscous failure takes place, whichever occurs first.

Self-diffusion within the grain boundary volume should be enhanced over that in grains. Perhaps one reason why this should be selective, i.e., observed for anion but not for cation diffusion [11-14], may be that the sublattice energy and defect population of the larger ion are more perturbed in the regions of disorder than are those of the smaller ion on account of the closer packing of the larger ions. The reason why the activation energy for self-diffusion over large distances in the boundary may not be measurably different from that in crystals in undoubtedly that the process series-connects vacancy movement in both islands and seaways, the energy level in the island centers being the lowest (and sensibly equal to that in grains) and hence dictating the activation energy of the series process.

Summary
    The above postulates and estimates have been arrived at largely by physicochemical reasoning. In the sections that follow, experimental evidence and theoretical background are assembled from a number of sources, which illustrate the mechanical behavior of polycrystal vis-à-vis single-crystal refractory oxides and test the consistency of the grain boundary model proposed.

ELASTIC BEHAVIOR

Basic Data
    In the preceding section, it was supposed that the elastic constants of grain boundaries are lower than those of grains, but also it was noted that the grain boundary volume fraction is very small. Observing that impurities will influence the melting point of the grain boundary domain, it was concluded that the influence of this domain on the mechanical properties would be observable (and impure systems distinguishable from pure systems) at high temperatures.

    We will investigate here only the interrelations among the single-crystal Young's modulus, polycrystal Young's modulus, grain size, and temperature.

    A number of recent measurements were compared for consistency up to moderately high temperatures. These included data on single crystals of: MgO by Chung and Lawrence [15], Al_2O_3 by Wachtman [16] and by Chang [17], and BeO by Bentle [18], and data on polycrystals of: MgO by Chung and Lawrence [15] and by

TABLE II
Young's Modulus of Single Crystals

Compound	E_0 (psi $\times 10^{-6}$)	A ($°C^{-1} \times 10^4$)	Linearity limit (°C)	Reference
MgO	44.4	2.00	1300	[15]
Al$_2$O$_3$	$\begin{cases} 60 \\ - \end{cases}$	$\begin{matrix} - \\ 0.94 \end{matrix}$	$\begin{matrix} - \\ 1400 \end{matrix}$	$\begin{matrix} [16] \\ [17] \end{matrix}$
BeO	56	-	-	[18]

Mitchell, Spriggs, and Vasilos [19], Al$_2$O$_3$ by Crandall, Chung, and Gray [20] and by Spriggs, Mitchell, and Vasilos [19,21], and BeO by Chang [17,22] and by Fryxell and Chandler [23]. Data on polycrystals were corrected as necessary by the present author to give elastic constants for zero porosity. Summaries of the data are given for single crystals in Table II, and for polycrystals in Table III. In the tables, the temperature-dependent E is given by the simple linear relationship,

$$E_T = E_0 (1 - AT) \tag{1}$$

or

$$- \Delta E = E_0 AT \tag{1a}$$

where E_0 is the value at 0°C and T is in degrees centigrade. The upper temperature limit of linearity is recorded, and for polycrystals the mean grain diameters are given, identified by the symbol $\bar{D}$.

Grain-Size Effect

Those studies which included a broad range of grain sizes were unable to discern a dependency on $\bar{D}$, as limited by the precision of measurement and the determination of preferred grain orientation. Also, within the absolute accuracy of measurement methods, at room temperature, $E_{polycrystal} = E_{single-crystal}$. These results are, of course, expected, since V_b/V_t is very small.

Young's Modulus Versus Temperature

The agreement of values for temperature coefficients for polycrystals, even well below $0.5 T_m$, among different workers is not striking.   Possible sources of discrepancy in making cross comparisons among workers include interaction effects of porosity and temperature; consistent instrument errors; and, in anisotropic materials (Al_2O_3 and BeO), interaction effects of thermal expansion and E, as well as the more troublesome variable of preferred grain orientation. Extrinsic (impurity) differences are also likely.

Of major significance, however, and illustrated by the tabulations, is that the Young's modulus computed from single-crystal data varies regularly with temperature to much higher limits than pertain to any of the polycrystals; and the upper limits for the latter vary widely. Single-crystal curves bend only very slightly downward above the limit of linearity (Chang [17] enumerates some of the reasons for this), while the polycrystal curves break sharply downward [15, 17, 19—23].

Conclusions — Role of Grain Boundaries and Impurities in Elastic Behavior

That the termination of the linear branch of the $E-T$ curve for polycrystals signals a grain boundary effect is undoubted. Close to the temperature of the break, an "unrelaxed" and a "relaxed" value of E appear, $E_u > E_r$, because a time-dependent, reversible, anelastic grain boundary relaxation [24] becomes important; and, in turn E_r becomes difficult to measure because anelastic relaxation and irreversible plastic creep phenomena may overlap.

The fact that the $E-T$ curves break at different temperatures for nominally the same material is clearly extrinsic, i.e., due to the presence of grain boundary impurities, as was illustrated by Chang [17] for several oxides, by Carniglia and Hove [22] using Chang's data on BeO, and later elaborated on by Carniglia [10] using microscopic evidence. As is well known, internal friction peaks, the onset of transient creep, and a break in the tensile strength–temperature curve all accompany the break in E versus T.

If the impurities responsible for changing the elastic properties of grain boundaries are distributed uniformly over all boundaries present, then their active volume is approximately constant irrespective of grain size (see Table I); consequently, to a first approximation, one would expect no grain-size effect either on the temperature of the $E-T$ break or on the E values above this temperature, within a single investigation. This expectation seemed borne out by a detailed examination of the results reported by Fryxell and Chandler [23] and by Spriggs et al. [19, 21].

TABLE III

Young's Modulus of Polycrystals (Corrected to Zero Porosity)

Compound	Impurity (wt. %)	Range of $\bar{D}$ (μ)	$E_0 \times 10^{-6}$ (psi $\times 10^{-6}$)	A (°C$^{-1} \times 10^4$)	Linearity limit (°C)	Reference
MgO	0.4 metals	4– 5	44.4	2.00	1000	[15]
	0.3 typ.	1–190	44.8	1.2	700	[19]
Al$_2$O$_3$	0.5 metals	3, 20	59.3	0.47	1000	[20]
	0.2 typ.	1–100	59.3	1.1	800	[19,21]
BeO	0.2 metals	30–40	55	1.1	1100	[17,22]
	0.5 typ.	2–80	56.2	1.01	1000	[23]

The impurity levels in the ceramics investigated are summarized in Table III. Knowledge of the total impurity levels is very crude; spectrographic evidence was used, from which weight-percent data are given as metals. In no case were anionic impurities reported, and in those marked "typ." no detailed analysis was reported at all in the references cited. Carniglia [10] has pointed out the significance of contamination of BeO at the several thousand parts per million level, and the particular importance of anionic contaminants with respect to other properties. The evidence of Table III is that none of these polycrystal materials was within a decade of adequate purity for intrinsic high-temperature elastic-property measurements, and that individual impurity levels must be known to far higher precision for an accurate assessment of grain boundary behavior.

FLOW BEHAVIOR

Grain Boundary Microshear

It is desired to test an argument developed previously, namely, that the activation energy of impurity-induced grain boundary shear processes should be amplitude-dependent.

Chang [25] studied the steady-state tensile creep of single-crystal and polycrystal alumina with and without the addition of 3 wt.% Cr_2O_3, and also determined the low-frequency internal friction of selected specimens in bending. The two types of measurement lie at opposite ends of the range of amplitudes which we intend to examine. His results are summarized in Table IV, where H_F is the activation energy of internal friction and H_C is the activation energy of steady-state creep. From the equality of activation energies for creep of single-crystal and polycrystal Al_2O_3 and the same value for polycrystal Al_2O_3 internal friction, Chang concluded that grain boundary "sliding" contributed undetectably to microdeformation. In terms of the postulate, sliding presumably did occur in polycrystal Al_2O_3, but showed the normal activation energy because of the participation of the "islands" in the shear process.

Segregation of Cr^{+3} was clearly responsible for the low activation energy of internal friction of the polycrystal $Al_2O_3-Cr_2O_3$ system, since the ruby single crystal showed no corresponding peaks and gave an activation energy for creep equal to that of undoped alumina. The model described in the first section suggests that a

TABLE IV

Grain Boundary Microshear in Al_2O_3

Alumina	H_F (kcal)	T (°C)	H_C (kcal)	T (°C)
Al_2O_3 (cryst.)	No peak	1100 − 1200	200	1800 − 1925
Al_2O_3 (poly.)	200	1100 − 1200	200	1510 − 1570
$Al_2O_3 - Cr_2O_3$ (cryst.)	No peak	850 − 950	−	−
$Al_2O_3 - Cr_2O_3$ (poly.)	47	860 − 910	200	1529 − 1565

Taken from Chang [25−27].

continuous impurity film has divided the grain boundary islands; hence, viscoelastic sliding in internal friction shows an activation energy influenced by the impurity. The value of the activation energy for (large-amplitude) creep of $Al_2O_3-Cr_2O_3$ is the same as that for single-crystal deformation, in accordance with the model, because grain deformation is involved.

Later, Chang and Graham [26] studied the internal friction and transient bending creep of polycrystal BeO, undoped and doped with 1 wt.% MgO. Again, internal friction of the undoped oxide gave the normal activation energy, while the BeO−MgO value was much lower. The data are summarized in Table V.

Transient creep, although of three to four decades smaller amplitude than steady-state creep, subjects the specimen to much higher stress and strain than does internal friction measurement. Anelastic strain of the BeO ceramics was about $1 \cdot 10^{-6}$. Chang and Graham [26] observed two characteristic segments of the

TABLE V

Grain Boundary Microshear in BeO

Beryllia	H_F (kcal)	T (°C)	H_{SL} (kcal)	H_{SH} (kcal)	T (°C)	H_C (kcal)	T (°C)
BeO (poly.)	115	1400	75	120	1150	115	1650
BeO−MgO (poly.)	40	1200	40	115	1000	−	−

Taken from Chang [25−27].

strain–time curves—an initial segment where the plot of log ϵ versus log t had a slope of roughly $\frac{2}{3}$, and a second segment with a slope of approximately $\frac{1}{3}$. These segments were identified, respectively, with grain boundary "slide" (along smooth segments) and with grain boundary "shear" (overcoming jogs and small angles, thus involving deformation extending into grains). These log–log plots are reproduced in Figs. 4 and 5 for BeO and BeO–MgO, respectively. The consistency of initial slopes is observed to be poor in Fig. 4 and good in Fig. 5.

Chang and Graham computed activation energies for the two regions of Fig. 5. The writer computed the corresponding values from Fig. 4. The results are compiled in Table V, where H_{SL} is for the initial sliding process and H_{SH} is for the shearing process. The value for steady-state creep H_C [27] is also given.

It is particularly noteworthy that, while H_F for undoped BeO measured 115 kcal/mole, H_{SL} for the same material, under higher stress, measured 75 kcal/mole. This activation energy for sliding seems too far below the value for shearing and occurs at too low a temperature to represent intrinsic behavior. It must be concluded that, even without doping, we are dealing with an impurity-induced phenomenon. The suggestion is that shear strain has coupled with thermal activation to distribute the impurity into the island regions, and the impurity film has become continuous. The normal activation energy for still larger-amplitude deformation is again observed, the transition from lower to higher activation en-

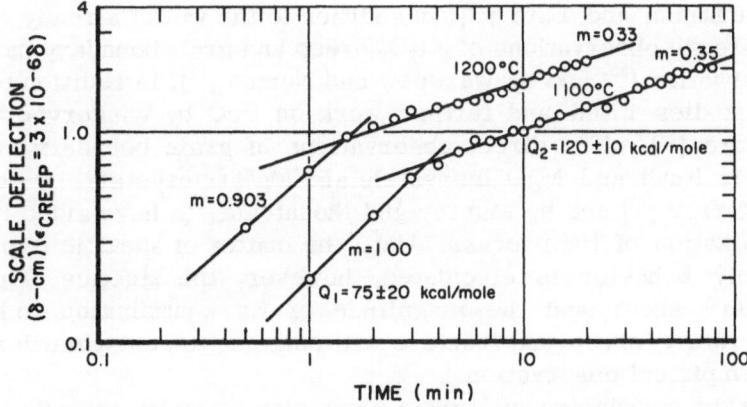

Fig. 4. Transient creep of undoped BeO, after Chang [26].

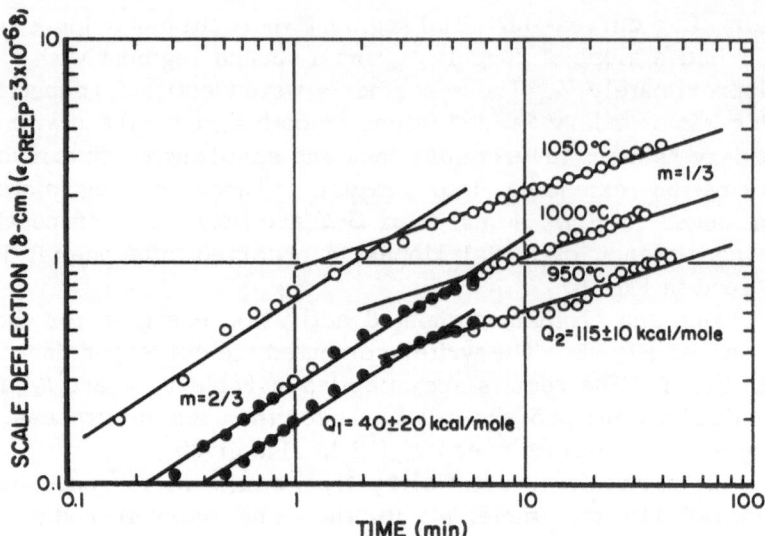

Fig. 5. Transient creep of MgO-doped BeO, after Chang [26].

ergy appearing where grain boundary jogs become a barrier to sliding, and within the region of recoverable deformation.

Creep

The concept of high-temperature grain boundary shearing processes due to impurities in refractory oxides appears now to be well established, through interpretation of Young's modulus and internal friction data for Al_2O_3 by Wachtman and Maxwell [28] and by Wachtman and Lam [29], in addition to the works already cited; and through observations of Al_2O_3 creep and grain boundary parting by Folweiler [30] and by Warshaw and Norton [31], in addition to the BeO studies cited and further work on BeO by Vandervoort and Barmore [32]. The direct observations of grain boundary movement in NaCl and MgO bicrystals and MgO tricrystals, by Adams and Murray [33] and by Murray and Mountvala [34], have aided in the visualization of the process. Until the matter of specific impurity dynamic behavior is elucidated, however, the kinetics of grain boundary shear and the magnitude of its contribution to high-temperature creep and fracture will continue to resist much more than empirical observation.

Grain boundaries and grain sizes play a significant role in the stable region of high-temperature creep, viz., wherein grain boundary shearing and grain deformation are coupled in magnitude. Creep obeying Nabarro–Herring vacancy diffusion kinetics, according to the following equation:

$$\dot{\epsilon} \propto \sigma/(\bar{D})^2 \qquad (2)$$

has been observed for polycrystal Al_2O_3, BeO, and recently by Vasilos, Mitchell, and Spriggs [35] for MgO. Even NaCl, which exhibits easy glide at room temperature, has apparently been forced to deform by vacancy diffusion through loading the polycrystal with a finely subdivided foreign phase [36]. As might be expected from the $(\bar{D})^{-2}$ dependency, transitions from the vacancy diffusion mode to the dislocation climb mode of creep have been either observed or inferred as a function of grain size in Al_2O_3 [31] and in MgO [35]. Transitions as functions of temperature, and as functions of strain rate if fracture does not occur, should also be observable.

Dislocations in MgO, Al_2O_3, and BeO

In order to complete the treatment of grain boundary effects on plastic deformation and to prepare for the following section on fracture, it is necessary to describe some features of dislocation dynamics. No attempt will be made to detail references to the vast literature on the subject, which includes highly informative observations on other nonmetal systems (e.g., NaCl, AgCl, and LiF) and on many metals.

Since the earlier work of Parker et al. [37], Stokes and co-workers [38-43], Pask and co-workers [44, 45], and numerous others (e.g., [46]) has clarified the dislocation behavior of MgO sufficiently, a comprehensive understanding of its crystal deformation is now possible. The following summary, although framed in general terms, serves to provide the information needed here.

MgO has numerous slip planes, the set most easily activated being {110}, but the {100} and {111} sets are apparent also. Although the critical resolved shear stress for the sets are different below very high temperatures ($\sim 1600°C$), work-hardening can compensate for the difference down to much lower temperatures. Wavy glide has been seen down to as low as 400°C. Therefore, the von Mises condition [47] of at least five independent operating slip systems may not be expected to be obtained below this temperature, although it could be obtained above it.

The yield stress of MgO single crystals, in addition to varying with the glide planes operative, also depends very significantly on the presence or absence of dislocation generators and on locking of dislocations by impurities. Diffusion which produces locking and dissociation phenomena appears significant at 600°C and above. Precipitated impurity particles, which affect the lattice resistance to the motion of unlocked dislocations, are not appreciably re-arranged or dissolved below 1100 to 1200°C. This is consistent with the finding that the yield stress and the strain-hardening rate decrease rapidly at about 1100°C and above. At low temperatures (below 400—600°C), the yield stress increases rapidly with decreasing temperature. Between 500 and 1100°C, the yield stress is nearly level, the $\sigma_c - T$ curve running about parallel to the $E - T$ curve.

Because of the sharp dependency of the observed yield stress of single crystals on surface condition, it is difficult to relate single-crystal data directly to polycrystals. The model of the grain boundary proposed in the first section suggests that generators are plentiful in the disordered regions, but that a band of dense, locked dislocations probably stands between the bulk of each grain and the boundary midplane. The lowest yield and flow stresses observed in single-crystal studies (of the order of several thousand psi) therefore should not apply. Probably the yielding of single crystals associated with unlocking of existing dislocations should reasonably represent the yield behavior of grains in a polycrystal; the room-temperature yield stress in bending obtained by Stokes [40] for this condition was about 30,000 psi.

As a consequence of the above observations, we should not expect polycrystal MgO to show appreciable ductility below 400—600°C under any circumstances, owing to failure to meet the von Mises condition for transmittal of deformation from grain to grain. The extent of ductility above this limit should depend in complex ways upon impurity and porosity content and deformation rate. The key to understanding of polycrystal ductility at high temperatures should lie in the critical resolved shear or flow stress in bound-aries relative to that in grains. The nearer these are to equality, the larger will be the displacements which can be accommodated homogeneously. The more unequal these are, the greater will be the disparity between grain boundary and grain deformation, with consequent opening of grain boundary cracks as the expected failure mechanism.

The experimental results of Stokes [40] on (impure) polycrystal

MgO are in accord with this view. Granular slip was seen at 700–800°C, but not below this range on account of genuine brittleness, nor above it due to grain boundary weakness and strain inhomogeneity. Fractures in the range 700–2000°C were predominantly intergranular, and only above 1800°C (presumably accompanied by homogenization of impurities) did even a few percent "ductility" appear.

In Al_2O_3 and BeO, both of which have a higher covalency of bonding and anisotropic crystal structure, the lattice resistance to dislocation motion may be expected to be much higher than in MgO, and directional, so that a wider disparity will exist between critical resolved shear stresses for different slip systems.

The major contributions to the fund of information on sapphire have recently been summarized by Conrad [48]. He refers to disclosures by Kronberg, Nye, Wachtman and co-workers and Gibbs and co-workers of the operation of basal slip at temperatures above about 900°C. Slip on prism planes has been reported only at 1500–1600°C, by Wachtman and Maxwell [49], Scheuplein and Gibbs [50], and Alford and Stephens [51]. Twinning as a deformation mode in alumina, at the lower temperatures ($\gtrsim 1000$°C) and high strain rates, was seen by Stofel and Conrad [52]. Below 900°C, even single crystals of Al_2O_3 appear to lack any evidence of ductility; above this temperature, strain rate is an important factor in determining ductile deformation versus brittle fracture.

Kronberg [53] showed the existence of yield drop in alumina single crystals between 1200 and 1700°C, and related in detail the initial yield stress, strain rate, and temperature for {0001} slip oriented 30° from the tensile axis. Typical yield stress was about 20,000 psi at very low strain rates ($\sim 0.001/$min) and at 1270°C; at lower temperatures or higher strain rates, fracture occurred without evidence of yielding.

Based on the above, we may expect polycrystal alumina to be sensibly brittle to temperatures of at least 1500°C and probably much higher than this, since it is not clear that the von Mises condition has ever been met and it is certain that it is not met below 1500°C. To achieve stable deformation between 1500 and 2000°C would require the utmost purity in order to reduce viscous behavior in grain boundaries.

Virtually all of the deformation work on BeO single crystals done to date in this country has been by Bentle and co-workers

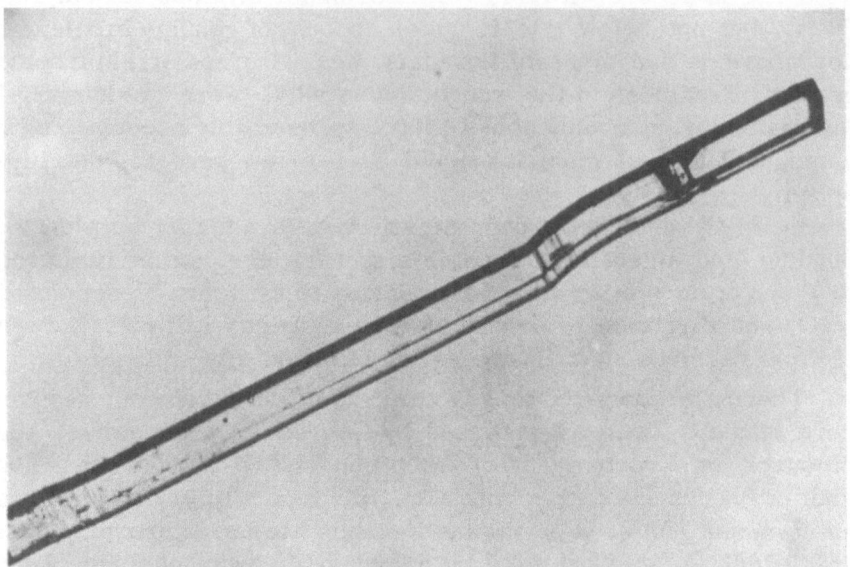

Fig. 6. BeO single crystal deformed under bending load at 1600°C, after Bentle [54].

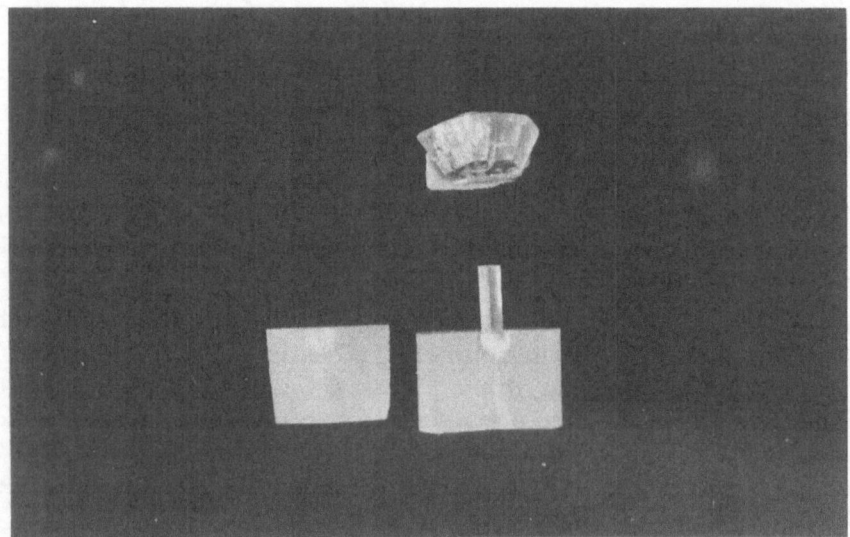

Fig. 7. BeO single crystal loaded in compression at 1600°C, undeformed, after Bentle [54].

[54]. Bending and compression of prism crystals having the basal plane oriented normal to the long axis have been investigated in the range 1000–2000°C. Brittle fracture always occurred below 1000°C. Basal slip was induced in bending experiments above 1000°C, as illustrated in Fig. 6, which shows a crystal deformed at 1600°C. Slip occurred only opposite the loading points, resulting in kinking. Resolved shear stress was about 5,000 psi, in approximate agreement with Kronberg's data on Al_2O_3 at this temperature if account is taken of differences in strain rate. Slip on prism planes has not been seen at comparable stresses, as might be anticipated from the behavior of sapphire.

A crystal loaded in compression at 1600°C is shown in Fig. 7. The resolved shear stress along pyramidal planes was about 50,000 psi, but no slip occurred. The loading fixtures—a highly imperfect BeO crystal above and Lucalox polycrystal alumina below (shown sectioned in the photograph)—were highly deformed.

It is safe to conclude that the deformation behavior of BeO is much like that of Al_2O_3. Beryllia polycrystals should be sensibly brittle to very high temperatures, and should show only small quantitative differences in yield behavior from that of alumina.

Conclusions – Ductility Criteria

It appears that most of the flow properties recorded for MgO, Al_2O_3, and BeO polycrystals, like the high-temperature elastic properties (see the preceding section), are influenced by extrinsic character, i.e., by impurities segregated at the grain boundaries.

Should one expect to be able to produce refractory oxide ceramics exhibiting useful amounts of ductility at temperatures of engineering interest? The answer is probably not, except for the possibility of limited ductility over a limited temperature range around $0.5T_m$ in the simplest and most ionic, least covalent lattices. What frequently passes for ductility in the engineering sense, viz., plastic macroscopic deformation, is largely due to grain boundary mechanical instability. It seems inevitable that efforts to bring additional slip systems into play in anisotropic lattices, by work-hardening, solution-alloying, or dispersion-alloying techniques, will exaggerate the grain boundary instability problem. With or without such efforts, the strengthening of refractory ceramic oxide grain boundaries against elevated-temperature viscous behavior should have high priority in new research in physical ceramics. Chemical purity in the range of four to five nines is the first order of business, perhaps followed by investigation of refractory particulate additives. The finest-possible grain sizing is desirable for stable deformation, because this diminishes the magnitude of lateral displacement which grains must endure in the course of sliding over one another; but unless higher purities are also achieved, the creep strength of such materials will be inordinately low.

FRACTURE BEHAVIOR AND SHORT-TIME STRENGTH

Crack Initiation, Growth, Propagation, and Stoppage Mechanisms

The extensive literature on initiation of cracks within crystals by dislocation interactions, involving massive plastic flow, need be only briefly sampled here [55−59]. It is certain that crack initiation can occur within ductile crystals, and the various flow and interaction mechanisms described in the literature have much in common when observed in a broad view. Single crystals of alumina and beryllia, however, always fracture at low temperatures without appreciable plastic deformation [53,54]; and, owing to the paucity of slip systems operating, glide band intersection is rare even at high temperatures.

Crack initiation at surfaces where glide bands are stopped has been demonstrated or inferred by Stofel and Conrad [52] in high-temperature tensile and bending experiments on Al_2O_3 single crystals, by Bentle [54] in high-temperature bending of BeO single crystals, by Stokes, Johnston, and Li [60] in deformation of NaCl and MgO, and by Miles and Clarke [61] in thermal shock fracture of MgO crystals.

Crack initiation by interaction of dislocations with grain boundaries has been demonstrated or inferred by Clarke and co-workers [61−63], by Westwood [64], and by Johnston, Stokes, and Li [65]. In all cases, these investigations dealt with the inherently ductile compounds, MgO and NaCl.

Parker [66] summarized by stating, in 1962, "The evidence now seems conclusive that cracks form as a result of dislocation interactions in regions where the motion of dislocations is impeded by some sort of barrier," and he noted further that examples of barriers are slip bands, sub-boundaries, and grain boundaries. However, it is not apparent for ceramic materials that fracture must always be precipitated by "initiation" of a crack, since unnumbered examinations of refractory polycrystals have disclosed pre-existing cracks and flaws, and since internal crack-initiation phenomena in ductile nonmetal single crystals have rarely been observed when surface flaws were present [67,68], i.e., without careful polishing of crystal surfaces. Furthermore, it is not yet clear how to interpret the low-temperature fracture behavior of single-crystal alumina or beryllia, nor how to describe microcrack initiation where the deformation of inherently ductile crystals is restricted by the surrounding grains.

To dispose first of the question of crack initiation without massive deformation, it must be assumed that stress risers and mechanical "weak links" (e.g., pores, foreign particles, grain boundary triple points, and grain boundary emergences at surfaces) are scattered throughout all refractory polycrystals. If initiation of a microcrack occurs at one of these and is accompanied or caused by massive slip, we know the yield stress has been reached. If only a minute amount of slip takes place, it may not be detectable and we may have to seek indirect means of observing that the yield stress was reached. Even if slip across the grain cannot occur, we can be confident that the system will use the most economical means available for moving atoms, which is to utilize thermal activation in addition to the applied and locally concentrated stress. Hence, dislocation activity is presumed to accompany crack initiation in any case. If the yield stress is too high for dislocations to communicate across the crystal, they will form an isolated net about the region of greatest deformation.

Such a situation evidently pertains to another phase of fracture, namely, the catastrophic propagation phase, wherein the time duration of stress at any one point is not long enough to drive dislocations across a grain. Figure 8, for example, is an electron micrograph of dislocation tangles adjacent to the room-temperature fracture surface of a BeO fragment, one of many obtained by Willis [69] and observed independently by Wilks [70]; see also evidence of slip traces in grains of fractured MgO as seen by Stokes [40] and of X-ray line broadening in fractured Al_2O_3 obtained by Guard and Romo [71].

There seems to be no reason, then, to assume qualitatively different crack-initiation mechanisms relative to the presence or absence of observed slip. The amount of energy absorbed in the whole process (over and above that of the new surfaces created), and the stress level relative to the yield stress, will be quantitatively distinguishing criteria.

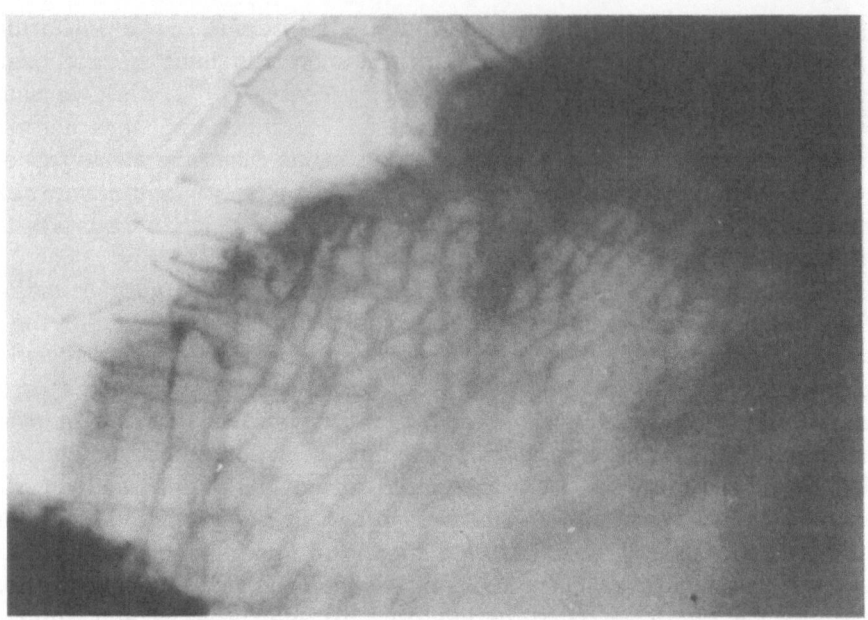

Fig. 8. Dislocation tangles in fractured BeO crystal, after Willis [69].

One should take precisely the same view of crack growth processes. The pre-existence of a crack does not alter the necessity
to create new surface by its extension (hence, fixing the minimum
energy absorbed); the crack itself constitutes a stress riser, and
the radius of its root together with its length determine the relation between the local stress and the applied stress; and the length
determines how much of the energy absorbed in its differential
extension can be supplied by elastic relaxation of the surrounding
medium. The density and extent of plastic deformation which
accompanies crack growth still determine how much energy is
absorbed over and above the surface energy. In general, it should
be easier to extend an existing crack than to start a new one, but
the phenomenology is the same, whether slip is extensive or negligible.

It follows, of course, that the yield stress fixes the approximate maximum stress necessary to initiate or grow a crack.

Quantitatively, we have two conditions for crack initiation:

$$M \cdot \sigma_a = \sigma_c \qquad (3)$$

and

$$dQ = \gamma_s dS + dW_p \tag{4}$$

where σ_a is the applied stress, σ_c is the stress necessary to activate that local flow process which results in crack initiation, and M is the "stress concentration" or "stress multiplication" factor; $\gamma_s dS$ is the energy of the new differential surface element created, dW_p is the plastic work done by all flow processes attending initiation, and dQ is, therefore, the net energy absorbed in the differential process. Similarly, for growth of a pre-existing crack,

$$M \cdot \sigma_a = \sigma_c \tag{5}$$

and

$$dQ = \gamma_s dS + dW_p - dE_r \tag{6}$$

where dE_r is the elastic relaxation energy released in the course of differential extension of the crack. The derivative dE_r/dS is an increasing function of c, which term may be identified for an enclosed crack with its half-length, or, for a crack intersecting the surface, with its length. Hence, the larger the crack (other terms remaining unchanged), the smaller dQ/dS becomes.

So long as $dQ/dS > 0$, some external agent must do work on the system. In test or experimental setups, this work is supplied by the loading device:

$$F_a dx = - dQ \tag{7}$$

i.e., a differential increase in macroscopic deformation takes place, accompanying the formation of new surface dS. When $dQ/dS \le 0$, then crack extension may proceed spontaneously, all the necessary energy coming from the elastic relaxation. This is the condition for catastrophic fracture.

These familiar facts have been restated here in order to lend emphasis to the following arguments:

(1) The relation of c to σ_a when $dQ/dS = 0$, as derived by Griffith [72], assumed a number of idealized conditions which do not necessarily apply, for example, that $dW_p/dS = 0$; that dS is oriented perpendicular to σ_a; that γ_s is invariant; and that $M \cdot \sigma_a = \sigma_c$ is irrelevant. The Griffith equation was admirably suited for the description of brittle fracture of glass, but has not been satisfactorily applied to crystalline aggregates. Subsequent modifica-

tions* by Orowan [73] and others [74, 75] appear to be mathematically satisfactory; nevertheless, most interpreters of the "critical-crack" criterion have been unable to resolve certain physical problems, for example, what is the real significance of grain-size dependency, how is "built-in" surface damage accounted for, and how can the growth of a crack be stopped after it has begun?

(2) The first step toward an understanding of polycrystal fracture is taken by recognizing the fundamental relevance of the condition $M \cdot \sigma_a = \sigma_c$ [equation (5)] as independent of Griffith's energy balance. This was imposed axiomatically previously. What is required is that sufficient local resolved stress exist for initiation of a deformation process which, in interacting with a crack tip, is mechanistically capable of extending it. Granted Griffith's energy condition, if local atomic flow is not activated, there will be no extension of the crack.

(3) The next step is to recognize that the applied stress must be resolved into the plane and direction of that particular local deformation process, the result of which is crack extension. The term M must contain, therefore, besides Orowan's radius ratio ρ/a, a function which is perhaps highly complex but which may for simplicity be identified with geometry and be called g.

$$M = \left(\frac{1}{g} \cdot \frac{a}{\rho} \right)^{1/2} \tag{8}$$

Furthermore, the particular value of σ_c which applies to a particular process at a particular location and mechanical state will be unique only to that situation; hence, σ_c is in general not single-valued for a given material. Out of ignorance, we can incorporate the variability of σ_c into that of g.

*Common forms of the modified Griffith criterion are

$$\sigma_f = \left[\frac{2E \cdot \gamma_s}{\pi c (1 - \nu)} \cdot \frac{\rho}{a} \right]^{1/2}$$

and

$$\sigma_f = \left[\frac{2E(\gamma_s + \gamma_p)}{\pi c (1 - \nu)} \right]^{1/2}$$

where σ_f is the applied stress at $dQ/dS = 0$, E is Young's modulus, γ_s is the surface energy, c is the critical crack length (or half-length) at $dQ/dS = 0$, ν is Poisson's ratio, ρ is the radius at the root of the crack, a is the average interatomic spacing in the material, and γ_p has the same significance as W_p used above. The significance of (ρ/a) in the first of these equations should be familiar [73]; it can be developed from either atomistic reasoning [75] or from the principles of continuum mechanics [76,77].

(4) Next, because of the essential independence of the activation criterion, the author prefers to employ neither the term ρ/a (or $g\rho/a$) nor the term $(\gamma_s + \gamma_p)$ as a "catchall" in the modified Griffith equation, but rather to show them both as follows:

$$\sigma_f = \left[\frac{2E}{\pi c(1 - \nu)} \cdot \frac{g\rho}{a} \cdot (\gamma_s + \gamma_p) \right]^{1/2} \qquad (9)$$

To appreciate the physical significance of this, suppose that a subcritical crack is growing perpendicular to the applied stress at high temperature in a polycrystal of anisotropic character, such as BeO, in which the critical stress τ_c for basal slip is probably an order of magnitude lower than for pyramidal or prismatic slip. Suppose that, as it approaches critical dimensions, the crack passes out of one crystal and into another whose basal plane lies parallel to σ_a and perpendicular to the crack. The resolved shear stress at the crack tip is a maximum normal to it [77], τ_c in that plane is now at a minimum, and there is no doubt as to basal shear in the new grain. The plastic deformation volume in BeO is severely limited, so that in any orientation γ_p is not very large relative to γ_s. The immediate effect on the crack radius ρ will be profound. Such a crack will inevitably stop. The same argument pertains to interaction of a crack with any grain boundary normal to it, if the postulate of the first section concerning in-boundary shear strength is correct.

At the same time, inclusion of a separate term γ_p permits us to attribute the proper role to plastic work and to crack-jogging processes independent of the sharpness of the tip.

(5) It is generally accepted that γ_p may be composed of a number of terms [75], to take care not only of coupled plastic work processes, but also of components of the new surface which are parallel to the applied stress. None of these terms is single-valued for a given material.

What requires emphasis in the work term, however, is that γ_s is not single-valued either. It is commonly recognized that γ_s is smaller on a crystal cleavage plane than in other crystallographic orientations, and cleavage on such planes is the preferred mode of crystal fracture (modified in detail by jogging mechanisms that produce the familiar steps and river patterns of fractogaphs). However, one of the postulates of the first section arises here, namely, that the cleavage energy of a grain boundary may be substantially less than that of a crystal. It was also postulated that r_c for grain boundary shear is less than that for crystal shear. Since r_c must be a component of σ_c for the complex of flow processes which causes boundary cleavage [and in paragraph (3) above, it was decided to include variation of σ_c into g], a decrease in g relative to its crystal value follows. This decrease will be a minimum for grain boundary orientation normal to σ_a, and a maximum for grain boundary orientation parallel to σ_a (on account of the resolution of r_c), thus helping to compensate for the opposite variation of γ_p with orientation of the crack plane. It follows, therefore, that a crack, once established in a grain boundary, should follow it for some distance even at the price of acute orientation to the applied stress. Since, in the first section, it was assumed that γ_s for grain boundaries varies from nearly the crystal value to much lower values depending on the degree of conjugate lattice mismatch, a whole spectrum of local crack behavior from indifference to the existence of grain boundaries to a strong preference for running in them should be seen.

On the average, consequently, grain boundaries should make themselves evident in fracture initiation (i.e., in exceeding the Griffith–Orowan condition). As Rice [1] stated regarding low-temperature bending of hot-pressed MgO bicrystals, "Failure appears to originate at boundary pores or flaws." Stofel and Conrad [78] noted that alumina bicrystals fracture preferentially on grain boundaries, and even saw persistence of cracking along twin boundaries. Other investigators [2,61–65] came to the same conclusion.

Having established what seems to be a workable model for crack initiation, growth, and catastrophic propagation, one finds at least some of the possible mechanisms for crack stoppage or impedance to be apparent as well. These are:

1. Crack blunting by plastic flow, i.e., increase in ρ.
2. Excess work absorbed by plastic flow, i.e., increase in γ_p.
3. Crack jogging in cutting through screw dislocations or other crystal defects, i.e., increase in γ_p.
4. Crack deflection on entering and following a grain boundary, i.e., increase in γ_p and g.
5. Crack blunting on striking a grain boundary at right angles, i.e., increase in ρ.

6. Change in σ_c, or crack deflection, due to dislocation pileup or work-hardening, i.e., increase in g.
7. Change in σ_c in passing from grain to grain with change in orientation of principal yield direction, i.e., increase in g.

Polycrystal Fracture Behavior and Strength

On examining fracture-strength data, one should expect to find many of the factors that influence the observed strength responsible for variability but not to vary the mean value for different specimens of a given material. The single most persistent question concerning the mean value of σ_f has been why and how it is related to grain size, and why and how it is (to a lesser degree) affected by surface finish, or surface "perfection."

If fracture is initiated by the growth of pre-existing flaws to critical size, the distribution of built-in crack lengths should have some effect on the mean strength, and it is not clear that this distribution should be necessarily related to the grain diameter. At least, however, if this were so, the separate effect of surface polishing would be understandable. If fracture begins with crack initiation independent of the existence of surface flaws, then a model can be constructed which relates strength to grain size; but surface finish would appear irrelevant. In addition, how does one deal with those grain-size studies [79] which have related the strength to powers of $-\frac{1}{6}$ or $-\frac{1}{3}$ or some power of mean grain diameter other than the more common $-\frac{1}{2}$ power (which appears to have some foundation in the behavior of metals)? The examination of these questions requires the examination of some data.

Four recent sets of bend-test data were selected for analysis. The first set was obtained by Bentle and Kniefel [54] on very high-purity ($\sim 99.98\%$), very high-density ($\gtrsim 99.7\%$), hot-pressed BeO. Each of the data points shown in Fig. 9 is the mean of ten specimens, ruptured in bending in a vacuum environment. The σ_f versus $(\bar{D})^{-\frac{1}{2}}$ plots seem to be linear (with random scatter), the extrapolated curves passing through the origin at all temperatures.

The second set was selected from the data of Fryxell and Chandler [23] on extruded, relatively impure BeO (AOX grade), which had been shown not to be troubled with grain orientation effects common to other extruded grades. Only their highest-density series (96.5%) was selected. Each of the data points shown in Fig. 10 is

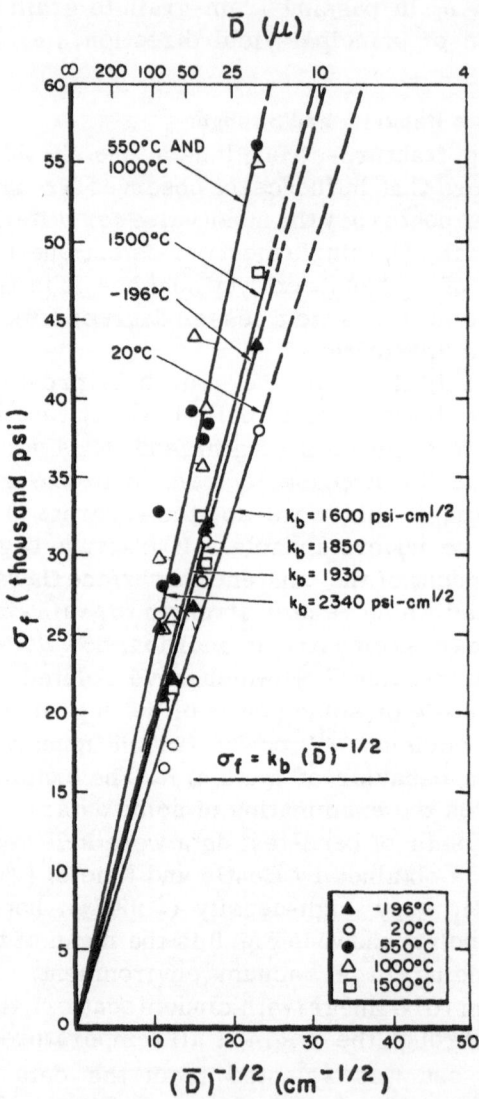

Fig. 9. Petch plot of BeO bending data, after Bentle [54].

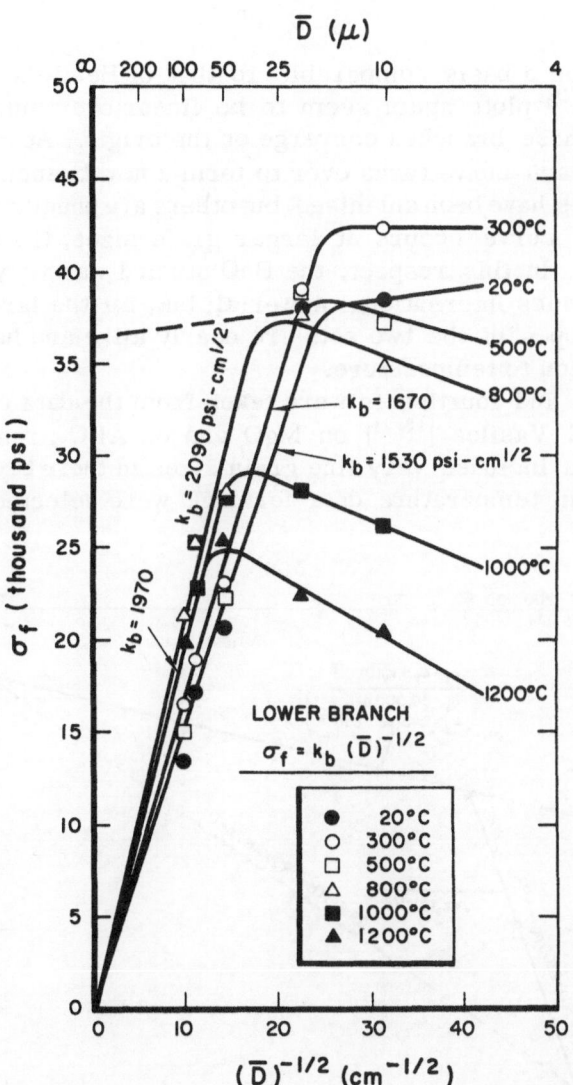

Fig. 10. Petch plot of BeO bending data, after Fryxell [23].

the mean of multiple specimens, ranging from 7 to 62 in number.
The author corrected the strength values by the relation

$$\sigma_0 = \sigma_P \ e^{3P} \tag{10}$$

to put them on a basis comparable to that of Bentle's data. The
σ_f versus $(\bar{D})^{-\frac{1}{2}}$ plots again seem to be linear over much of their
length, and these branches converge on the origin. At grain sizes
below 50μ, each curve turns over to form a new branch. Some of
these branches have been surmised, but others are unequivocal. The
break in the curve occurs at larger grain sizes, the higher the
temperature. In this respect, the BeO studied was very different
from the Atomics International material; but, for the largest grain
sizes, the slopes of the two sets are nearly alike and have a very
similar relation to temperature.

The third and fourth sets were taken from the data of Spriggs,
Mitchell, and Vasilos [19, 21] on MgO and on Al$_2$O$_3$, respectively.
These studies included very fine grain sizes in their broad range.
Only the room-temperature data for MgO were selected, because

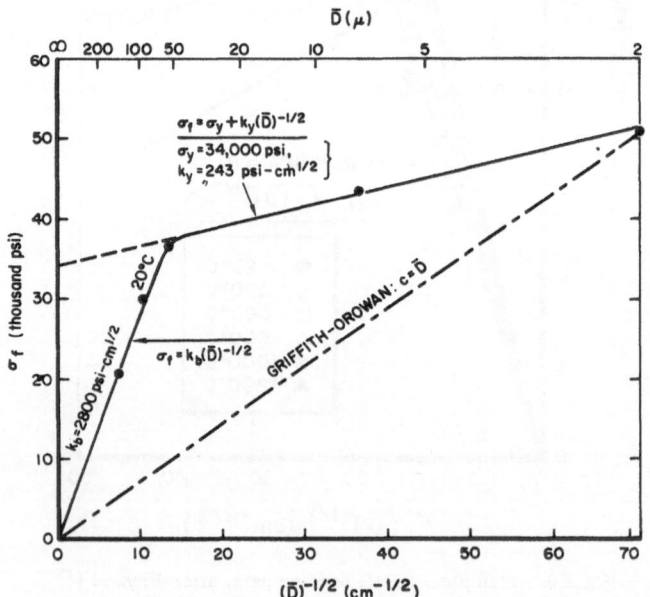

Fig. 11. Petch plot of MgO bending data, after Spriggs [19].

the higher-temperature strengths appeared too inconsistent for this type of analysis. All the Al_2O_3 data were taken.

When plotted (with a small correction of the strength for density made by the writer), the MgO data appeared as in Fig. 11, and the Al_2O_3 data as in Fig. 12. Again, two-branched curves are present, but for these materials the upper branch is quite reminiscent of the behavior of brittle metals. The slopes of the lower convergent branches are similar to those for BeO but somewhat higher, and for Al_2O_3 the slopes vary nearly monotonically with temperature (cf., the reversals at low temperatures for BeO).

Since the observations by Petch [80] on iron and steel at low temperatures, it has been repeatedly confirmed [81-84] that the body-centered-cubic metals in the region both above and below their brittle–ductile transition temperature exhibit a yield stress related to the mean grain size (the "Petch equation"):

$$\sigma_y = \sigma_i + k_y (\bar{D})^{-\frac{1}{2}} \tag{11}$$

Petch [80], Stroh [85], Cottrell [86], Johnson [84, 87], and others have attempted theoretical elaborations on this equation with varying success, but there is general agreement that σ_i is a grain-size-independent term related to the resistance afforded by a crystal lattice to the passage of dislocations; and $k_y(\bar{D})^{-\frac{1}{2}}$ describes the back-stress resulting from dislocation pileup at grain boundaries. Petch and co-workers [82, 83] showed that the tensile fracture stress followed a similar law:

$$\sigma_f = \sigma_i + k_f (\bar{D})^{-\frac{1}{2}} \tag{12}$$

where σ_i was about the same for fracture as for yield, but $k_f \neq k_y$. Low [88] showed that, irrespective of grain size,

$$\sigma_f \text{ (tension)} = \sigma_y \text{ (compression)}$$

Johnson [87] stated, "Yielding is a necessary prerequisite of brittle fracture," and cited data on molybdenum showing that, at a given temperature, $\sigma_f - \sigma_y = \text{constant} \neq f(\bar{D})$.

It is, thus, not unreasonable to assume behavior of the form of the Petch equation for ceramic materials, and a number of workers

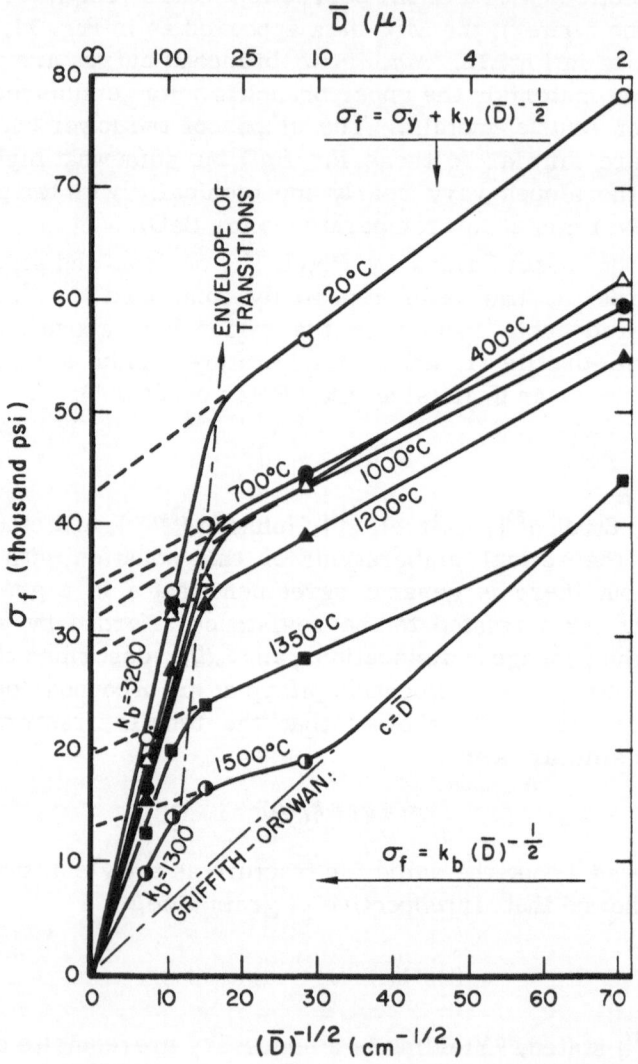

Fig. 12. Petch plot of Al_2O_3 bending data, after Spriggs [19,21].

have done so. We shall henceforth use the notation

$$\sigma_f = \sigma_y + k_y(\overline{D})^{-\frac{1}{2}}$$ (13)

to describe tensile or bending fracture strength which appears to involve yielding; we shall not yet, however, attempt to attach physical significance to σ_y or k_y; we only state that these are proportionality constants, and that finding a significantly large value of σ_y indicates that fracture was initiated by a yield phenomenon.

By all available criteria, the upper (i.e., fine-grained) branch of the curve for MgO (Fig. 11) describes fracture in consequence of yielding. Equally certainly, the lower branch does not; lacking a σ_y term, it seems to indicate pure elastic failure. This means that pre-existing cracks start catastrophic propagation without elongation, and we are immediately led to ask two questions: (1) Why is the average length of the critical crack proportional to grain size and (2) why, in spite of the lack of observable slip in MgO polycrystals at room temperature, is the proportionality constant k_b so high?

For purposes of comparison, an elastic-failure criterion based on the Griffith–Orowan equation and the assumptions $c \simeq \overline{D}$ and $\gamma_p \simeq \gamma_s$ has been sketched in Fig. 12. If the crack-tip radius ρ is used as a "catchall" instead of γ_p, the corresponding assumption is that $\rho \simeq 2a$. The value of k_b obtained from the MgO data corresponds to either $\rho \simeq 30a$ or $\gamma_p \simeq 30\gamma_s$, or some division of the effect between these. One is led to question how the large value of k_b could come about by a large value of the plastic work term without the data having yield characteristics, so that one is hence led to prefer the concept of a "blunt" crack.

The following proposition is, therefore, proffered: The magnesia tested contained "built-in" cracks, of length equal to the grain diameter, and consequently the tips were lying in grain boundaries and blunted to an average radius of about 10 to 30 atom spacings. Whenever the critical fracture stress corresponding to this crack length (or grain diameter) and shape was less than the yield strength, a crack was slightly elongated by minute flow processes; it emerged thereby into a second grain and sharpened, immediately triggering catastrophic fracture. Whenever the corresponding fracture stress was above the yield stress, the precritical crack elongation and sharpening occurred by yielding, so that no fracture strength appreciably higher than σ_y ever appeared.

This postulate is phenomenologically perfect with respect to Fig. 11, but it is very difficult to believe that all built-in flaws extend from the surface of a ceramic to the first underlying grain boundary—none longer and none shorter. Suppose instead that damage to the surface exists to a variety of depths, but that such cracks are usually stopped and blunted by the first grain boundary on which they impinge. Then many such cracks would possess the condition $c = D$, and be blunt, while others would have $c < D$ and be sharp; and some of the latter would become critical first in the course of a rising stress, depending on the ratio ρ/c. For these, cleavage would proceed in the surface grains at stresses below the fracture stress, but only to the nearest grain boundary, where they would come to rest, blunted.

Of the brittle fracture of metals, Allen [90] remarked: "Cracks that stop at grain boundaries have been observed... When they are observed, they are always fully formed and of lengths equal to the diameters of the grains in which they occur." Orowan [73] addressed himself to the question of grain boundary stoppage of cracks in steels, as did Low [81]. Johnson and Shaw [91] noted that in molybdenum cracks often come to rest in the region of "forced slip" (i.e., dense, tangled dislocations) adjacent to the grain boundary. Hahn et al. [92] observed that, in iron and steel stressed at low temperatures, "Cleavage fracture is preceded by discrete bursts of deformation ... accompanied by audible clicks," identified as twinning, restrained by grain boundaries. They concluded that "a cleavage crack that attains a size critical with respect to the host grain is not, at the same time, (necessarily) critical with respect to the overall aggregate."

Parikh [93] has now reported microscopic evidence of the formation and growth of cracks of $c < D$ in the surface grains of Lucalox Al_2O_3 bent to only 50–75% of the fracture stress at room temperature. That these are grain cleavages initiated and grown elastically is indicated by the very low stress levels relative to the yield stress of sapphire; and this occurrence is reasonable in the light of new information on the impurity and porosity characteristics of Lucalox also reported by Parikh. Irrespective of the mechanism of formation, it is clear that these cracks did not become critical with respect to the overall aggregate; hence, the largest of them must have become blunted and stopped at grain boundaries before the actual initiation of catastrophic failure.

The crack-tip radius is invariant with $\bar{D}$ (see Fig. 9–12). If the radius is formed by grain boundary shear, then r at ρ is r_c .

If $\sigma_a \simeq \sigma_f$ and $c \simeq D$, then near the crack tip

$$M\sigma_f = \tau$$

and

$$M_\rho \sigma_f \simeq \tau_c$$

However,

$$M_\rho \propto (c/\rho)^{1/2} \propto (D/\rho)^{1/2}$$

and

$$\sigma_f \propto (1/D)^{1/2}$$

Substitution of

$$K \cdot (D/\rho)^{1/2} (1/D)^{1/2} \simeq \tau_c$$

yields

$$\rho \simeq K/\tau_c$$

A constant ρ thus implies that $\tau_c \neq f(\bar{D})$ which would be expected for large grain sizes and small shear amplitude.

Available mechanisms for grain boundary stoppage of cracks in refractory ceramics were included in the listing developed earlier in this section. To these may be added the following conjecture: When a surface crack enters a grain boundary and stops, the relative ease of shear in the boundary plane may give rise to a "J" or "T" configuration at the crack end, which could isolate the cracked grain acoustically from the conjugate grain and, thus, impede the transmission of the elastic wave (which accompanies a running crack) into the next grain. This configuration should be sought via direct microscopy, and, in the meantime, we should continue to look for other evidence of in-boundary shear in testing of the model of the first section. But why does this process occur one grain deep into the body and not two or three or more ? The answer would seem to lie in the fact that boundaries of surface grains are on the average approximated by a hemispherical shell, while all deeper levels correspond to a full sphere. Thus, a cleaved grain fragment at the surface is bounded on the average by only a quarter-segment of a sphere, all deeper-lying fragments by a hemisphere. Surface grains are thereby much less constrained against rocking or rotation in the process of cleavage than are underlying grains. Verification or correction of this model by direct experimental evidence is one of the greatest of the present requirements of physical ceramics. Its validity will be assumed for want of an alternative even remotely as plausible, in the face of the behavior seen in Figs. 9–12.

If pre-existing cracks, either through their origin or by extension of cleavage at low stresses, all penetrate from the surface to
the first underlying grain boundary, what is the effect of improving
surface finish? This may be complex, because the polishing
process may introduce new damage while removing the old; but let
us assume an ideal case in which only improvement occurs. If it
does not introduce new cracks, removal of the surface by polishing
will reduce the average distance from the surface down to those
grain boundaries which terminate (actually or potentially) existing
cracks, and, hence, change the relationship between $\bar{c}$ and $\bar{D}$. It
must be remembered that built-in surface cracks constitute only one
of a number of structural flaws present (e.g., pores and precipitates), which are all parallel-connected as potential initiators of
fracture. There is therefore no inference, even ideally, that polishing can effect improvement in strength ad infinitum. Only minutely
controlled polishing experiments which compare isolated effects of
impurity state, of porosity and its location, of grain size, and of
fracture in the "elastic" versus "yield" zones will add measurably
to our present understanding. Meanwhile, the venerable concept
of grain refinement in the first few layers from the surface appears
as valid as ever.

We have established $\sigma_y + k(\bar{D})^{-\frac{1}{2}}$ as the upper limit of the strength
of a ceramic polycrystal. There could be a region in which the
strength exceeds this, however. If the Griffith–Orowan line
sketched in Fig. 11 is extended to the right until it crosses the σ_y
curve, then we must have $c > \bar{D}$ at fracture, even if $\rho = a$. Consequently, we might see the influence of work-hardening, as appreciable crack elongation by a yield process precedes fracture. If the
work-hardening coefficient is high enough, a new branch of the
$\sigma_f - (\bar{D})^{-\frac{1}{2}}$ curve should appear which is again linear and extrapolates
through the origin. This concept is introduced schematically in
Fig. 12 in the 1350 and 1500°C data of Spriggs et al. for alumina;
however, the data points do not yield information on this question.
The curves of Fig. 12 are generally well-behaved relative to Fig. 11
and the model developed from it, in all other respects.

The BeO data of Bentle (Fig. 9) show only an "elastic-fracture"
zone. This seems not unreasonable, in view of the fact that $18\,\mu$
was the smallest grain size investigated. However, it is noted that
the beryllia curves of Fig. 10 and the alumina curves of Fig. 12
turn over at larger grain sizes than this. A comparison with
Fryxell and Chandler's data (Fig. 10) suggests that major im-

purities in BeO may be responsible for two effects: (1) the introduction of yield fracture where none existed in Fig. 9, and (2) the
inversion of the slope of the yield branches relative to those of
Fig. 12. It is unlikely that that latter effect is due to porosity, since
the materials of Figs. 10 and 12 are of comparable pore fraction.
A useful postulate is that the inverted slope of the yield branches
of Fig. 10 is due to gross grain boundary sliding; the effect is
quite the same as that observed for internal friction by Hanna and
Crandall [34], who found this property to increase with decreasing
grain size in a series of MgO polycrystals.

One may be led to question whether the Griffith–Orowan line
would necessarily uphold the strength at very small grain sizes
as illustrated shcematically in Fig. 12, since the σ_y curve and the
Griffith–Orowan curve can be expected practically to intersect
only at high temperatures, where impurity-induced, viscous grain
boundary behavior may ordinarily be anticipated. Sliding as a
cause of fracture initiation has, of course, been shown clearly, for
example, in the experiments of Adams and Murray [33] and Murray
and Mountvala [34].

To investigate this matter further, the present author replotted
(again, making small density corrections where required) the
compressive failure strength of MgO polycrystals and of Lucalox
polycrystal Al_2O_3 as reported by Evans [95]. Evans took as the
fracture strength the point at which the σ–ϵ curve showed initial
yielding. His MgO data are shown in Fig. 13, where it is seen that
the elastic failure mode persists at 1000°C over the full range of
grain sizes studied. The data at 1300°C seem to be as well plotted
on a two-branched curve as by Evans' own method (loc. cit.), while
the data at 1600°C seem to be upheld by the Griffith–Orowan curve
and, hence, do not show two-branch characteristics. The curves
of Fig. 13, drawn in accord with the fracture processes postulated
in this treatise, thus seem to be consistent with Evans' data; however, again there is a lack of truly definitive information on this
question. If the postulate of work-hardening is accepted, then
plastic yield within the grains of MgO at 1300 and 1600°C is implied
as initiating failure, rather than grain boundary viscous slide; this
difference between the present interpretation of Evans' data and
the findings of Murray and Mountvala [34] could be ascribed to compressive versus tensile loading. Murray's MgO bicrystals were
very impure, however, and this fact must be taken into consideration.

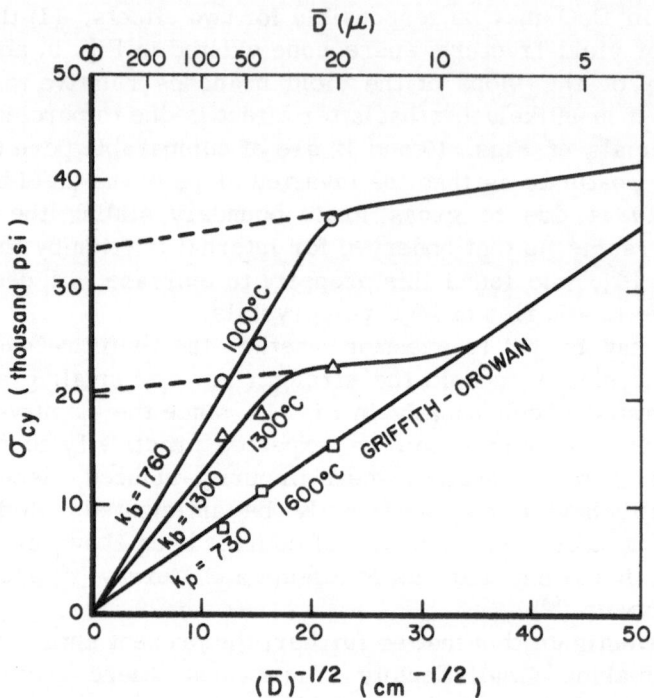

Fig. 13. Petch plot of MgO compressive data, after Evans [95].

Evans' data for Lucalox in compression are replotted in Fig. 14. Viscous behavior seems to be the rule, even at 1000°C. The material analysis given indicated significant amounts of sodium, calcium, and silicon, as well as a major amount of magnesium impurity. Some viscous sliding at 1000°C is thus possible, in the light of the very high stress levels sustained; and sliding at 1600°C and above is to be expected. This time the Griffith–Orowan curve seems not to uphold the strength at all, as would be expected from the fact that strain-hardening does not accompany viscous shear.

We have not proved unequivocally that work-hardening during plastic growth of cracks can increase the slope of the $\sigma_t - (\bar{D})^{-\frac{1}{2}}$ curve where it crosses the Griffith–Orowan curve, but it is certain that no such phenomenon will be evident when impurity-induced viscous grain boundary sliding is in effect. A negative slope of the $\sigma_t - (\bar{D})^{-\frac{1}{2}}$ curve indicates viscous behavior.

Finally, we come to a most important question concerning the plastic yield phenomenon affecting the strength: What is yielding when viscous behavior is not evident? We are prone to assume granular yield, but have not tested this assumption.

For MgO, the room-temperature value of σ_y (Fig. 11) of 34,000 psi is in excellent agreement with the single-crystal value of 30,000 psi [40] cited in the second section in preparation for this comparison. (Ku and Johnston [96] have recently reported a much lower stress for straight bicrystal boundaries, however.) We may conclude that crystalline yield is involved (although not necessarily exclusively) in MgO fracture at low temperatures, and that grain boundary yielding or sliding or both will become progressively more important with increasing temperature, at a rate determined by the melting point and detailed solution behavior of impurities.

For Al_2O_3 and BeO, all the indications are that the yield branch of the $\sigma_f - (\bar{D})^{-\frac{1}{2}}$ curve at low temperatures is extrinsic. We have no independent knowledge of the yield stress of BeO crystals, but expect it to be very high and similar to that of sapphire. The

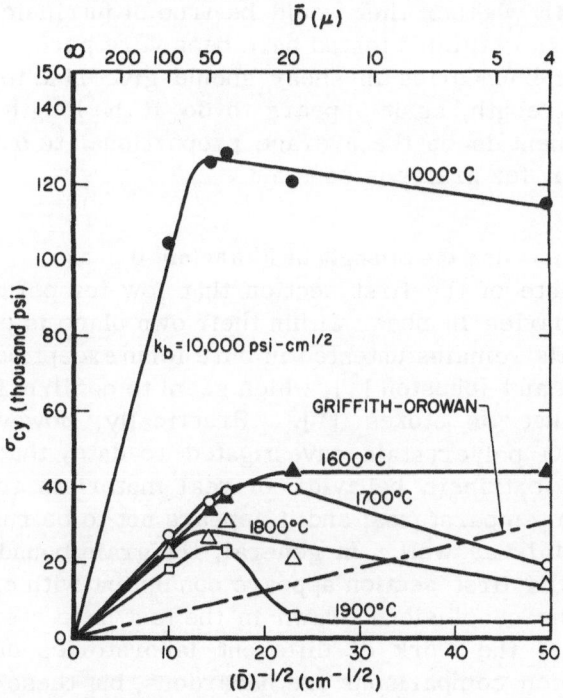

Fig. 14. Petch plot of Al_2O_3 compressive data, after Evans [95].

absence of a yield branch even at high temperatures in Fig. 9 for
99.98% pure BeO and its presence even at room temperature in
Fig. 10 for a very impure grade support the interpretation that
both yielding and sliding in polycrystal BeO are grain boundary
phenomena.   The yield strength of grain boundaries of high-
purity BeO has not yet been seen, because no sufficiently fine-
grained ceramic has yet been investigated.

For Al_2O_3, we have the following evidence—first, the high
yield strengths of sapphire as seen by Kronberg [53], reaching
20,000 psi at 1270°C under the most favorable strain rates and
increasing at a very steep slope with decreasing temperature;
the appearance of compressive yield of polycrystal Al_2O_3 at 120,000
psi at 1000°C in Evans' work [95], which at least sets a lower limit
to the extrapolation of Kronberg's data; and the appearance of a
yield branch in Spriggs' data (Fig. 12) even at room temperature,
with all values of σ_y below 45,000 psi.

The inescapable conclusion for polycrystal Al_2O_3 is that yield
as well as slide is predominantly a grain boundary phenomenon.
It is not clear whether this would be true of intrinsic behavior,
since no alumina ceramics tested have been very pure.

Yielding of boundaries in shear should give rise to a $k_y (\bar{D})^{-\frac{1}{2}}$
term in the strength, as it appears to do, if the length of a grain
boundary segment is on the average proportional to $\bar{D}$, that is, if
the shape factor for grains is constant.

Conclusions – Improving the Strength of Refractory Oxides

The postulate of the first section that low-temperature yield
of grain boundaries in shear within their own plane is easier than
yield of crystals remains untested in pure form except possibly for
the data of Ku and Johnston [96], which seem to confirm it for MgO
relative to that of Stokes [40].   Practically, however, for all
Al_2O_3 and BeO polycrystals investigated to date, that postulate
describes the extrinsic behavior of real materials from low to
relatively high temperatures; and it appears not to be ruled out for
MgO polycrystals as well.   In general, the grain boundary model
presented in the first section appears consistent with experience.

Many of the conclusions drawn in the text have resulted from
comparisons of the work of different laboratories on different
materials.   Such comparisons are hazardous, but these are all we
have from which to work in a comprehensive analysis of fracture.

Many of these comparisons should be repeated in a single laboratory, under conditions of optimum control. Until this can be done,
the following general conclusions concerning the amenability of
refractory oxide ceramics to improvement of strength seem sup-
portable.

Grain refinement and maximum feasible density have tradi-
tionally been deemed major virtues, and the present analysis
emphasizes these. Furthermore, the appearance of a yield phenomenon ascribed partly or solely to grain boundaries, together
with the frequent observation that at high temperatures the elongation to fracture increases with decreasing grain size, accents
fine-grained materials still more emphatically. At the same time,
the existence of this yield phenomenon even at low temperatures
focuses the need for strengthening on the grain boundaries, whose
stiffness relative to that of grains is all-important to both low-
temperature and high-temperature strength, as well as to deforma-
tional stability.

First, in order to obtain the maximum strength of the material,
it will be necessary to constitute oxide ceramics at grain sizes (at
least in the surface layers) small enough that yield, rather than
elastic fracture, occurs; the finer, the better. Second, within this
grain-size limit, it will be necessary to raise the yield stress,
specifically of the grain boundaries themselves. The simplest
improvement to be made in this respect concerns purity. The
work of Bentle on BeO (and the supporting preparative work at
Atomics International) illustrates the fallacy of the term "high
purity" which has accompanied so much of the past effort in oxides.
If purity of four to five nines can be achieved in MgO, Al_2O_3, and
BeO ceramics, together with full density and fine grain size, both
the low-temperature and high-temperature strength levels presently
being observed will surely be dwarfed.

Suggestions of other techniques for strengthening the grain
boundaries are now given physical significance: Some progress can
be expected from dispersion or precipitation alloys, if attention is
given to the size, spacing, and segregation of the particulate phase,
relative expansion coefficients, thermal stability, etc. On the other
hand, the relative futility of raising the crystal yield stress alone,
or of reasoning that yield stress is important in the grain-size
region of elastic failure, is now apparent.

Investigation of ways of increasing the yield stress through
subsequent treatment should go hand in hand with optimization of

microstructures by forming processes. Surface modification seems not to have been exhaustively explored as yet; and not all the details of metallurgical treatments have been investigated or understood. The most exciting possibility seems to be the use of nuclear radiation. Its usefulness in raising the yield strength of crystals and perhaps of grain boundaries is emphasized by the evidence of yield phenomena presented here; but radiation can be employed for more sophisticated purposes than simply creating defects. It can be used to effect nucleation and precipitation reactions in the solid state, for example, and perhaps to swell surface grains as an unorthodox form of prestressing. As an example of the most straightforward use of irradiation, Sambell and Bradley [97] have reported that the shear stress of MgO single crystals was increased from 2900 psi to about 390,000 psi by fast neutron irradiation between $2 \cdot 10^{19}$ and $2 \cdot 10^{20}$ nvt. This whole area of research should have high priority.

The "100,000-psi refractory oxide" is at hand, as a laboratory specimen [98]. It remains to exploit known principles of physical ceramics, and to develop the compositions and the means of preparing the lattice, grain, and boundary structures dictated by these principles, to put even stronger ceramics into production.

GENERAL CONCLUSIONS

1. In broad terms, the grain boundary model proposed in the first section is found to be consistent with the available data on the mechanical behavior of refractory oxide polycrystals.

2. The yield and flow characteristics of grain boundaries vis-à-vis those of grains are found to be of great importance in understanding the deformation behavior and the short-time strength of MgO, Al_2O_3, and BeO ceramics, and the temperature dependence of these. The evidence indicates that short-time tensile fracture may be elastic (at coarse-grain sizes), or due to a yield phenomenon (at fine-grain sizes) most often attributable to the grain boundaries, or due to viscous flow (at high temperatures) uniquely attributable to grain boundaries. The latter two processes were found to be extrinsic, i.e., due to impurities; apparently, they are also responsible for most of the observed plastic deformation of these oxides, and lead to creep rupture and the observed limitations of apparent ductility at high temperatures.

3. The evidence is convincing that purity levels of oxides in the region of two to three nines are entirely inadequate for either optimum performance or the obtaining of intrinsic behavior data. It is probable that no intrinsic polycrystal deformation or fracture data have been obtained to date, with the possible exception of one study of beryllium oxide. At least four- and preferably five-nines purity or better is required.

4. The applicability of the Griffith–Orowan energy criterion and of the Petch rule to fracture of these oxides, and the significance of these criteria in terms of physical phenomena, are verified over a wide range of conditions and materials. Some modifications for more detailed study are suggested.

ACKNOWLEDGMENTS

The author is deeply indebted to the many individuals engaged in the mainstream of scientific productivity for the data, original conclusions, and origins of ideas which have made this synthetic work possible. In particular, discussions with R. W. Guard and with G. G. Bentle and their review of the manuscript have been most helpful. This treatise is based in part on work supported by Fuels and Materials Branch, Division of Reactors, U. S. Atomic Energy Commission, under Contract AT(11-1)-GEN-8.

REFERENCES

1. R. W. Rice, Boeing Co., unpublished manuscript, presented at 15th Pacific Coast Regional Meeting, American Ceramic Society, Seattle, Washington, October 1962.
2. J. J. Gilman, "Mechanical Behavior of Crystalline Solids," Natl. Bur. Std. Monograph 59:79 (1963).
3. W. M. Lomer and J. F. Nye, Proc. Roy. Soc. (London) A212:576 (1952).
4. S. Amelinckx and W. Dekeyser, Solid State Physics, Vol. 8, Academic Press (New York), 1959, p. 325.
5. N. F. Mott, Proc. Phys. Soc. (London) 60:391 (1948).
6. W. D. Kingery, Introduction to Ceramics, John Wiley and Sons (New York), 1960, p. 199.
7. D. H. Bowen and F. J. P. Clarke, Phil. Mag. 8:1257 (1963).
8. S. B. Austerman, presented at the International Conference on Beryllium Oxide, Sydney, Australia, October 1963, to be published. See also USAEC Reports NAA-SR-8056 (1963), NAA-SR-8235 (1963), and NAA-SR-8361 (1963).
9. P. J. Jorgensen and J. H. Westbrook, J. Am. Ceram. Soc. 47:332 (1964).
10. S. C. Carniglia, presented at the International Conference on Beryllium Oxide, Sydney, Australia, October 1963, to be published.
11. M. O. Davies, J. Chem. Phys. 38:2047 (1963).
12. Y. Oishi and W. D. Kingery, J. Chem. Phys. 33:480 (1960).
13. A. E. Paladino and W. D. Kingery, J. Chem. Phys. 37:957 (1962).

14. S. B. Austerman, presented at the International Conference on Beryllium Oxide, Sydney, Australia, October 1963, to be published. See also USAEC Reports NAA-SR-3170 (1958), NAA-SR-5893 (1961), NAA-SR-6427 (1961), and NAA-SR-7637 (1962).
15. D. H. Chung and W. G. Lawrence, J. Am. Ceram. Soc. 47:448 (1964).
16. J. B. Wachtman, Jr., W. E. Tafft, D. G. Lam, Jr., and R. P. Stinchfield, J. Res. Nat. Bur. Std. 64A:213 (1960).
17. R. Chang, in: W. W. Kriegel and H. Palmour III (eds.), Mechanical Properties of Engineering Ceramics, Interscience Publishers (New York), 1961, p. 209.
18. G. G. Bentle, Atomics International, 1964, to be published.
19. J. B. Mitchell, R. M. Spriggs, and T. Vasilos, "Microstructure Studies of Polycrystalline Refractory Oxides," USN Report RAD-TR-63-2 (1963).
20. W. B. Crandall, D. H. Chung, and T. J. Gray, in: W. W. Kriegel and H. Palmour III (eds.), Mechanical Properties of Engineering Ceramics, Interscience Publishers (New York), 1961, p. 349.
21. R. M. Spriggs, J. B. Mitchell, and T. Vasilos, J. Am. Ceram. Soc. 47:323 (1964).
22. S. C. Carniglia and J. E. Hove, J. Nucl. Mater. 4:165 (1961).
23. R. E. Fryxell and B. A. Chandler, J. Am. Ceram. Soc. 47:283 (1964).
24. C. Zener, Elasticity and Anelasticity of Metals, University of Chicago Press, 1948, p. 150.
25. R. Chang, J. Nucl. Mater. 1:174 (1959). See also USAEC Report NAA-SR-2770 (1958).
26. R. Chang and L. J. Graham, "Transient Creep and Associated Grain Boundary Phenomena in Polycrystalline Alumina and Beryllia," USAEC Report NAA-SR-6483 (1961).
27. R. Chang, "Creep of Polycrystalline BeO at High Temperatures and Low Stresses," USAEC Report NAA-SR-2458 (1958).
28. J. B. Wachtman and L. H. Maxwell, USAF Report WADC-TR-57-526 (1957).
29. J. B. Wachtman and D. G. Lam, J. Am. Ceram. Soc. 42:254 (1959).
30. R. C. Folweiler, J. Appl. Phys. 32:773 (1961).
31. S. I. Warshaw and F. H. Norton, J. Am. Ceram. Soc. 45:479 (1962).
32. R. R. Vandervoort and W. L. Barmore, J. Am. Ceram. Soc. 46:180 (1963).
33. M. A. Adams and G. T. Murray, J. Appl. Phys. 33:2126 (1962).
34. G. T. Murray and A. J. Mountvala, "The Role of the Grain Boundary in the Deformation of Ceramic Materials," USAF Report ASD-TDR-62-225, Part 2 (1963); see also Part 1 (1962).
35. T. Vasilos, J. B. Mitchell, and R. M. Spriggs, J. Am. Ceram. Soc. 47:203 (1964).
36. R. L. Coble and A. E. Paladino, Massachusetts Institute of Technology, 1964, to be published.
37. A. E. Gorum, E. R. Parker, and J. A. Pask, J. Am. Ceram. Soc. 41:161 (1958).
38. R. B. Day and R. J. Stokes, Honeywell Research Center, 1964.
39. R. B. Day and R. J. Stokes, "Research Investigation of Mechanical Properties of Selected High-Purity MgO," USAF Quarterly Report No. 1 under RTD Contract AF 33(615)-1282 (1964).
40. R. J. Stokes, "Thermal-Mechanical History and the Strength of MgO Single Crystals," USAF Report HR-64-258 (1964).
41. R. J. Stokes and C. H. Li, J. Am. Ceram. Soc. 46:423 (1963).
42. R. J. Stokes, Trans. AIME 224:1227 (1962).
43. R. J. Stokes, T. L. Johnston, and C. H. Li, Phil. Mag. 6:9 (1961).
44. S. M. Copley and J. A. Pask, "Plastic Deformation of MgO Single Crystals up to 1600°C," J. Am. Ceram. Soc. 48:139 (1965).
45. C. O. Hulse, S. M. Copley, and J. A. Pask, J. Am. Ceram. Soc. 46:317 (1963).
46. A. E. Gorum and J. W. Moberly, J. Am. Ceram. Soc. 45:316 (1962).
47. R. von Mises, Z. Angew. Math. Mech. 8:161 (1921).
48. H. Conrad, "The Mechanical Behavior of Sapphire," J. Am. Ceram. Soc. 48(4):195-201 (1965).
49. J. B. Wachtman and L. H. Maxwell, J. Am. Ceram. Soc. 40:377 (1957).
50. R. Scheuplein and P. Gibbs, J. Am. Ceram. Soc. 43:458 (1960).
51. W. J. Alford and D. L. Stephens, J. Am. Ceram. Soc. 46:193 (1963).
52. E. Stofel and H. Conrad, Trans. AIME 227:1053 (1963).

53. M. L. Kronberg, J. Am. Ceram. Soc. 45:274 (1962).
54. G. G. Bentle and R. M. Kniefel, Atomics International, 1963, to be published.
55. C. Zener, Fracturing of Metals, ASM, Cleveland (1948).
56. A. N. Stroh, Proc. Roy. Soc. (London) A223:404 (1954) and A232:548 (1955).
57. R. J. Stokes, T. L. Johnston, and C. H. Li, Phil. Mag. 3:718 (1958).
58. J. Washburn, A. E. Gorum, and E. R. Parker, Trans. AIME 215:230 (1959).
59. A. S. Argon and E. Orowan, Nature 192:447 (1961).
60. R. J. Stokes, T. L. Johnston, and C. H. Li, Trans. AIME 218:655 (1960).
61. G. D. Miles and F. J. P. Clarke, Phil. Mag. 6:1449 (1961).
62. F. J. P. Clarke, R. A. J. Sambell, and H. G. Tattersall, Trans. Brit. Ceram. Soc. 61:61 (1962).
63. F. J. P. Clarke, R. A. J. Sambell, and G. D. Miles, Trans. Brit. Ceram. Soc. 60:299 (1961).
64. A. R. C. Westwood, Phil. Mag. 6:195 (1961).
65. T. L. Johnston, R. J. Stokes, and C. J. Li, Phil. Mag. 7:23 (1962).
66. E. R. Parker, "Mechanical Behavior of Crystalline Solids," Natl. Bur. Std. Monograph 59:1 (1963).
67. F. J. P. Clarke, R. A. J. Sambell, and H. G. Tattersall, Phil. Mag. 7:393 (1962).
68. H. G. Tattersall and F. J. P. Clarke, Phil. Mag. 7:1977 (1962).
69. A. H. Willis, Atomics International, 1963, to be published.
70. R. S. Wilks, "The Observation of Dislocations in BeO by Transmission Electron Microscopy," UKAEA Report AERE-R4436 (1963).
71. R. W. Guard and P. C. Romo, "X-Ray Microbeam Studies of Fracture Surfaces in Alumina," J. Am. Ceram. Soc. 48:7 (1965).
72. A. A. Griffith, Trans. Roy. Soc. (London) 221:163 (1920).
73. E. Orowan, Repts. Prog. Phys. XII:185 (1948).
74. N. F. Mott, Eng. 16:2 (1948).
75. J. Friedel, Fracture, John Wiley and Sons (New York), 1959, p. 498.
76. C. Inglis, Trans. Inst. Nav. Archit. (London) 55:219 (1913).
77. I. N. Sneddon, Proc. Roy. Soc. (London) A187:229 (1949).
78. E. Stofel and H. Conrad, J. Metals 14:87 (1962).
79. W. D. Kingery and R. L. Coble, "Mechanical Behavior of Crystalline Solids," Natl. Bur. Std. Monograph 59:103 (1963).
80. N. J. Petch, J. Iron Steel Inst. 174:25 (1953).
81. J. R. Low, Symposium on Relation of Properties to Microstructure, Am. Soc. Metals, 1954, p. 163.
82. A. Cracknell and N. J. Petch, Acta Met. 3:186 (1955).
83. J. Heslop and N. J. Petch, Phil. Mag. 1:866 (1956).
84. A. A. Johnson, J. Less Common Metals 2:241 (1960).
85. A. N. Stroh, Advan. Phys. 6:418 (1958).
86. A. H. Cottrell, Trans. Met. Soc. AIME 212:192 (1958).
87. A. A. Johnson, Phil. Mag. 7:177 (1962).
88. J. R. Low, IUTAM Colloquium on Deformation and Flow of Solids, Springer-Verlag (Berlin), 1956, p. 60.
89. N. J. Petch, Fracture, John Wiley and Sons (New York), 1959, p. 54.
90. N. P. Allen, Fracture, John Wiley and Sons (New York), 1959, p. 123.
91. A. A. Johnson and B. J. Shaw, Nature 183:1541 (1959).
92. G. T. Hahn, B. L. Averbach, W. S. Owen, and M. Cohen, Fracture, John Wiley and Sons (New York), 1959, p. 91.
93. N. M. Parikh, "Studies of the Brittle Behavior of Ceramic Materials," USAF Report ASD-TR-61-628, Part III (1964), p. 17.
94. R. Hanna and W. B. Crandall, "Dissipation of Energy by the Grain Boundaries," ASTIA Report 274956 (1962).
95. P. R. V. Evans, "Studies of the Brittle Behavior of Ceramic Materials," USAF Report ASD-TR-61-628, Part II (1963), p. 164.
96. R. C. Ku and T. L. Johnston, Phil. Mag. 9:231 (1964).
97. R. A. J. Sambell and R. Bradley, Phil. Mag. 9:161 (1964).
98. R. M. Spriggs, private communication, 1964.

Chapter 25

Fractographic Evidence of Multiple Slip in Deformed Hot-Pressed Spinel

Dong M. Choi* and Hayne Palmour III

North Carolina State University
Raleigh, North Carolina

Spinel is a refractory oxide ceramic which, in principle, should meet the Taylor–von Misés criterion of five slip modes, a necessary condition for the achievement of polycrystalline plasticity without loss of structural integrity. Crystallographically, its (111)[110] slip systems are multiple and quite comparable to those of a face-centered-cubic metal. Fractographic examinations of a spinel specimen (> 99.9% pure) previously subjected to deformation in three compressive stress cycles at 1600°C have revealed several microscopic features characteristic of polycrystalline plasticity which experimentally support the expectation of multiple slip processes in spinel ceramics.

INTRODUCTION

As Copley and Pask [1] have pointed out, few ceramic materials have structures which provide the five independent modes for crystalline slip that are required for plastic deformation of poly-crystalline masses. Furthermore, in terms of Carniglia's analysis [2], grain size effects and grain boundary localizations of any impurities present are likely to prevent observations of plasticity, even in material intrinsically capable of slip, by promoting either brittle fracture or viscous grain boundary shearing under stress-temperature conditions substantially lower than those required for yielding by glide of dislocations. Consequently, experimental evidences of polycrystalline plasticity in ceramics are understandably not common and are expected to be confined only to those having suitable isotropic structures, fine grain sizes, and high levels of purity. Spinel ceramics may be almost uniquely qualified for such a role.

The cubic structure of spinel yields (111)[110] as the slip plane and slip direction for dislocations. They are multiple, i.e., repeated

*Present address: E. I. duPont de Nemours Company, Wilmington, Delaware.

about the center of symmetry, as in a face-centered metal [3-5]. Magnesium aluminate spinel of high purity has been synthesized, and its mechanical properties have been investigated during a four-year study at this laboratory.

In our study, attention has been directed to the influence on mechanical behavior of process parameters, microstructures, and purity [6]. For example, though the test specimens were comparable in density and microstructure, a tenfold difference in flow stress was reported for compressive deformation at 1600°C between two spinel grades having dissimilar purity levels, viz., about 99.5% for the weaker and > 99.9% for the stronger. Above about 1350°C, flow stresses (rather than fracture stresses) for high-purity spinel are observed which depend sensitively upon both temperature and strain rate, as is the case for ductile metals. The character of the high-temperature stress—strain diagrams and the kinetic data now being developed are definitely indicative of ductility in this refractory ceramic material.

However, it is not easy to find direct evidence for the presence and mobility of dislocations in spinel. For example, it is very difficult to develop chemical etch pits, even in single crystals [7], and, in polycrystalline spinel ceramics, etchants having sufficient strength to cause etch-pitting also badly over-etch grain boundaries. Furthermore, cutting and polishing of spinel specimens for reflected light microscopy introduces much cold-work, so that subsequently an etchant principally develops dense rows of dislocation pits along scratch marks, to the point where such artifacts almost obscure any useful evidence about the bulk. In this report, optical replication fractography has been used in preference to conventional polish-and-etch photomicrographic techniques.

Those familiar with dislocations will realize that slip traces attributable to the displacement of crystalline material by individual dislocations generally are not resolvable by optical microscopy and result only in fine textural differences at electron microscopic magnifications (e.g., see Fig. 29 of Palmour et al. [3]). In optical replication fractography, therefore, one must look for textural features associated with groups of once-mobile dislocations; i.e., evidence of plasticity, if present, is most likely to be observable in the form of slip bands containing many dislocations.

Textural differences resolvable by fractography and attributable to slip bands probably arise due to local interactions between stress fields associated with dislocations in the band and stress fields at the advancing crack tip. Such interactions would bring about local deflections of the fracture wave, resulting in topographic variations characteristic of the stress fields and, hence, of the dislocations or slip bands.

In very fine-grained material, visible slip bands are not likely to form, since individual grains can be deformed sufficiently by the motion of relatively few dislocations, operating almost singly. Thus, a search for slip bands as prima-facie evidence of plastic deformation is more likely to be successful in relatively coarse-grained material.

EXPERIMENTAL TECHNIQUE

Replicas of ultrasonically cleaned fracture surfaces were made by applying a drop of an acetone-diluted replicating solution,* allowing the solvent to evaporate, stripping the replica with transparent tape, and vapor-depositing a reflective thin film (≈ 400 A) of chromium metal at an incident angle of about 45°. The replicas were examined and photographed with a Reichert Model MeF Universal microscope † using incident illumination from a zirconium arc source.

SPECIMEN HISTORY

The spinel specimen initially was prepared by rate-controlled hot-pressing from co-precipitated, high-purity starting material by methods described elsewhere [6]. The bulk density was greater than 99% of theoretical, and the average grain size was approximately $2-5\mu$. Figure 1 illustrates the initial microstructure as viewed by optical replication fractography; it shows the uniform, equiaxed texture characteristic of spinel.

The particular specimen was utilized as an anvil pad at high temperature in vacuo, transmitting compressive loads from a tungsten push rod to a smaller test specimen of the same spinel material. It was subjected to three successive cycles, each involving heating to 1600°C and applying a stress of 20,000–25,000 psi to the area directly beneath the test specimen. After the third cycle, it was found to be fractured approximately parallel to the load axis. As Fig. 2 illustrates, it had undergone considerable

*A product of Ladd Industries, Inc., Roslyn Heights, New York.

†A product of Optische Werke C. Reichert, Vienna, Austria.

Fig. 1. Microstructure of dense, fine-grained, high-purity spinel prior to deformation. Chromium-shadowed optical replication fractograph.

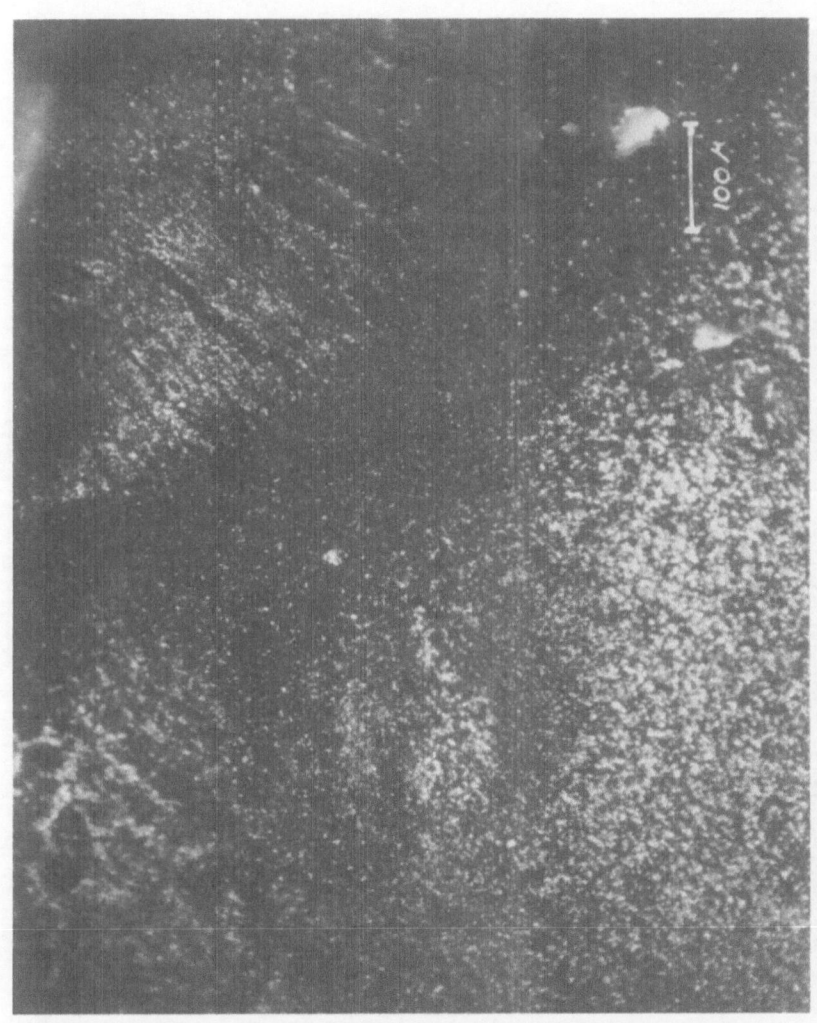

Fig. 2. Grain growth in polycrystalline spinel after three successive cycles of compression at 1600°C. Vertical load was concentrated at upper right; grain growth occurred in regions radially disposed about the strained portion. Chromium-shadowed optical replication fractograph.

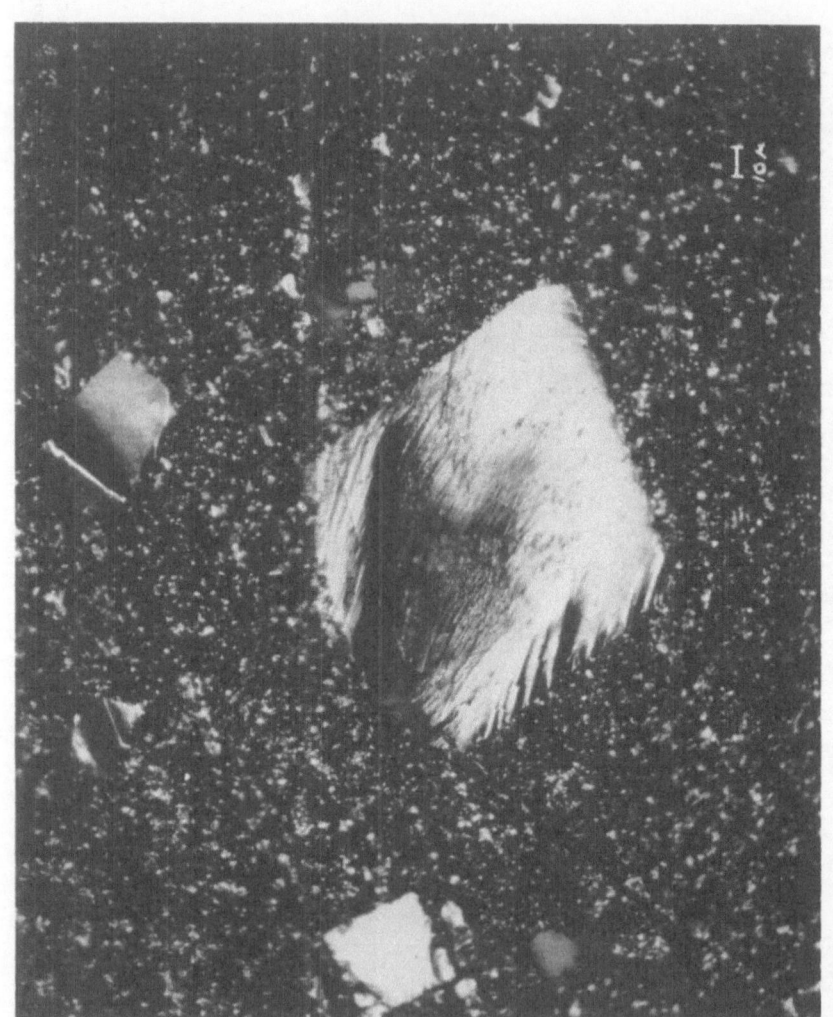

Fig. 3. Idiomorphic large grains in fine-grained spinel matrix after three successive cycles of compression at 1600°C. Chromium–shadowed optical replication fractograph.

grain growth in some regions that bore a distinct radial relationship to the localized region where compressive stress had been
greatest.

Elsewhere in the specimen, different examples of grain growth
could be noted, often yielding giant grains in an almost unchanged,
fine-grained matrix. These large grains, as illustrated in Fig. 3,
tended to take on euhedral shapes (e.g., cubes and tetrahedrons) and
are presumed to have grown under the infleunce of small, local
impurity concentrations in a manner akin to liquid-phase sintering
[8].

EVIDENCES OF PLASTICITY

Strain-Anneal

The rather consistent grain growth observed in areas radially
equidistant from the highly stressed and somewhat indented test
area (Fig. 2) suggests that the onset of grain growth was in some
way associated with the strain gradient and that strain-annealing
processes were involved. Commonly observed in deformed and
heated metals, these processes allow excess energy stored in
grains during plastic straining to contribute to the driving force for
subsequent grain growth when the temperature is high enough to
permit the required mass transport.

Loss of Strength with Increasing Grain Size

The spinel anvil block withstood the first two heat-and-stress
cycles without visible deterioration and technically completed its
third mission insofar as loading the test specimen was concerned,
although the anvil itself sustained a form of columnar fracture,
producing the surfaces examined here. A progressive loss of
strength (leading to brittle fracture only on the third cycle) is
suggested and is considered to be an illustration of the adverse
effect of increased grain size upon fracture strength, given by the
Stroh–Petch relationship [9, 10]:

$$\sigma_f = \sigma_y + Kd^{-\frac{1}{2}} \tag{1}$$

Dislocation interactions with other dislocations and particularly with
grain boundaries generate pile-up stresses, which increase in a
way proportional to the average grain diameter d. When these
stresses become high enough, cracks are generated and extend to
propagable length, ultimately inducing brittle behavior at the grain-

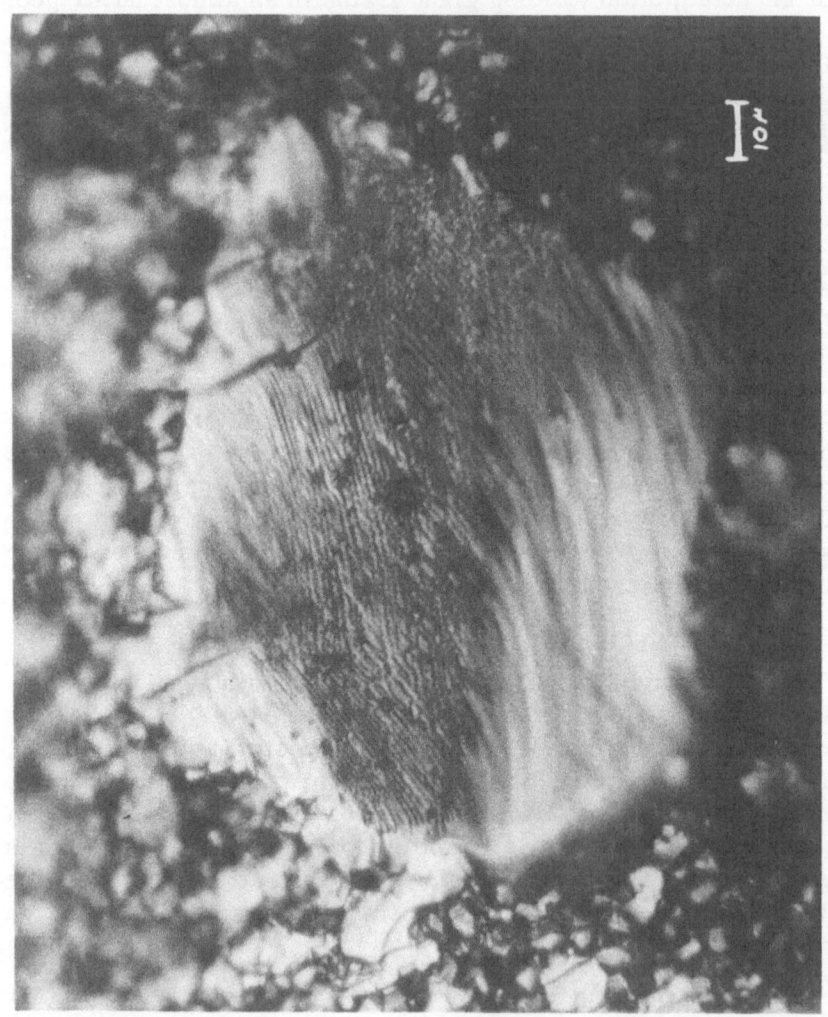

Fig. 4. Intersecting slip bands in idiomorphic large grain of spinel. Chromium–shadowed optical replication fractograph.

size-dependent fracture stress σ_f as a consequence of prior yielding σ_y and work-hardening $Kd^{-\frac{1}{2}}$ in polycrystalline material.

Observation of Intersecting Slip Bands

In the largest grain illustrated in Fig. 3, finely textured gradations in height, which form intersecting linear patterns, are detectable on the fracture surface. As pointed out in an earlier section, they are considered to have been generated by interactions between the fracture wave and stress fields of slip bands induced in the grain by plastic deformation. The intersections of several sets of differently oriented slip bands (at least four are discernible) are shown even more clearly in a different large grain (Fig. 4). This figure attests to extensive plastic alteration of the grain, including some evidences of very localized shearing around the three crack-like markings near the top. Although slip band markings are not detectable in the much smaller grains of the matrix, it is noteworthy that the matrix microstructure is still intact and free of cracks (except for the one large columnar fracture which "prepared" these viewing surfaces), so that plastic deformation of the matrix grains commensurate with that indicated by the larger ones need not be ruled out.

SUMMARY

Microscopic evidence obtained from optical replication fractographs has been presented to show that high-purity polycrystalline spinel undergoes microstructural alteration during cyclic compressive straining at high temperatures. It has been pointed out that such alterations, including grain growth and grain deformation (as well as the final initiation of brittle fracture), are consistent with well-established concepts of plastic deformation involving slip of dislocations by several independent modes, together with interaction of dislocations with barriers created by other dislocations and by grain boundaries. A two-dimensional section through a single large grain, which shows four (or more) intersecting sets of slip bands, has been cited as experimental confirmation of multiple slip at high temperature in this isotropic, face-centered, ductile, polycrystalline ceramic.

ACKNOWLEDGMENTS

Support of this research by the U. S. Army Research Office, Durham, North Carolina is gratefully acknowledged. We are indebted to Dr. A. E. Lucier for assistance in replica preparation and to Dr. W. W. Kriegel and R. Douglas McBrayer for their helpful discussions.

REFERENCES

1. S. M. Copley and J. A. Pask, this volume, Chapter 13.
2. S. C. Carniglia, this volume, Chapter 24.
3. Hayne Palmour III, Dong M. Choi, Lawrence D. Barnes, R. Douglas McBrayer, and W. W. Kriegel, "Deformation in Hot-Pressed Polycrystalline Spinel," in: H. H. Stadelmaier and W. W. Austin (eds.), Materials Science Research, Vol. 1, Plenum Press (New York), 1963, pp. 158-197.
4. Hayne Palmour III, W. W. Kriegel, and R. D. McBrayer, "Research on Growth and Deformation Mechanisms in Single-Crystal Spinel," Technical Documentary Report ASD-TDR-62-1086, Contract AF 33 (616)7820, North Carolina State University, Raleigh, North Carolina, February 1963.
5. J. Hornstra, "Dislocations, Stacking Faults, and Twins in the Spinel Structure," Phys. Chem. Solids 15:311 (1960). See also "Dislocations in Spinels and Related Structures," in: H. H. Stadelmaier and W. W. Austin (eds.), Materials Science Research, Vol. 1, Plenum Press (New York), 1963, pp. 88-97.
6. W. W. Kriegel, H. Palmour III, and D. M. Choi, "The Preparation and Mechanical Properties of Spinel," in: P. Popper (ed.), Proceedings of a Symposium on Special Ceramics, 1964 (British Ceramic Research Association, Stoke-on-Trent, July 1964), Academic Press (London), 1965.
7. R. D. McBrayer, Hayne Palmour III, and Povindar K. Mehta, "Chemical Etching of Defect Structures in Alumina-Rich Spinel Single Crystals," J. Am. Ceram. Soc. 46 (10):504-505 (1963).
8. J. E. Burke, "Grain Boundary Effects in Ceramics," in: H. H. Stadelmaier and W. W. Austin (eds.), Materials Science Research, Vol. 1, Plenum Press (New York), 1963, pp. 69-81.
9. A. N. Stroh, "The Formation of Cracks as a Result of Plastic Flow," Proc. Roy. Soc. 223(A):404 (1954).
10. N. J. Petch, "The Cleavage Strength of Polycrystals," J. Iron Steel Inst. 174:25 (1953).

PART V. Surface and Environmental Contributions to Mechanical Behavior

N. M. PARIKH, Presiding

Illinois Institute of Technology Research Institute
Chicago, Illinois

Chapter 26

Origin of Calcite Decomposition Nuclei

Eugene Nicholas Kovalenko

General Atomic Division of General Dynamics Corporation
San Diego, California

and Ivan B. Cutler

University of Utah
Salt Lake City, Utah

The thermal decomposition of natural calcite single crystals was found to occur at surface defects at which well-formed decomposition nuclei eventually developed. The occurrence of the nuclei was strongly dependent on the history of the specimen surface. An initial heat treatment in CO_2 would anneal out many surface defects; however, longer annealing drastically increased the defect concentration by allowing impurity precipitation throughout the crystal. Closed dislocation loops appeared to surround the pockets of precipitation. Strong evidence is presented to suggest impurity-associated dislocations are primarily responsible for nucleation.

INTRODUCTION

Because the effect of physical imperfections on chemical reactivity is of continuing interest, the authors elected to study the effect of surface defects on calcite thermal decomposition. Two papers [1,2] on calcite suggested its selection. Hyatt [1,3] calculated a decomposition site density of less than $10^{10}/cm^2$, and Keith and Gilman [2,4] demonstrated plastic deformation by identifying dislocation etch pits.

Two major questions arose—are dislocations the origin of decomposition nuclei, and, if so, will plastic deformation accelerate decomposition The first query is investigated herein; a discussion of the second has been presented elsewhere [5]. Other aspects of this reaction have also been studied. [6-11].

Five experiments were designed to investigate the following aspects of nucleation: (1) sensitivity to surface history, (2) dependence on CO_2 partial pressure, (3) simultaneous decomposition of matching cleavage faces, (4) dependence on plastic deformation and short-term annealing, and (5) dependence on long-term annealing.

EXPERIMENTAL PROCEDURE

A Unitron HHS-3 hot-stage unit with Polaroid attachment was used in this study. The hot-stage interior was modified as described elsewhere [5]. Small rods of optical grade calcite* were cleaved from rhombs whose ends had been squared with a diamond saw, lapped, and lightly etched in HCl. Gilman's [12] crystal-cleaving technique was modified so that specimens were cleaved with one end embedded in a small amount of clay. This minimized mechanical damage to freshly cleaved surfaces by securing the cleaved halves. Great care was taken to avoid even touching a fresh surface.

Because the effects of different aspects of specimen history on nucleation were investigated, procedures varied widely. The procedures of the first two experiments were slight variations of that used to measure the growth kinetics of decomposition nuclei [5]. The other three were more specific in design. Specimens were nucleated in a mixed nitrogen–CO_2 gas stream. A total flow rate of 10 cc/sec was usually maintained except where rapid CO_2 partial pressure changes† made constant flow impractical. Specimens were photographed during the course of each reaction.

Surface History

A one-day age experiment was performed to demonstrate the importance of considering the effect of surface history on decomposition nucleation. Two rods were cleaved from a freshly cleaved plate to give adjacent (not matching) faces. One was placed in the hot stage and heated to 745°C in 10% CO_2. After a 30-min soaking, it was allowed to nucleate in 5% CO_2. The other rod was left exposed to laboratory room conditions for about 24 hr and then given the same treatment.

CO_2 Partial Pressure

A freshly cleaved specimen was heated to 745°C, soaked in 10% CO_2, and nucleated in 5% CO_2. After a few nuclei had developed

---

*Iceland spar from Wards Natural Science Establishment.
†CO_2 partial pressures were calculated from the flow rates of gases before mixing.

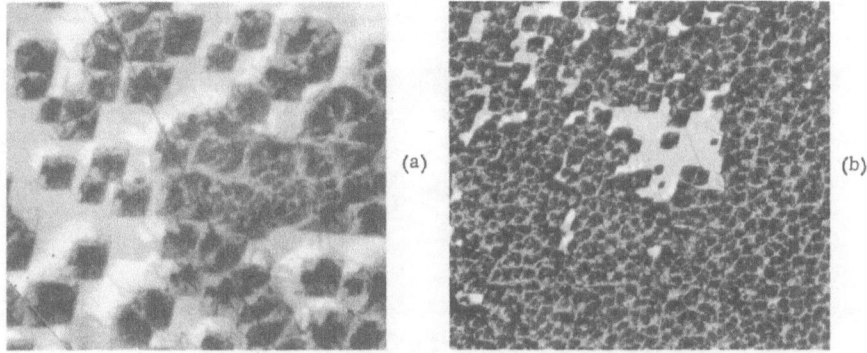

Fig. 1. One-day aging, adjacent cleavage faces nucleated: (a) fresh; (b) after 24-hr exposure to laboratory room conditions. 150 ×.

to considerable size in a relatively step-free region of the specimen surface, the CO_2 flow was reduced stepwise from 5% to 4, 3, and 2%, with each step being held about a minute, and raised again to 5%. It was then dropped to 1% for a few seconds and raised to 100% to stop the reaction.

Simultaneous Decomposition of Matching Cleavage Faces

Two freshly cleaved matching faces placed side by side were simultaneously heated at 590°C in 1% CO_2 for about 1 hr, cooled to

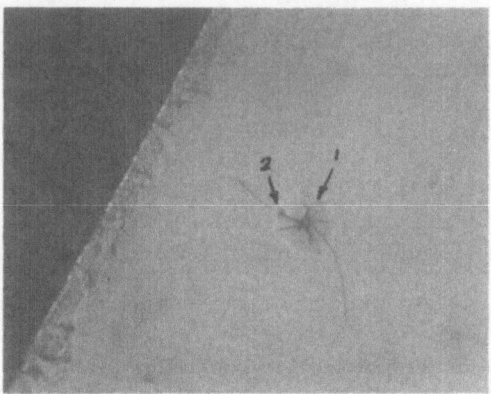

Fig. 2. Nucleation by cleavage whisker, smaller pit (2) nucleated later than larger (1). Note decomposition at crystal edge. 150×.

Fig. 3. Nucleation preferentially along cleavage steps. 75 ×.

532°C, and nucleated in pure nitrogen. The reaction was stopped after $2\frac{1}{2}$ hr. This experiment was then duplicated except that one face was lightly etched in formic acid to establish the dislocation etch-pit pattern.

Plastic Deformation and Short-Term Annealing

Keith and Gilman's glass-rod experiment was reproduced so that the effect of plastic deformation on decomposition could be observed. The effect of a short-term anneal was also investigated.

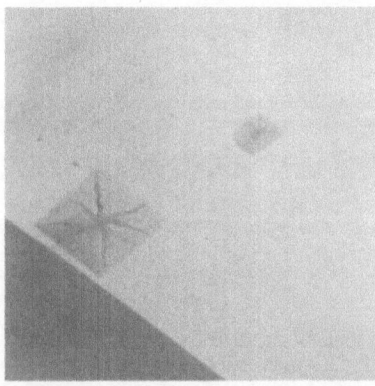

Fig. 4a. Effect of reducing CO_2 flow. Nucleation and growth in 5% CO_2. 150×.

Fig. 4b. Same as Fig. 4a, but in 2% CO_2 after step reductions to 4 and 3%. 150×.

Fig. 4c. Same as Fig. 4a, but in 1% CO_2. 150×.

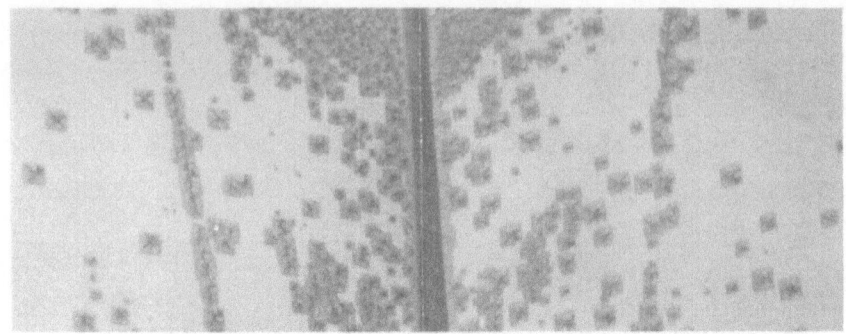

Fig. 5a. Matching, as-cleaved faces simultaneously decomposed. 150 ×; reduced 15% for reproduction.

A freshly cleaved right prism measuring 0.2 by 0.2 by 0.8 cm was kept from moving in the specimen holder by a small ceramic bar placed across its upper surface and normal to its long axis. With the upper part of the hot-stage unit secured to the lower part, the bar bridged the bottom of the hot-stage furnace and applied a small bending stress to the specimen.

After soaking at about 525°C for several minutes in 3% CO_2 (T = 695°C), the specimen was impressed with the blunted end of a thin glass rod.   This was done by inserting the rod through the viewing port, from which the quartz window had been temporarily removed (the lens housing was also removed), and firmly hand-

Fig. 5b. Matching cleavage faces simultaneously decomposed.   Left face etched before decomposition.   Arrows indicate several dislocation pits. 150 ×; reduced 15% for reproduction.

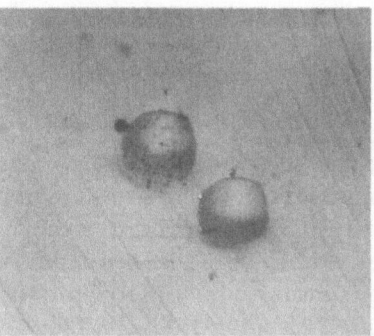

Fig. 6a. Preferential nucleation at fresh impression (made by glass rod) on left. 75 ×.

pressing it into the surface. The specimen was then annealed at 650°C for 9 hr, again cooled to 525°C, and, after the CO_2 flow was turned off, again impressed. The run was allowed to continue until the first decomposition nuclei became visible. This took about $2\frac{1}{2}$ hr, after which time the power was turned off without recarbonating the nuclei. The prism was etched, cleaved along the major rod

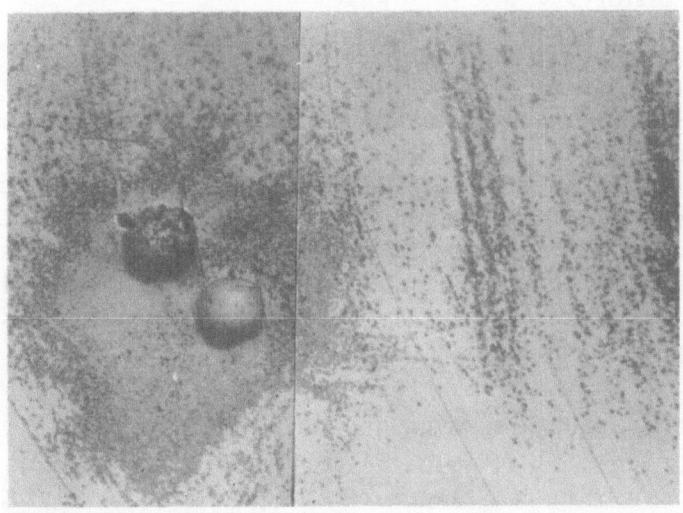

Fig. 6b. Nucleation at impressions made by a glass rod. Specimen cooled and quick-etched. Vertical lines to the right of impressions are nuclei. 75 ×.

axis normal to the impressed region, and again etched to compare matching faces.

Long-Term Annealing

A rhomb was cleaved in half to obtain matching faces. One of the halves was annealed for 60 hr at 780°C in 100% CO_2, nucleated for a few minutes in 10% CO_2, and cooled without recarbonating. The unannealed half was etched for 1 min in room-temperature formic acid. The annealed, partially decomposed half was quick-etched in ice-cold formic acid. Matching cleavage faces were compared. The two halves were again cleaved to make two sets of matching faces, given identical etches, and compared.

RESULTS AND DISCUSSION

Surface History

Figure 1 shows the number of decomposition nuclei on the aged specimen is much greater than that on the fresh specimen. Nucleation even appears sensitive to surface contact of a cleavage whisker (Fig 2) and also seems generally to prefer cleavage steps (Fig. 3). These results show nucleation to be very sensitive to surface history.

CO_2 Partial Pressure

The number and induction time of decomposition nuclei are functions of CO_2 partial pressure, as revealed in Figs. 4a–4c. A variety of sites of varying activity is indicated, the chemical potentials of which should be calculable by more deliberate quantitative work using this approach. Note that nucleation has preferentially occurred around larger nuclei and in well-defined crystallographic directions. Also, the shapes of well-defined nuclei parallel the host-lattice morphology.

Simultaneous Nucleation of Matching Cleavage Faces

This approach revealed no correlation between etch pits and decomposition nuclei. In general, the population density of dislocation etch pits in as-cleaved, as-received crystals was considerably lower ($< 10^3/cm^2$) than that of decomposition nuclei. Furthermore, simultaneous decomposition of as-cleaved matching cleavage faces failed to show a one-to-one correlation between densities of nuclei, as shown in Fig. 5a, although there is a regional correlation. Figure 5b shows no correlation at all between faces. Thus, comparison of matched faces using etching and decomposition methods fails to demonstrate that dislocations are responsible for nucleation. It does not follow, however, that dislocations are not the cause of nucleation. Factors, such as etching out of shallow defects, surface annealing, and dislocation migration, may be involved in the lack of correlation.

Plastic Deformation and Short-Term Annealing

Figures 6a–6c show that decomposition is sensitive to fresh surface damage and that surface damage can be significantly annealed. Again, however, there seems to be little or no correlation between decomposition nuclei and the dislocation etch-pit pattern.

A comparison of matching faces (Fig. 7) after cleaving a specimen through the impressed region reveals what appear to be impurity strata intersecting the original surface. These strata correspond to the lines of nuclei in Fig. 6b. A general one-to-one etch-pit pattern corresponds to the region around the impressions, but not to the strata region. These results demonstrate that

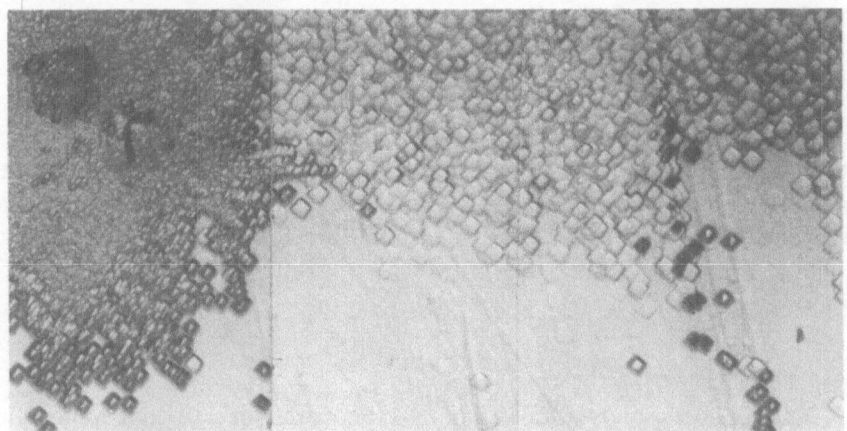

Fig. 6c. Nucleations at impressions made by a glass rod. Specimen etched 20 sec. Pointed-bottom etch pits are at dislocations. 75×; reduced 15% for reproduction.

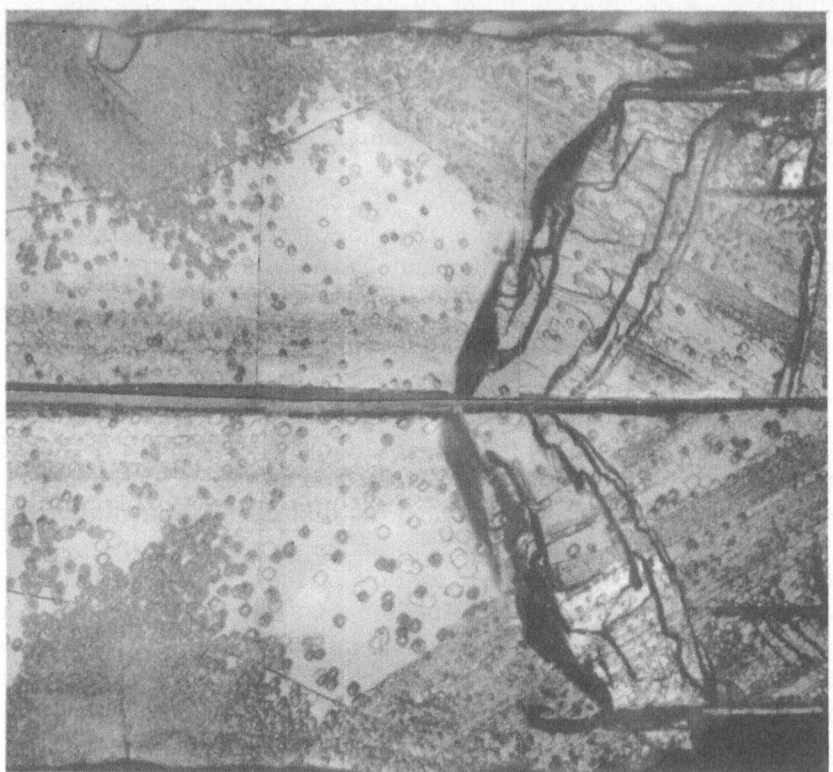

Fig. 7. Matching cleavage faces etched after cleaving through impressed region. Compare with Fig. 6c. Apparent impurity strata intersect impressed surface in region of flat-bottomed pits. 75 ×; reduced 20% for reproduction.

nucleation generally has not occurred at fresh dislocations, but rather at sites associated with impurities. This does not mean that dislocations are not generally responsible for nucleation, but that, if they are, they are associated with impurities, i.e., they are pinned.

Long-Term Annealing

The results shown in Figs. 8a–10b demonstrate that annealing significantly increases the dislocation density, and these results support the hypothesis that impurity-associated dislocations are the primary nucleant. Figures 8a and 8b show a close correspondence

Fig. 8a. Matching cleavage faces—as-cleaved and etched. 45 ×.

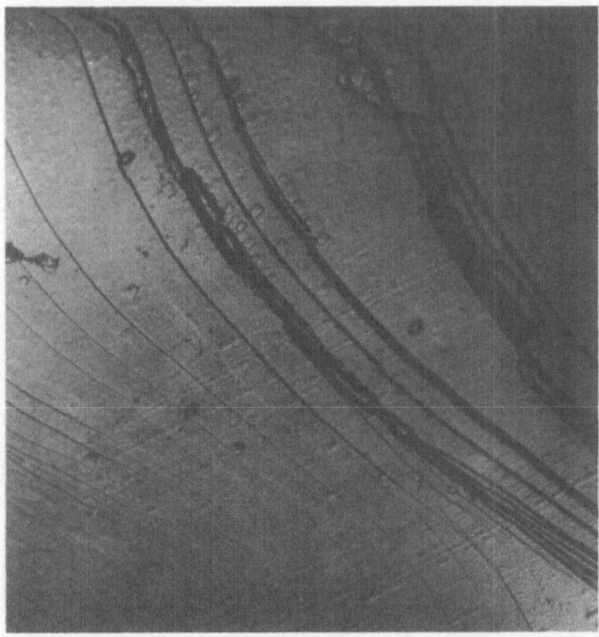

Fig. 8b. Matching cleavage faces—annealed for 60 hr, nucleated, and quick-etched. 45 ×.

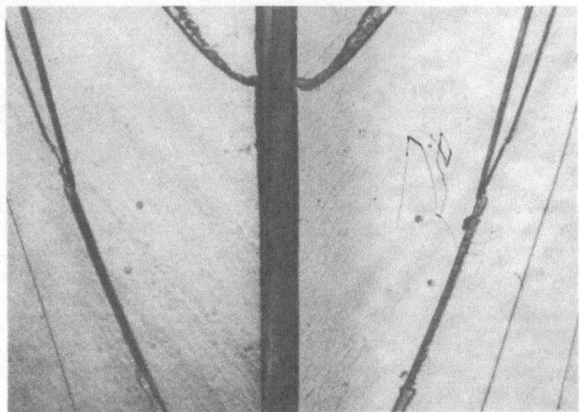

Fig. 9a. Sets of matching cleavage faces from specimen of Fig. 8a. As-received, cleaved, and etched 5 sec. 150 ×.

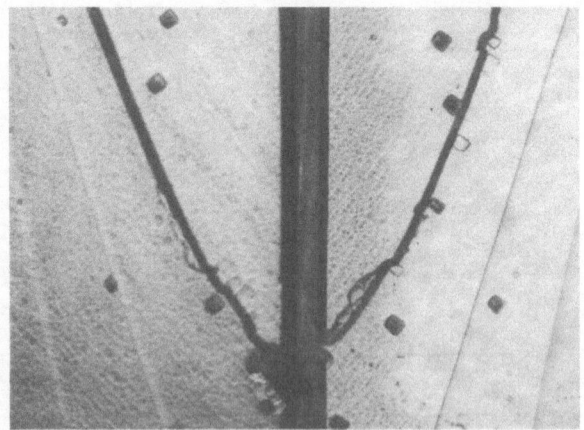

Fig. 9b. Same as Fig. 9a, but etched 60 sec. 150 ×.

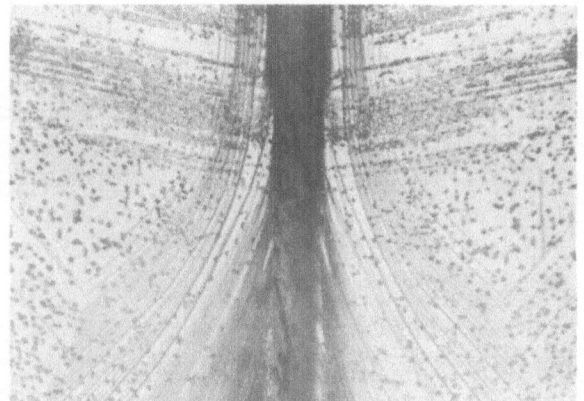

Fig. 9c.   Sets of matching cleavage faces from specimen of Fig. 8b. Annealed 60 hr,
cleaved, and etched 5 sec. 150×.

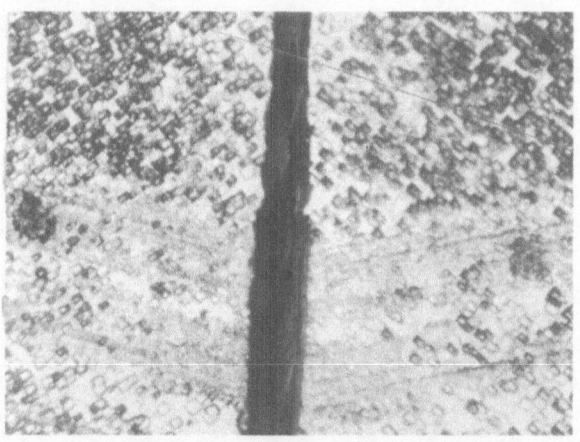

Fig. 9d. Same as Fig. 9c, but etched 60 sec. 150×.

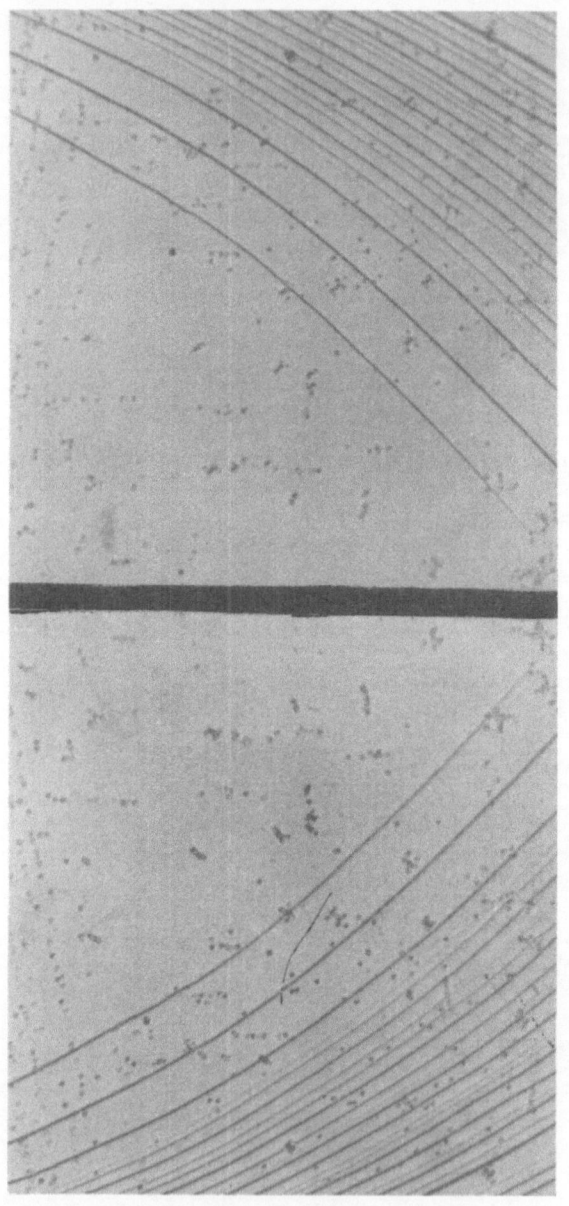

Fig. 10a. Matching cleavage faces of annealed, quick-etched specimen. 375 ×.

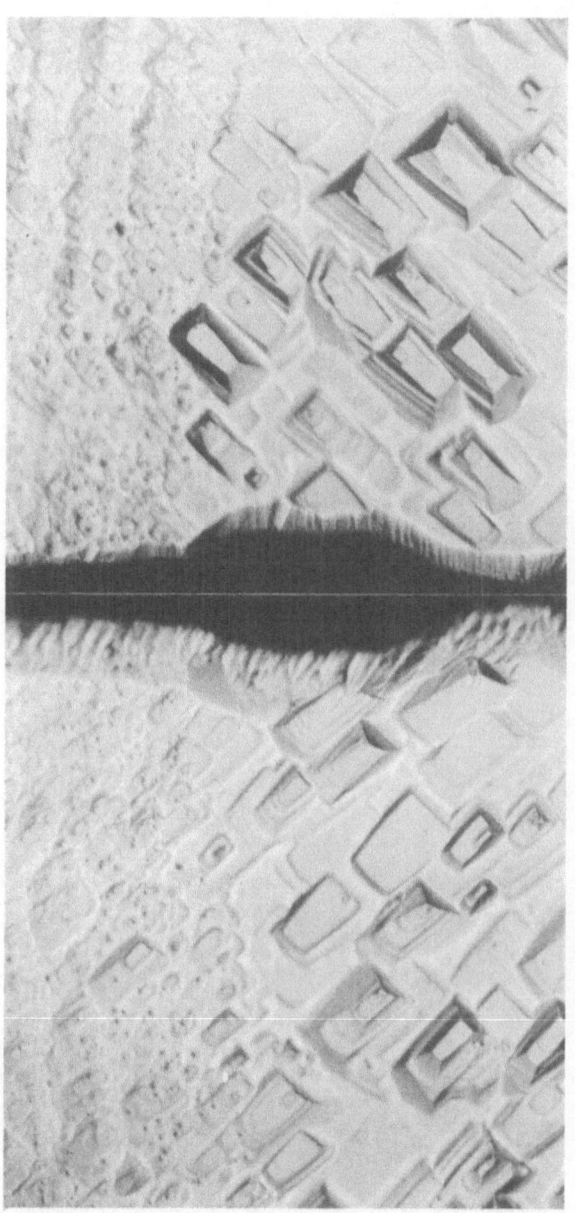

Fig. 10b. Matching cleavage faces of annealed, 60-sec etched specimen. 375 ×.

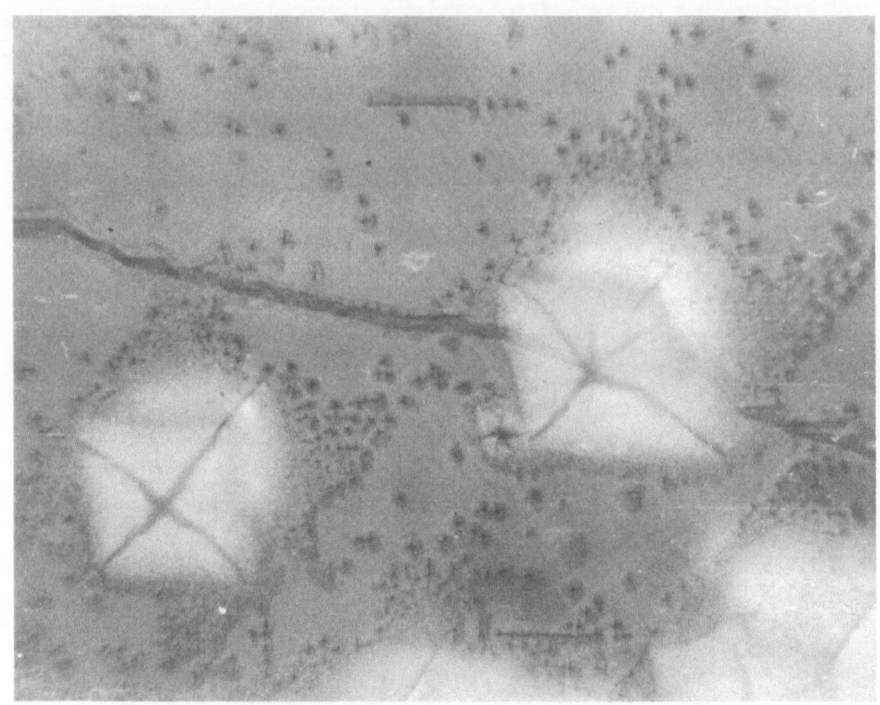

Fig. 11a. Further nucleation around advanced nucleus initiated by sharply reducing CO_2 flow. 650 ×.

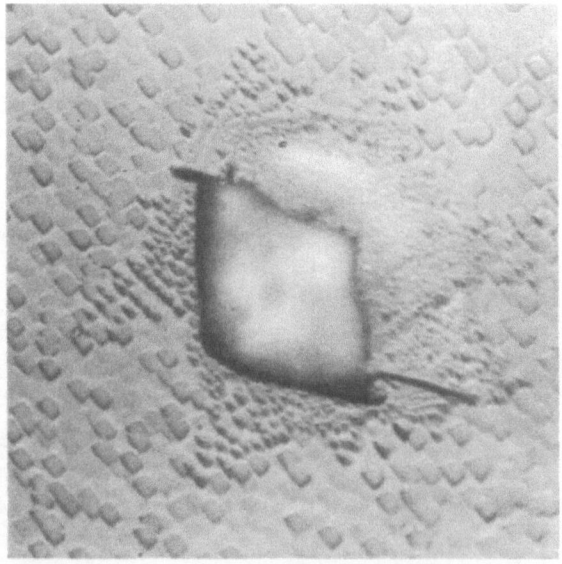

Fig. 11b. Dislocation etch-pit pattern around advanced decomposition nucleus. 500 ×.

between the nucleation pattern in an annealed cleavage face and the multidirectional, granular pattern of its unannealed, etched, matching face. This granular appearance is probably due to stratified inclusions. Nucleation started abruptly, without a significant induction time. That these inclusions coagulate in time during annealing and are surrounded by closed dislocation loops is shown in Figs. 9a–10b. The one-to-one correlation of etch pits on the quick-etched face disappears as the surface is etched below the corresponding dislocation loops which intersect the surface.

Perhaps the strongest evidence to support the hypothesis of primary nucleation by dislocations is illustrated in Fig. 11. The unusually close correspondence of the nucleation pattern to the dislocation etch-pit pattern cannot easily be explained otherwise. Fresh dislocations presumably are so mobile at reaction temperatures that they are less likely to provide a stable base for crystallization of the decomposition product than impurities, except under severe enough decomposition conditions. Once a stable nucleus has been formed, it apparently maintains its own steadily multiplying dislocation network around its periphery. It would seem, then, that the observed decomposition nucleation can be attributed to the presence of dislocations.

CONCLUSIONS

The following conclusions with respect to natural calcite crystals are drawn from the foregoing discussion:

1. The as-cleaved surface is very sensitive to mechanical damage, which can serve to nucleate decomposition under proper conditions.
2. A short-term anneal under recarbonating conditions can significantly reduce surface damage.
3. Longer-term annealing causes precipitation of impurities to occur throughout the crystal lattice, some of it at the crystal surface.
4. Dislocation loops surround each precipitated inclusion; thus, long-term annealing increases, rather than decreases, the dislocation density.
5. The precipitation of impurities at the surface appears to be a primary source of decomposition nucleation.

6. A pattern of fresh dislocations exists around each decomposition nucleus.
7. The population of decomposition nuclei decreases with increasing CO_2 overpressure; therefore, the final size of stable CaO nuclei increases with increasing CO_2 overpressure.
8. At low decomposition rates, such as were used in this investigation, nucleation generally does not occur at fresh dislocations, although, at higher decomposition rates or under severe energy fluctuations, it does.

Thus, calcite decomposition nucleation was found to be a function of the condition of the surface and the decomposition conditions. Under the normal conditions of this investigation, primary nucleation appeared to be associated with impurities precipitated at the crystal surface. Under more severe decomposition conditions, fresh dislocations also served as nucleation sites. In view of these and other observations, it was strongly inferred that dislocation, particularly impurity-pinned dislocations, were generally responsible for nucleation.

REFERENCES

1. E. P. Hyatt, Ph. D. Thesis, University of Utah, 1957.
2. R. E. Keith and J. J. Gilman, General Electric Company Report GE-59-RL-2207, 1959.
3. E. P. Hyatt, I. B. Cutler, and M. E. Wadsworth, J. Am. Ceram. Soc. 41: 70 (1958).
4. R. E. Keith and J. J. Gilman, Acta Met. 8: 1 (1960).
5. E. N. Kovalenko, I. B. Cutler, and M. E. Wadsworth, "Growth Kinetics of Calcite Decomposition Nuclei," presented at 66th Annual Convention of the American Ceramic Society, 1964.
6. J. M. Thomas, G. D. Renshaw, and C. Roscoe, Nature 203: 72 (1964).
7. T. R. Ingraham, Can. J. Chem. Eng. 41: 170 (1963).
8. W. E. Garner, Chemistry of the Solid State, Butterworth's (London), 1955, pp. 214-231.
9. A. J. Sedman and A. J. Owen, Trans. Faraday Soc. 58: 2033 (1962).
10. E. Cremer and W. Nitsch, Z. Elektrochem. 66: 697 (1962).
11. H. Hashimoto, Chem. Abstr. 57: 6861e (1962).
12. J. J. Gilman, in: B. L. Averback et al. (eds.), Fracture, Technology Press, Massachusetts Institute of Technology (Cambridge), 1959, p. 194.

Chapter 27

Fracture Surface Energy of Soda-Lime Glass

S. M. Wiederhorn

National Bureau of Standards
Washington, D. C.

The fracture surface energy of soda-lime glass was measured at temperatures of 77, 195, and 300°K in various media using the double-cantilever
cleavage technique. Values obtained for the fracture surface energy were
3.20 J/m² in nitrogen (*l*), 3.10 J/m² in toluene (*l*) — CO_2 (*s*), and 2.83 J/m²
in dry nitrogen (*g*). During the experiment, slow crack motion was always
observed prior to catastrophic failure of the specimens. The crack motion
was complex, depending on the stress at the crack tip and the concentration
of water in the medium surrounding the crack. Experimental results will be
discussed with respect to several different mechanisms of crack growth.

INTRODUCTION

The tremendous difference that exists between the actual strength
of ordinary glass, 10,000 psi, and the theoretical strength, $1-5 \cdot 10^6$
psi, was explained by Griffith [1] in 1921 as being due to small flaws
in the surface of the glass. These flaws act as stress concentrators
and severely weaken the glass. Glass fractures when the stress at
a flaw tip exceeds the intrinsic strength of the glass. The breaking
stress equation derived by Griffith for an elliptically shaped crack
in an elastic medium is

$$S = \sqrt{2E\gamma/\pi C} \tag{1}$$

Thus, the breaking stress S is a function of Young's modulus E, the
half-length of the crack C, and the surface energy γ. Providing
these three parameters are known, it should be possible to predict
the strength of any piece of glass.

Most methods of measuring the surface energy of solids depend
on high atomic mobility within the solid at elevated temperatures
near the melting point [2]. At low temperatures, i.e., 0.5 T_m, these
methods are inapplicable, and one must resort to a quantitative
fracture technique to measure the surface energy. All variations of

503

the quantitative fracture technique take advantage of the appearance of a surface energy term in the Griffith equation. Clearly, one could determine the surface energy of a solid using equation (1), providing the fracture stress, flaw dimension, and Young's modulus were measured. Since the surface energies measured by this technique may not be identical to the true surface free energy of the solid, due to possible irreversible effects occurring at the crack tip, surface energies determined by this method are termed fracture surface energies in this paper. In the past, three investigators have used the quantitative fracture technique to measure the fracture surface energy of glass. Their work will be reviewed in the discussion section of this paper.

The technique used in our experiment, the double-cantilever cleavage technique, was originally developed by Gilman [3] and later modified by Westwood and Hitch [4], and has been used successfully to measure the fracture energy of several different crystalline materials [3, 4]. The shape of the samples used in this technique is shown in Fig. 1. A crack of length L_0 is introduced into the specimen by cleavage or by some other technique, and the critical force P_{cr} necessary to move the crack is measured. The fracture energy γ_0 is then calculated using the following equation, which was derived originally by Westwood and Hitch [4]:

$$1/\gamma_A = 1/\gamma_0 + (aE/4\gamma_0 G) (t/L_0)^2 \qquad (2)$$

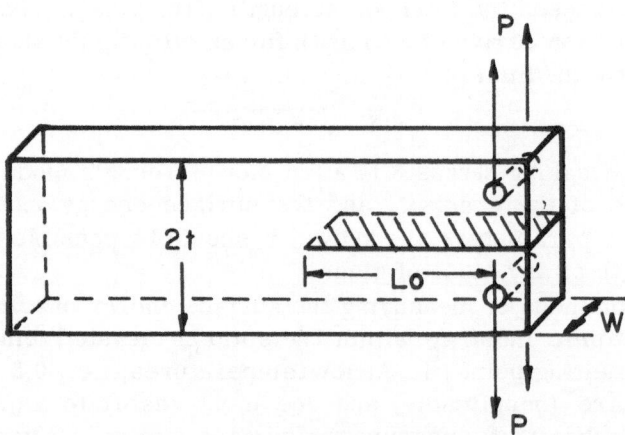

Fig. 1. Sample configuration.

where

$$\gamma_A = 6P_{cr}^2 \ L_0^2/EW^2t^3 \tag{3}$$

and where a is an experimental constant, E is Young's modulus, and G is the shear modulus. If one plots $1/\gamma_A$ against $(t/L_0)^2$, a straight line results and the intercept of the straight line with the ordinate gives $1/\gamma_0$. The slope of the straight line may be used to calculate the empirical constant a. As the crack becomes very long, $L_0 \gg t$, $\gamma_A \to \gamma_0$.

Berry [5] analyzed the technique and developed equations of motion for the crack, giving the velocity v_L and the acceleration a_L:

$$v_L = \frac{v_s t}{2L\sqrt{3}} \left(1 - \frac{L_{cr}}{L}\right)\left(1 + \frac{2L_{cr}}{L}\right)^{1/2} \tag{4}$$

$$a_L = \frac{v_s^2 t^2}{12L^3}\left(1 - \frac{L_{cr}}{L}\right)\left(\frac{5L_{cr}^2}{L^2} - \frac{L_{cr}}{L} - 1\right) \tag{5}$$

where v_s is equal to $(E/\rho)^{1/2}$. L is the crack length at any time t and L_{cr} is the critical crack length which is defined by the following equation:

$$L_{cr} = \left(\frac{E\gamma w^2 t^3}{6P^2}\right)^{1/2} \tag{3a}$$

From equations (4) and (5), it can be shown that a crack under constant load P will not propagate if its length is less than L_{cr}. If the crack length is greater than L_{cr}, the crack will propagate and accelerate rapidly, leading to a sharp transition between the stationary and moving states. For example, in the present experiment, velocities of about 10^{-1} m/sec would be expected after the crack had propagated 10^{-6} m and 1 m/sec after it had propagated 10^{-5} m. By comparing equations (3) and (3a), we see that the initial crack length L_0 is equal to the critical crack length L_{cr} when the applied load P is equal to the critical force P_{cr} for crack motion.

In the present study, the double-cantilever cleavage technique was used to measure the fracture surface energy of soda-lime glass in five different environments. Experimental observations are discussed with reference to different theories of crack motion and static fatigue in glass.

EXPERIMENTAL PROCEDURE

The specimens used were annealed soda-lime glass microscope slides having dimensions of 75 by 25 by 1.5 mm and a composition reported by the manufacturer to be 72.13% SiO_2, 14.12% Na_2O, 7.2% CaO, 3.97% MgO, 1.70% Al_2O_3, 0.45% K_2O, 0.24% SO_3, and 0.044% FeO. To test the effect of specimen thickness on the measured fracture surface energy values, a set of runs was made with 1.0-mm thick slides of the same composition. The slides were annealed at 528°C (the reported annealing point) for 30 min and then cooled at a rate of 2°C/hr until the temperature was less than 300°C, at which point the furnace was turned off. The density and the modulus of elasticity were measured on 10 slides from each box of 72 slides used in our experiments and were found to be $2.48 \cdot 10^3$ kg/m^3 and $7.34 \cdot 10^{10}$ N/m^2, respectively. Holes were drilled into the ends of the microscope slides using a $\frac{1}{16}$-in. diamond drill. A crack of length L_0 was introduced into the slide by inscribing a scratch into the surface of the slide with a diamond stylus and then inducing a crack to propagate along the scratch. A light stress was then applied to the ends of the slides to straighten the crack front. The surfaces of the cracks so formed were perpendicular to the faces of the slide and the leading edge of the crack was at an angle varying between 60 and 90° to the faces of the slide. This angle was due to the slight unevenness of loading at the grips. Cracks varied in length between 1.2 and 3.5 cm with only a few having lengths greater than 3.5 cm. Before testing, the slides were annealed at approximately 350°C for one hour in a nitrogen atmosphere (0.001% H_2O) to remove excess moisture from the crack tip and to release residual stresses along the scratch.

The tests were run on model TM-M-L Instron testing machine using a CTM (metric) load cell. A mounting jig, identical to the one used by Westwood [4], was attached to the movable crossarm of the machine. The sample was attached to the jig by two small hooks, which imposed little constraint on the sample being tested. There was, nevertheless, a small unavoidable torsion which caused the crack front to lie at an angle to the slide face as mentioned previously. Crack motion was viewed through a small 20× microscope that was attached to a cathetometer. The crack length could be measured to ±0.01 cm using the microscope and cathetometer. Other dimensions of the specimens were obtained with a micrometer and vernier calipers. The height and length were measured to ±0.01 cm, and the sample width w was measured to ±0.003 cm.

In order to measure the crack velocity, a filar eyepiece was attached to the microscope, and a potentiometer was attached to the filar eyepiece so that motion of the cross hairs in the eyepiece could be converted directly into a potential difference, which could then be registered on a strip recorder.

Fracture surface energy measurements were conducted in five different environments: (1) liquid nitrogen, 77°K; (2) a mixture of toluene (l)–$CO_2(s)$, 195°K; (3) tank nitrogen gas containing 0.001 mol.% water vapor, 300°K; (4) tank nitrogen gas dried over liquid nitrogen, 300°K; and (5) air with 1.3 − 1.6 mol.% water vapor, 300°K. Measurements that were made in nitrogen gas were conducted in a test chamber through which a continuous stream of gas flowed. Samples were annealed and tested in the chamber so that they never came in contact with the air. Samples tested in other media were similarly annealed in the test chamber, but were removed from the chamber before testing. It was possible to view crack motion in liquid nitrogen by attaching to the end of the microscope a small tubular window, which projected below the level of the liquid nitrogen. Crack motion could not be observed in toluene (l) − $CO_2(s)$ because of ice formation. Toluene was selected as a test medium because of its low dielectric constant [6] and because it dissolved little water at low temperatures [7].

As will be discussed later in the paper, initial crack motion was due to statistical fluctuations and to the presence of water vapor at the crack tip. Consequently, the critical crack length and load had to be measured in such a way as to eliminate the effects of the initial crack motion. One method was to load the specimen rapidly enough so that the critical load was reached before the crack could move very far. The critical load was then equal to the maximum load before failure, and the critical crack length was equal to the initial crack length before testing. In the second method, a constant load was applied to the specimen. The load was maintained constant by inserting a door spring between the sample and the load cell and presetting the Instron machine to the desired load. Thus, slight deflections due to crack motion did not change the load. The velocity of the crack was measured and the critical length was determined as the length at which the crack suddenly accelerated to failure. The critical load was equal to the constant applied load.

For tests in toluene (l) − $CO_2(s)$, only the first method of loading was used. For tests in air, the second method was slightly modified because of the large distance traveled by the crack prior to fracture.

Instead of using a constant load, the sample was continuously loaded at a low rate of deformation, 0.05 cm/min. The crack motion was observed under the microscope, and the critical crack length was determined as the length at which the crack suddenly accelerated to failure. The critical load was the maximum load before failure.

In all media except air, the rate of deformation was 1.0 cm/min (12,000 g/min). In air, two different rates of deformation were used — 0.05 cm/min (600 g/min) and 2.0 cm/min (24,000 g/min).

At least twelve samples were used for each fracture energy determination. The critical load and the critical crack length were substituted into equations (2) and (3) to determine the fracture energy and the empirical constant a. A plot of $1/\gamma_A$ versus $(t/L_{cr})^2$ was constructed, and a straight line was fitted to the data points by the method of least squares. The fracture energy γ_0 was determined from the y-intercept of the straight line, and a from the slope of the line.

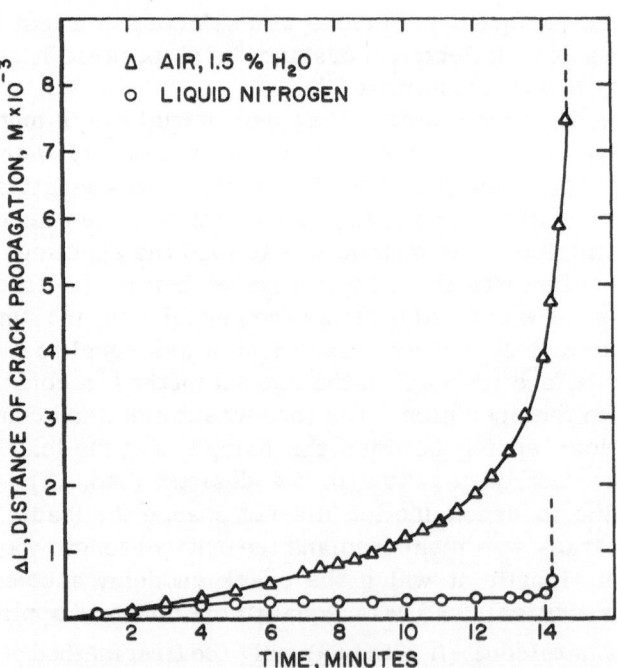

Fig. 2. Crack propagation to failure under constant load, in air and in liquid nitrogen.

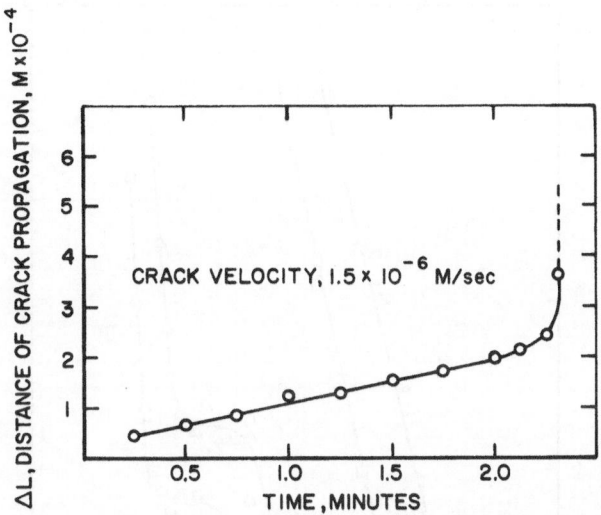

Fig. 3. Crack propagation to failure under constant load, in liquid nitrogen.

RESULTS

Fracture always occurred in two stages—an initial stage during which the crack velocity was relatively low, less than $5 \cdot 10^{-4}$ m/sec and a catastrophic stage during which the velocity was high and increased very rapidly with time. The slow initial stage of fracture is shown graphically in Figs. 2 and 3, where the distance traveled by a crack under constant load is plotted against time. The shape of the curve depended on the moisture content of the surrounding medium.

In moist air with 1.5 mol.% water vapor (Fig. 2), the crack velocity gradually increased with time from $8.4 \cdot 10^{-7}$ to $1.0 \cdot 10^{-4}$ m/sec prior to the catastrophic stage of crack motion, which could be identified by sudden failure of the sample (dotted lines in Figs. 2 and 3). A gradual increase in crack velocity and a relatively long distance of propagation prior to rapid failure, $7.5 \cdot 10^{-3}$ m (Fig. 2) were characteristic of samples tested in moist air.

The initial stage of crack motion in a dry atmosphere, such as liquid nitrogen or gaseous nitrogen dried over liquid nitrogen, could be characterized by a relatively constant velocity, $1.5 \cdot 10^{-6}$ m/sec, and short distances of motion, $3 \cdot 10^{-4}$ m (Fig. 3). The crack velocity was constant in liquid nitrogen or in dry nitrogen gas,

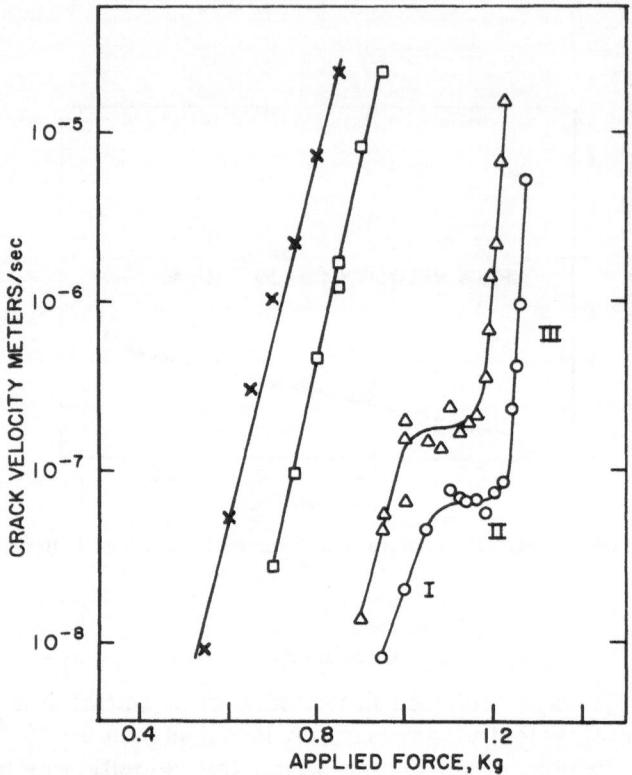

Fig. 4. Variation of crack velocity with applied force, tested in different media. X = water; □= moist air containing 1.5 mol.% water vapor; △=nitrogen gas containing 0.001 mol.% water vapor; O= nitrogen gas dried over liquid nitrogen.

and the transition between the initial and catastrophic stages of fracture occurred rapidly (less than 15 sec) for 60–70% of the cases studied.

Additional information on crack motion during the initial stage of fracture was obtained by studying the load dependence of the crack velocity with respect to various atmospheric conditions (Fig. 4). The behavior of slides tested in water and in moist air is similar and is probably due to water at the crack tip. The crack velocity is exponentially dependent on the load, and the slopes of these curves are 9.1 kg^{-1} for water and 11.7 kg^{-1} for moist air. The curve for water is shifted to lower loads with respect to the curve for air, probably reflecting the effect of a greater water concentration at the crack tip of the sample tested in water.

In dry nitrogen gas, three distinct regions of crack propagation were observed during the initial stage of fracture. At low velocities (region I), the crack velocity depended exponentially on the stress. Measured slopes were 10.0 kg^{-1} for tank nitrogen (0.001% H_2O) and 7.7 kg^{-1} for dried nitrogen [tank nitrogen dried over N_2 (l)]. These slopes are close to those obtained in water and moist air and suggest the same mechanism for fracture. The curves in region I are shifted to higher loads with respect to that for air, reflecting the effect of a lower concentration of water.

In region II (Fig. 4), the crack velocity deviates from its initial exponential dependence on load and becomes nearly independent of load at velocities of $2 \cdot 10^{-7}$ m/sec in tank nitrogen and $7 \cdot 10^{-8}$ m/sec in dried nitrogen. Region II occurs at a lower velocity in the dried nitrogen which contains less water vapor than the tank nitrogen.

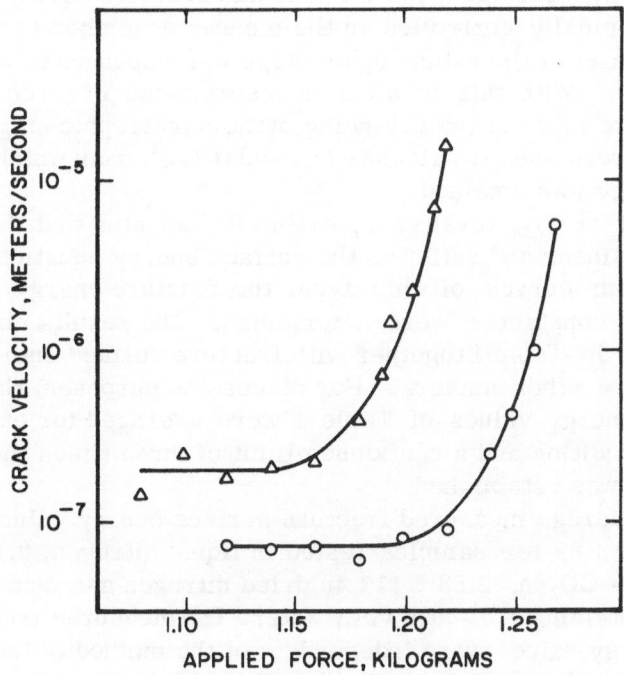

Fig. 5. Stress dependence of crack velocity on applied load in the transition from region II to region III of Fig. 4.

Region III occurs at higher loads than regions I or II. The velocity again appears to be exponentially dependent on the stress, but the dependence is greater and the slopes are 38 kg^{-1} for the tank nitrogen and 53 kg^{-1} for dried nitrogen. The fact that the slopes in region III differ from those in region I suggests a different mechanism for crack-motion. It is of interest that these slopes represent only average values, since the curves in region III exhibit a small positive curvature which can be demonstrated by expanding the P axis (Fig. 5). It is not known whether the curvature is intrinsic to the mechanism of propagation or was caused by a slight lengthening of the crack during the test; however, the existence of such curvature is quite important to our theoretical discussion of the results.

Crack propagation during the catastrophic stage of fracture was not studied quantitatively because of the high velocities involved. Samples failed quite suddenly after the onset of the catastrophic stage, and it is probable that velocities were as high as those observed by Schardin [8]. We feel that fracture during this stage was dynamically controlled in the manner described by Berry and that the onset of the catastrophic stage corresponded to the Griffith conditions. With this in mind, measurements of force and crack length were made at the beginning of the catastrophic stage of fracture and were used to calculate $1/\gamma_A$ and $(t/L_{cr})^2$, from which the fracture energy was obtained.

A plot of $1/\gamma_A$ versus $(t/L_{cr})^2$ (Fig. 6) indicates that the data obtained is linear and satisfies the surface energy equation, equation (2). From curves of this type, the fracture energy γ_0 and the empirical constant a were determined. The results obtained are presented in Table I together with fracture surface energy results obtained by other authors. For discussion purposes, the fracture surface energy values of Table I were averaged for each atmospheric condition and a confidence limit of three times the standard deviation was established.

The average measured fracture surface energy values (in J/m^2) were 3.20 ± 9% for samples tested in liquid nitrogen, 3.10 ± 10% in toluene $(l)-CO_{2}(s)$, 2.83 ± 11% in dried nitrogen gas, and 2.62 ± 14% in air containing 1.3 − 1.6 vol.% water. The measured fracture surface energy values were independent of the method of testing. The differences observed for values obtained in nitrogen (l), toluene $(l)-CO_{2}(s)$, and nitrogen (g) are believed to be due to differences in temperature and not environment, as the water concentration in each

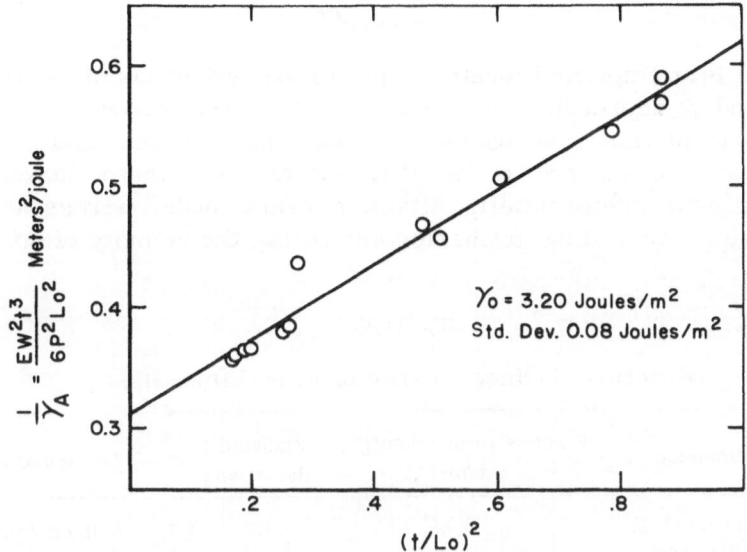

Fig. 6. Fracture surface energy of soda-lime glass, tested in liquid nitrogen.

of these environments was low and the tests were run so as to minimize the effect of the remaining water. Furthermore, it was felt that the glass was inert to toluene $(l) - CO_2(s)$ and nitrogen, and that the surface energy would not be altered due to chemical reaction with the glass. It was expected that tests in air would yield the same values of the fracture surface energy as the tests in nitrogen; however, a difference of about 7.5% was observed. This difference may have been due to residual water and might be eliminated by increasing the loading rate.

DISCUSSION OF RESULTS

Crack Motion

Slow crack motion has been observed previously by several authors [9–11]; however, relationships between the loading force and crack motion were not obtained by them and, consequently, it is difficult to compare our work with theirs. An experimental arrangement similar to our own was used by Murgatroyd [9], who obtained velocities ranging from $1.7 \cdot 10^{-4}$ to $8.3 \cdot 10^{-6}$ m/sec, which fall within the range we have measured.

Crack motion in water or moist air and in region I in dry nitrogen can be expressed empirically by the following relationship:

$$v = v_0\, e^{\beta P} \tag{6}$$

where v_0 is an empirical constant which probably depends on environment and β represents the slopes of the curves shown in Fig. 4. The form of this relationship is reminiscent of the equation derived by Charles and Hillig [12], who assumed that a corrosive medium will preferentially attack a crack under stress at the crack tip. According to Charles and Hillig, the velocity of corro-

TABLE I

Fracture Surface Energy of Soda-Lime Glass

Environment	Fracture surface energy, J/m^2	Standard deviation	α	Test condition
Air, 1.3−1.6 vol. %	2.58	0.130	1.31	0.05 cm/min
Water vapor, 300°K	2.66	0.110	1.16	0.05 cm/min
	2.62	0.140	1.18	2 cm/min
Nitrogen (g), dried	2.85	0.100	1.28	Constant load
over liquid nitrogen,	2.81	0.140	1.13	Constant load
300°K	2.82	0.070	1.11	1 cm/min
Toluene (l)−CO_2(s),	3.02	0.120	1.65	1 cm/min
195°K	3.18	0.090	1.64	1 cm/min
Nitrogen (l), 77°K	3.25	0.080	1.67	Constant load
	3.13	0.070	1.74	Constant load (slide 1.0 mm thick)
	3.20	0.080	1.67	Constant load
	3.21	0.140	1.61	1 cm/min
Water	0.966			Corrected data of Berdennikov
Ethyl alcohol	1.77			
Isoamyl plus isobutyl alcohol	2.12			
Nitrobenzene	2.58			
Vaseline oil	3.22			
Vacuum	4.06			
Air	10−1.8 (preferred value, 4.1)			Roesler
Air	1.70			Shand

sion normal to an interface can be expressed by the following equation:

$$v = v_0' \exp - [E*(0) + \Gamma V_m/\rho - \sigma V*]/RT \tag{7}$$

The first term of the exponential, $E*(0)$, represents the activation energy for dissolution in the absence of stress. The second is a molar free-energy term associated with the curvature of the surface. Γ is the surface free energy of the glass, ρ is the radius of curvature of the crack tip, and V_m is the molar volume. The third term represents the stress dependence of the activation energy for surface dissolution; σ is the stress on the surface and $V*$ is the activation volume. Since $E*(0)$ is a constant, equation (7) may be rewritten as follows:

$$v = v_0 \exp(V*\sigma - \Gamma V_m/\rho)/RT \tag{8}$$

If we now define an effective stress at the tip of the flaw as $\sigma_0 = \sigma - \Gamma V_m/V*\rho$, equation (8) may be expressed as follows:

$$v = v_0 \exp \sigma_0 \, V*/RT \tag{9}$$

This is identical to our empirical equation (6), providing σ_0 is proportional to the applied load P. If this is indeed the case,* it follows that the constant β is proportional to the activation volume of the reaction. Since the slopes of the curves of water, moist air, and dry nitrogen (Fig. 4, region I) are nearly equal, the activation volumes will be nearly equal, and it is reasonable to assume that the reaction mechanism is the same in all cases. Therefore, if the observed crack motion in water is due to water corrosion at the crack tip, it follows that crack motion in moist air and nitrogen, region I, is also due to water corrosion. The displacement of the different curves in Fig. 4 along the P axis reflects the effect of varying water concentration at the crack tip. As would be expected, a crack having a lower concentration of water at the crack tip requires a greater stress to reach the same velocity than one having a greater concentration of water at the crack tip.

The theory of Charles and Hillig contains the explicit assumption that the reaction rate of the corrosive medium with the glass

*In order to demonstrate a proportionality between σ_0 and P, an Inglis-type relationship [13] is necessary to relate the applied force P to the stress σ at the crack tip. To the author's knowledge, a complete elastic solution to the cleavage problem and, consequently, an Inglis-type relationship is not available.

surface is controlled by a thermally activated process at the interface and not by the rate of transport of reactant to the interface. This assumption is probably valid in media that contain a high concentration of reactant, e.g., water(l); however, in more dilute media, the assumption might not be valid and the rate of transport of corrosive medium to the reaction site might well be the controlling step of the reaction. In dry gaseous nitrogen, for example, the water vapor forms only a small percentage of the gaseous environment and must travel through the inert nitrogen to reach the crack tip. If the rate of transport to the crack tip is slower than the thermally activated process at the crack tip, then the rate of reaction will be transport-controlled. Since transport processes, such as diffusion and condensation, are not expected to depend on the applied stress, the reaction rate and thus the crack velocity would be expected to be stress-independent. This is exactly the behavior observed in region II of crack propagation in nitrogen; consequently, we believe that crack propagation in region II is transport-controlled.

In region III (Fig. 4), the slopes and, therefore, the activation volumes differ considerably from those obtained in region I. It is, therefore, reasonable to conclude that the mechanism of crack propagation is different. In other words, crack propagation in region III should not depend on water vapor concentration, but on intrinsic properties of the material being tested. It is also clear that crack propagation in region III does not satisfy the equations derived by Berry [5] because of the low velocities observed. One means of explaining crack motion in this region is to assume that crack motion is controlled by the occurrence of thermodynamic fluctuation at the crack tip. Of the several fluctuation mechanisms proposed previously [14-20], the one by Hillig [20] can be most easily compared with our experimental results.

Hillig assumed (following previous practice [3, 21]) that the force separation law for separating two atoms in the material is given by the expression $\sigma = \sigma_u \sin \pi\epsilon/\epsilon_0$, where σ is the local stress, σ_u is the maximum stress to separate the atoms, and $\epsilon_0/2$ is the strain at maximum stress. From this expression, he derived the following equation for the work that a fluctuation would have to perform to separate a single atom pair:

$$E_i = 2\lambda^2 \gamma \left[2(\sigma_u - \sigma)/\sigma_u\right]^{1/2} \tag{10}$$

where λ is the interatomic spacing of the material and γ is the surface energy.

To obtain a law of crack motion, Hillig assumed that a crack of length L, having a radius of curvature of atomic dimensions, was subject to an applied stress σ. If the fluctuation at the crack tip were too localized, the atoms constituting the broken bond would be constrained by neighboring atoms and the broken bond would re-form. Consequently, fracture would not occur. To overcome this objection, Hillig assumed that the fluctuation had to be large enough to cause the remaining bonds on the crack front to be at the critical stress σ_u after the fluctuation, thus enabling the whole line of bonds to rupture. The crack velocity would then depend on the frequency of fluctuations.

To apply Hillig's ideas to the present discussion, we have assumed a crack of length L_0 and a width of w. Assume a fluctuation of width dw to occur at the crack tip (Fig. 7a). The force supported by the crack tip prior to the fluctuation is $(w/\lambda)\,\lambda^2\sigma$, where σ is the local stress at the crack tip. After the fluctuation the force supported by the crack tip is $[(w-dw)/\lambda]\,\lambda^2\bar{\sigma}$ where $\bar{\sigma}$ is the new stress at the crack tip. Since the force supported before and after the fluctuation remains the same, $(w/\lambda)\,\lambda^2\sigma = [(w-dw)/\lambda]\,\lambda^2\bar{\sigma}$. When $\bar{\sigma} = \sigma_u$, the entire crack length will jump forward one atomic spacing. The critical fluctuation size is then equal to

$$dw = w(\sigma_u - \sigma)/\sigma_u \tag{11}$$

The fluctuation area is λdw, and the required fluctuation energy is

$$\lambda\,dw\,E_i = \lambda\,w\gamma\,[2(\sigma_u - \sigma)/\sigma_u]^{3/2} \tag{12}$$

The frequency of fluctuation is, therefore, $f = (w/\lambda)\,\nu_0\,e^{-\lambda w E_i/kT}$, where ν_0 is a frequency factor of the order of $\nu_0 = 10^{13}$ and w/λ is an approximation of the number of ways the fluctuation can be arranged on the crack front. Since the crack front jumps forward a distance λ each time a fluctuation occurs, the crack velocity will be

$$v = w\,\nu_0\,\exp\{-\lambda\,w\gamma[2(\sigma_u - \sigma)/\sigma_u]^{3/2}/kT\} \tag{13}$$

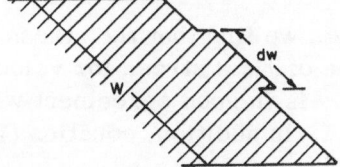

Fig. 7a. Crack surface containing a fluctuation on its perimeter.

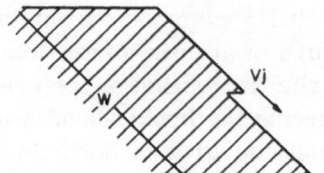

Fig. 7b. Crack surface containing a jog on its perimeter.

In our experiments, a crack propagating in dry nitrogen at a velocity of 10^{-6} m/sec would travel about 1% of its initial length before catastrophic failure ensued. If we substitute this velocity into equation (13) together with $\nu = 10^{13}$ sec^{-1}, $\lambda = 1.6 \cdot 10^{-10}$ m, $\gamma = 3.0$ J/m^2, $w = 1.5 \cdot 10^{-3}$ m, and $T = 300°$K, we arrive at a value of 10^{-5} for $(\sigma_u - \sigma)/\sigma_u$. It follows that the stress must be within 0.001% of the critical stress before a velocity of 10^{-6} m/sec would be observed. If L_u is the critical length at failure, it can be shown that $(L_u - L)/L_u = (\sigma_u - \sigma)/\sigma_u$, providing one assumes that the stress at the crack tip is directly proportional to the applied force. Since the crack propagates 1% of its initial length before failure, we can conclude that $(\sigma_u - \sigma)/\sigma_u$ is only within 1% of the critical stress when a velocity of 10^{-6} m/sec is observed and, therefore, Hillig's theory cannot be used to explain experimental observations in region III.

A different approach may be used, which more successfully explains our experimental observations. Assume that the crack front is not straight, but is actually jogged by one atomic spacing (Fig. 7b). The bond existing at the jog will be most likely to break, since it is the most strained bond along the crack front. The energy for the rupture of this bond should be the same as that derived by Hillig [see equation (10)] for a single bond, and the frequency of bond rupture at the jog will be

$$f = \nu_0 \exp \left\{- 2\lambda^2\gamma[2(\sigma_u - \sigma)/\sigma_u]^{1/2}/kT\right\} \tag{14}$$

The velocity of jog motion is $f\lambda/w$, and the velocity of crack motion is

$$v = \lambda^2\nu_0/w \exp \left\{- 2\lambda^2\gamma[2(\sigma_u - \sigma)/\sigma_u]^{1/2}/kT\right\} \tag{15}$$

Substituting the same values used above, we find that the stress at the crack tip must be within 0.7% of the critical stress for velocities of 10^{-6} m/sec to be observed. This is in good agreement with the experimentally observed value of 1%. In addition, equation (15) predicts a maximum crack velocity in region III of $\lambda^2 \nu_0 / w = 1.7 \cdot 10^{-4}$ m/sec, which is approximately the limiting velocity actually observed in region III. By rearranging equation (15) and taking the first and second derivatives of log v with respect to σ, it can be shown that a plot of ln v versus σ should yield a curve having positive slope and positive curvature, demonstrated by the curves of Fig. 5. This point should be further demonstrated experimentally since the curves of Fig. 5 were not taken with the intention of determining the curvature, and the observed curvature may be due to slight changes in crack length. Equation (15) was integrated graphically to obtain the relationship between L and t shown in Fig. 8. Although the theoretical curve exhibits more curvature than the experimental curve (Fig. 3), both curves are fairly flat over most of their length and both indicate rapid acceleration prior to catastrophic failure. The difference between the curves probably reflects the fact that the theoretical description of crack propagation in region III is only an approximation to the actual crack behavior.

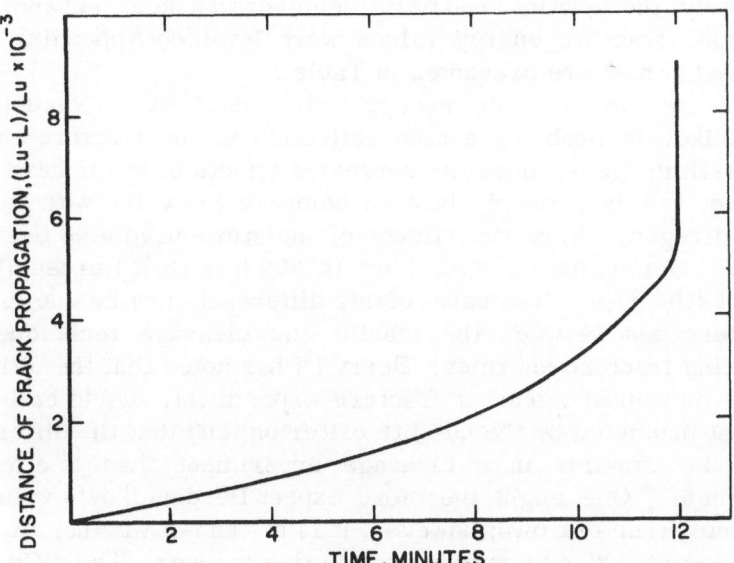

Fig. 8. Crack propagation to failure under constant load, calculated from equation (15). Failure time arbitrarily selected as 12 min.

In conclusion, the experimental results of region III can be largely explained by a theory which assumes that crack propagation in this region is controlled by motion of jogs on the crack front. This theory predicts a maximum crack velocity and the correct order of magnitude for the velocity of crack motion. Further studies should be carried out to confirm the crack velocity dependence on the applied stress.

Fracture Surface Energy Values

It is of interest to compare our experimental values of the fracture surface energy with others in the literature. Perhaps the most extensive previous study was by Berdennikov [22], who measured the effects of several environments on the fracture energy of soda-lime glass. His experiments were conducted using tensile-type apparatus in which he measured the stress necessary to initiate crack motion in a thin plate of glass containing a crack of known dimensions. By quickly reducing the stress after crack motion was initiated, Berdennikov prevented complete failure in the glass plate and was able to make several measurements on each. The fracture surface energy was calculated using a Griffith-type equation. Unfortunately, the equation used by Berdennikov was incorrect and, as a result, his fracture energy values were low (see Appendix). The corrected values are presented in Table I.

The fracture surface energy value obtained in vacuum by Berdennikov is probably a true reflection of the fracture energy of soda-lime glass, since the corrosive effects of water have been removed. It is probably best to compare his value with ours in dried nitrogen, where the effects of moisture have also been removed. Our value of 2.83 J/m^2 is 30% less than Berdennikov's value of 4.06 J/m^2. The cause of this difference may be due in part to differences between the tensile and cleavage techniques of measuring fracture energies. Berry [5] has noted that the ultimate stress, measured during a fracture experiment, should be larger than that predicted by the Griffith criterion, and that the difference should be smaller in a cleavage experiment than in a tensile experiment. One might therefore expect Berdennikov's values to be greater than our own; however, it is not known whether or not a difference of 30% can be explained in this manner. The difference in values may also have been due in part to differences in glass composition and to the presence of nitrogen in our experiments.

A different method of measuring the fracture surface energy was developed by Roesler [10], who applied a constant force to a flat-ended punch, inducing a conically shaped crack to grow from the edge of the punch into the glass. The crack was relatively stable and its dimensions could be easily measured. Roesler calculated the fracture surface energy from the applied force and the crack dimensions. His values depended on the duration of load and ranged from 10 J/m^2 for a load duration of a few seconds to 1.8 J/m^2 for loading over long periods. The high values were attributed to anelastic effects at the crack tip, while the low values were attributed to fracture surface energy lowering caused by chemical changes and adsorption of vapors at the crack tip. Roesler arbitrarily selected a 15-min value of the fracture energy (4.1 J/m^2) as a good approximation of the true fracture energy of glass. Our fracture surface energy value in dry nitrogen falls within the range of values obtained by Roesler. A more quantitative comparison between our data and Roesler's data is not possible because of the wide range of values obtained by Roesler.

A third method of measuring the fracture surface energy of soda-lime glass was used by Shand [23], who introduced cracks of known dimension into the face of a glass sheet and then broke the sheets in flexure. The fracture surface energy was calculated from the crack dimensions and the critical load necessary for fracture after a loading time of one second. Tests were carried out in air at room temperature with the humidity uncontrolled. A fracture surface energy value of 1.7 J/m^2 was obtained, but this value was undoubtedly too low, due to the presence of moisture in the air. If a shorter load duration were used, the measured breaking stress and the calculated fracture surface energy would have been higher. For example, the extrapolated load for breakage to occur in 0.1 sec yields a fracture energy value of 3.02 J/m^2.

At high temperatures, where the glass has become fluid, it is possible to measure its surface energy by several standard techniques [24]. The measured surface energy is observed to decrease linearly with increasing temperature. A room-temperature value may be obtained from a linear extrapolation of the high-temperature data back to room temperature. Following this procedure, Griffith [1] obtained a surface energy value of 0.560 J/m^2 for soda-lime glass. This value is 20% of our measured fracture surface energy value of 2.83 J/m^2. As suggested by Morey [24], identical values would not be expected. At high temperatures, the components of the glass possess high mobility and the glass surface is capable of attaining its equilibrium structure through chemical changes and atomic rearrangement. Since atomic mobility is low at room temperature, such changes do not occur at surfaces newly created by fracture. Therefore, the fracture surface does not attain its equilibrium configuration during fracture and the measured fracture energy will thus be greater than the surface energy obtained by extrapolation from elevated temperatures.

Morey distinguishes between two different surface energies in a multicomponent system—a static surface energy, which is obtained after the surface has reached a state of thermodynamic equilibrium, and a dynamic surface energy which refers to the instantaneous value for a newly formed surface. The surface energy quoted by Griffith is undoubtedly equivalent to the true static surface energy of glass, while the fracture energy values obtained by the present author would be equal to the dynamic value, providing no irreversible mechanical effects occur at the crack tip during fracture. The difference between the dynamic and static values of the surface energy in glass is not known, but these values have been measured in water containing various surfactants [25]. For example, in water containing 0.284 moles/liter amyl alcohol, Hiss [25] found a dynamic value of 0.0549 J/m^2 and a static value of 0.0348 J/m^2. If glass behaves in a similar manner, a value of about 1.1 J/m^2 would be expected for the dynamic surface energy. Our value of 2.83 J/m^2 is about 2.5 times as large as this value.

A theoretical estimate of 1.75 J/m^2 for the surface energy of silica glass was obtained by Charles [21], who assumed it to be equal to the sublimation energy of the material associated with the surface. From this estimation, a value of 1.76 J/m^2 is obtained for the glass used in our experiments if we assume that the surface energy of the glass is proportional to its Young's modulus [3]. Our experimental value for the fracture surface energy is about 1.6 times as great as the theoretical estimate. Despite the crudeness of the theoretical estimate, it does seem that our value is high. Part of the difference in value might be due to nonconservative effects at the crack tip, such as heat generation and plastic flow.

Recently, Marsh [26] has demonstrated plastic flow around indentations and scratch marks in glass. If plastic flow did occur around a crack tip, the two fracture surfaces produced during propagation would not match. It was felt that the mismatch would be particularly severe at the point at which the crack started to move. Comparison of the two fracture surfaces using a Zeiss interference microscope showed that the surfaces matched exactly, to within the limits of height resolution of the microscopy, 500 A (Figs. 9–11). This condition does not preclude the possibility of plastic deformation on a scale smaller than 500 A, although it does eliminate the possible occurrence of gross plasticity at the crack tip. Other observations on the fracture surface topology are interesting. The

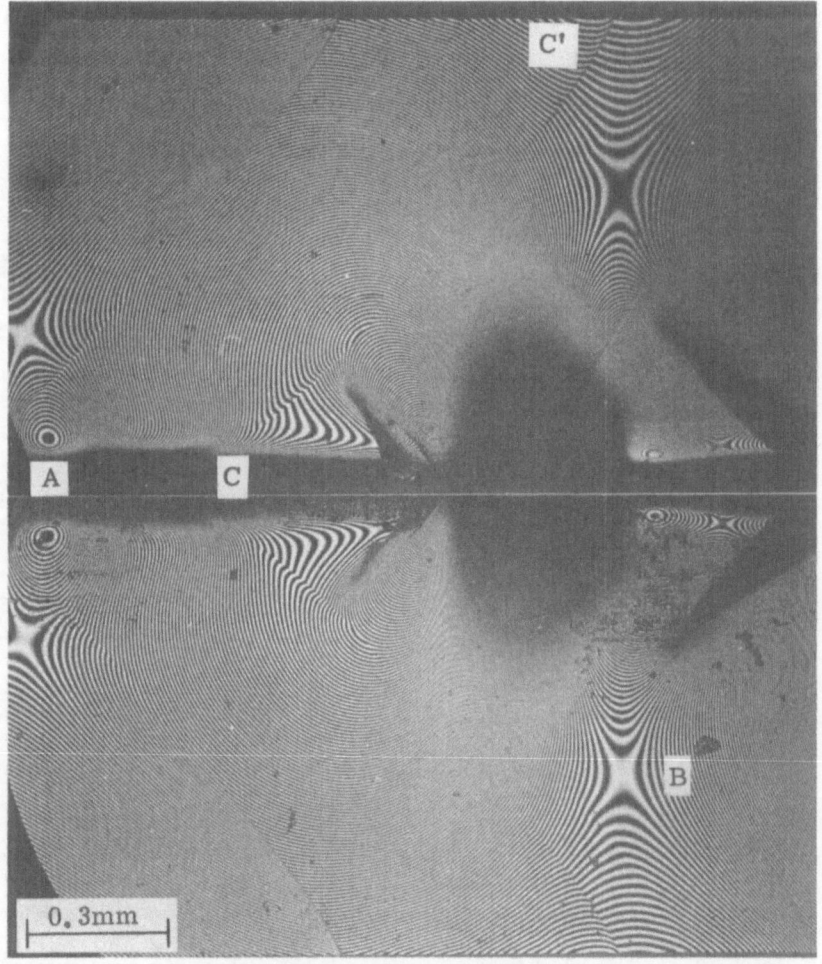

Fig. 9. Topology of matched fracture surfaces: (A) hill—valley, (B) saddle point, (C) and (C') ridge-grooves.

crack surface is not flat, but conchoidal as might be expected, containing hills, valleys, and saddle points (Fig. 9). In agreement with the observations of Murgatroyd [9], ridges appeared on the fracture surface each time the crack was stopped and restarted (Fig. 10).

Despite the fact that our value of the fracture energy is high, it may be the correct one to use in estimating the ultimate strength of glass, since irreversible effects occurring at the crack tip during fracture should be included in the surface energy term of the Griffith equation [27]. A theoretical estimate of the strength of

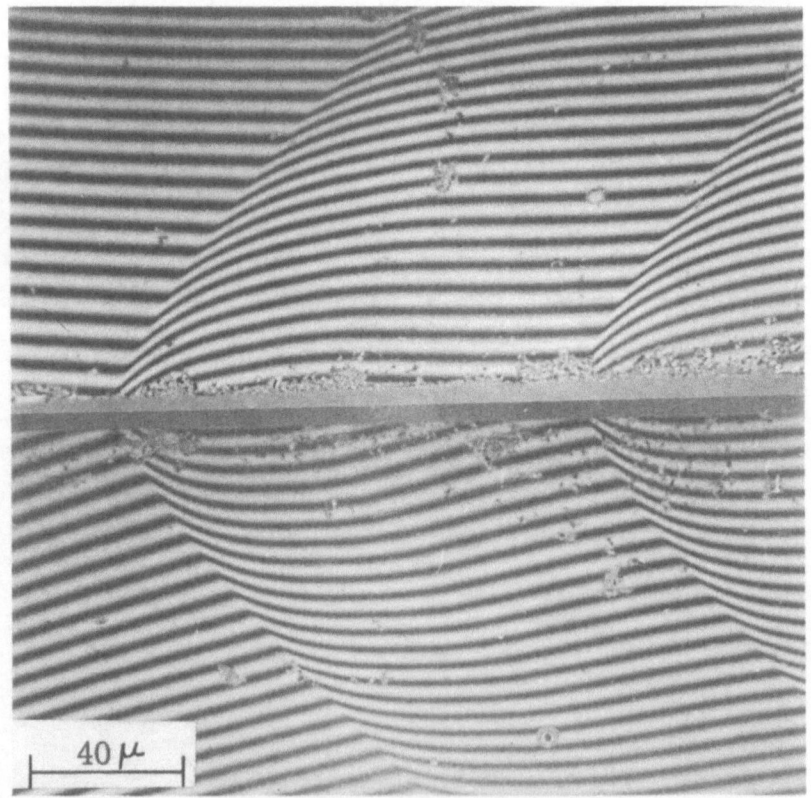

Fig. 10. Topology of matched fracture surfaces in the vicinity of a ridge-groove. The crack propagation was stopped and restarted along the ridge-groove and the direction of crack propagation was from left to right.

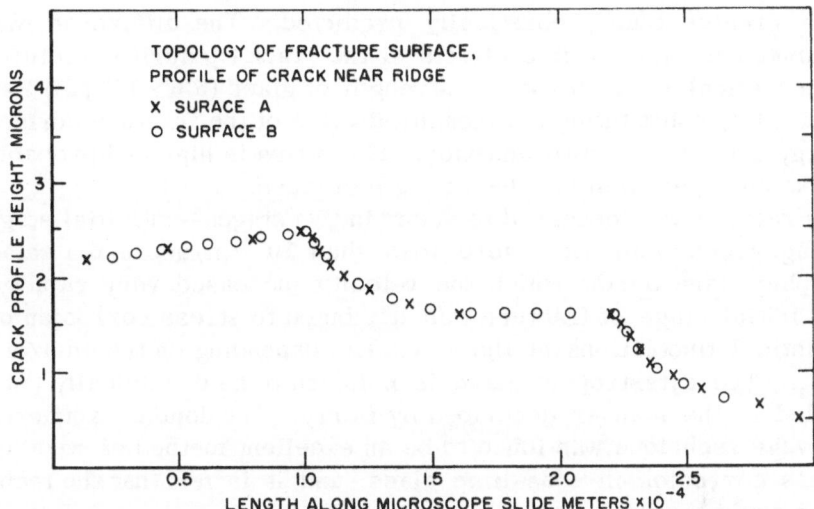

Fig. 11. Profile of matched crack surfaces near a ridge-groove. This is the same area as shown in Fig. 10. The crack propagation was from left to right.

glass can be obtained by substituting our fracture surface energy value in the Griffith equation and permitting the crack length to approach a value equal to the atomic spacing of glass, $1.6 \cdot 10^{-10}$ m. Although the strength value obtained will undoubtedly be high, it should establish an upper limit for the theoretical strength of glass. Following this procedure, we arrive at a value of $4.1 \cdot 10^{10}$ N/m^2 or $5.6 \cdot 10^6$ psi. This value is 3.8 times the value measured by Marsh [28] ($1.46 \cdot 10^6$ psi) on soda-lime glass fibers. Thus, our estimate is considerably higher than the highest measured value of the strength of soda-lime glass. A closer comparison is obtained if we compare our estimate of the strength of glass with that of Náray-Szabó and Ladik [29], who arrived at a value of $3.5 \cdot 10^6$ psi for fused silica. If it is assumed that the theoretical strength is proportional to the square root of the Young's modulus, the strength of soda-lime glass should also be $3.5 \cdot 10^6$ psi, which is about 60% of our estimated values.

SUMMARY

The double-cantilever cleavage technique was used to measure the fracture surface energy of soda-lime glass in several different environments. Although the values obtained were reasonable, they were greater than theoretically predicted. The difference was attributed to irreversible effects at the crack tip during fracture. A theoretical estimate of the strength of glass ($5.6 \cdot 10^6$ psi) was obtained by substituting the measured value of the fracture surface energy into the Griffith equation. This value is high and probably marks an upper limit for the strength of glass.

Fracture was observed to occur in two stages—an initial stage during which velocities were less than 10^{-4} m/sec and a catastrophic stage during which the velocity increased very rapidly. The initial stage of fracture was attributed to stress corrosion or statistical fluctuations at the crack tip, depending on the environment. The catastrophic stage is believed to be dynamically controlled in the manner described by Berry. The double-cantilever cleavage technique was found to be an excellent method of studying stress corrosion in soda-lime glass, and it is felt that the technique could be applied to other materials.

ACKNOWLEDGMENT

The author would like to express his gratitude to J. B. Wachtman, Jr., for his lively interest and helpful suggestions during the course of this work.

APPENDIX

Berdennikov [22] derived the following equation to calculate the fracture surface energy of glass:

$$\gamma = S_{cr}^2 \ C\pi(1 + \nu)/8E \tag{1}$$

where γ is the fracture surface energy, S_{cr} is the critical stress for crack propagation, C is the half-length of the crack, ν is Poisson's ratio, and E is Young's modulus. This equation differs from the one originally derived by Griffith [1] for the same system:

$$\gamma = S_{cr}^2 \ \pi C/2E \tag{2}$$

The question is then raised as to which of the two equations is correct.

In a recent review of the two-dimensional crack problem, Sneddon [30] demonstrated that the equation obtained by Griffith can be derived by six different methods; therefore, Griffith's equation is probably the correct one. The error in Berdennikov's equation occurred because an incorrect strain energy term, obtained by Wolf [31], was used in its derivation.

The fracture surface energy values presented by Berdennikov can be corrected by multiplying them by a factor of $\frac{8}{2}(1 + \nu) = 3.22$. The corrected values are presented in Table L

REFERENCES

1. A. A. Griffith, "Phenomena of Rupture and Flow in Solids," Trans. Roy. Soc. (London) A221:163–198 (1921). See also A. A. Griffith, "Theory of Rupture," Proc. First Intern. Cong. Appl. Mechanics, Delft, 1924, pp. 55–63.
2. J. C. Fisher and C. G. Dunn, "Surface and Interfacial Tensions of Single-Phase Solids," in: W. Shockley, J. H. Hollomon, R. Maurer, and F. Seitz (eds.), Symposium on Imperfections in Nearly Perfect Crystals, Pocono Manor, John Wiley and Sons (New York), 1952.
3. J. J. Gilman, "Direct Measurements of Surface Energies of Crystals," J. Appl. Phys. 31:2208–2218 (1960). See also J. J. Gilman, "Cleavage, Ductility, and Tenacity in Crystals," in: B. L. Averbach, D. K. Felbeck, G. T. Hahn, and D. A. Thomas (eds.), Proc. Intern. Conf. on Mechanisms Fracture, Swampscott, Massachusetts, John Wiley and Sons (New York), 1959, pp. 193–224.
4. A. R. C. Westwood and T. T. Hitch, "Surface Energy of {100} Potassium Chloride," J. Appl. Phys. 34:3085–3089 (1963).
5. J. P. Berry, "Some Kinetic Considerations of the Griffith Criterion for Fracture, Part I," J. Mech. Phys. Solids 8:194–206 (1960).
6. "Handbook of Chemistry and Physics," 37th edition, Chemical Rubber Publishing Co., 1956, p. 2325.
7. W. F. Linke (ed.), "Solubilities of Inorganic and Metal Organic Compounds, Seidel," fourth edition, Vol. 1, D. Van Nostrand Co. (Princeton), 1958, p. 1135.
8. H. Schardin, "Velocity Effects in Fracture," in: B. L. Averbach, D. K. Felbeck, G. T. Hahn, and D. A. Thomas (eds.), Proc. Intern. Conf. on Mechanisms of Fracture, Swampscott, Massachusetts, John Wiley and Sons (New York), 1959, pp. 297–330.
9. J. B. Murgatroyd, "The Significance of Surface Marks on Fractured Glass," J. Soc. Glass Technol. 26:155–171 (1942).
10. F. C. Roesler, "Brittle Fractures near Equilibrium," Proc. Phys. Soc. (London) 69B:981 (1956).
11. W. C. Levengood and W. H. Johnston, "Kinetics of Slow Fractures in Glass," J. Chem. Phys. 26:1184–1185 (1957).
12. R. J. Charles and W. B. Hillig, "The Kinetics of Glass Failure by Stress Corrosion," Symposium on the Mechanical Strength of Glass and Ways of Improving It, Union Scientifique Continentale du Verre, 1961.
13. C. E. Inglis, "Stresses in a Plate Due to the Presence of Cracks and Sharp Corners," Trans. Inst. Naval Architects (London) 55:219 (1913).
14. A. Smekal, "The Nature of the Mechanical Strength of Glass," J. Soc. Glass Technol. 20:432–448 (1936).
15. E. F. Poncelet, "A Theory of Static Fatigue for Brittle Solids," in: Fracturing of Metals, Am. Soc. Metals, 1948, pp. 201–227.
16. E. Saibel, "The Speed of Propagation of Fracture Cracks," in: Fracturing of Metals, Am. Soc. Metals, 1948, pp. 275–281.
17. P. Gibbs and I. B. Cutler, "On the Fracture of Glass Which Is Subjected to Slowly Increasing Stress," J. Am. Ceram. Soc. 34:200–206 (1951).

18. D. A. Stuart and O. L. Anderson, "Dependence of Ultimate Strength of Glass Under Constant Load on Temperature, Ambient Atmosphere, and Time," J. Am. Ceram. Soc. 36:416–424 (1953).

19. S. M. Cox, "A Kinetic Approach to the Theory of the Strength of Glass," J. Soc. Glass Technol. 32:127–146 (1948).

20. W. B. Hillig, "Sources of Weakness and the Ultimate Strength of Brittle Amorphous Solids," in: J. D. Mackenzie (ed.), Modern Aspects of the Vitreous State, Vol. 2, Butterworth and Co. (London), 1962, pp. 152–194.

21. R. J. Charles, "A Review of Glass Strength," in: J. E. Burke (ed.), Progress in Ceramic Science, Vol. 1, Pergamon Press (New York), 1961, pp. 1–38.

22. W. P. Berdennikov, "Messung der Oberflächenspannung von Festen Körpern," Soviet Phys. Z. S. 4:397–419 (1933); Zhur. Fiz. Khim. 5(2–3):358 (1934). (A summary of these articles is given in: Surface Energy of Solids by V. D. Kuznetsov, Her Majesty's Stationary Office, London, 1957, pp. 224–234.)

23. E. B. Shand, "Correlation of Strength of Glass with Fracture Flaws of Measured Size," J. Am. Ceram. Soc. 44:451–455 (1961).

24. G. W. Morey, "The Properties of Glass," second edition, Reinhold (New York), 1954, p. 191.

25. H. Freundlich, "Colloid and Capillary Chemistry," translated from the 3rd German edition by H. S. Hatfield, Dutton (New York), 1922, p. 52.

26. D. M. Marsh, "Plastic Flow in Glass," Proc. Roy. Soc. 279A:420–435 (1964).

27. E. Orowan, "Energy Criteria of Fracture," Welding J. Res. Suppl. 34:157s–160s (1955).

28. D. M. Marsh, "Plastic Flow and the Mechanical Properties of Glass," Technical Report No. 161, Tube Investments Research Laboratories, Hinnton Hall, Cambridge, 1963.

29. I. Náray-Szabó and J. Ladik, "Strength of Silica Glass," Nature 188:226–227 (1960).

30. I. N. Sneddon, "Crack Problems in the Mathematical Theory of Elasticity," Nonr 486(06), File No. ERD-126/1, May 15, 1961, pp. 13–44.

31. K. Wolf, "Zur Bruchtheorie von A. Griffith," Z. Math. Angew. Mech. 3:107-112 (1923).

Chapter 28

Chemical Polish and Strength of Alumina

A. G. King

Norton Company
Worcester, Massachusetts

Grinding polycrystalline alumina objects with abrasives introduces structural damage into the surface which weakens the specimen. Solution of the surface by a borax fusion removes this surface damage and materially increases the transverse rupture strength.

INTRODUCTION

An important area of materials research is the development of stronger and more durable ceramics. A principal factor which influences the strength of fully dense ceramics is the physical condition of the exterior surface. The literature contains many references to studies concerning the condition of the surface and the effects of chemical polishing on the strength of such materials as silicon, magnesium oxide, alkali halides, and glasses. In a technological sense, alumina is one of the more important refractory oxides, as it combines many useful physical and chemical properties which favorably permit its utilization in a wide variety of environments.

EXPERIMENTAL

The technique of chemically polishing alumina depends upon its solubility in a fusion of borax. The solution of the ceramic surface proceeds at an appropriate rate and in such a manner that structural discontinuities, such as dislocations and grain boundaries, are not preferentially etched. The effect of this solution process is to produce a smoothed and highly reflective surface on the alumina ceramic.

Figure 1 shows the apparatus which was used for these experiments. It consists of a small, rotating electrical resistance

Fig. 1. Rotary furnace used for chemical polishing.

furnace which contains a removable platinum crucible 3 $\frac{1}{2}$ in. by 1 $\frac{1}{2}$ in. in diameter. Approximately 50 g of borax glass are melted in the crucible, and the temperature is adjusted by means of an immersion platinum–platinum 10% rhodium thermocouple. Satisfactory polishes are produced between about 800 and 900°C. Temperatures above 900°C produce an etched, rather than a polished, surface. The furnace is tilted to an appropriate angle so that the sample tumbles freely in the melt. The alumina sample is preheated to the fusion temperature to avoid thermal shock of the specimen, and then it is dropped into the crucible. It is advantageous to reverse the rotation direction periodically to assure even solution of the ceramic on all surfaces. After approximately 5 min, the rotation is stopped and the specimen is removed from the crucible with nickel tongs. After cooling, the adhering borax glass is dissolved with dilute acid.

The ceramic material used in this study was hot-pressed alumina containing 0.5% MgO, which was added to retard grain growth. The powder was hot-pressed so as to produce a fully

dense structure at 3.98 g/cm^3 with an average grain diameter of
about 3 μ . This material was cut and ground with resinoid bonded
diamond wheels into specimens which measured 0.187 by 0.187 by
0.500 in.   These specimens were polished as described above, but
at variable times, and then broken in crossbending over a 0.33-in.
span using three-point loading.   The size of the specimen was
measured with a micrometer to the nearest ten-thousandth of an
inch before and after polishing. These data were used to compute
solution rates.

Three specimens were polished at each of three temperatures
for varying lengths of time so as to remove 6, 12, and 25-μ depths
of surface. The polishing time for each depth was variable because
of the increase in solution rate with increase in temperature.

DISCUSSION

The Nature of the Surface

Since we are largely dependent upon grinding as a process for
precisely shaping ceramics, the effects of this process on the
strength are naturally of interest. Material is removed from the
specimen by scratching and cutting. The surface that is produced
is extensively damaged.   Moreover, metallographic polishing
methods inevitably leave a few isolated large scratches. When
such a surface is etched in a fusion of potassium pyrosulfate, the
scratches develop etch pits along their length, as shown in Fig. 2.

The configuration of the etch pits indicates that they are
associated with structural defects which have a crystallographic
regularity.   When a surface such as the one shown in Fig. 2 is
alternately chemically polished and etched, the density of pits along
the scratches decreases, indicating that the amount of damage decreases with distance away from the scratch. This behavior suggests that the etch pits are the termini of dislocations on the sample
surface.   The character of the damage produced by grinding is
probably similar.

Figure 3 is a photograph taken with reflected light by a metallo-
graph of a ground ceramic and one of similar character which was
chemically polished. It is evident from the photograph that a great
improvement in surface topography is produced by the chemical-
polishing technique.   The latter is highly reflective, and the term
"polish," when used to describe the surface, is not used inaccurately.

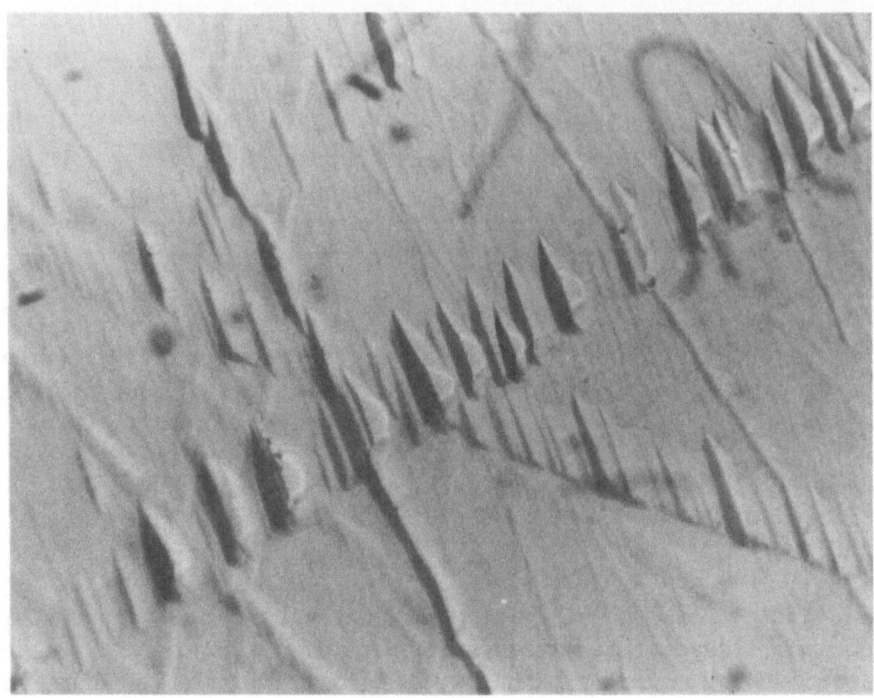

Fig. 2. Symmetrical etch pits developed in scratches on sapphire crystal. 450×.

Some surface irregularities are still present after polishing; at least on the polycrystalline alumina examined, there was no instance of a flaw-free surface having been produced. The principal surface flaw is a symmetrical fan-shaped structure, such as the one shown in Fig. 4. These configurations have no relationship with any internal structure, but seem to result from an etching process that occurs at the ceramic–borax-glass interface during cooling. Due to a difference in thermal expansion coefficient between borax and alumina, the boundary surface is stressed after the solidification of the flux. The amount of etching that occurs is a function of the flux composition. In particular, the volatile content of the borax seems to be important. Fluxes which have been held at a high temperature for a considerable time produce better polishes than those which use fresh borax. Since the alumina content of the flux has no effect in this regard, it was tentatively concluded that the volatile content of the flux, probably traces of water, causes stress corrosion phenomena to occur at the flux–ceramic interface during cooling of the polished specimen.

Solution Kinetics

The solution rates of these specimens at temperatures of 800, 850, and 900°C varied with the length of polishing time. The data indicated that the initial solution rates were higher. For longer polishing intervals, the rates leveled out to a steady-state condition for each temperature, as shown in Fig. 5.

Figure 6 shows that a straight line is obtained when the steady-state solution rates are plotted against polishing temperature for the materials and temperature range studied. As would be expected, the rates are higher for the higher temperatures.

Strength of Chemically Polished Specimens

The transverse strength of the specimens was determined for each polishing temperature at three values of material removal. The data (Table I) were essentially structureless, except for what appears to be random variation, but with one major exception. All of the polished specimens were stronger by about 35% than the

Fig. 3. Comparison of surface textures between diamond-wheel ground specimen on right, and chemically polished specimen on the left. Reflected light at normal incidence. 50×.

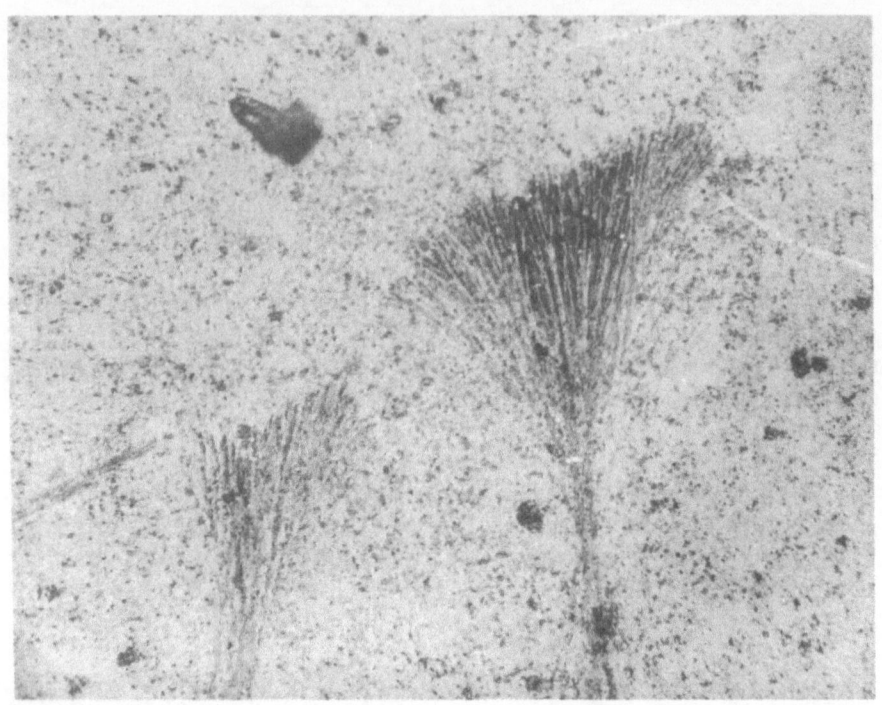

Fig. 4. Stress corrosion which develops on chemically polished surface during cooling
100× .

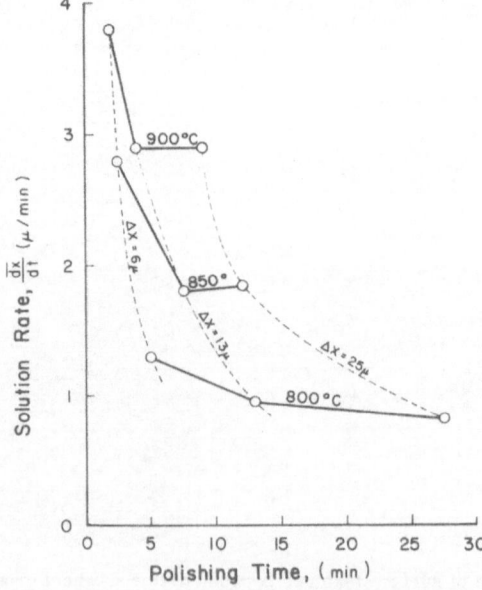

Fig. 5. Average solution rates as a function of polishing time at three temperature

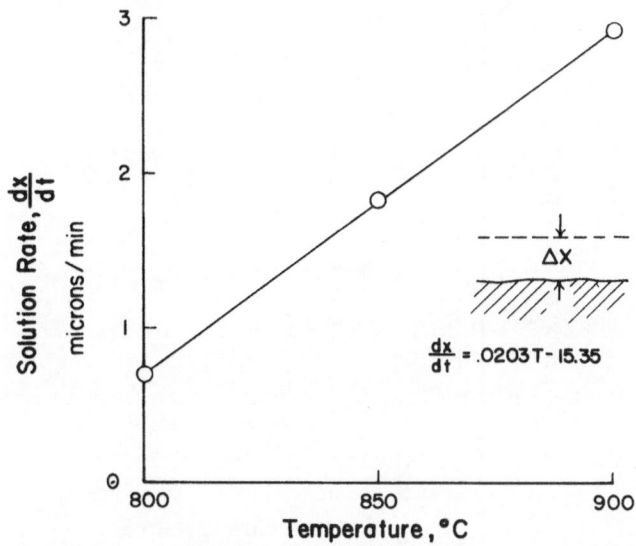

Fig. 6. Steady-state solution rate as a function of temperature.

TABLE I

Dependence of Transverse Strength upon Depth of Removal of Surface Layers by Chemical Polishing

| Temperature of molten borax (°C) | Transverse strength (psi×10⁻³) | | | | Average for given temperature |
| | Depth of removed surface layer (μ) | | | | |
	0	6	13	25	
800	66.8	95.9	100.3	104.0	
	73.1	86.4	90.5	66.3	
	77.2	89.4	86.6	102.0	
Average	72.4	90.6	92.5	90.8	91.3
850	79.7	107.6	76.1	81.9	
	61.6	103.6	99.1	99.1	
	83.7	92.6	99.3	84.4	
Average	75.0	101.3	91.5	88.4	93.7
900	87.9	82.8	101.6	92.9	
	45.1	103.2	117.1	87.6	
	73.5	99.1	106.3	87.4	
Average	68.8	95.0	108.3	89.3	97.6
Average for given depth	72.1	95.6	97.4	89.5	

In the figure: $\dfrac{dx}{dt} = .0203\,T - 15.35$

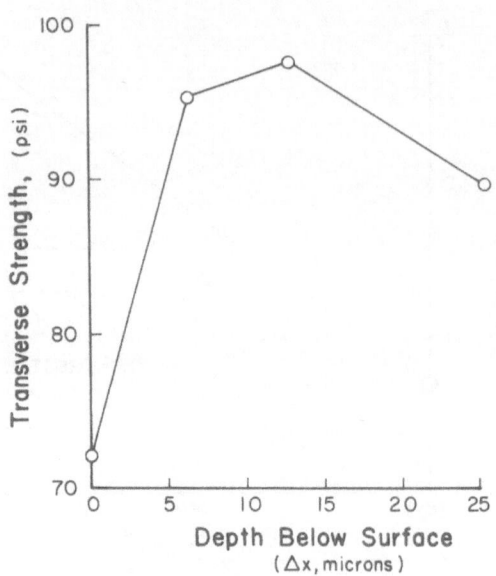

Fig. 7. Variation in transverse strength with depth of solution from the ground surface.

Fig. 8. Surface finishes of ceramic tools. Left — Ground with resinoid bonded diamond wheels. Right — chemically polished.

ground samples. Figure 7 shows the average values of strength as increasingly larger amounts of material are dissolved from the surface. The data at zero depth represent the strength of the ground, unpolished samples. These data and those on solution kinetics (Fig. 5) both indicate that the depth of the grinding damage is no greater than about $6\,\mu$. A microscopic study of these specimens indicates that they have an average grain size of about $3\,\mu$, with occasional grains (in fair abundance) as large as $6\,\mu$. It is concluded, therefore, that the grinding damage is stopped at the first grain boundary within the body of the ceramic. It is also evident that the damage that is present within this layer is responsible for the crack nucleation which normally results in fracture as the material is stressed. The strength of these materials can be limited by the character of the surface damage which is introduced in grinding (in this study, 72,000 psi). In order to produce precision-shaped alumina ceramics stronger than 70,000 psi, it is necessary either to remove the damage by solution or to develop ceramics less sensitive to the particular damage which results in fracture. The latter approach is probably limited due to the inherent lack of ductility in aluminum oxide. Ceramic materials containing flaws which propagate at stresses below 70,000 psi, of course, would not be benefited by chemical polishing. The maximum strength that has been observed on chemically polished sam-

TABLE II

Tool−Life Improvement With Chemical Polishing (Time in Seconds)

	Ground cutting edges	Chemically polished edges
A	53	267
B	66	116
C	116	404
D	90	188
E	76	167
F	117	193
Average	86	222

ples is 155,000 psi. This value probably represents the lower strength limit of a well-polished surface.

Application and Utility

To date, the only application of this polishing technique to an industrial process has been made to increase the durability of ceramic turning tools. Figure 8 shows two ceramic tools; the one on the right was polished using this technique.

When a polished tool such as the one shown in Fig. 8 is used to turn hardened steel, the tool life can be increased threefold. Some typical data are shown in Table II, which gives a comparison between polished tools and ground tools having similar edge configurations. The machining conditions were selected to give short tool-life values for testing purposes.

Chemically polished ceramic tools also have been used to machine cast iron using a workpiece containing four longitudinal slots which impart four severe impacts to the tool on each revolution. Polished tools can function up to 830 sfpm at a feed of 0.0165 in./rev without failure.

CONCLUSIONS

The process of chemically polishing alumina ceramics with a borax fusion removes surface damage which is introduced during grinding. This damage extends into the ceramic for just one grain diameter beneath the scratch root. The removal of this damage results in an increase of strength of 35% in the cases described. Other samples have shown flexural strengths up to 155,000 psi. This increase in strength can be utilized in certain applications to increase the durability and impact resistance of the ceramic. The future development of strong and precisely shaped polycrystalline ceramics will require the use of solution techniques to remove surface damage.

ACKNOWLEDGMENT

Appreciative acknowledgment is made to E. F. Reiner for his skilled assistance in preparing the samples and in performing many of the tests.

Chapter 29

Influence of Surface Conditions on the Strength of Ceramics

S. A. Bortz

Illinois Institute of Technology Research Institute
Chicago, Illinois

The effects of mechanical treatment (grinding, lapping, and vibratory polishing), chemical machining, and atmospheric conditions on the mechanical strength of alumina are described and analyzed.

INTRODUCTION

This research effort was concerned with both mechanical and chemical surface treatment of oxide ceramics and the effects on mechanical strength. An attempt was made to study various factors influencing the resistance of brittle materials to fracture, which is measured quantitatively by some critical value of stress. Analysis is complicated by the fact that fracture does not occur by one simple process for all materials and conditions, but rather a given material may fail by different mechanisms depending on the stress level, strain rate, previous history, and environmental condition.

Brittle fracture requires crack nucleation as well as crack propagation; therefore, we have considered the most likely origins of microcracks as well as the stress required for fracture. This likely source of failure is the exposed surface of the material.

Surface microcracks in polycrystalline materials are commonly due to the difference in the thermal expansion coefficient between the phases present in the body, a difference which gives rise to boundary stresses. Stresses arising at the surface during the cooling of samples from the firing temperature can also be a source of surface checking. However, it is usually assumed that mechanical abrasion of the surface and chemical attack are the major sources for crack development. Under the action of a stress field, these pits act as points of stress concentrations which lower the apparent strength of the material compared to its unflawed state. This phenomenon has been extensively studied in glasses; it is known that when surface irregularities are removed, the apparent strength of the glass is increased. Thus, the purpose of this study was to determine if the strength of oxide ceramics could be improved by the removal of surface microcracks.

Three surface-treatment techniques were studied for their effect on mechanical strength: (1) mechanical surface treatment, (2) chemical machining, and (3) surface diffusion. Mechanical polishing should remove high areas and reduce surface irregularities, but it is doubtful whether this polishing could eliminate the sharp ends of deep pits unless considerable surface material was also removed. The danger existed of pulling out whole grains, which might cause greater irregularities; also, thermally induced stresses might be produced due to heat generated when grinding. Chemical machining appeared to offer some promise, since this treatment should round out the sharp ends of the surface irregularities and, thus, reduce stress concentration in these areas. Surface diffusion of a second phase can produce a residual compressive stress condition, which, in turn, will produce an apparent increase in the strength of the matrix material.

EXPERIMENTAL PROCEDURE

Mechanical Surface Treatment

Wesgo AL-995 was obtained in the as-fired condition. The samples were divided into groups according to treatment: (1) fine-ground; (2) fine-ground and optically polished; (3) fine-ground, optically polished, and lapped; and (4) vibratory-polished. This latter treatment consisted of placing samples in a rubber-lined chamber with a kernel-sized abrasive (alumina grains) and gently revolving the entire mass in a detergent lubricant. The purpose of this treatment was to round the sharp corners of the as-fired specimens to eliminate the edge stress concentrations. In order to assess the effect of edge-removal using this technique (which, it is alleged, might reduce strength due to mechanically induced flaws), the corners of one group of specimens were edge-ground, while, for another group, the edges were both ground and lapped.

Profilometer readings for as-received and ground specimens

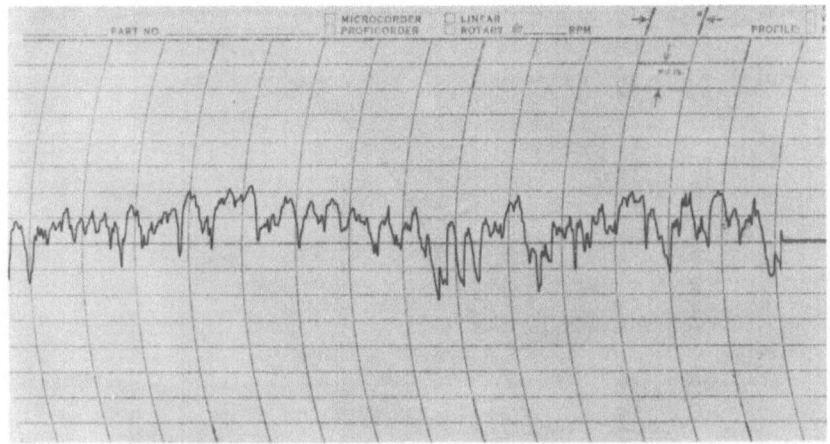

Fig. 1a. Profilometer reading of a ground surface, 40 rms.

showed about the same magnitude of surface readings, i.e., 40 rms. Lapping removed most of the surface irregularities. (See Figs. 1a and 1b.)

Lucalox, a purer form of Al_2O_3 produced by General Electric Company, was also studied. This material was received in the ground condition and was subjected only to optical lapping as a mechanical treatment.

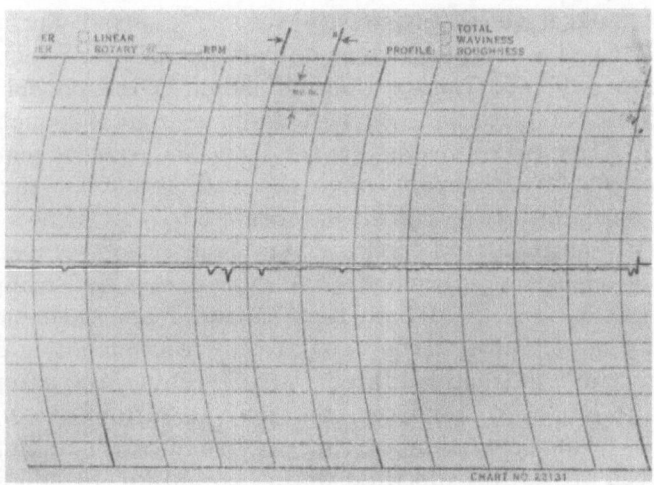

Fig. 1b. Profilometer reading of a lapped surface, 2 rms.

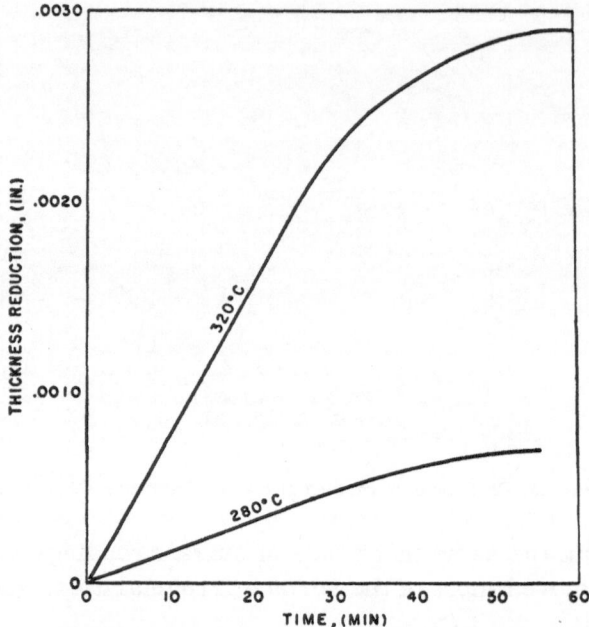

Fig. 2. Etching rate of orthophosphoric acid on alumina.

Chemical Machining

Chemical-machining experiments were made using as-fired, unground specimens. To reduce the possibility of thermal shock, it was imperative that experiments be carried out somewhat below 400°C. Two etchants were chosen for this treatment—orthophosphoric acid and fuming sulfuric acid. Orthophosphoric acid is a successful material-removal agent over a reasonable period of time. The experiments showed that the rate of material removed at 320°C is three times the rate at 280°C. (See Fig. 2.)

The samples were slowly preheated to 300°C (the test temperature) in an insulated container and held at this temperature for $\frac{1}{2}$ hr to ensure equilibrium. At this point, the samples were transferred to the acid bath that had also been preheated to the required temperature. The temperature was manually controlled to within ±10°C of that specified. After etching, the samples were removed to a heated, insulated container and allowed to cool slowly to 100°C. When this temperature was reached, the samples were transferred to boiling distilled water, boiled for 10 min to remove the acid, then cooled in water. After cooling, the samples were washed in methyl alcohol to remove any residual material and then washed in ether.

A photomicrograph of an as-fired surface of Wesgo material is shown in Fig. 3. The surface flaws can be readily observed and consist of several sharp-ended pits or surface cracks. A surface treated with H_3PO_4 shows chemical attack, which appears to occur rather uniformly over one grain depth (Fig. 4). The appearance of the surface is very rough; more irregularities for crack nucleation are present than for the unetched surface.

Fuming sulfuric acid removes very little surface material. The specimens treated with this etchant were processed in a manner similar to those treated with H_3PO_4. After a 3-hr exposure at 280°C, less than 1 mil of surface material was removed. The surface of the H_2SO_4-treated specimens appears to be smoother; however, the texture under the surface has changed from the original as-fired condition, having a porous look (Fig. 5). This is a surface phenomenon; the interior of the specimen is the same as the as-fired material (Fig. 6).

Material subjected to both treatments has the same appearance

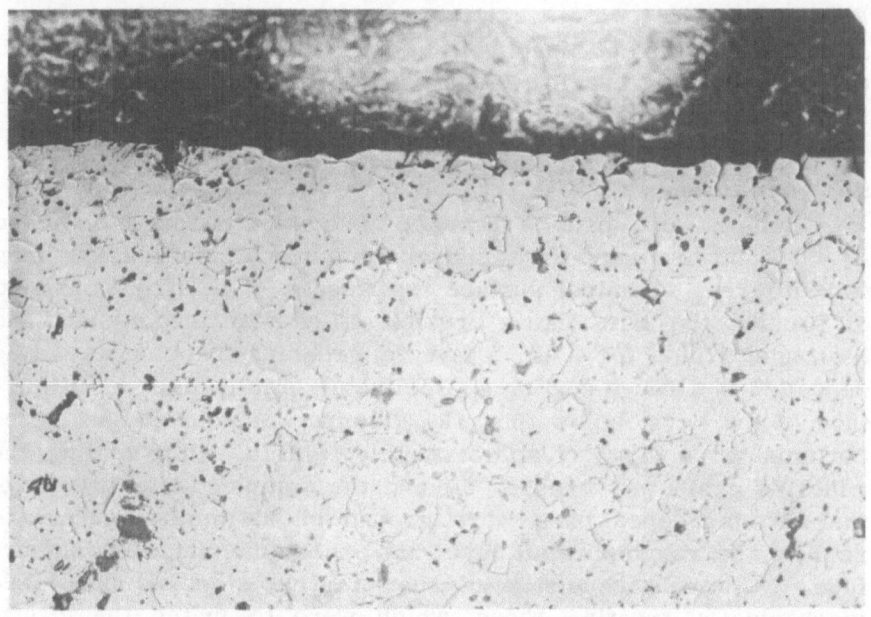

Fig. 3. Alumina specimen, unattacked surface. 200×; reduced 35% for reproduction.

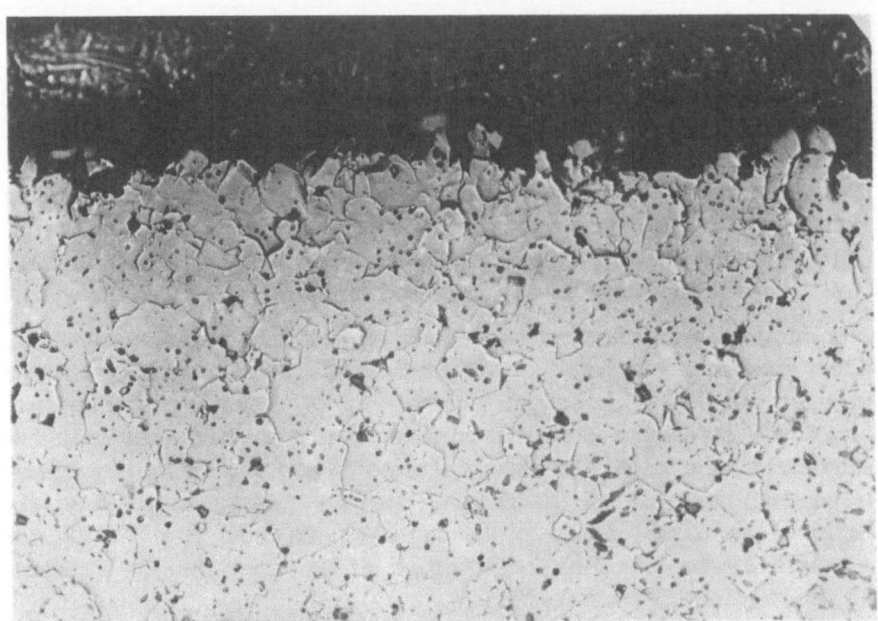

Fig. 4. Alumina treated in H_3PO_4 for 1 hr at 300°C. 200×; reduced 35% for reproduction.

as the H_2SO_4-treated specimens (Fig. 7). Both the Wesgo AL-995 (3% porosity) and Lucalox (0% porosity) were used in the chemical-treatment experiments.

Surface Diffusion

Attempts were made to increase the ambient strength of Wesgo AL-995 and Lucalox bars by inducing a surface compression using two different chemical surface treatments. The first consisted of packing the bars into a crucible filled with chromic acid and heating at 1700°C for 4 hr. A layer of greenish-black chromia was deposited in a rough coating approximately 5 mils thick on the surface of the bars; below this, the alumina was colored pink by the chromia for a depth of approximately 15 mils. After cooling, the adhering grain was brushed off and the samples tested. Some of the chromia-doped bars were ground on the tension surface to remove the rough chromia layer and expose the pink alumina surface. Controls consisted of heat-treating standard bars, and mechanically treating those which were to match the ground chromia-doped bars.

Vacuum Testing

Previous experimentation [1] has shown that an adsorbed surface film of water and the accompanying hydroxide formation has a greater influence on the strength of Al_2O_3 than surface texture. In order to verify the presence of a surface water film or hydroxide layer, transmission spectroscopy measurements were undertaken. Water and a hydroxide layer were identified on the surface of the specimens.

The strength of both Wesgo AL-995 and Lucalox were determined after chemically adsorbed water had been driven off by heating the specimens to above 400°C in vacuum and testing them in a vacuum environment. The specimens were placed in a vacuum chamber containing a small furnace with an Inconel test jig; the chamber was then sealed and evacuated to 0.1 mm Hg. The specimen was held at temperature for 1 hr, then allowed to cool to room temperature and tested in situ. The cycle was repeated for each new test.

RESULTS AND DISCUSSION

Mechanical-test data were obtained from bend tests on 4-in. long specimens of $^1/_4$-in.^{2} cross section. Mechanical-test data for

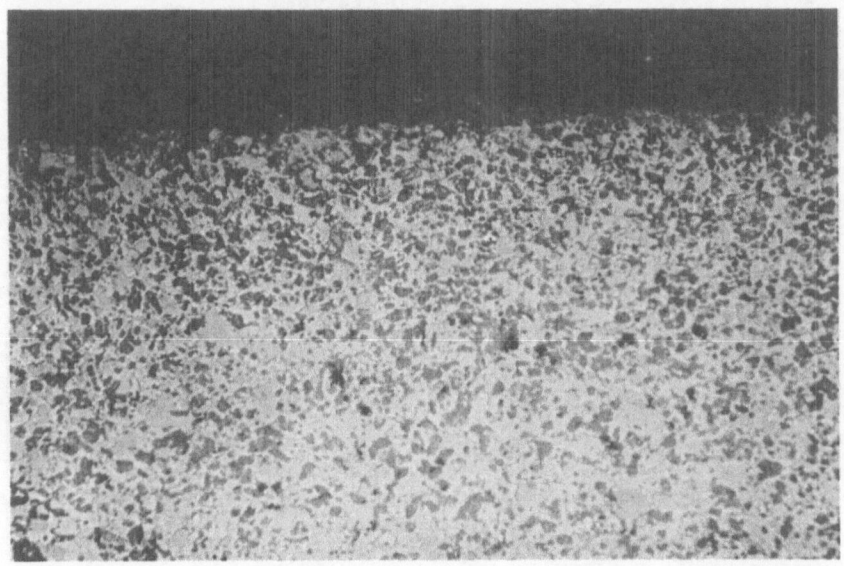

Fig. 5. Alumina treated with H_2SO_4 for 3 hr. 200×; reduced 30% for reproduction.

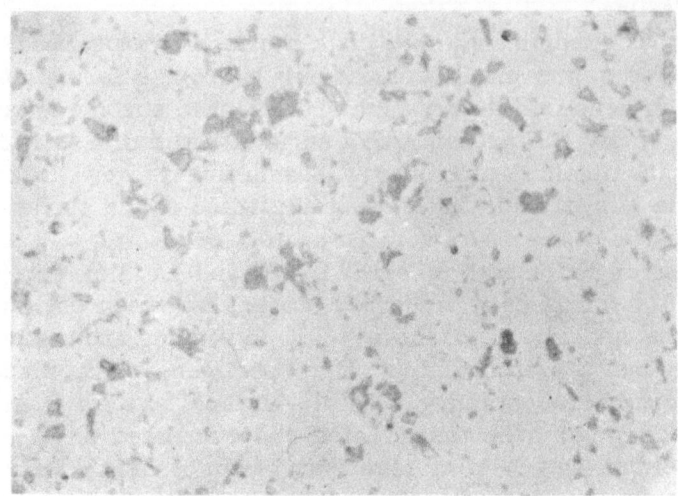

Fig. 6. Interior of sample treated with H_2SO_4. $500\times$.

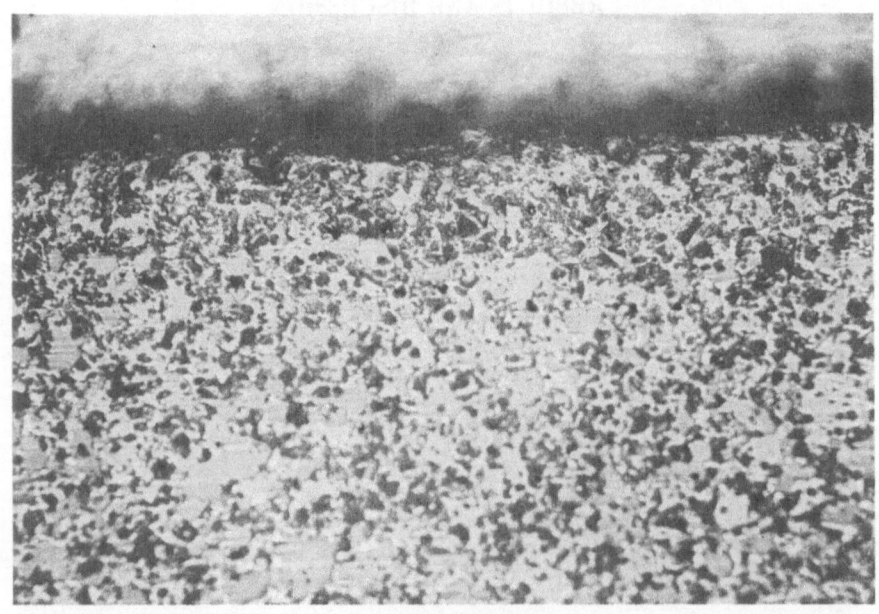

Fig. 7. Alumina treated with H_3PO_4 for 1 hr and H_2SO_4 for 2 hr. $200\times$; reduced 35% for reproduction.

Wesgo AL-995 are listed in Table I and the data for Lucalox in Table II. Grinding the surface of the Wesgo material lowered the strength, as compared to as-fired material. Polishing and lapping the surface did not appear to affect the strength to any great extent. The variability of the material tends to wash out any small strength changes. The data correspond to results obtained in previous work [2].

Tumbling the specimens with kernel-sized alumina balls produced the highest strength of the mechanically treated specimens. The as-fired material, subjected to low-power vibratory polishing, tested 22.5% higher than the as-fired controls. The ground material, subjected to vibration, tested 33% higher than the ground control specimens. Statistical examination of the results indicates that these differences are significant and that there is an apparent improvement in mechanical properties due to mechanically tumbling the Wesgo AL-995. The data also reveal that there is an optimum amount of tumbling in order to obtain maximum strength.

The specimens were originally tumbled to round the corners in order to reduce stress concentrations due to the sharp edges of the specimens, and to provide a uniform surface finish. When the increased strength was observed, a group of specimens was edge-ground; another group was edge-lapped to observe any behavior similar to the vibrated specimens. The data from the edge-ground specimens exhibited no significant strength changes from the as-fired controls. This evidence suggests that the increased strength is not due to rounding of the corners.

Wesgo AL-995 tested in vacuum, after heat treatment to drive off adsorbed moisture, has higher strength than the as-fired control sample. The increase is slightly higher than that achieved by tumbling the specimens (26%). These results are consistent enough to state that removal of adsorbed surface moisture increases the strength. Gregg [3] has reported that the adsorbed water film on alumina, calcined at 1000°C, still covers approximately one-third of the total surface area, and, at room temperature, the layer is several molecules thick. The temperature at which this chemically adsorbed film just covers the surface with a layer one molecule thick is approximately 400°C. Rupture of this film may be the cause of the increased strength noted for Wesgo alumina heated to 500°C in vacuum.

A traditional claim is that mechanical abrasion of the surface of ceramic materials is a major source of crack development.

TABLE I

Surface Treatment and Strength of Wesgo AL-995 Alumina

Number of samples	Treatment	Thickness of material removed (in.)	Average flexural strength (psi)	Standard deviation (psi)	Coefficient of variation (%)
30	As-fired		34,400	3985	11,6
18	Ground		28,600	3320	11,6
15	Polished		32,500	5000	15.0
3	Lapped		32,200	1100	3.5
12	Tumbled (Sweco)		42,100	4600	11.0
5	Ground, tumbled $^1/_2$ hr		34,150	3585	10.5
5	Ground, tumbled 1 hr		38,100	2635	6.9
5	Ground, tumbled 2 hr		37,270	2950	7.0
5	Ground, tumbled 3 hr		34,200	5920	17.3
5	As-fired, ground edges		35,000	2600	7.5
10	Ground, lapped edges		31,037	5920	19.0
6	As-fired, 500°C, vacuum 1 hr		41,700	1710	4.0
9	As-fired, 700°C, vacuum 1 hr		43,500	4300	9.9
1	As-fired, 1000°C, vacuum 1 hr		43,400	—	—
10	As-fired, H_3PO_4	0.0030	28,360	1618	5.7
10	As-fired, H_2SO_4	0.0002	32,790	3740	11.4
10	As-fired, H_3PO_4, 1 hr	0.0040	28,290	2180	7.7
6	As-fired, chromia-doped		26,650	1320	5.0
3	Ground, chromia-doped		26,470	2810	10.5
2	Borax-doped		14,250	1356	9.5

TABLE II

Surface Treatment and Strength of Lucalox Alumina

Number of samples	Treatment	Average flexural strength (psi)	Standard deviation (psi)	Coefficient of variation (%)
18	As-received, ground	24,200	2820	11.6
5	As-received, lapped	28,800	2195	7.6
10	As-received, ground H_3PO_4	23,500	2321	9.7
5	As-received, ground H_2SO_4	25,840	1350	5.1
3	Vacuum-ground	22,100	2130	9.5
5	Vacuum-ground, 500°C, 1 hr	16,850	2670	32.0
4	Vacuum-ground, 700°C, 1 hr	23,100	2696	11.6
3	Chromia-doped	19,750	3460	17.0
3	Chromia-doped, polished	23,486	3060	13.0
3	Borax-doped	13,370	2790	21.0

Early speculation was that the tumbling of the specimens should lower its mechanical strength. However, both Petch [4] and Parikh [5] claim to have found some evidence of room-temperature microplastic behavior or dislocation movement. The data obtained for the tumbled specimens tend to confirm this evidence. It is suggested that the energy imparted to the specimens during the tumbling processes causes something similar to work-hardening to occur at the surface. This effect is readily observed in metals under similar treatment. This is a possible explanation of strengthening due to tumbling.

Treatment with H_3PO_4 reduced the tested strength approximately 17% below that of the control group. The photomicrograph (Fig. 4) shows extensive grain boundary attack which leads to the introduction of a more extensive microcrack system than the as-fired material (Fig. 3). This would naturally reduce the strength of the material treated in this manner.

Surface treatment with fuming H_2SO_4 does not remove much material, but it does change the surface of the as-fired material. The surface of the specimens appears much smoother after treatment than before. However, strengths are somewhat below the as-fired specimens, which can be explained by increased surface porosity. Treating the specimens with H_3PO_4 first and then attempting an H_2SO_4 treatment does not improve the apparent strength of the Wesgo AL-995. A similar observation can be made for ground Wesgo AL-995. Apparently, H_2SO_4 cannot remove enough material to reduce the irregularities of a coarse surface.

Diffusion of chromia into the surface of the Wesgo bars lowered the strength. Surface diffusion into alumina with borax drastically reduced the strength of Wesgo. Lapping the surface of Lucalox specimens increased the strength of this material over that achieved by the controls (19%). This confirms previous data [6].

Experiments to measure material removal of Lucalox in H_2SO_4 and H_3PO_4 indicated that little or no removal occurred when treated in the same manner as AL-995. The main difference between the two materials is that Lucalox is denser with little or no porosity and has a smaller amount of impurities added to control grain growth. (The strength of the acid-treated Lucalox specimen is approximately the same as that of the ground specimens.)

Vacuum testing of Lucalox after heat treatment has little effect on strength. As a matter of observation, the data for heat treatment at 500°C appears to lower the strength considerably. Chromia diffusion into the surface apparently has little effect on the strength of Lucalox. Borax diffusion into the surface of this material drastically reduces its strength, as would be expected from the etched appearance of the surface. A limited supply of Lucalox specimens prevented tumbling experiments from being performed.

CONCLUSIONS

The reactions of Wesgo AL-995 and Lucalox to surface treatment do not appear to be the same. Chemical treatment lowers the strength of Wesgo, vacuum treatment increases it, grinding lowers it somewhat, while polishing and lapping raise the strength to the level of the control specimens. Chemical and vacuum treatment do not affect the strength of Lucalox, but mechanical polishing increases it. Chromia diffusion into the surface of these two materials indicates a strength drop in the Wesgo AL-995, but little or no change in the Lucalox. Borax diffusion drastically lowers the strength for both materials.

The differences observed in the two materials must be due to impurities [7] and the percentage of pores present in these two materials, since the principal constituent of each is alumina. The Wesgo contains approximately 0.25% each of MgO and SiO_2, while the Lucalox contains 0.25% MgO and a very small amount of silica; in fact, the presence of any appreciable amount of silica prevents full densification of Lucalox [8].

The apparent increase in strength of the Wesgo AL-995 due to vibratory tumbling of specimens provides additional evidence that some room-temperature strain hardening within the grain or at the grain boundary is possible. The reaction appears to be similar to shot-peening in metals.

It is apparent that a more basic study of the microstructure of these materials is required to explain the phenomena and the material differences which have been observed during these experiments.

REFERENCES

1. N. A. Weil (ed.), "Studies of Brittle Behavior of Ceramic Materials," ASD-TR-61-628, April 1962, Task 9.
2. N. A. Weil (ed.), "Studies of Brittle Behavior of Ceramic Materials," ASD-TR-61-628, April 1962, Task 1.
3. S. J. Gregg, Proc. Chem. Soc. Symp. on Chemisorption at Kerle, Butterworth's (London), 1965.
4. N. M. Parikh (ed.), "Studies of Brittle Behavior of Ceramic Materials," ASD-TR-61-628, Part III, June 1964, Task 6, pp. 139-141.
5. Ibid., Section III, pp. 36-45.
6. N. A. Weil (ed.), "Studies of Brittle Behavior of Ceramic Materials," ASD-TR-61-628, April 1962, Task 1, p. 19.
7. Ibid., p. 16.
8. R. F. Vines et al., J. Am. Ceram. Soc. 41:304-309 (1956).

Chapter 30

Complex-Ion Embrittlement of Silver Chloride

A. R. C. Westwood, D. L. Goldheim, and E. N. Pugh

Research Institute for Advanced Studies
Martin Company
Baltimore, Maryland

Previous studies have revealed that when polycrystalline AgCl is deformed in aqueous solutions containing complex ions of high negative charge, e.g., $AgCl_4^{-3}$, $Ag(SCN)_4^{-3}$, and $Ag(S_2O_3)_3^{-5}$, the fracture mode changes from ductile and transcrystalline (as in air) to brittle and intercrystalline. In the present paper, it is demonstrated that: (1) embrittlement in chloride environments can be prevented by the presence, in solution, of inhibitor ions such as K^+, Cs^+, Zn^{+2}, Cd^{+2}, and Hg^{+2}; (2) embrittlement can be induced by complex ions of high positive charge; and (3) monocrystals can be embrittled, providing they contain a pre-existing crack. Studies of monocrystal fracture surfaces have revealed that the fracture process is discontinuous. For this and other reasons discussed, it is concluded that embrittlement cannot be explained on the basis of a dissolution-dependent mechanism, but is more likely to be associated with the adsorption of complex ions of high charge in the vicinity of strained surface bonds. It is suggested that the charge on the complex induces a localized redistribution of the shared electrons constituting the bond, effectively reducing its strength and causing the bond to break at an abnormally low stress level.

INTRODUCTION

When polycrystalline AgCl is deformed in air, it is ductile and fracture is transcrystalline. However, when deformed in concentrated aqueous solutions of NaCl [1−4], HCl, LiCl, NaBr, NaSCN [3,4], LiBr, and $Na_2S_2O_3$ [3], it behaves in a relatively brittle manner, and fracture is intercrystalline. Somewhat surprisingly, polycrystalline AgCl is not embrittled when tested in concentrated solutions of KCl or CsCl [3]. The effects of applied stress and some of the above environments on the time to failure of polycrystalline AgCl specimens tested in dead loading are illustrated in Fig. 1. Of particular interest in this figure are: (1) the fact that embrittlement does not occur in 2N NaCl, but does occur in solutions of concentration $\geq$ 3N, for which $AgCl_4^{-3}$ is the predominating complex

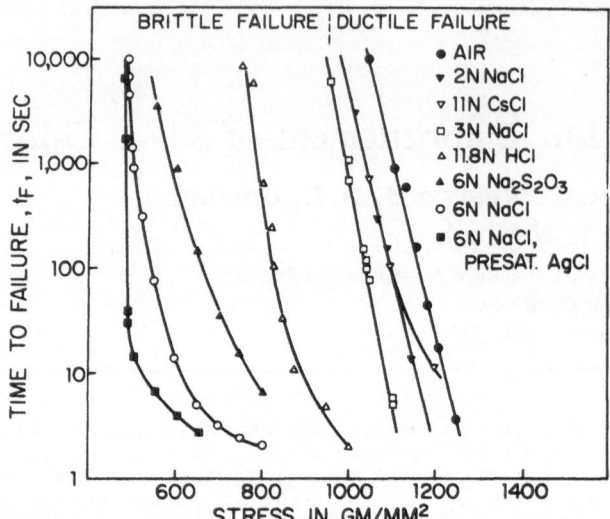

Fig. 1. Effects of applied stress and environment on the time to failure t_F of polycrystalline AgCl at room temperature. Note variation in degree of embrittlement with concentration of aq. NaCl and also effect of presaturating 6N NaCl with silver chlorocomplexes.

species in solution [5]; (2) the stress sensitivity of the phenomenon: The stress range over which time to failure t_F is reduced from approximately 10^4 to 10 sec is remarkably small; and (3) the effect of presaturating 6N NaCl with silver chlorocomplexes: At a stress of 500 g/mm^2, this procedure reduces t_F from approximately 10^3 to approximately 10 sec.

It is known that the solubility of AgCl in aqueous solutions of halogen or pseudohalogen salts (e.g., cyanides and thiocyanides) increases markedly with halogen-ion or pseudohalogen-ion concentration. This increase is associated with the formation of soluble complex ions such as $AgCl_4^{-3}$, $Ag(SCN)_4^{-3}$, and $Ag(S_2O_3)_3^{-5}$. Thus, it might be suspected that intercrystalline embrittlement in these environments is related in some way to specific dissolution at grain boundaries, perhaps in the vicinity of piled-up groups of dislocations [6]. However, if this were so, embrittlement might also be expected to occur in solutions of KCl and CsCl, since the solubility of AgCl in these environments is not significantly different from that in LiCl or NaCl at similar normalities [5,7,8] (Fig. 2) and the same complexes are formed. The effects of presaturation [2,3] (Fig. 1) also are difficult to account for in terms of a dissolution model. An alternative and more likely explanation is that embrittlement is associated with the adsorption and interaction of complex ions of high negative charge at regions of stress concentration in the lattice, for example, at or near grain boundaries against which dislocations are piled up [2,3,9]. A similar possibility involving adsorption at strained bonds has been discussed in connection with the phenomenon of embrittlement by liquid metals [10-12]. Another possible mechanism involves the adsorption of simple anions or cations at strained bonds, followed by replacement of the strained bonds by strain-free adsorbed-ion bonds [4].

The present work was undertaken to clarify the mechanism of embrittlement and the nature of the fracture process, to provide a better understanding of the role of the grain boundary in the phenomenon, and to determine whether embrittlement could be induced by complex ions of high positive charge. Possible means of preventing embrittlement were investigated also, and it will be demonstrated that the presence, in solution, of appropriate concentrations of ions such as K^+, Cs^+, Zn^{+2}, Cd^{+2}, and Hg^{+2} completely inhibits embrittlement by negatively charged complexes.

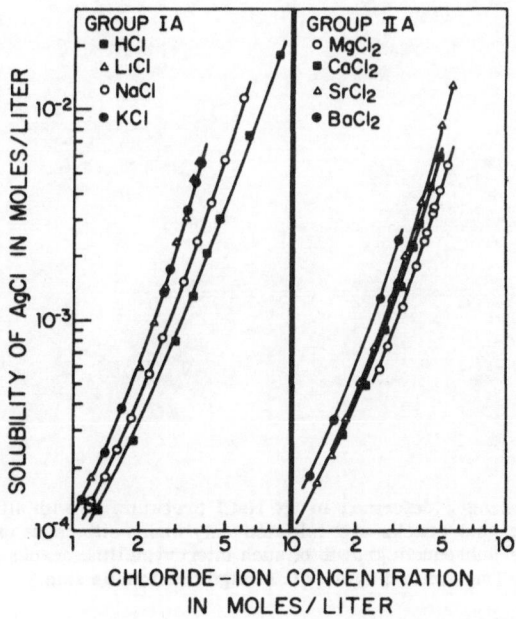

Fig. 2. Solubility of AgCl in various chloride solutions. Data of Kratohvil et al. [5], Forbes [7], and Kendall and Sloan [8].

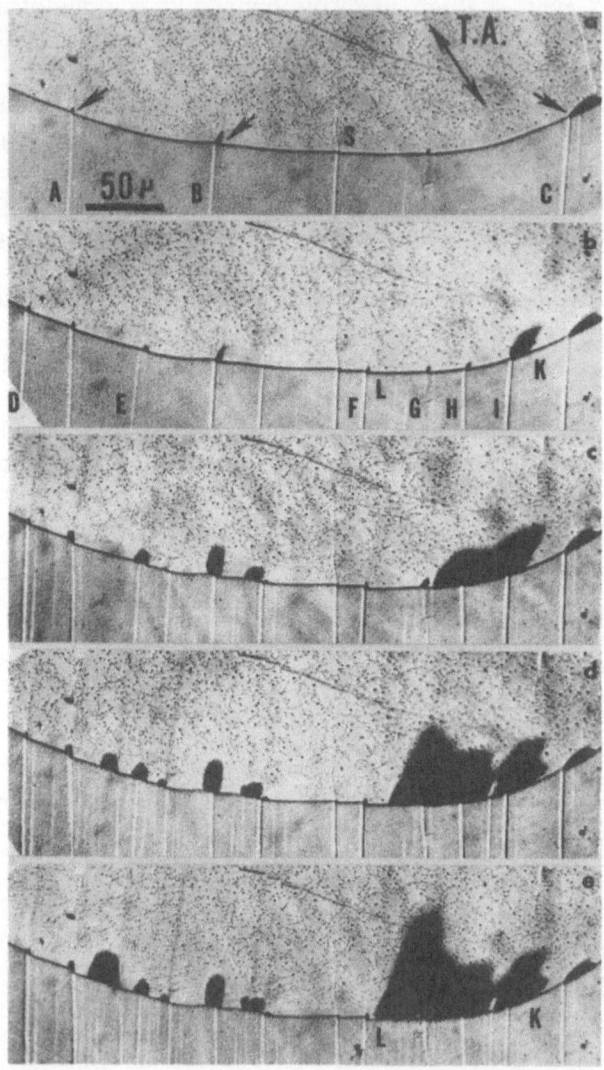

Fig. 3. Polycrystalline AgCl deformed in 6N NaCl presaturated with silver chlorocomplexes, demonstrating that cracks are initiated only where slip bands are arrested at a grain boundary. The subsequent growth of such intercrystalline cracks is illustrated in parts (b) through (e). Transmitted light. (T. A. signifies tensile axis.)

EXPERIMENTAL

Cast and rolled polycrystalline AgCl sheet, 1 mm thick, was obtained from the Harshaw Chemical Company, and tensile specimens, approximately 45 by 12 mm in size and having gage dimensions of 10 by 3 mm, were stamped from this material. Following chemical polishing for 1 min in boiling 10N NH$_4$OH, specimens were annealed at 300°C for 2 hr, and then air-cooled to room temperature. This heat treatment produced equiaxed grains 0.5–1.0 mm in diameter, with 2–3 grains across the thickness of the specimen. Gage dimensions were determined optically, and specimens were repolished for 1 min immediately before testing. Specimens for metallographic study were annealed for 16 hr at 450°C to produce grains 1–3 mm in diameter extending through the thickness of the sheet. Monocrystal tensile specimens were stamped from sheets sawed from a large Harshaw monocrystal, chemically polished, and then annealed at 450°C. Specimens containing a Na$^+$-doped region were prepared by heating crystals in contact with NaCl powder at 230°C for 16 hr.

Tensile tests were performed at room temperature in air and in various aqueous solutions, utilizing a static loading technique. The variable determined was time to failure t_F, under specific conditions of applied stress and environment. Coarse-grained and monocrystal specimens also were tested in a manually operated stressing jig which allowed study of crack initiation and propagation by means of a low-power binocular microscope.

RESULTS

Observations on the Fracture Process

Polycrystals. Earlier work suggested that, in the presence of an embrittling environment, crack initiation occurs where dislocations are held up against a grain boundary of suitably large misorientation [3]. However, this was first clearly demonstrated during the present studies [13]. For example, a polycrystalline specimen was lightly deformed in air and the boundary illustrated in Fig. 3(a) was revealed as potentially susceptible to embrittlement because slip bands, such as A, B, and C, were arrested by the boundary without inducing slip in the neighboring grain [3]. A drop of 6N NaCl solution, presaturated with silver chlorocomplexes, was placed over the boundary, the stress increased momentarily, and then the drop of NaCl solution removed by flushing with

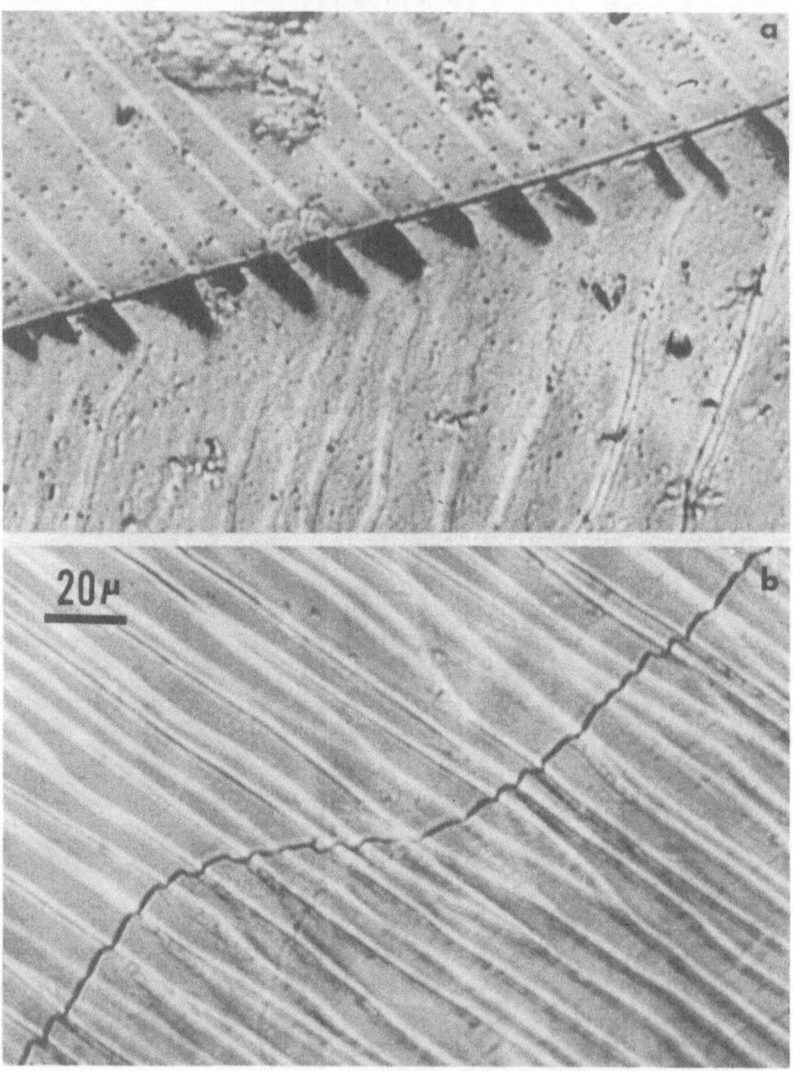

Fig. 4. Polycrystalline AgCl specimen deformed in 6N NaCl presaturated with silver chlorocomplexes. (a) Cracks are initiated where slip is arrested by a grain boundary of suitable orientation. (b) Cracks are not formed when the boundary is such that stress concentrations associated with arrested slip bands are relieved by slip in the neighboring grain. Transmitted light.

distilled water. After drying the specimen in an air blast, examination by transmitted light revealed small cracks lying in the grain
boundary [indicated by arrows in Fig. 3(a)] at the point of intersection of each of the arrested slip bands. This sequence was
repeated several times, with the results illustrated in Fig. 3(b)–(e).
It can be seen that cracks have formed at each of the newly arrested
slip bands, e.g., D–I in Fig. 3(b), and nowhere else. Crack propagation occurred first where the grain boundary lay approximately
perpendicular to the tensile axis [T.A. in Fig. 3(a)].

Figure 4(a) presents similar observations on another specimen.
In this case, the arrest of slip bands originating in the upper grain
led to the formation of intercrystalline cracks. Analysis of the slip
distribution in the lower grain of this figure revealed that it was not
directly related to that in the upper grain, and that the correlation
suggested by Fig. 4(a) was fortuitous.

Figure 4(b), and also Fig. 3(a) at S, demonstrate that cracks
are not formed when the stress field associated with an arrested
slip band is relieved by slip in the neighboring grain. It can be
concluded that, as in liquid-metal embrittlement phenomena [14],
the presence of dislocations piled up against some stable obstacle
is a prerequisite for crack initiation.

Once initiated, an intercrystalline crack propagates in a
relatively brittle manner provided that the grain boundary containing
it is oriented approximately perpendicular to the tensile axis. Thus,
in Fig. 3(b)–(e), no significant increase in slip band density occurred
in the vicinity of K–L during propagation of the crack over this
distance. Cracks lying in boundaries oriented approximately parallel
to the tensile axis, however, tend to blunt rather than propagate.
For example, note the faint slip markings in the left-hand side of
the upper grain in Fig. 3(d) and (e). Each band is associated with
a boundary crack.

It was observed that, in polycrystalline AgCl, crack propagation
was totally confined to grain boundaries. No examples of transcrystalline failure were noted.

Monocrystals. Un-notched monocrystals of pure AgCl are not
significantly embrittled in any of the otherwise active environments [3,4], and this observation led to the speculation that intercrystalline failure might be related to some inherent property of
grain boundaries in AgCl, for example, to the high density of
mobile interstitial silver ions present there [15]. Recently, however,

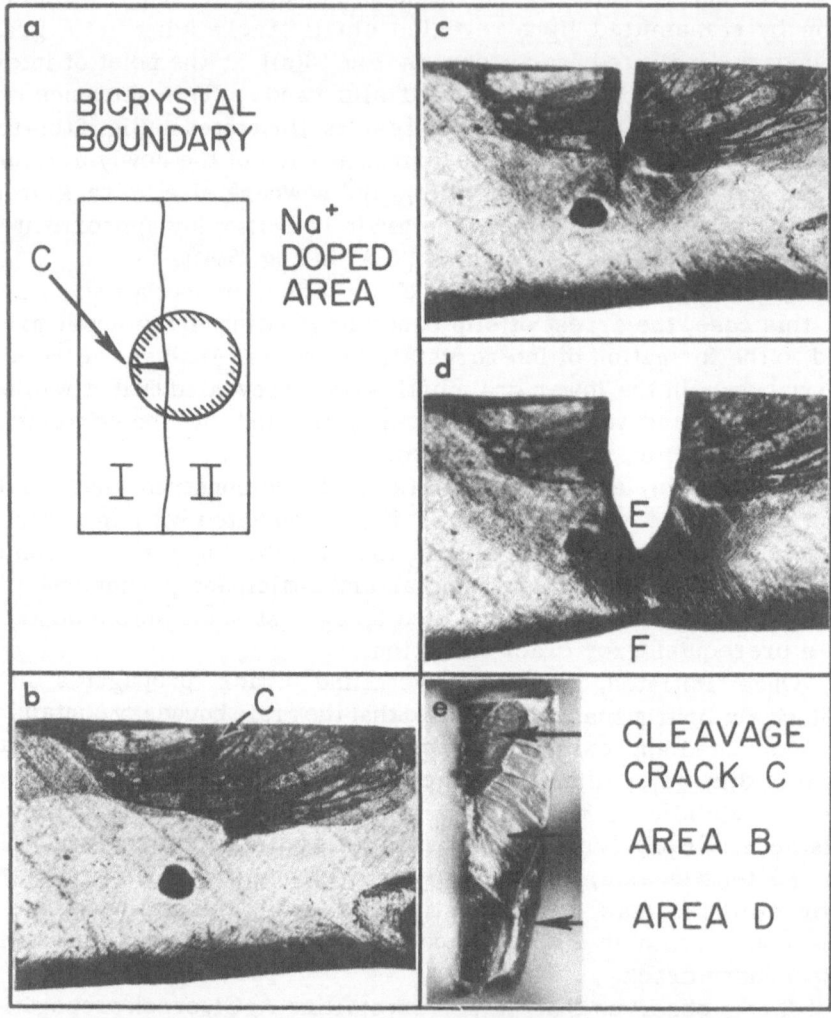

Fig. 5. (a) Schematic of bicrystal from which test specimen was taken, illustrating doped region and cleavage crack C. (b) Monocrystal I containing cleavage crack C. (c) Same area after straining in presaturated 6N NaCl. Crack has propagated in a relatively brittle fashion into undoped area of crystal. It was then propagated to a point just below dark indicator spot. (d) Crystal after further deformation in air. Ductile failure occurring by necking in region E-F. (e) Resulting fracture surface. Note cleaved area C, relatively brittle crack propagation through area B, and ductile failure in area D. Magnification about 16×; reduced for reproduction 30%.

Levine and Cadoff [16] have reported that deeply notched monocrystals of AgCl are brittle when tested in concentrated aqueous
NaCl solutions, but ductile in air. They have also suggested that
cracks can be formed at the interface between Na$^+$-doped and
undoped regions of a monocrystal, and that such cracks can be
propagated across the crystal only in the presence of an active
environment. Essentially similar experiments have been performed
in the present investigation, and Figs. 5 and 6 illustrate some of
the results obtained. In the sequence of Fig. 5, a AgCl bicrystal
partially doped with sodium ions, as indicated in Fig. 5(a), was
stressed in tension along a direction parallel to the boundary until
a cleavage crack C appeared in the doped region of one of the
component crystals. A solution of 6N NaCl, presaturated * with
silver chlorocomplexes, was then spread along the length of the
boundary, and a tensile stress applied across it. This caused the
bicrystal to separate into two monocrystals. Monocrystal I,

*Throughout the text, the terms "presaturated" and "unpresaturated" refer to the
relative concentration of complexed silver ions present in the environment at the onset
of the test.

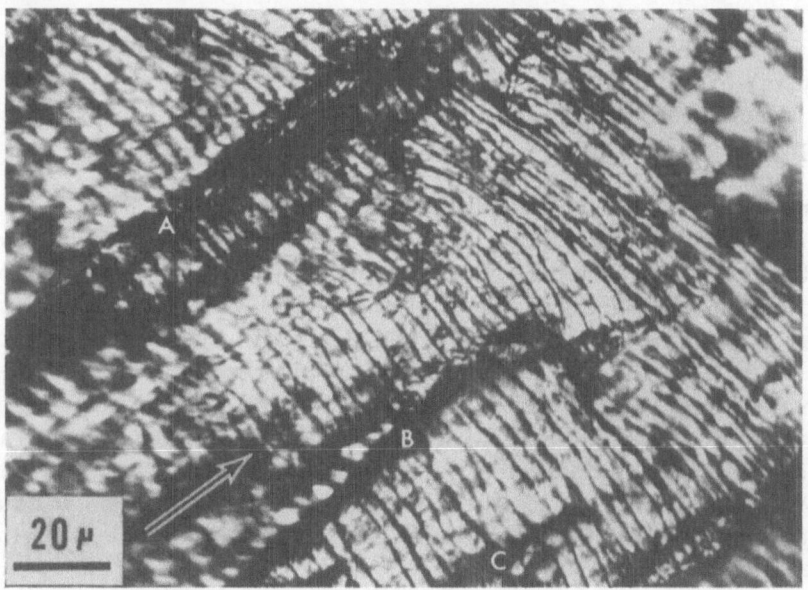

Fig. 6. Fracture surface from area B of Fig. 5(e). A, B, and C are tear marks, and the
direction of crack propagation is indicated by the arrow. Failure occurred in a discontinuous fashion, alternating between cleavage and plastic relaxation in the vicinity of the
crack tip (see Fig. 13).

containing the cleavage crack C, is shown in Fig. 5(b). A drop of presaturated 6N NaCl was then placed over the area illustrated, and a tensile stress applied perpendicular to the direction of the crack. The crack then grew in a relatively brittle manner in a direction perpendicular to the stress axis, and into the undoped area of the monocrystal [Fig. 5(c) and area B of Fig. 5(e)]. The NaCl solution was then removed by washing with distilled water, the specimen carefully dried, and stressing continued in air. The crack immediately became blunt, and the specimen deformed in a ductile manner, necking down to a chisel edge between E and F in Fig. 5(d) [area D of Fig. 5(e)]. A micrograph illustrating part of the fracture surface in area B is presented as Fig. 6. Lines A, B, and C are tear marks, such as those seen in area B of Fig. 5(e). Of particular interest, however, are the fine surface steps oriented approximately perpendicular to the direction of crack propagation. Such a structure reveals that, in the presence of an embrittling environment, crack propagation occurs discontinuously, probably alternating between cleavage and plastic relaxation in the vicinity of the crack tip. This microstructure will be discussed in a section to follow.

Other observations suggest that the susceptibility of a mono-crystal to embrittlement in presaturated 6N NaCl is a function of its orientation with respect to the tensile axis. This work is still in progress.

Embrittlement by Group-II Metal Chloride Solutions

In view of the interesting variation in embrittling behavior of solutions of the Group-IA metal chlorides [3], tests were performed in aqueous solutions of the Group-II metal chlorides. Data from tests performed in solutions of the Group-IIA chlorides are shown in Fig. 7. While the solubility of AgCl in these environments is very similar to that in Group-IA chloride solutions at a given normality [5] (Fig. 2), with the exception of $CaCl_2$, saturated solutions of these chlorides were much less effective as embrittling agents than solutions of the Group-IA metal chlorides. It can be seen from Fig. 7 that, again with the exception of $CaCl_2$ solutions, the degree of embrittlement increases with the period of the metallic element of the salt. It seems likely that this variation and also the smaller degree of embrittlement produced by the Group-IIA metal chlorides in comparison with the Group-IA chlorides (HCl, LiCl, and NaCl) both are related to differences in the chemical nature of

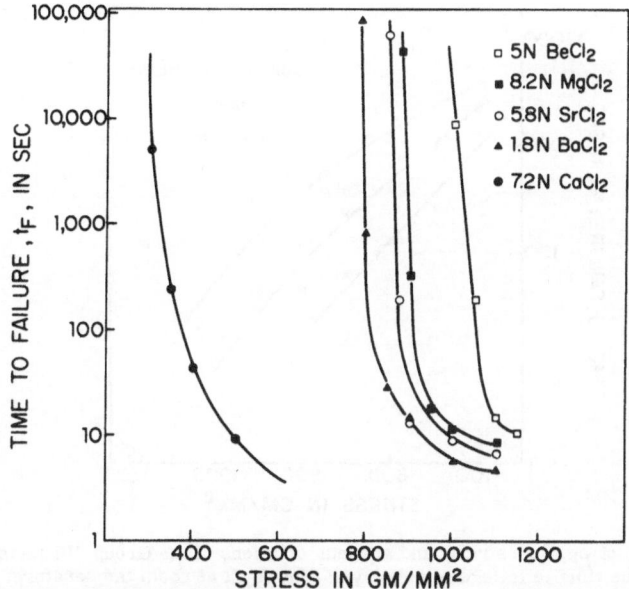

Fig. 7. Effects of applied stress and aqueous solutions of the Group–IIA metal chlorides on the time to failure of polycrystalline AgCl at room temperature.

the Group–IIA and Group–IA metal ions. In particular, Group–IA cations exhibit little tendency to form complex ions, whereas the lighter elements in Group IIA do form weak complexes [17]. For example, it is possible that magnesium ions might compete with silver ions in solution for chloride ions, and, thus, restrict or retard the formation of the complex $AgCl_4^{-3}$. In contrast with this effect, however, a 7.2N $CaCl_2$ solution proved to be one of the most embrittling environments yet tested, inducing brittle failures at stresses as low as 350 g/mm^2 in 250 sec (cf. Fig. 1). This anomalous behavior presently is not understood.

Data from experiments with the Group–IIB metal chlorides $ZnCl_2$ and $CdCl_2$ are given in Fig. 8. $ZnCl_2$ is extremely soluble in water, and, thus, in view of the high concentration of chlorine ions potentially available (approximately 66N), marked embrittlement was expected in saturated solutions of this salt. In fact, though intercrystalline failures at reduced stresses were obtained, the degree of embrittlement recorded was by no means as spectacular as might have been predicted. Moreover, the form of the t_F versus stress curve was not that usually observed in embrittling chloride environments (cf. Figs. 1 and 8), but instead was rather similar to

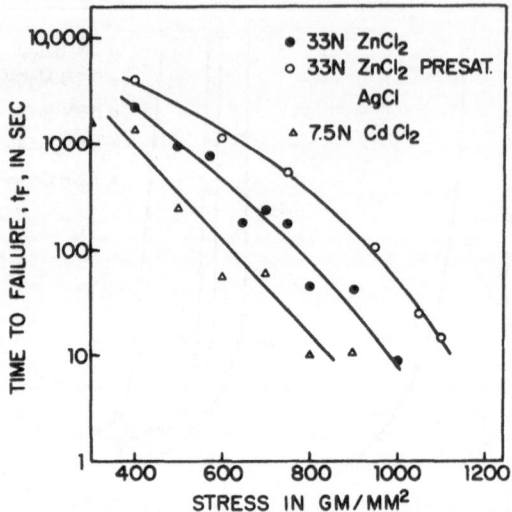

Fig. 8. Effects of applied stress and aqueous solutions of the Group-IIB metal chlorides on the time to failure of polycrystalline AgCl at room temperature.

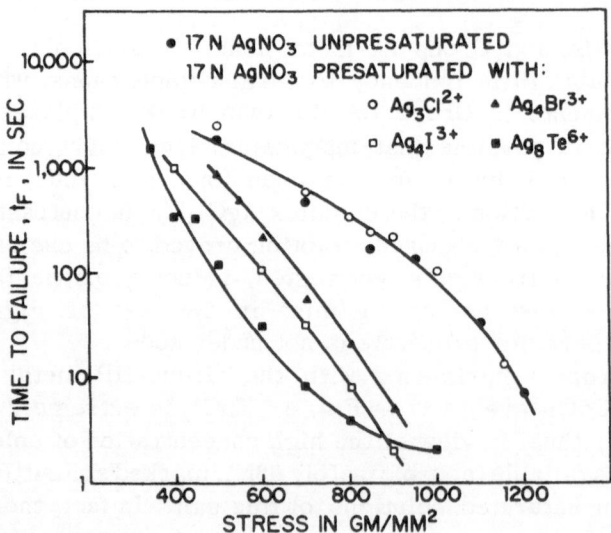

Fig. 9. Effects of applied stress and aqueous solutions of $AgNO_3$ containing various positively charged silver chlorocomplexes on the time to failure of polycrystalline AgCl at room temperature. Note that the degree of embrittlement increases with charge on the complex, and that the form of the curve obtained in the presence of Ag_8Te^{+6} is similar to that obtained in the presence of negatively charged complexes (cf. Figs. 1 and 12).

that obtained in 17N $AgNO_3$ solutions (Fig. 9). At first it was considered that the high solubility of AgCl likely in solutions of such high chloride-ion concentration might be counteracting the embrittlement process via Joffe's effect [18,19]. Accordingly, tests were performed in $ZnCl_2$ solutions presaturated with AgCl. However, the time to failure at a given stress in presaturated $ZnCl_2$ was found to be greater than that in unpresaturated solutions of similar normality (Fig. 8). This observation is not in accord with the Joffe-effect hypothesis. It also is directly at variance with the effect of presaturating other embrittling chloride environments, such as LiCl and NaCl, for which t_F is reduced [2,3] (Fig. 1). On the other hand, 17N $AgNO_3$ environments exhibit no significant effect of presaturation with AgCl [3] (Fig. 9).

In concentrated $AgNO_3$ solutions, AgCl dissolves to produce positively charged complexes, such as Ag_2Cl^+ and Ag_3Cl^{+2} [20]. Therefore, in view of the similarities in behavior of polycrystalline AgCl in solutions of $ZnCl_2$ and $AgNO_3$, the possibility was considered that the zinc ion might be successfully competing for the chloride ions present in solution, causing some reaction of the following type to occur:

$$ZnCl_2 + 3AgCl \rightleftharpoons ZnCl_4^{-2} + Ag_3Cl^{+2} \qquad (1)$$

In such circumstances, embrittlement might result from the presence of the positively charged complex Ag_3Cl^{+2}, since previous work has demonstrated that embrittlement of polycrystalline AgCl does not occur in the presence of negatively charged complex ions of charge less than -3 [3]. Such a reaction appears feasible in view of the known tendency of the Group-IIB metal ions to form relatively stable halide complexes [21]. If this hypothesis were correct, the addition of $ZnCl_2$ to presaturated 6N NaCl might be expected to decrease the effectiveness of this solution as an embrittling agent by reducing the $AgCl_4^{-3}$ ions present to complexes of lower charge, for example,

$$ZnCl_2 + AgCl_4^{-3} \rightleftharpoons ZnCl_4^{-2} + AgCl_2^{-1} \qquad (2)$$

Such an inhibition effect has been observed (Fig. 10), providing support for the proposed competitive action of zinc ions in solution. It is interesting to note that here two embrittling environments have been combined to produce an inert environment.

Tests performed in 7.5N $CdCl_2$ solutions exhibited a similar stress dependence to those from $ZnCl_2$ solutions, but the degree of

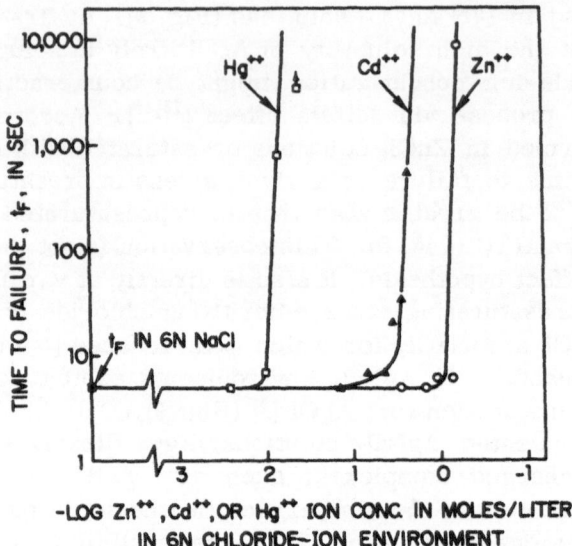

Fig. 10. Inhibition of embrittlement in 6N NaCl solutions by the addition of Group-IIB cations. The applied stress was 625 g/mm^2.

embrittlement was greater (Fig. 8). Similarly, additions of $CdCl_2$ and $HgCl_2$ to presaturated 6N NaCl also inhibited embrittlement in this solution, and preliminary experiments indicate that the order of effectiveness for inhibition is Zn < Cd < Hg (Fig. 10). It is known that the order of thermodynamic stabilities of the Group-IIB metal chloride complexes also is Zn < Cd < Hg [21].

TABLE I

Behavior of Group-IA Ions in Water

	Li$^+$	Na$^+$	K$^+$	Rb$^+$	Cs$^+$
Crystal radii, A	0.60	0.95	1.33	1.48	1.69
Hydrated radii (approx.), A	3.40	2.76	2.32	2.28	2.28
Hydration numbers (approx.)	25.3	16.6	10.5	—	9.9
Hydration energies, kcal/mole	123	97	77	70	63
Ionic mobilities (at infinite dilution and 18°C)	33.5	43.5	64.6	67.5	68

Taken from F. A. Cotton and G. Wilkinson [24].

Other Examples of Inhibition Phenomena

It has been mentioned that solutions of the Group-I chlorides HCl, LiCl, and NaCl embrittle polycrystalline AgCl, whereas KCl and CsCl do not. In an earlier publication [3], it was suggested that this variation in behavior might be associated with the formation, in solution, of uncharged complexes of the type Cs_2AgCl_3 [22]. This remains a possible explanation, for Jones and Penneman [23] have reported that ions corresponding to the solids $KAg(CN)_2$ and $K_3Ag(CN)_4$ are formed in aqueous solutions of KCN, and the cyanides are usually referred to as pseudohalogens, or halogenoids, because of their similarity in behavior to the halides. However, other explanations also may be suggested. For example, it is known that for the Group-IA elements, the larger the cationic radius, the smaller the hydrated radius, hydration number, and hydration energy (Table I), and the greater the strength or bonding to ion-exchange resins [24]. In other words, K^+, Rb^+, and Cs^+ are less effectively electrostatically screened by their water shells than Li^+ and Na^+.

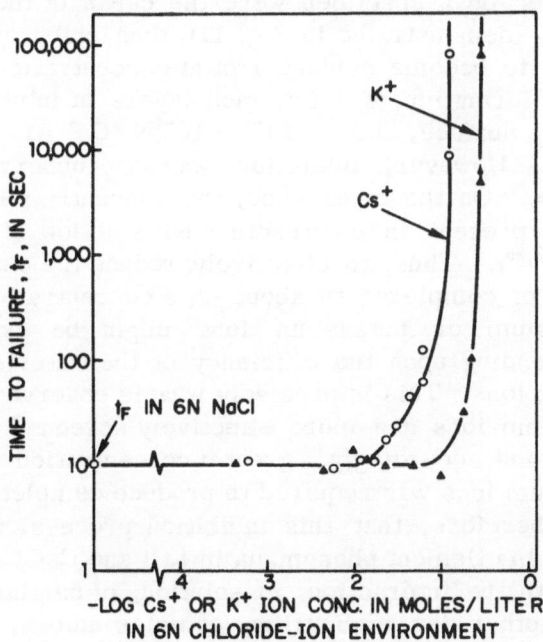

Fig. 11. Inhibition of embrittlement in a 6N chloride environment resulting from the replacement of sodium ions by potassium or cesium ions. The applied stress was 600 g/mm².

Thus, some interaction between the K^+ and Cs^+ and complex ions of high negative charge might be expected to occur in aqueous solutions. The result of such an interaction would be to alter (make more positive) the effective charge on the complex ion and to increase its apparent size. Either of these two factors is likely to reduce the ability of the complexion to cause embrittlement.

Another possible mechanism which has been suggested to explain the absence of embrittlement in these environments involves specific adsorption of potassium or cesium ions on to the surface of AgCl specimens, physically impeding adsorption of the complex ion and, thus, preventing embrittlement [25].

On the basis of any of these possible explanations, it would appear that replacement of the sodium ions in a 6N NaCl solution by either potassium or cesium ions, while maintaining the chlorine-ion concentration constant, should inhibit embrittlement in the resulting solution. Evidence supporting this prediction is presented in Fig. 11. If adsorption of the inhibitor ions Cs^+ or K^+ on the surface of the AgCl specimen were the cause of the significant increases in t_F (demonstrated in Fig. 11), then inhibition might have been expected to become evident first at concentrations sufficient to ensure the formation of a few monolayers of inhibitor ions on the specimen surface, i.e., $\sim 10^{-6} \rightarrow 10^{-5}$N Cs^+ or K^+ in these experiments. However, inhibition was not observed at such concentrations. On the other hand, the concentration of $AgCl_4^{-3}$ complex ions present in a presaturated solution of 6N NaCl is about 10^{-2}N [5,26]. Thus, to effectively reduce the charge on this concentration of complexes to about -2, a concentration of $1-10 \times 10^{-2}$N of cesium or potassium ions might be expected to be required, depending upon the efficiency of the water screening of these inhibitor ions. This is precisely what is observed. Further-more, potassium ions are more effectively screened than cesium ions (Table I), and, accordingly, a greater concentration of potassium ions than cesium ions was required to produce complete inhibition. It appears, therefore, that this inhibition process, and also the absence of embrittlement phenomena in KCl and CsCl solutions, is associated with the interaction, in solution, of partially screened cations with otherwise embrittling complex anions, and results from either: (1) a reduction in the effective negative charge on the complex, (2) an electrostatic screening of complex, (3) an increase in the effective size of the complex, or (4) some combination of these factors.

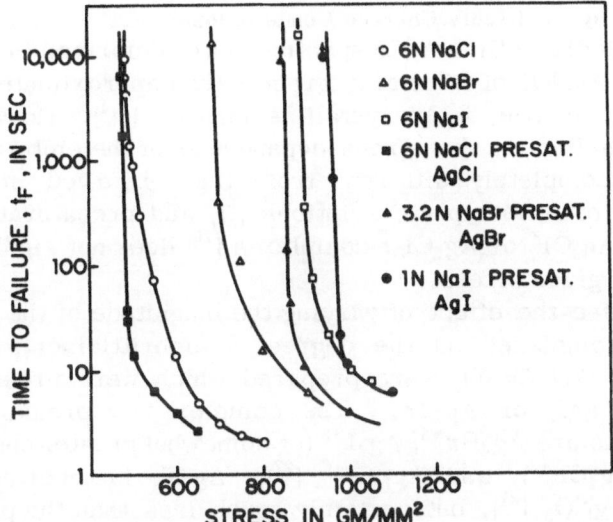

Fig. 12. Effects of applied stress and environment on the time to failure t_F of poly-crystalline AgCl at room temperature. The presaturated environments contain approximately equal concentrations of the respective tetrahalide silver complexes.

Embrittlement in Other Environments Containing Negatively Charged Complex Ions

Experiments were also performed in other halogen- and pseudo-halogen-containing environments with the object of clarifying the embrittlement mechanism. It was found that for 6N solutions, NaCl was more embrittling than NaBr, and NaBr more than NaI (Fig. 12). Since it is known that the thermodynamic stability of the silver tetrahalide complexes increases in the order Cl < Br < I [5,26], tests were performed in environments in which the complexes $AgCl_4^{-3}$, $AgBr_4^{-3}$, and AgI_4^{-3} were present in approximately equal concentrations, namely, 6N NaCl presaturated with AgCl, 3.2N NaBr presaturated with AgBr, and 1N NaI presaturated with AgI. These data also are shown in Fig. 12. The degree of embrittlement induced by these environments again was Cl > Br > I. Though the situation certainly is complicated by the fact that AgCl specimens were used throughout, it appears possible that the degree of embrittlement is inversely related to the stability of the complex. With this in mind, it is interesting to note that the embrittlement observed in saturated (~17N) NaSCN solutions has been found to be less than that observed in saturated (~8.5N) NaBr solutions, but greater than that in saturated (~12.7N) NaI solutions. The order of stabilities of the complexes involved is $AgBr_4^{-3} < Ag(SCN)_4^{-3} < AgI_4^{-3}$ [5].

Embrittlement by Positively Charged Complex Ions

When polycrystalline AgCl specimens are deformed in aqueous solutions of $AgNO_3$ of normality greater than approximately 0.2N, they fail in a brittle, intercrystalline manner [3,4]. However, as illustrated in Fig. 9, the stress dependence of the embrittlement process is completely different from that observed in halide, thiocyanate, or thiosulphate solutions [3], and presaturating 17N $AgNO_3$ with Ag_2Cl^+ or Ag_3Cl^{+2} complexes [20] does not significantly affect t_F at a given stress.

To examine the effect of varying the magnitude of the positive charge on complexes on the degree of embrittlement, aqueous solutions of 17N $AgNO_3$ were prepared which were presaturated with AgBr, AgI, or Ag_2Te. The complex ions predominant in such solutions are Ag_4Br^{+3}, Ag_4I^{+3} (of somewhat greater concentration than Ag_4Br^{+3}), and Ag_8Te^{+6} [27]. Ag_2Te is reportedly very soluble in $AgNO_3$ [27], but certainly much less than the predicted 1 mole/liter went into solution, and the actual concentration present was not determined. The results obtained using these environments are given in Fig. 9. It can be seen that the degree of embrittlement increases with charge on the complex ion, and that, in the presence of complexes of charge +6, a t_F versus stress curve similar to that obtained in the presence of complex ions of high negative charge was produced.

Other Observations

Comparative tests were made on monocrystals of NaCl deformed in three-point bending in saturated aqueous NaCl, and in saturated aqueous NaCl presaturated with silver chlorocomplexes. No significant variation in deformation or fracture behavior was observed.

DISCUSSION

Several important conclusions can now be drawn concerning the nature and characteristics of the embrittlement phenomena described. It has been demonstrated that grain boundaries are not a prerequisite for embrittlement, as might have been thought on the basis of earlier work [3,4]. Suitably oriented monocrystals can be embrittled, providing that a sharp crack or notch is present to serve as a stress concentrator of sufficient magnitude. Thus, intercrystalline embrittlement is not dependent upon some property specific to grain boundaries in AgCl, e.g., a high local concentration of interstitial silver ions. A grain boundary of appropriate orientation merely serves as a stable obstacle to dislocation motion, facilitating crack initiation.

That crack propagation proceeds in a discontinuous fashion [13] (Fig. 6) confirms the view [2,3] that embrittlement does not involve a dissolution-dependent mechanism. On the other hand, this observation and the demonstrated need for either a highly stressed grain boundary or the presence of a sharp notch to ensure embrittlement provide further support for the hypothesis that embrittlement involves adsorption and interaction of some surface-active species present in the environment at regions of strain in the lattice [2-4].

TABLE II

Relation between Complex Ion Stability and Embrittlement

Environment	Predominant complex-ion species	$\log_{10}K$ *	Embrittlement
10N NH$_4$OH	Ag(NH$_3$)$_2^+$	$K_{21} = 3.8$	No
< 0.1N AgNO$_3$	Ag$_2$Cl$^+$	$K_{12} = 2.0$	No
17N AgNO$_3$	Ag$_3$Cl^{+2}	$K_{13} = 0.25$	Yes
17N AgNO$_3$ presat. AgBr	Ag$_4$Br^{+3}	$K_{14} = 0.38$	Yes
6.2N Na$_2$S$_2$O$_3$	Ag(S$_2$O$_3$)$_3^{-5}$	$K_{31} = 0.28$	Yes
Satd. aq. NaCN	Ag(CN)$_4^{-3}$	$K_{41} = \sim -0.5$	Yes [4]
0.1N NaCl	AgCl$_2^-$	$K_{21} = 1.9$	No
1N NaI presat. AgI	AgI$_4^{-3}$	$K_{41} = 0.45$	Yes
17N NaSCN	Ag(SCN)$_4^{-3}$	$K_{41} = 0.28$	Yes
3.2N NaBr presat. AgBr	AgBr$_4^{-3}$	$K_{41} = 0.14$	Yes
6N NaCl presat. AgCl	AgCl$_4^{-3}$	$K_{41} = -0.4$	Yes

Increasing embrittlement (arrow pointing downward)

*Taken from J. Bjerrum, G. Swarzenbach, and L. G. Sillen [28]. K is the stability constant.

Westwood et al. [3] have proposed that embrittlement is associated with the adsorption of complex ions of high charge in the vicinity of strained surface bonds. Alternatively, Levine et al. [4,16] have suggested that the active species are simple silver or chloride ions, rather than complex ions. However, if this view were correct, then embrittlement might be expected to become evident at concentrations far below the observed 2–3N for NaCl solutions [3, 4], or 0.1N for AgNO$_3$ solutions [4]. Such embrittlement has not been observed. Moreover, the experimental evidence appears overwhelmingly in favor of the hypothesis that complex ions play an important role in the embrittlement process [1–3]. For example, embrittlement does not occur until certain highly charged complex ions become the predominant species in solution, e.g., AgCl$_4$$^{-3}$ in > 3N NaCl, and is inhibited when any reaction or interaction occurs in solution which affects either the structure or charge of the complex ions. Furthermore, the degree of embrittlement: (1) increases with concentration of the critical complex species present in the environment, (2) increases with charge on the complex ion, and (3) appears to be related to the stability constant of the complex ion predominating in the environment. Table II [28] illustrates the latter correlation.

It has been noted that the environmental embrittlement of AgCl exhibits several interesting similarities with liquid-metal embrittlement phenomena [2–4]. These include: (1) the remarkable stress sensitivity of both phenomena [2,10,28] (Figs. 1, 7, and 12); (2) the tendency for intercrystalline failure to occur unless the material contains planes of extremely low surface energy, i.e., cleavage planes; (3) the absence of embrittlement phenomena in monocrystals unless a crack pre-exists in the specimen [14] or can be readily initiated by dislocation–dislocation or dislocation–precipitate particle interactions [30]; and (4) the relatively high speed at which crack propagation can occur in both phenomena [3,10]. In view of these similarities, it does not seem unreasonable to suspect that an essentially similar mechanism involving adsorption and interaction at strained bonds [10–12] is involved. To be more specific, when a highly charged complex ion adsorbs at the site of a surface ion of opposite charge, the distribution of shared, or bonding, electrons between this ion and its neighbors is likely to be considerably perturbed. If the perturbation is such that cohesion is reduced locally, then, in the presence of an applied tensile stress, embrittlement will result. The degree of embrittlement is likely to depend upon the magnitude and distribution of the charge on the adsorbing complex. Both positively and negatively charged complexes will produce local perturbations, and, therefore, both are likely to constitute potentially embrittling species. However, the nature of the perturbations induced will be different. For this reason, the characteristics of the embrittlement phenomena resulting from the adsorption of positively or negatively charged complexes are likely to vary. The degree of embrittlement produced in a given environment also will be a function of the probability of adsorbing an embrittling complex ion at a site while the latter is in a suitably strained state. Thus, the degree of embrittlement should be directly related to the concentration of active complex ions present in the environment, as is observed [3]. Furthermore, for a given concentration of complex ions of equivalent charge, e.g., $AgCl_4^{-3}$, $AgBr_4^{-3}$, and AgI_4^{-3} in Fig. 12, the degree of embrittlement should be related to the distribution of charge on the complex, since this will determine the perturbation produced during adsorption. Thus, it appears reasonable that iodide complexes are less efficient embrittling agents than chloride complexes (Fig. 12), for it is known that the iodide complexes exhibit a much greater degree of covalency than chloride complexes, either as a result of an increased degree of $\pi(d)$ bonding [31,32] or because of the more polarizable nature of the iodide ligands [33]. Increasing the stress on a specimen will increase the probability of a given bond being in a suitably strained state when an embrittling complex ion adsorbs in its vicinity. Thus, t_F should be inversely related to stress, as observed. However, the stress sensitivity previously commented upon is not understood.

It should be emphasized that the present hypothesis does not postulate that adsorption is strain-induced, i.e., occurs only at strained bonds, as previously suggested for liquid-metal embrittlement [10]. It seems more likely that adsorption of the critical complex species at a strained bond is a purely statistical event, governed by factors such as temperature, concentration of the species, and diffusion rate. However, it is postulated that bond fracture occurs only when the bond is simultaneously acted upon by the applied stress and the perturbing forces of the charge on the complex.

The apparent relationship between degree of embrittlement produced by a given environment and the stability constant of the predominating complex species in that environment is most intriguing.

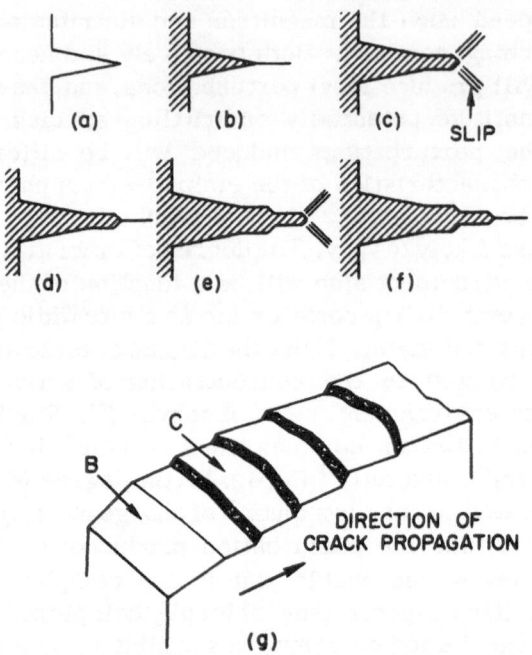

Fig. 13. Proposed mechanism for formation of fracture surface illustrated in Fig. 6.

Such a correlation might arise if the act of embrittlement also involves some change in the embrittling complex species, namely, reduction to the next lower complex. When a complex ion is adsorbed and interacts strongly with a surface ion, it seems likely that the complex itself will be raised to an excited state. Such an interaction might be expected to catalyze any tendency that this complex might possess to dissociate, for example,

$$AgCl_4^{-3} \rightarrow AgCl_3^{-2} + Cl^-$$

On the basis of the present hypothesis, a possible explanation for the surface structure illustrated in Fig. 6 can now be postulated. This is shown schematically in Fig. 13. Figure 13(a) represents the cleavage crack C initiated in the doped area of the crystal. Following immersion of this crack in a presaturated solution of 6N NaCl, a tensile stress was applied perpendicular to its length. While the crack tip was in a highly stressed state, adsorption and interaction of $AgCl_4^{-3}$ complex ions with the ions constituting the crack tip occurred, causing a significant reduction in cohesion across the tip. Accordingly, the bonds there fractured, and the crack propagated by cleavage in a relatively brittle manner [Fig. 13(b)]. The accelerating crack soon left behind the relatively slowly diffusing complex ions, but, because of the inherent ductility of AgCl at room temperature, did not achieve the critical velocity for completely brittle propagation [34]. Indeed, after traveling a distance of some $3-4\mu$, it was decelerated and arrested as a result of the generation of dislocations in the vicinity of the crack tip [Fig. 13(c)]. The embrittling complexes then diffused to the new position of the stressed crack tip and the cycle was repeated [Figs. 13(d)–(f)]. The fracture surface likely to result from such a process is illustrated in Fig. 13(g) (cf. Fig. 6). It can be seen that the concept of a cycle involving embrittlement, crack acceleration away from the embrittling species, and deceleration via plastic relaxation appears consistent with the fractographic observations.

Finally, it was found that no reduction in mechanical properties of NaCl monocrystals occurred when these were tested in solutions containing a relatively high concentration of $AgCl_4^{-3}$ complex ions, even though cracks can be readily initiated in this material via dislocation interaction. This observation appears to be in accord with the mechanism just proposed, which postulates that embrittlement results from an adsorption-induced variation in the distribution of the bonding electrons in a strained bond. Since bonding in NaCl is purely ionic,* there are no bonding electrons per se to be affected by an adsorbed complex.

ACKNOWLEDGMENTS

It is a pleasure to acknowledge stimulating discussions with Dr. Ruth Aranow and Dr. M. H. Kamdar. We also appreciate the financial support received from the U.S. Office of Naval Research under Contract No. Nonr–4162(00).

REFERENCES

1. T. L. Johnston and E. R. Parker, Rept. on Contract No. N7–ONR–29516 NR 031–255, Jan. 1957.
2. A. R. C. Westwood, E. N. Pugh, and D. L. Goldheim, Phil. Mag. 10: 345 (1964).
3. A. R. C. Westwood, D. L. Goldheim, and E. N. Pugh, "Dislocations in Solids," Disc. Faraday Soc. 38: 147 (1964).
4. E. Levine, H. Solomon, and I. Cadoff, Acta Met. 12: 1119 (1964).
5. J. Kratohvil, B. Tezak, and V. B. Vouk, Arkhiv. Kem. (English Transl.) 26: 191 (1954).
6. A. R. C. Westwood, RIAS Rept. No. 126, Aug. 1962.

*Bonding in AgCl is estimated to be 30% ionic and 70% covalent [35].

7. G. S. Forbes, J. Am. Chem. Soc. 33: 1937 (1911).

8. J. Kendall and C. H. Sloan, J. Am. Chem. Soc. 47: 2306 (1925).

9. A. R. C. Westwood, RIAS Rept. No. 162, Jan. 1964.

10. A. R. C. Westwood and M. H. Kamdar, Phil. Mag. 8: 787 (1963).

11. A. R. C. Westwood, in: Fracture of Solids, Interscience (New York), 1963, p. 553.

12. A. R. C. Westwood, Phil. Mag. 9: 199 (1964).

13. A. R. C. Westwood, D. L. Goldheim, and E. N. Pugh, Acta. Met. 13: 695 (1965).

14. M. H. Kamdar and A. R. C. Westwood, in: Environment-Sensitive Mechanical Behavior, Gordon and Breach (New York), to be published.

15. P. E. Goddard and F. Urbach, J. Chem. Phys. 20: 1975 (1952).

16. E. Levine and I. Cadoff, discussion to A. R. C. Westwood, D. L. Goldheim, and E. N. Pugh, "Dislocations in Solids," Disc. Faraday Soc. 38: 188 (1964).

17. F. A. Cotton and G. Wilkinson, Advanced Inorganic Chemistry, Interscience (New York), 1962, p. 329.

18. A. Joffe, N. W. Kirpitschewa, and M. A. Lewitsky, Z. Physik. 22: 286 (1924).

19. A. R. C. Westwood, Materials Science Research, Vol. 1, Plenum Press (New York), 1963, p. 114.

20. K. H. Leiser, Z. Anorg. Allgem. Chem. 304: 296 (1960).

21. F. A. Cotton and G. Wilkinson, op. cit., p. 471.

22. C. Brink and C. H. McGillivray, Acta Cryst. 2: 158 (1949).

23. L. H. Jones and R. A. Penneman, J. Chem. Phys. 22: 965 (1954).

24. F. A. Cotton and G. Wilkinson, op. cit., p. 321.

25. E. Matijevik, private communication.

26. K. H. Leiser, Z. Anorg. Allgem. Chem. 292: 97 (1957).

27. K. H. Leiser, Z. Anorg. Allgem. Chem. 305: 255 (1960).

28. J. Bjerrum, G. Swarzenbach, and L. G. Sillen, Stability Constants, Chemical Soc. (London), 1958.

29. L. S. Bryukhanova, I. A. Andreeva, and V. I. Likhtman, Soviet Phys.-Solid State (English Transl.) 3: 2025 (1962).

30. H. Nichols and W. Rostoker, Acta Met. 9: 504 (1961).

31. K. B. Yatsimirski and V. P. Vasilev, Instability Constants of Complex Compounds, Pergamon Press (London), 1960.

32. S. Ahrland, J. Chatt, and N. R. Davies, Quart. Rev. (London) 12: 265 (1958).

33. C. K. Jorgensen, Inorganic Complexes, Academic Press (New York), 1963, p. 52.

34. J. J. Gilman, C. Knudsen, and W. P. Walsh, J. Appl. Phys. 29: 601 (1958).

35. L. Pauling, The Nature of the Chemical Bond, 2nd edition, Cornell Press (Ithaca), 1940, p. 73.

Chapter 31

Role of Interfacially Active Metals in the Apparent Adherence of Nickel to Sapphire

Willard H. Sutton and Earl Feingold

Space Science Laboratory, General Electric Company
King of Prussia, Pennsylvania

The effects of 1 at. % of interfacially active metals (chromium, titanium, and zirconium) on the wetting and adherence of nickel to sapphire (α-Al_2O_3) were investigated. Sessile-drop tests were conducted in order to determine the contact angle and the interfacial energy. The apparent adherence of the solidified nickel drops was determined by a simple shear test. The structure of the nickel and the Ni—Al_2O_3 interface was characterized by optical microscopy, X-ray and electron diffraction, X-ray fluorescence analysis, electron probe X-ray microanalysis, and microhardness measurements. Each active metal had a distinct and reproducible effect on the thickness, the composition, and the microstructure of the interfacial zone, which is discussed in detail. The pure nickel and the chromium-doped specimens exhibited the greatest apparent adherence to the sapphire, whereas the titanium-doped and zirconium-doped specimens chemically degraded (weakened) the surface of the sapphire so that premature failure occurred. A model was developed to explain the effects of the various additives upon bond strength and upon failure modes.

INTRODUCTION

The properties of most metal—ceramic systems depend to a large extent on the wetting and bonding between the component phases. Thus, the properties of many practical systems, such as the ceramic-to-metal seals in electron tubes, cermets, enamels and protective coatings on metals, and structural composites, are governed by the nature and type of bonds at the grain boundaries and by the interfacial zones between the metallic and ceramic constituents.

Although there are a wide variety of techniques used to join metal and ceramic members [1−4], many of the fundamental principles are not well understood. This arises primarily from the fact that such systems are highly complex, and that trace impurities can have profound effects. As a result, it has been extremely difficult to design and perform critical experiments which clearly elucidate

TABLE I

Composition of Ni and Al_2O_3 Specimens Used in this Investigation

Specimen	Major impurity or additive (ppm)		Elements sought but not detected
Base metal, high-purity nickel*	Al	< 0.10	Ag, Na, Sn, Zn, Sb, As,
	Ca	< 0.10	Ba, Be, Bi, Cd, Cb, Ga,
	Cr	< 0.50	Ge, Au, Mo, Pt, K, Sr,
	Co	< 0.50	Te, W, V
	Cu	< 0.10	
	Fe	< 0.10	
	Pb	< 0.50	
	Mg	< 0.90	
	Mn	< 0.50	
	Si	< 0.40	
	Ti	< 0.50	
	B	< 0.20	
	Zr	< 0.50	
	Total	< 4.90	
Sapphire (α-Al_2O_3) substrate†	Mg	20	Ca, Cr, Co, Cu, Cd, Fe, Pb,
	Si	80	Mn, Ni, Ag, Na, Sn, Ti, Zn,
	Total	100	Sb, As, Ba, Be, Bi, B, Cb,
			Ga, Ge, Au, Mo, P, Pt,
			K, Sr, Te, W, V, Zr
Doped nickel‡ (desired concentration)			
High-purity nickel + 1 at. % chromium	Cr = 1.35 at. % (12,000 ppm)		
High-purity nickel + 1 at. % titanium	Ti = 0.99 at.% (8090 ppm)		
High-purity nickel + 1 at. % zirconium	Zr = 1.00 at. % (15,500 ppm)		

*Triply zone-refined, Materials Research Corporation.
†Linde Crystal Products Division, Union Carbide Corporation.
‡Additives uniformly diluted by a zone levelling process; chromium was 99.9% pure, titanium and zirconium were 99.95% pure, Materials Research Corporation.

or define the critical parameters. Furthermore, the physical properties are dependent upon structures and internal stresses at atomic, microscopic, and macroscopic levels.

The adherence of nickel to sapphire (single-crystal α-Al_2O_3) will be considered on the basis of adherence (interfacial bonding) on an atomic or microscopic level and adherence on a macroscopic level. In order to develop a model for elucidating the mechanism of adherence, an experimental investigation was undertaken whereby the factors affecting both the wetting and bonding* were examined. In this investigation, precautions were taken to utilize very pure materials, to carefully clean them, and to heat them in vacuo to the melting point of nickel until an equilibrium contact angle† was achieved. Shear tests were performed on the solidified drops in order to determine the relative adherence of the drops to the sapphire. The effects of various additives (indium, aluminum, copper, chromium, titanium, and zirconium) in concentrations of about 1 at. % in nickel were investigated and reported previously [5]. It was found that chromium, titanium, and zirconium had the greatest effects on the wetting (contact angle θ), on the interfacial energy γ_{SL}, and on the bond (adhesive) strength between the nickel and the sapphire. These results are in agreement with those of earlier studies, where the effects of chromium and titanium on the wetting of nickel to Al_2O_3 were investigated [6-8]. Since the chromium, titanium, and zirconium additives caused the greatest changes in the structure of the Ni—Al_2O_3 interface, they were considered as being interfacially active. The nature of these changes (e.g., chemical reactions and diffusion) and their effects on the adherence of nickel to Al_2O_3 are the subject of this paper.

EXPERIMENTAL PROCEDURE

Specimen Preparation

Sessile-drop specimens were prepared in an apparatus which has been fully described elsewhere [5]. These drops consisted of

---

*Although wetting implies bonding (or interaction of electrons) on an atomic scale, in this paper wetting and bonding are treated as two different phenomena on a micro-macroscopic scale. The term wetting is used when a liquid phase is present, and the term bonding (adherence) is used when all phases are solid.

†The contact angle θ for a sessile drop is measured in the liquid drop and lies in a plane which passes through the axis of symmetry. It is the angle between the substrate surface and the tangent to the surface of the drop at the point where the drop contacts the substrate. Wetting is assumed to occur for contact angles less than 90°.

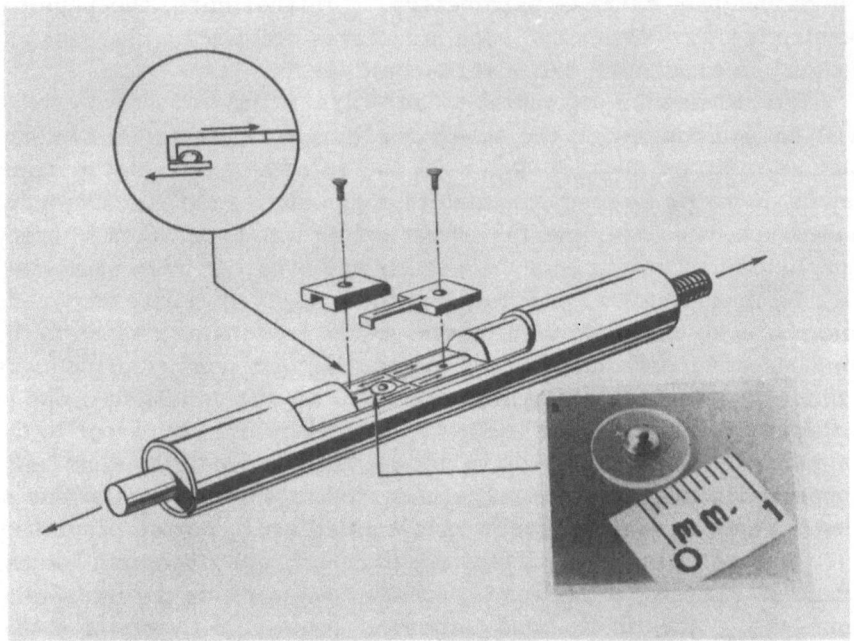

Fig. 1. Schematic diagram of device used for determining sessile-drop shear strength.
Inset: Photograph of solidified metal drop on α-Al_2O_3 plaque.

high-purity nickel containing 1 at.% of chromium, titanium, or
zirconium, which were melted on single-crystal substrates of
α - Al_2O_3. Chemical analyses of the samples are shown in Table I.
The impurity level of chromium, titanium, and zirconium in the
pure nickel and sapphire is well below that of the doped nickel
specimens. All specimens were heated to $1500 \pm 20°C$ for a
period of 2 min* under a pressure of 5×10^{-6} torr or lower. After
the 2-min period, the power was shut off, and the specimens were
allowed to cool rapidly to room temperature. Specimens were
next placed in a special jig (shown in Fig. 1) for shearing off the
nickel drops. Shear tests† were conducted in an Instron testing
machine [5]. Duplicate specimens were prepared under similar

*During this period, photographs were taken from which contact angles were determined.

†The shear tests provided a measure of the apparent shear strength between the nickel
drop and the sapphire plaque. The word apparent is used because failure occurred, in
some cases, through the sapphire (cohesive failure).

conditions for examination by X-ray, optical, and electron microscopic techniques.

Optical Microscopy

Sessile-drop specimens were individually mounted in bakelite
metallographic mounts using conventional metallographic hot-
pressing methods.   Each specimen thus mounted was then ground
in order to expose the metal–sapphire interface and adjacent
regions; the mode of preparation is shown schematically in Fig. 2.
The sectioned specimens were subsequently polished and electro-
etched.   Polishing techniques were developed which resulted in
minimal sapphire pull-out.   Electroetching was done in a 1 vol. %
solution of HCl.

Electron Microscopy

The specimens were purposely somewhat over-etched in order
to enhance detail in the interfacial region and to provide a step
discontinuity at the interface, as shown in Fig. 2(B). The step so.

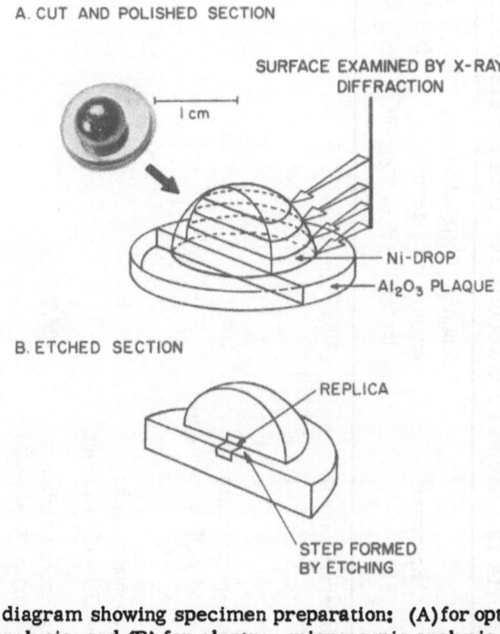

Fig. 2. Schematic diagram showing specimen preparation: (A) for optical microscopy and
X-ray analysis, and (B) for electron microscopic analysis (replica).

TABLE II

Results of Sessile-Drop Tests and Shear Tests Performed on Ni–Al$_2$O$_3$ Specimens
(All data for $1500 \pm 20°C$ and 5×10^{-6} torr except for shear tests, which were conducted at room temperature.)

Specimen	Contact angle (θ, degrees)	Nickel surface tension (γ_{LV}, ergs/cm^2)	Ni–Al$_2$O$_3$ interfacial energy (γ_{SL}, ergs/cm^2)	Work of adhesion (W_{AD}, ergs/cm^2)	Ni–Al$_2$O$_3$ apparent shear strength (psi)	Type of fracture
High-purity nickel	100.7	1770	1290	1390	17,800	Essentially interfacial separation
High-purity nickel +1.35 at,% chromium	107.7	1670	1470	1170	20,000+*	Essentially interfacial separation
High-purity nickel +0.99 at,% titanium	94.5	1770	1090	1640	4,800	Fracture in Al$_2$O$_3$ substrate
High-purity nickel +1.00 at,% zirconium	136.2	1770	2220	500	0†	Fracture in Al$_2$O$_3$ substrate

*Exceeded limit of test device.
†No shear strength; plaque cracked under drop on cooling.

produced enabled an examination to be made of three important regions in the specimens as indicated from top to bottom: (1) a transverse metal-drop section, (2) a lateral surface of the sapphire underlying the metal drop, and (3) a transverse sapphire section. Platinum-shadowed carbon replicas were prepared from primary cellulose acetate replicas of the surfaces of interest [see Fig. 2(B)]. The replicas were examined in an Hitachi HU-11 electron microscope and electron photomicrographs were produced.

Electron Probe X-Ray Microanalysis

Electron probe X-ray microanalyses were made on specimen sections which had been polished and etched. Analyses were made in a Cambridge electron probe instrument.

X-Ray Diffraction

Precision lattice parameters of the metal drops were obtained from back-reflection diffraction photographs of surfaces (parallel to the surface of the sapphire plaque) which were generated by lightly filing the metal drops. The graphical method of Taylor, Sinclair, Nelson, and Riley was employed in the determination of lattice parameters. Powder X-ray diffraction (Debye–Scherrer) photographs were made of the filings described in the preceding section. The crystallographic orientations of the single-crystal sapphire plaques were determined through the use of a diffractometer mounted single-crystal orienter.

EXPERIMENTAL RESULTS

Sessile-Drop and Shear Tests

The results from the sessile-drop tests (contact angle, nickel surface tension, Ni–Al_2O_3 interfacial energy, and work of adhesion at 1500°C) are summarized in Table II along with the results from the shear tests conducted on the solidified nickel drops at room temperature. It can be seen that all the additives affected the contact angle of the pure nickel; 1 at. % zirconium increased the angle to 136°, while 1 at. % titanium lowered θ to just below 95°. Thus, it is evident that small amounts of interfacially active elements* can markedly change the interfacial energy

*In this discussion, interfacially active elements refer to those which concentrate at the Ni(l)–Al_2O_3 (s) boundary and which lower the interfacial energy (γ_{SL}). The zirconium additive is an exception and increases γ_{SL}.

and the wetting, without appreciably altering the surface tension of the nickel (with the exception of the chromium additive). The work of adhesion (W_{AD}) *gives some indication of the adherence of the liquid drop to the substrate, but does not necessarily predict the adherence of the solidified drop. Table II also shows that the additives have very significant effects on the $Ni-Al_2O_3$ shear strengths.

$Ni-Al_2O_3$ Interfaces

High-Purity Nickel – Al_2O_3 Specimens. Observations of the base of the solidified metal sessile drop, which were made by viewing

*The work of adhesion W_{AD} represents the free-energy change for separation of a unit area of interface into liquid and solid surfaces, and can be expressed as the Young-Dupré equation: $W_{AD} = \gamma_{SL} (1 + \cos \theta)$, where γ_{SL} is the solid-liquid interfacial energy and θ is the contact angle.

A.

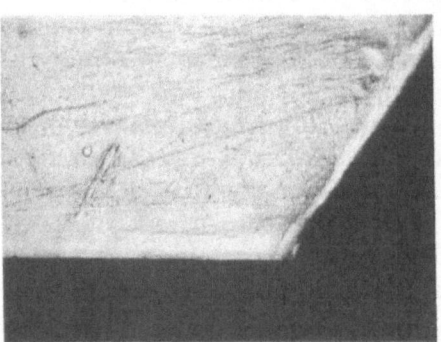

B.

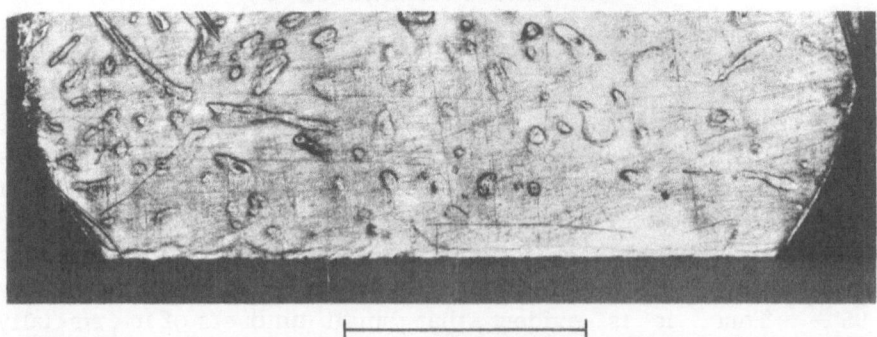

|⟵   1 mm (B)   ⟶|
0.25 mm (A)

Fig. 3. Cross-sectional photographs of a solidified drop of high-purity nickel on an α-Al_2O_3 substrate. Upper photo A is an enlarged portion (4×) of lower photo B.

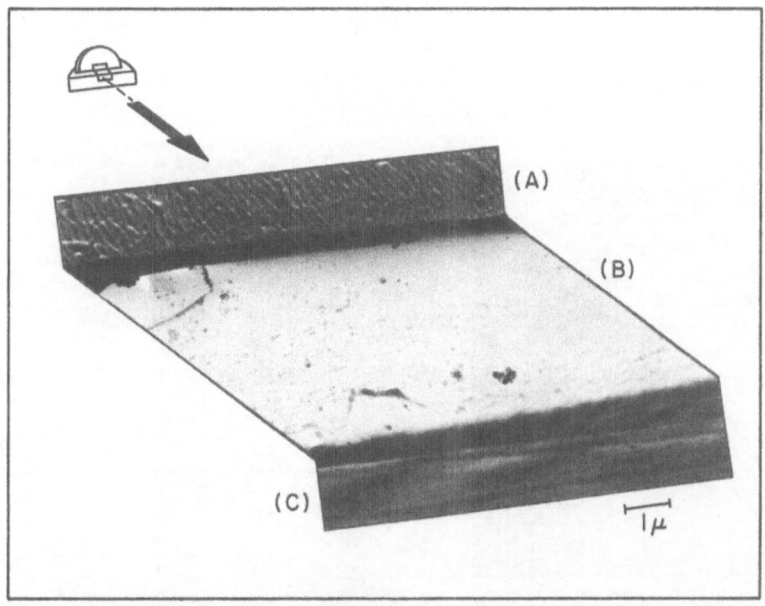

Fig. 4. Electron photomicrograph of replica of: (A) etched section of high-purity nickel drop, (B) exposed surface of the α-Al_2O_3 plaque, and (C) polished section of plaque.

through the transparent sapphire plaque, revealed a mirror-smooth $Ni-Al_2O_3$ interface. Photomicrographs of a polished and etched cross section of the solidified high-purity nickel drop on a sapphire plaque are shown in Fig. 3. The upper photograph is an enlarged portion of the lower one, and reveals in greater detail the $Ni-Al_2O_3-$vapor interface, where a slight chemical reaction (etching of the Al_2O_3 surface) occurred. This indicates that the composition of the surrounding vapor (5×10^{-6} torr) may play a dominant role. No etching could be seen at the $Ni-Al_2O_3$ interface at regions inward from the periphery of the drop. It is suspected that residual oxygen in the vapor was the source of the reaction at the three-phase boundary, where NiO could form and react with the substrate to form $NiAl_2O_4$. However, the extent of the reaction was too small to be detected by X-ray diffraction methods. Figure 3(B) also indicates that the interior of the high-purity nickel drop contains more than one phase, although X-ray analysis did not detect any new phases. Spectroscopic analysis of a solidified high-purity nickel drop revealed that the impurity content was below 5 ppm. Thus, it is suspected that the "second phase" is due to columnar solidification and orientation of the nickel during cooling. It was also ap-

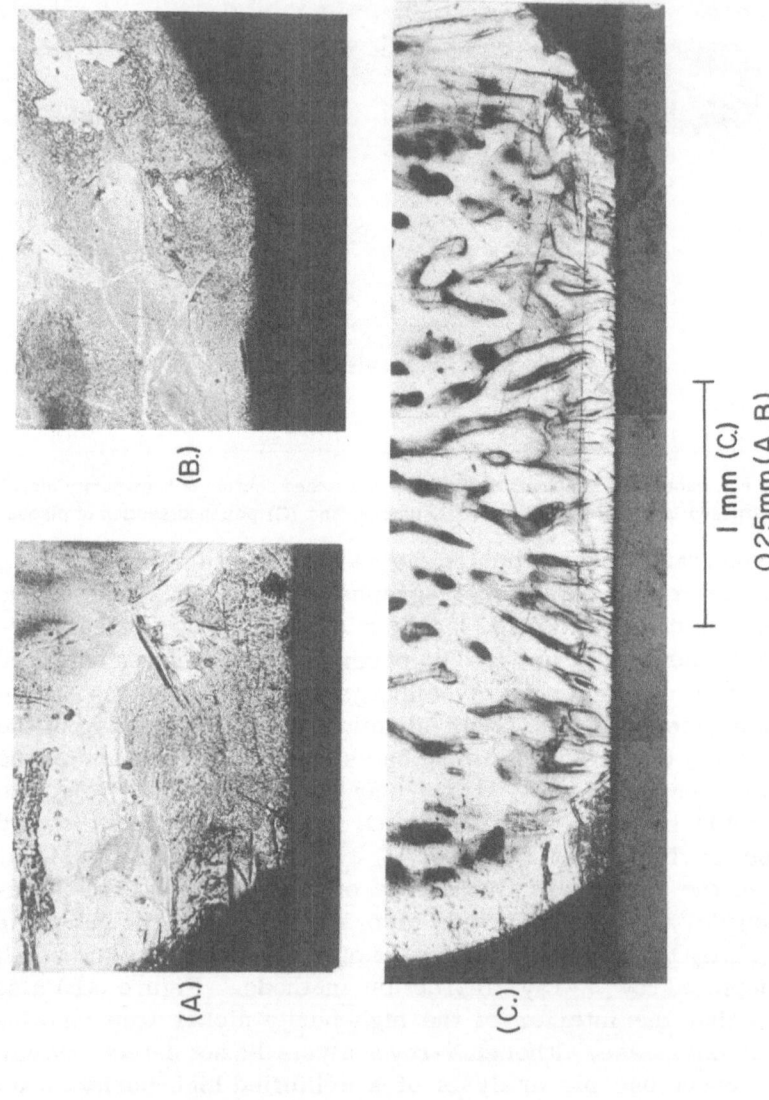

Fig. 5. Photomicrographs of the cross section of a solidified sessile drop of nickel–chromium (1 at. % chromium) on an α-Al_2O_3 substrate. Photos A and B are enlarged portions (4×) of photo C.

parent from visual observations and from the cooling curve, which was recorded automatically, that the nickel drop was supercooled before it solidified. This was evidenced through a sharp and sudden increase in thermal emittance (due to the latent heat of solidification) at temperatures well below the melting point of nickel (1455°C).

Figure 4 is an electron photomicrograph of a replica made in the interfacial region of the sectioned nickel–sapphire specimen which had been polished and etched. The upper portion (A) of this figure corresponds to a vertical section through the metal drop; the light band (B) is the horizontal surface of the sapphire plaque underlying the metal drop; the lower portion (C) is a vertical section through the sapphire. It can be seen that there was a slight etching or erosion of the sapphire in the region directly under the high-purity nickel.

Ni-Cr – Al₂O₃ Specimens. In comparison to the high-purity nickel specimens, chemical reaction at the Ni–Al₂O₃ interface was more pronounced as shown in Fig. 5. Chemical reaction at the three-phase boundary, (Ni-Cr–Al₂O₃–vapor), is also more pronounced as shown in Fig. 5 (A) and (B). The columnar structure of the etched

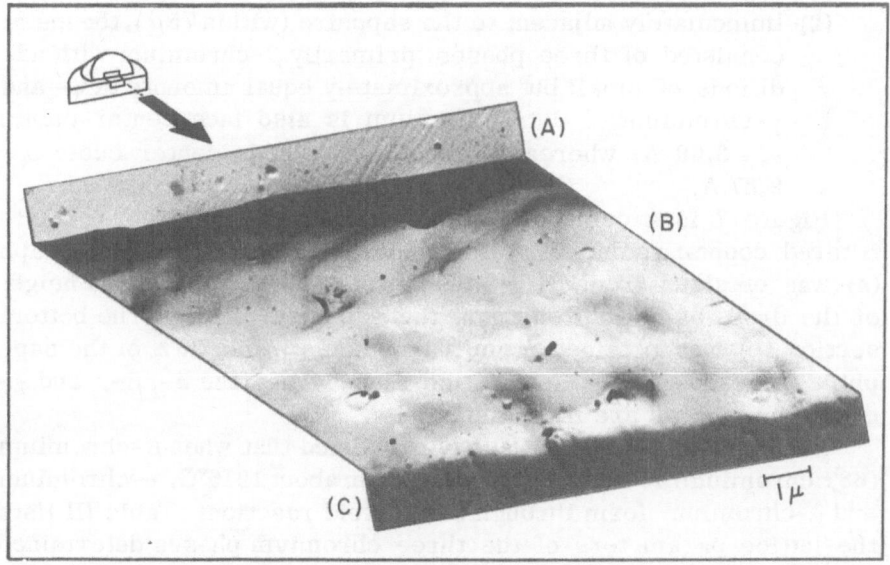

Fig. 6. Electron photomicrograph of replica of: (A) etched section of nickel–chromium drop, (B) exposed interfacial surface of α-Al₂O₃ plaque, and (C) polished section of plaque.

nickel [Fig. 5(C)] is again probably due to the nucleation–solidification process which occurred when the molten nickel was cooled. It appears that nucleation occurred first at the Ni–Al$_2$O$_3$ interface where the chromium was concentrated, and then solidification of the nickel proceeded from this interface towards the top of the molten drop.

The extent of chemical reaction at the Ni–Al$_2$O$_3$ interface is shown more clearly in Fig. 6. The light diagonal band (B) representing the sapphire surface underlying the nickel–chromium drop shows that the interface is no longer smooth and that the sapphire plaque had reacted with the molten nickel–chromium drop. X-ray diffraction analysis of the material from the interface showed that a new crystalline phase had formed. X-ray fluorescence and electron probe X-ray microanalysis showed a very much enriched concentration of chromium at the interface. X-ray diffraction analysis of filings produced from specific levels in the metal drop revealed the following information:

(1) At $\frac{9}{10}$ of the height of the drop (as measured from the surface of the sapphire plaque), the metallic phase was essentially pure nickel (γ-chromium).* It was face-center-cubic with a lattice parameter $a_0 = 3.51$ A ($a_0 = 3.524$ A for pure nickel).

(2) Immediately adjacent to the sapphire (within 75μ), the metal consisted of three phases, primarily β-chromium with additions of small but approximately equal amounts of α- and γ-chromium. Beta-chromium is also face-center-cubic, $a_0 = 3.69$ A, whereas α-chromium is body-center-cubic $a_0 = 2.87$ A.

Figure 7 is a composite X-ray diffraction photograph (nickel-filtered copper radiation). The upper section of this photograph (a) was obtained from filings taken at the level of $\frac{9}{10}$ of the height of the drop, as measured from the sapphire plaque. The bottom section (b) was obtained from filings taken within 75μ of the sapphire surface. The diffraction lines from the α-, β-, and γ-chromium phases are indicated.

Abrahamson and Grant [9] have reported that when β-chromium (68% chromium) is rapidly quenched from about 1215°C, α-chromium and γ-chromium form through a eutectoid reaction. Table III lists the lattice parameters of the three chromium phases determined

---

*In this paper, γ-chromium refers to the γ- (nickel-rich) solid-solution field in the chromium–nickel phase diagram [29].

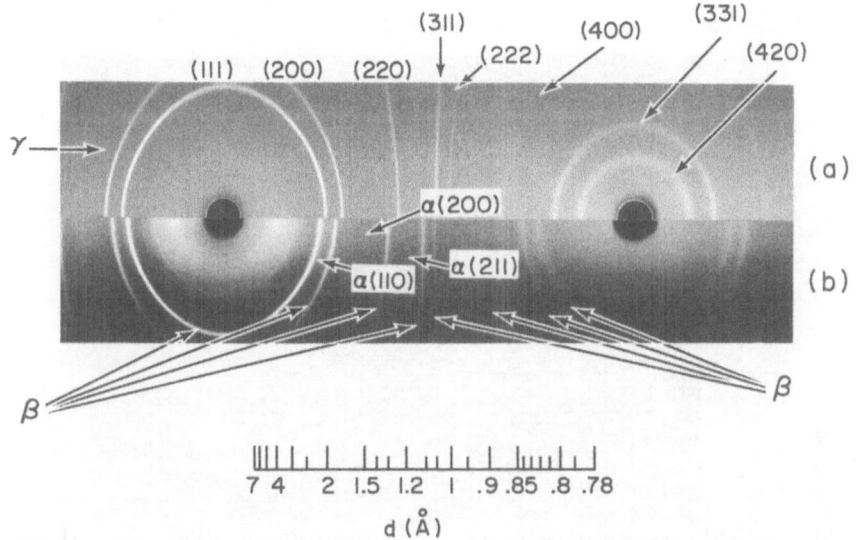

Fig. 7. X-ray diffraction photographs (nickel-filtered copper radiation) obtained from solidified nickel–chromium sessile drop. (a) From near top of drop and (b) within 75 μ of sapphire surface.

in this study as compared with the values given by Abrahamson and Grant [9]. These findings provide further evidence that chromium had indeed concentrated in the region of the sapphire–metal interface. (Bechtoldt and Vacher [10] have more recently determined that the maximum solubility of nickel in chromium is 38 at. %.)

Ni-Ti – Al₂O₃ Specimens. This system, like that of Ni-Cr–Al₂O₃, showed that chemical attack had occurred in the region underlying the metal drop. A thin dark blue film was observed to have been formed between the metal drop and the sapphire plaque. Figure 8 presents optical photomicrographs of the sectioned and etched drop.

The upper portion (A) of the electron photomicrograph shown in Fig. 9 represents the sapphire surface underlying the metal drop. The bottom (B) is a section through the sapphire plaque. Evidence of rather severe roughening of the surface of the sapphire substrate by the molten nickel–titanium drop is shown in this figure. Electron beam X-ray microanalysis also showed that titanium had concentrated in the vicinity of the interface, particularly within 5 μ of the sapphire. This is the region in which the thin blue film previously mentioned was observed.

(A)

(B)

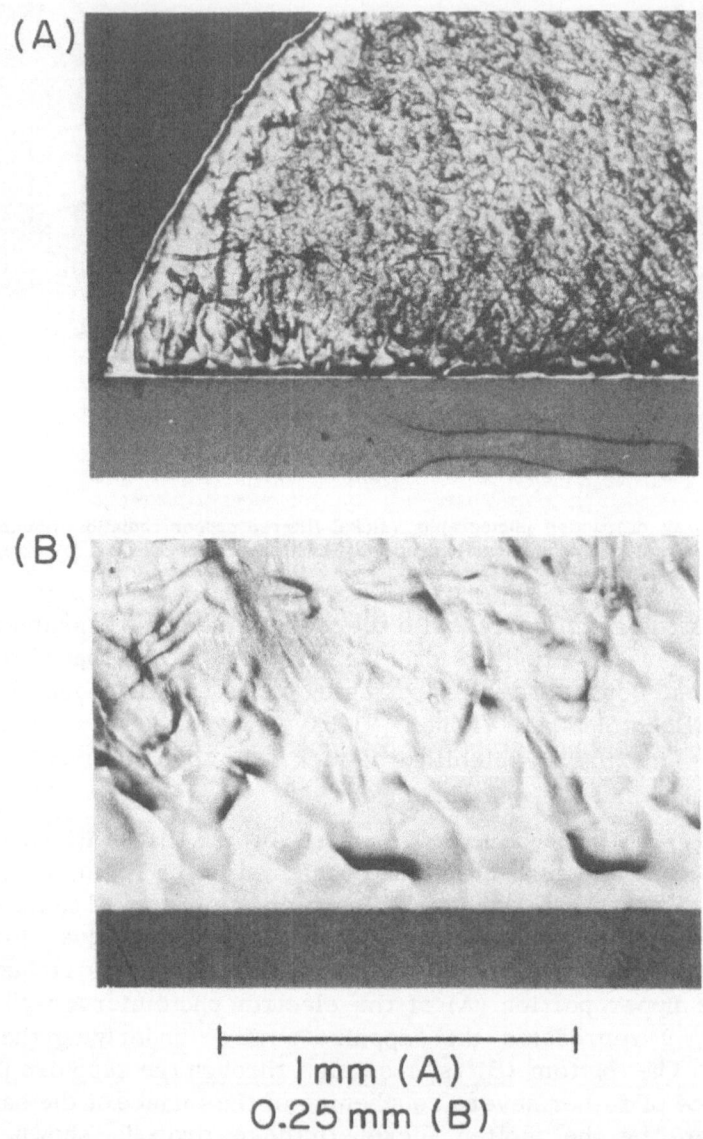

├─────────────────────────────┤
1 mm (A)
0.25 mm (B)

Fig. 8. Photomicrographs of the cross section of a solidified sessile drop of nickel–titanium (1 at.% titanium) on an α-Al_2O_3 substrate. (B) is an enlarged portion (4×) of (A).

TABLE III

Comparison of Chromium Lattice Parameters

Phase	Value, this work (A)	Value, [9] (A)
Alpha	2.87	2.88
Beta	3.69	3.68
Gamma	3.51	3.52

The thin dark blue interfacial film was carefully removed in HCl from a portion of the solidified drop from which the sapphire had been broken away. This film was examined by transmission electron diffraction and by X-ray diffraction. Figure 10 shows an electron photomicrograph of a self-supporting portion of this film.

Particles with an average area of about $0.25 \mu^2$ in a clear (void of contrast detail) matrix can be observed. Selected area electron diffraction revealed that the particles were polycrystalline whereas the clear portions were noncrystalline. Unlike the results reported by Armstrong et al. [7] and by Clarke [8], no evidence of the formation of a-Ti_2O_3 was found. The d-spacings as obtained from the resulting diffraction patterns of the "new" phase which had formed in the interface region best compared with the standard (ASTM) diffraction data for Ti_3O_5. However, the intensities of the experimental diffraction lines differed considerably from the standard (ASTM) Ti_3O_5 data. It was suspected that the new oxide phase possessed a defect structure.

Ni-Zr – Al_2O_3 Specimens. This system exhibited the highest degree of sapphire degradation. The series of optical photomicrographs of Fig. 11 reveals several interesting features. A new phase is observed to have formed in the interface region, and is a porous structure about 45μ thick. Approximately $\frac{2}{3}$ of this structure extended into the sapphire. The sapphire underlying the new phase was found to be modified, in that it contained many (approximately $10^{11}/cm^3$) small spherical cavities which were approximately 0.1–3μ in diameter. The central band (B) in the electron photomicrograph of Fig. 12 shows the great extent to which the sapphire surface beneath

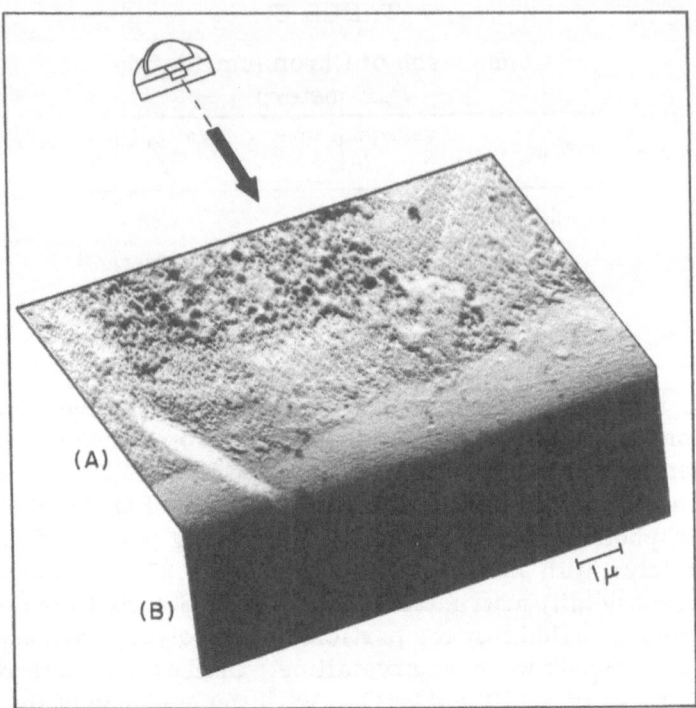

Fig. 9. Electron photomicrograph of replica of: (A) exposed interfacial surface of a-Al_2O_3 plaque which had reacted with nickel–titanium drop and (B) polished section of plaque.

the drop had deteriorated. Sectional views of the new phase in the nickel and of the sapphire plaque are given in Fig. 12 (A) and (C), respectively.

X-ray diffraction analysis of the new phase proved it to be ZrO_2. This was expected, since thermodynamically the reduction of Al_2O_3 by zirconium is favored. The reaction [11] is

$$2Al_2O_3 + 3Zr \xrightarrow{\text{1800° K}} 4Al + 3ZrO_2 - \Delta F \tag{1}$$

where the free energy of reaction ΔF equals approximately 18 kcal. The mechanism for mass transport of Al^{+3} and O^{-2} from below the reaction region and the subsequent formation of spherical cavities in the Al_2O_3 is probably due to a type of Kirkendall effect [12].

Effects of the Additives on the Structure of the $Ni - Al_2O_3$ Interface

Electron Probe X-Ray Microanalysis. X-ray fluorescence and electron probe X-ray microanalysis showed that the concentration of each additive had become enhanced in the $Ni-Al_2O_3$ interfacial zone (see Fig. 13). In addition to the enhanced interfacial concentration, which was as much as 50 to 70 times greater than in the interior portions of the nickel drop, the additives reacted with and diffused into the Al_2O_3 substrate. Figure 13(C) shows that chromium had diffused about 5 μ into the Al_2O_3, while titanium [Fig. 13(B)] and zirconium [Fig. 13(A)] diffused as much as 75 μ into the Al_2O_3. (The ZrO_2 layer that formed extended into both the nickel and substrate.) However, zirconium (as ZrO_2 or as Zr in the Al_2O_3 lattice) extended at least 150 μ beyond the base of the solidified nickel drop. The crosshatched portion of the curve [Fig. 13(A)] represents the thickness of the sponge-like ZrO_2 layer (shown in Figs. 11 and 12), while the portion to the right in Fig. 13(A) represents the modified

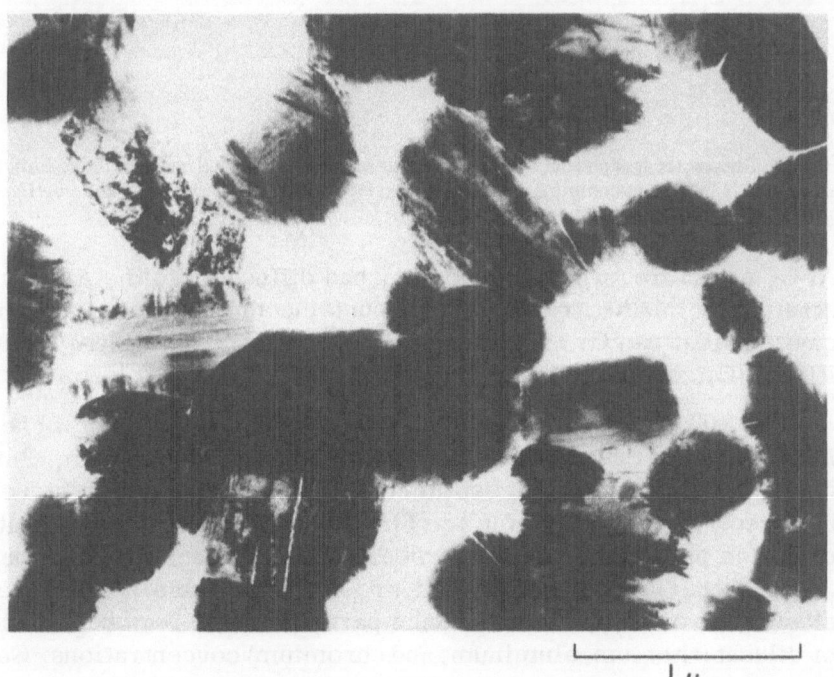

$| \mu$

Fig. 10. Transmission electron photomicrograph of the blue titanium–oxide layer which was removed from interfacial region between nickel–titanium–sapphire.

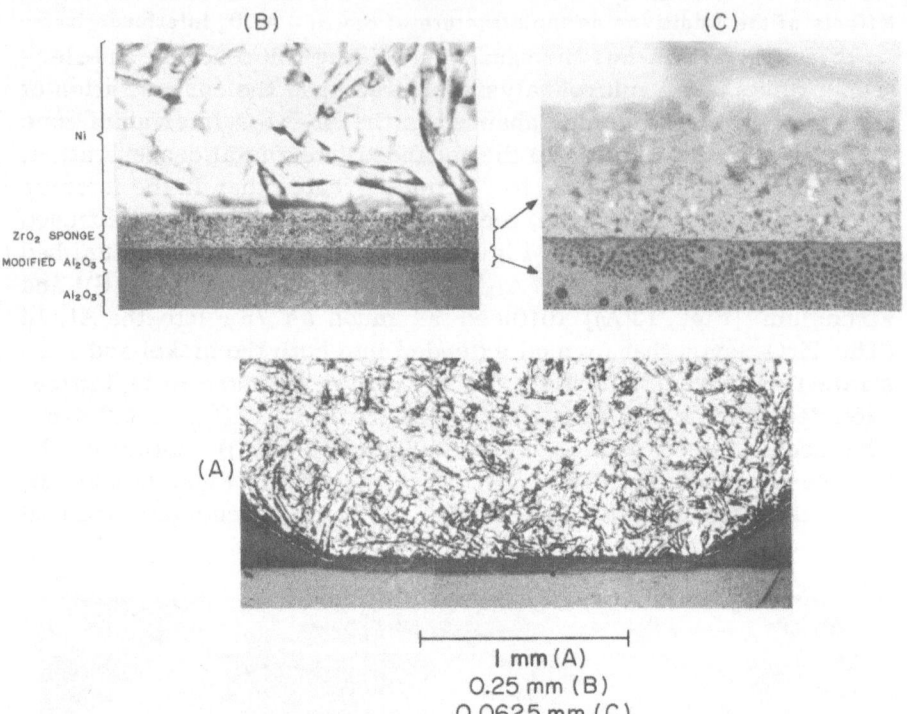

Fig. 11. Photomicrographs of the cross section of a solidified sessile drop of nickel–zirconium (1 at.% zirconium) on an α-Al_2O_3 substrate. (B) is an enlarged portion (4×) of (A). (C) is an enlarged portion (4×) of (B).

Al_2O_3 structure, where zirconium had diffused into the Al_2O_3 substrate. It is this region which contains the many fine spherical cavities that can be clearly seen at the top of the sapphire layer in Fig. 11(C).

The Nickel Lattice Parameter. The lattice parameter of nickel a_0 is sensitive to the addition of foreign atoms such as titanium, aluminum, and chromium. The solid solubility of zirconium in nickel is believed to be lower than 0.3 at.% [13,14], and no detectable alteration of lattice parameter has been reported. For diluent additions below 10 at.%, the lattice parameters are shown in Table IV. Figure 14 shows the variations in the lattice parameters of α-nickel in terms of diluent (titanium, aluminum, and chromium) concentrations. Comparisons are made in Table V of the concentrations of titanium and chromium as obtained from the metal specimens used in this study through X-ray lattice parameter measurements and by spectroscopic analysis. The agreement between X-ray and spectroscopic analysis is surprisingly good considering that, in the X-ray analysis, it was assumed that the nickel specimens contained only titanium and chromium additions. Figure 15 presents lattice parameter profiles of the α-nickel phases of the solidified metal sessile drops. The right-hand ordinate of these plots represents the region near the top of the metal drop, whereas the left ordinate indicates the position of the surface of the sapphire substrate. The arrows on the right ordinate show the lattice parameter values of the various specimens prior to the sessile-drop experiments.

The shapes of the resulting lattice parameter profile curves in Fig. 15 are related to certain uncontrollable factors, such as concentration eddies (within the drops immediately prior to solidification) and solidification stresses. However, certain trends can be recognized. The outer regions of the drop, which cool much more rapidly than the interior, provide what is believed to be the most useful and reliable information.

Nickel–titanium: The decrease in lattice parameter, -0.0009 A, at the top of the drop corresponds to a concentration change from an initial value of 1.2 to 0.9 at.%titanium. A simple calculation can

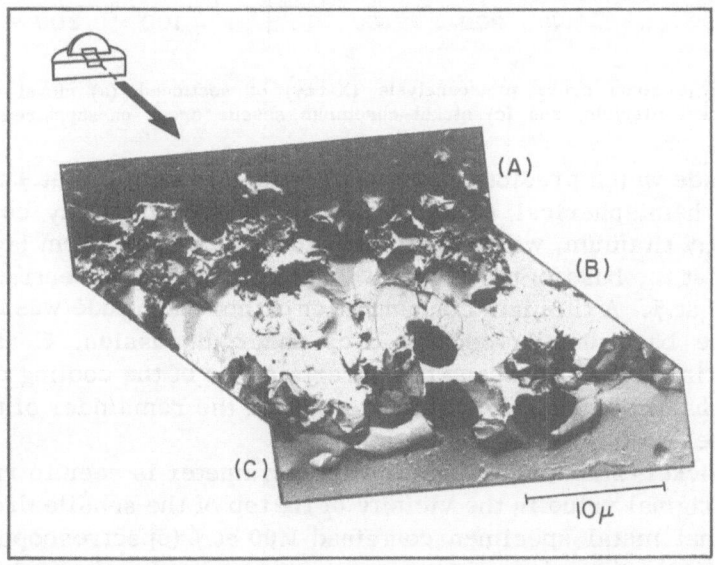

Fig. 12. Electron photomicrograph of replica of: (A) etched section of nickel–zirconium drop, (B) exposed interfacial surface of α-Al$_2$O$_3$ plaque, and (C) polished section of plaque.

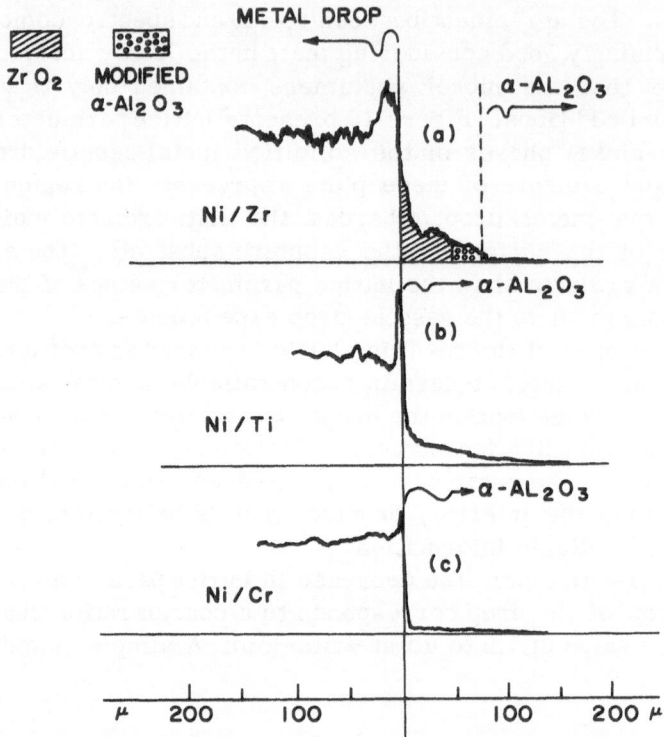

Fig. 13. Electron probe microanalysis (X-ray) of sectioned: (a) nickel—zirconium, (b) nickel—titanium, and (c) nickel—chromium sessile drops on sapphire substrates.

be made which predicts that a uniform decrease of 0.3 at.% titanium in a hemispherical drop of 3-mm radius, originally containing 1.2 at.% titanium, will be sufficient to result in a uniform layer 10 μ thick at the base of the drop in which the titanium concentration will be 60 at.%. A titanium concentration of this magnitude was detected in the base of the sessile drop under discussion. Considering experimental errors and the uncertainities of the cooling process, the changes in lattice parameters over the remainder of the drop are believed to be insignificant.

Nickel—zirconium: The lattice parameter is seen to rise over the original value in the vicinity of the top of the sessile drop. The original metal specimen contained 1.00 at.% (spectroscopic determination) zirconium. As discussed previously, the reduction of Al_2O_3 by zirconium is favorable under the conditions of the sessile-drop experiment. Since the solid solubility of zirconium in nickel

TABLE IV

Effect of Additives on the Changes in the Lattice Parameter of Ni

Diluent	Change in a_0 (A/at. % diluent)
Titanium	0.0035
Aluminum	0.0017+
Chromium	0.0010

is less than 0.3 at.% [14], zirconium was readily available for reaction at the sapphire interface. Four atoms of aluminum are released into the metal drop for each three atoms of zirconium which react with Al_2O_3 to form ZrO_2. Therefore, an excess of 1.05 at.% zirconium in the system could result in an overall aluminum concentration in the drop of 1.4 at.%. Figure 14 shows that aluminum dilates the a-nickel lattice. Thus, the migration of aluminum atoms from the sapphire plaque into the nickel–zirconum drop can explain the noted increase in a_0.

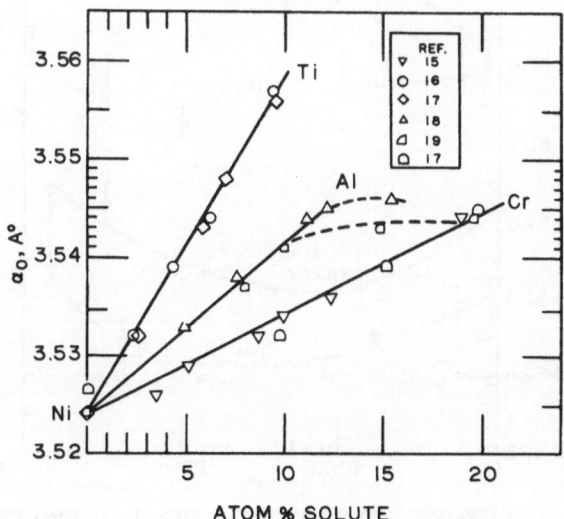

Fig. 14. Effect of concentration of titanium, aluminum, and chromium on the lattice parameter of nickel.

TABLE V

Diluent Concentration in Nickel–Chromium and
Nickel–Titanium Specimens

Specimen (diluent)	a_0 (A) (for nickel)	Composition (at. %)	
		X-ray	Spectroscopic
Titanium	3.5285	1.2	0.9
Chromium	3.5273	3.2	1.35

Nickel–chromium: The decrease of a_0 in the vicinity of the top of this sessile drop (compared to that of the untreated specimen) corresponds to a decrease in the chromium concentration of about 1 at.%.   As explained in the nickel–titanium case, a uniform decrease of diluent concentration of this magnitude in the metal drop is sufficient to account for a sixty-fold or greater increase in concentration in the vicinity of the metal–oxide interface.

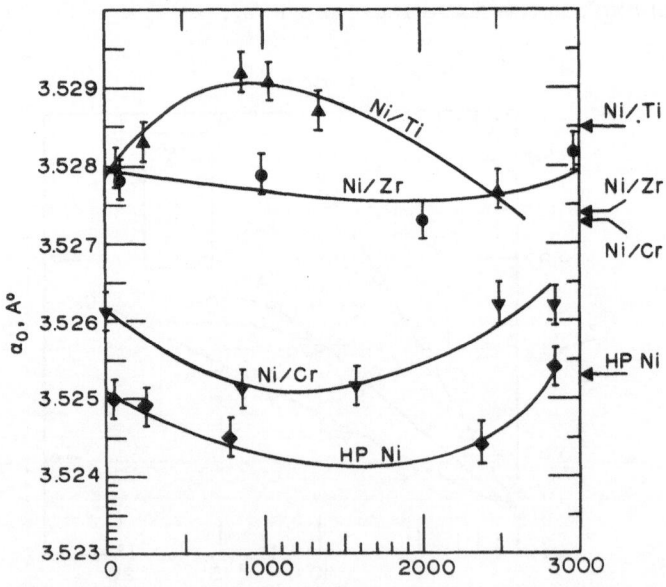

DISTANCE FROM SAPPHIRE SURFACE, MICRONS

Fig. 15. Lattice parameter profile of nickel for various compositions as a function of distance from the sapphire interface.

High-purity nickel: There was no significant alteration of a_0 at the surface of this solidified sessile drop. The X-ray lattice parameters, therefore, suggest further evidence that the Al_2O_3– metal interface provides a driving force for the segregation of the active metals titanium, chromium, and zirconium from molten nickel containing these diluents.

Microhardness. Additions of 1 at.% of zirconium, titanium, or chromium to high-purity nickel were found to reduce the Knoop microhardness (200-g load) values by as much as 17% as shown in Table VI. Microhardness profile data obtained from polished sessile-drop sections are presented in Fig. 16. The average decreases in microhardness values in the metal drop sections as compared to the value for high-purity nickel before treatment (dashed line in Fig. 16) are listed in Table VII.

Quantitative explanations of the data in the metal drop regions presented in Fig. 16 cannot be given since these data were obtained from highly strained (unannealed) metal drops. However, certain qualitative features at the interface regions can be related to X-ray, electron probe, and microscopic findings. The relatively gradual slope of microhardness in the vicinity of the nickel-zironium– sapphire interface is related to the fact that a wide band of ZrO_2 (ZrO_2 microhardness about 1000) exists in this region. The next most gradual slope of hardness versus displacement in the vicinity of the interface is that for the nickel–chromium specimen. This is the region in which a chromium-rich phase, β-chromium, was found to exist. The interfacial microhardness slopes from the nickel–titanium and high-purity nickel specimens are the steepest and indicate that any new phases which are present are concentrated

TABLE VI

Knoop Microhardness Values of Nickel Specimens

Diluent	At. %	Knoop microhardness	Δ Microhardness (%)
High-purity nickel	—	239	—
Zirconium	1.00	237	−0.8
Titanium	0.9	201	−16
Chromium	1.35	191	−17

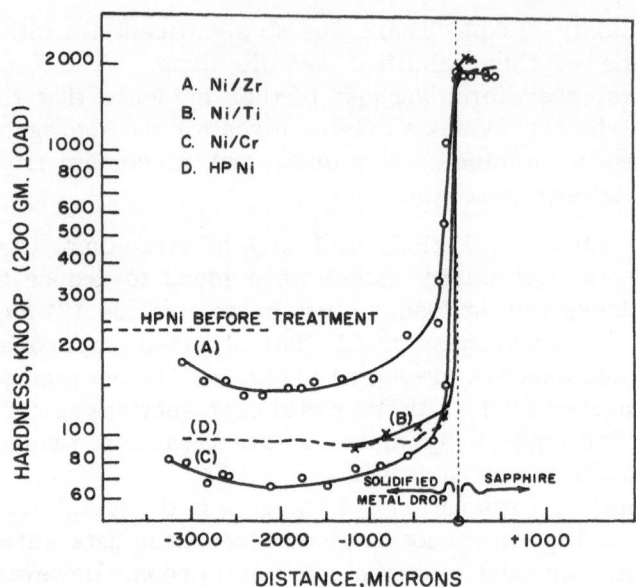

Fig. 16. Microhardness profiles of cross sections through solidified (A) nickel–zirconium, (B) nickel–titanium, (C) nickel–chronium, and (D) high-purity sessile drops on sapphire substrates.

very close to the sapphire substrate. Indeed, the nickel–titanium specimen was found to have an extremely thin oxide film at the sapphire interface.

Crystal Orientation of Sapphire Plaques

The crystal orientations in terms of the angle $\phi_{n,c}$ between the $<00l>_{hex}$ or c-axis direction of the hexagonal unit cell of α-Al_2O_3

TABLE VII

Decrease in Microhardness of Nickel Specimens After Sessile-Drop Test

Specimen	Average decrease in microhardness (%)
Nickel – zirconium	33
High-purity nickel	58
Nickel – titanium	58
Nickel – chromium	71

and the direction of the normal to the flat plaque surfaces were measured by X-ray diffraction methods. (See Fig. 17.) Crystal orientations for each of the plaques utilized in the characterization studies are listed in Table VIII. Although these data show a fairly wide spread in crystal orientation (55 ± 15°), other plaques with different orientations were found to produce results which were qualitatively consistent with the findings of this report.

DISCUSSION

Since the adherence between metals and oxides depends on numerous interrelated factors, it does not seem likely that any simple or universal theory can be used to explain or predict the magnitude of the apparent adherence for all metal–alloy–oxide combinations. For example, the work of adhesion has been used to provide some guidelines for predicting the magnitude of bond strengths between metals and aluminum oxide [20,21]. However, such predictions are generally based on the adherence between a liquid and solid phase where no chemical interactions occur. For solid–solid systems, the work of adhesion is not a reliable tool for predicting adherence since chemical reactions [20,22,23], internal stresses [24], and processes responsible for nucleation and propagation of fracture [25] are of far greater influence.

Thus, in discussing adherence between solid metals and oxides, at least four factors should be considered: (1) the composition and purity of the phases, (2) the joining history, (3) the geometrical and size effects, and (4) the failure modes. These factors will be discussed as they apply to the Ni–Al$_2$O$_3$ system.

Composition of the Phases

Experimental results show that nickel adherence to sapphire is greatly influenced by small additions of an active metal.

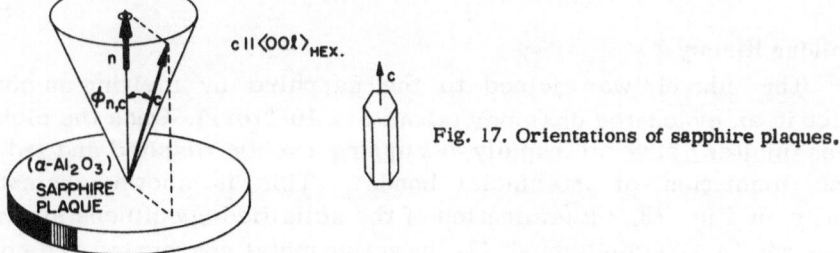

Fig. 17. Orientations of sapphire plaques.

TABLE VIII

Crystal Orientations

Composition of drop	$\phi_{n,c}$ (degrees)
High-purity nickel	64.3
Nickel−chromium	69.9
Nickel−zirconium	40.8
Nickel−titanium	57.8

In this study, considerable care was taken to minimize the effect of unwanted impurities through a careful choice of materials, and also essentially the same thermal history was provided for each $Ni-Al_2O_3$ specimen. Furthermore, since the amount of active metal was added in relatively low concentrations (about 1 part per 100), the residual stresses due to thermal expansion mismatch and to geometrical or size effects probably were of the same order of magnitude for each of the samples. Thus, the diluent species (chromium, titanium, or zirconium) was the major variable investigated. The results clearly indicate that each additive had a distinct and pronounced effect on the wetting (contact angle), on the interfacial energy, on the crystal structure of the nickel, on the composition and microstructure of the interfacial region, on the hardness profile of the nickel and the interface, on the adherence of the nickel to the sapphire, and on the failure modes which caused fracture. Visual examination of the base of the nickel drop through the transparent sapphire substrate clearly revealed different types of texture for each additive. The interface between pure nickel and sapphire was mirror-smooth, while that of the chromium-doped specimen was somewhat dull and roughened. The titanium and zirconium interfaces were very dull and roughened.

Joining History

The nickel was joined to the sapphire by melting on contact in an evacuated chamber (about 5×10^{-6} torr). Once the nickel was molten, several rapidly occurring events resulted and led to the formation of interfacial bonds. This is shown schematically in Fig. 18. Examination of the solidified specimens showed that within a 2-min period: (1) the active metal segregated and concentrated at the interface, (2) a chemical reaction occurred at the Ni(l)–Al$_2$O$_3$ (s) interface, (3) new phases formed in the interfacial regions, and (4) the active metal additive diffused into the sapphire substrate. In Fig. 18(D), the sapphire plaque is shown notched at the edges of the nickel drop in order to indicate the region where the greatest roughening (reaction) occurred.

The driving force for the preferential adsorption (chemisorption) of the solute atoms (chromium, titanium, and zirconium) has been pointed out by Kurkjian and Kingery [6] to be dependent on the free energies of formation of the oxides. Thus, metals, such as chromium and titanium, having more stable oxides than the solvent, nickel, will tend to be preferentially adsorbed at the Ni(l)–Al$_2$O$_3$(s) interface.

The depth of the interfacial zone depends on the nature of the diluent in the nickel. In the case of the chromium-doped nickel, the new phases at the interface were primarily metallic. Since the free energy of oxide formation of chromium is more positive than that of aluminum, no reaction with the Al$_2$O$_3$ substrate was predicted. X-ray analysis revealed no Cr$_2$O$_3$, although some specimens which had been heated to 1500°C for 30 min exhibited a slight green coloration at the interface, which is characteristic of Cr^{+3} ions.

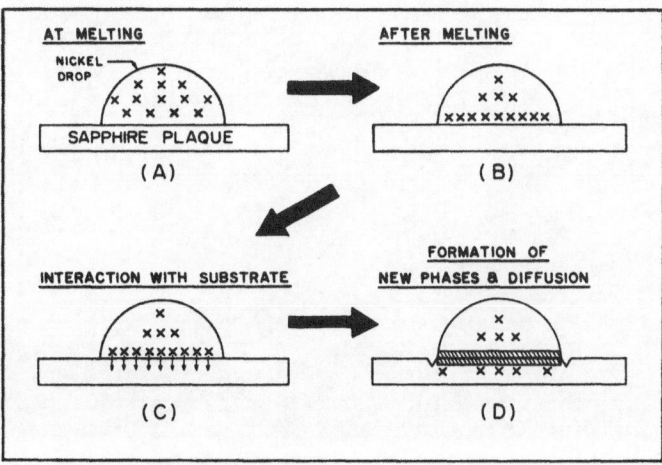

Fig. 18. Events occurring in the bonding of nickel doped with active metal X to sapphire. (A) homogeneous doped nickel drop on sapphire at melting, (B) after melting, active metal segregates and concentrates at interface, (C) active metal interacts with sapphire surface, and (D) new crystalline phases are formed, also active metal diffuses into sapphire plaque.

A.

B.

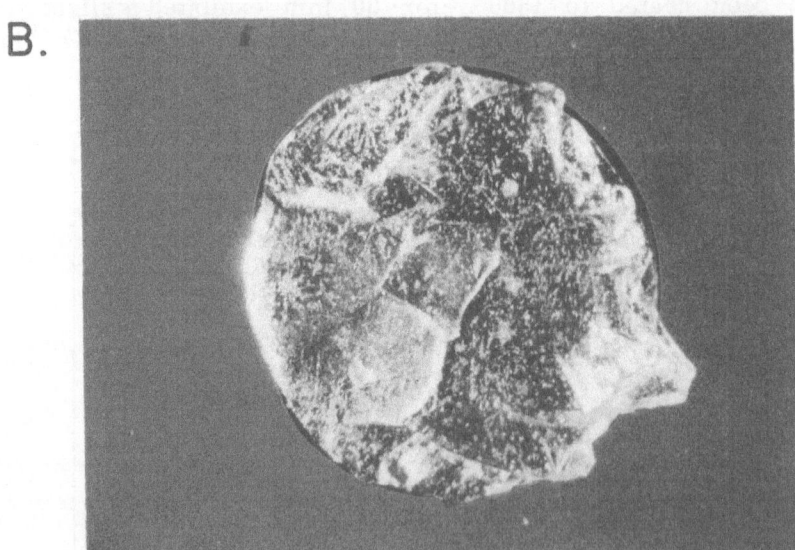

Fig. 19. Photomacrographs of the base of nickel drops sheared off sapphire plaques (15×). (A) Specimen of high-purity nickel showing interfacial separation (note slip lines in nickel). (B) Specimen of high-purity nickel containing titanium showing adherence of Al_2O_3 to the drop; fracture occurred through the sapphire plaque.

The colored layer was extremely thin. No new phases could be detected by X-ray analysis, and it was assumed that the oxidation of chromium to Cr^{+3} resulted from the residual oxygen in the nickel or in the system.

The titanium-doped specimens produced a very thin interfacial layer which appeared to consist of crystallites of nonstoichiometric Ti_3O_5 embedded in an amorphous matrix. Since $\Delta F_{Ti_3O_5}$ is more positive than $\Delta F_{Al_2O_3}$, the source of oxygen here also was probably from the residual gases in the system (available as oxygen, NiO, or H_2O).

Of the oxides of the three additives, only ΔF_{ZrO_2} is more negative than $\Delta F_{Al_2O_3}$, and, hence, zirconium would be expected to rob the sapphire of its oxygen. This was observed. A relatively large amount of ZrO_2 was formed in the interfacial region. Also, unique for the zirconium specimen, numerous microcavities appeared in the sapphire, and spectroscopic analysis indicated a significant increase in the aluminum content of the nickel.

Geometric and Size Effects

The factors influencing the adherence between a metal and oxide depends on the geometry (morphology, mechanical keying and curvature) and on the dimensions of the two phases. On a submicroscopic level, adhesion depends primarily on the number, the type, and the directions of the bonds at the interface. Thus, the adherence of thin metallic films on oxide surfaces [26] and thin oxidized layers on metallic surfaces [27] is primarily dependent on van der Waals interactions and on chemical bonds. However, adherence on a micro-macroscopic level, such as that observed in metal-to-ceramic seals, depends primarily on chemical interactions. These are subsequently modified by bulk phenomena, such as the presence of critical flaws and strain concentrations, and by the behavior of the brittle phase. For example, chemical reactions at the surface of the brittle (oxide) phase are likely to weaken it, so that failure is likely to initiate and propagate through this phase.

Failure Modes

Essentially two types of failure modes were observed—fracture through the interfacial zone and fracture through the brittle phase (Al_2O_3). Examples of these fractures are shown in Fig. 19. The roughening of the sapphire surface in the pure nickel and chromium-doped nickel specimens was the least severe. Con-. sequently, they exhibited interfacial fracture [Fig. 19(A)] and had greatest measured shear strengths. On the other hand, the titanium- and zirconium-doped specimens severely attacked the sapphire surface, which resulted in fracture through this brittle phase [Fig. 19(B)] at lower stress levels.

In summary, the adherence of nickel to Al_2O_3 was found to be determined by the combined effects of the factors discussed previously. The net result is that the measured strength of adherence (shear strength) is governed primarily by two competing effects: (1) the enhancement of interfacial bonding, and, at the same time, (2) the weakening of the sapphire plaque due to surface roughening. These effects are produced by chemical reactions and by diffusion which occur at the interface. Maximum adherence results when chemical reaction and diffusion are sufficient to ensure adequate interfacial bonding, but not so severe that the sapphire is weakened by excessive roughening. These two competing effects are illustrated in Figs. 20 and 21. In Fig. 20, the strength of the sapphire is shown to decrease with increasing surface roughness or with

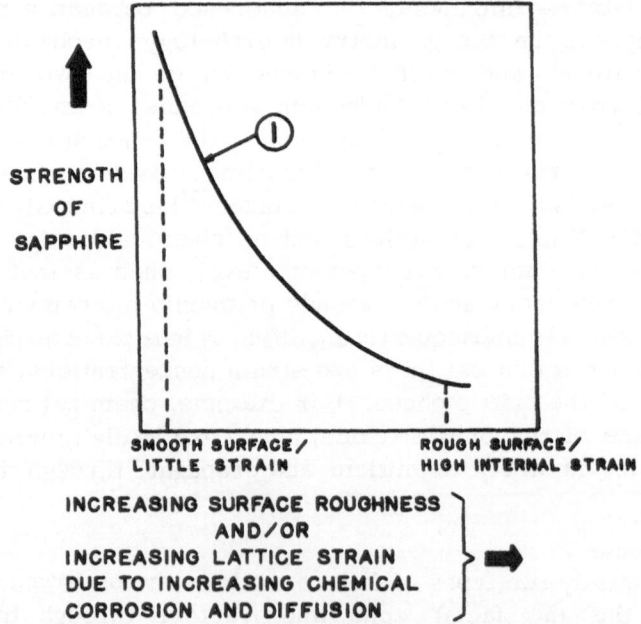

Fig. 20. Effect of chemical reactions and diffusion on weakening sapphire as a function of increasing surface roughening or lattice strain.

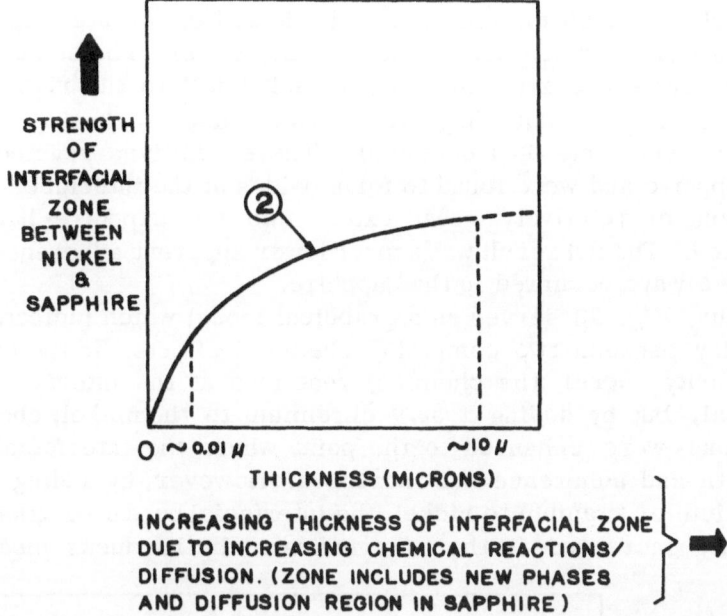

Fig. 21. Effect of chemical reactions and diffusion on increasing interfacial bond strength between nickel and sapphire as a function of increasing thickness of the interfacial zone.

increasing internal stresses or both. In Fig. 21, the interfacial bond strength is shown to increase with increasing thickness of the interfacial zone. This thickness is a measure of the extent of chemical reactions and includes the region where new phases exist as well as the diffusion depth into the sapphire. With little or no chemical interactions, there should be minimal bonding strength between the two phases. However, as the extent of the reactions increases, the interfacial bond strength will increase until a maximum value is achieved. The combined results of these two competing effects are illustrated by the dark curve in Fig. 22. The maximum adherence between nickel and sapphire occurs where curves 1 and 2 intersect. This point indicates the optimum degree of chemical reaction and diffusion. To the left of the maximum point on the curve, the sapphire is not adversely weakened, and failure occurs through the interface. Two data points are included in this region: (1) high-purity nickel where the sapphire interface is smooth and where the thickness of the interfacial zone is probably on the order of a few hundred angstroms, and (2) chromium-doped nickel where the interfacial zone is about

5 μ thick. The chromium specimen showed the greatest apparent shear strength, which was in excess of 20,000 psi. The strengths of the titanium- and zirconium-doped nickel fell to the right of the maximum point, indicating that considerable weakening of the sapphire substrate had occurred. These additives reacted with the sapphire and were found to form oxides at the interface. Also, diffusion of relatively great extent into the sapphire had also occurred. The net result was a much lower apparent adherence, and failure always occurred in the sapphire.

Thus, Fig. 22 serves as a graphical model which indicates the interplay between two competing chemical effects. In the case of high-purity nickel, the chemical reactions at the interface were minimal, but by adding 1 at. % chromium to the nickel, chemical reactions were enhanced to the point where the interfacial bond strength and adherence was increased. However, by adding 1 at. % zirconium or titanium to nickel, the interfacial bond strengths were probably increased further (as indicated by hardness measure-

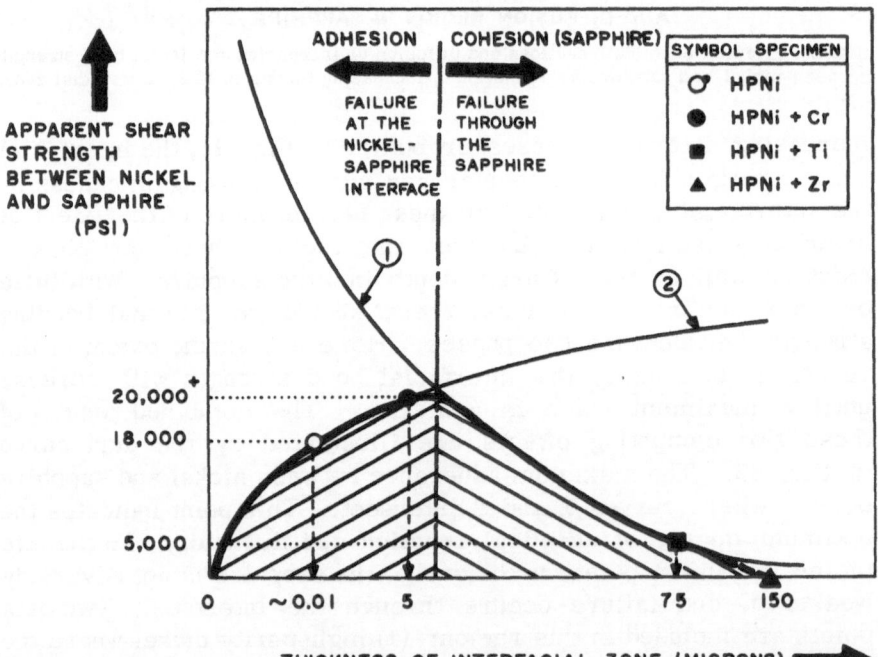

Fig. 22. Apparent shear strength between nickel and sapphire as a result of two competing effects: (1) weakening of sapphire, and (2) enhancement of interfacial bonds, both of which are the result of chemical reactions and diffusion.

ments), but these same chemical reactions had a much greater dele-
terious effect on the sapphire, causing surface deterioration. The
net result was a low apparent shear strength. This model indicates
that by reducing the extent of the chemical reactions and diffusion
at the interface (through a lowering of the concentration of the
zirconium and titanium additives), the adherence between the doped
nickel and sapphire should increase. This has been demonstrated
in recent studies [28] in which the concentrations of titanium and
zirconium in nickel were reduced by an order of magnitude. The
strength of these specimens was in excess of 10,000 psi, even
though failure still occurred in the brittle sapphire phase.

CONCLUSIONS

The adherence of nickel to sapphire depends on factors including:
(1) the composition and purity of the phases, (2) the joining history,
(3) geometric and size effects, and (4) failure modes. These factors
govern the thickness, composition, and microstructure of the interfacial zone which in turn determines the adhesive strength. The
importance of these factors was demonstrated in the experimental
studies which showed that small amounts (1 at.%) of chromium,
titanium, and zirconium in high-purity nickel could produce major
changes in both the wetting and the adherence of nickel to sapphire.
When the nickel first became molten, these active metal additives
segregated and concentrated at the interface. Once there, they
reacted to form new phases and also diffused into the sapphire
substrate. The new phases and the diffusion depths were determined
by the nature of the additives and by the thermal (joining) history of
the specimens.

A graphical model was proposed on the basis of the failure
modes for showing a relationship between the apparent shear
strength and the extent of chemical reactions and diffusion at the
nickel–Al_2O_3 interface. The apparent shear strength was the result of two competing chemical effects: (1) enhancement in interfacial bonding, and (2) weakening of the brittle sapphire phase by
roughening its surface or by introducing strain through lattice
diffusion or both. The net result was illustrated by a curve which
showed that adherence increased with increasing chemical reaction or diffusion (as measured by interfacial thickness) until a
maximum point was reached; then, the interfacial reactions proceeded to such an extent that they weakened the sapphire sub-

strate, and adherence decreased with increasing severity (thickness) of reaction. High-purity nickel and chromium-doped nickel did not react severely with sapphire and exhibited relatively high shear strengths; failure in these cases occurred in the interfacial zone. On the other hand, in the case of the titanium- and zirconium-doped nickel, chemical reactions were much more severe and the measured shear strength was low; failure occurred through the sapphire.

Thus, in order to maximize the adherence between nickel and sapphire (and other metal—ceramic combinations), it is necessary to control the interfacial chemical reactions to the point where the two competing effects are optimized. This can be accomplished by controlling the composition and concentration of the additives in the metal and ceramic phases and by controlling the joining parameters, such as temperature, time, and composition (pressure) of the surrounding atmosphere.

ACKNOWLEDGMENTS

The authors wish to thank H. W. Rauch for reviewing the manuscript, M. Birenbaum for conducting the sessile-drop tests, J. Higgins for determining surface energies, C. Miglionico for the electron microscope and microhardness work, T. Harris for X-ray diffraction work, and R. Jakas for conducting the shear tests. The authors are also grateful to Drs. M. Schwartz and S. Nash and Mrs. M. Schuler of the U.S. Army Frankford Arsenal for the electron probe data, and to the U.S. Army Materials Research Agency for permission to publish the data used in this paper, which is based on work performed under contract DA 19-066-AMC-184(X).

REFERENCES

1. R. J. Bondley, "Metal Ceramic Brazed Seals," Electronics 20: 97-99 (1947).
2. G. R. Van Houten, "A Survey of the Bonding of Cermets to Metals," Welding J. (N.Y.) 37 (12): 588-569s (1958).
3. Symposium on Ceramics and Ceramic—Metal Seals, School of Ceramics, Rutgers University, New Brunswick, New Jersey (April 21, 1953); published in Ceramic Age, Vol. 63 (Feb.-Sept. 1954).
4. W. H. Kohl, Materials and Techniques for Electron Tubes, Reinhold Publishing Corp., (New York) 1960, pp. 470-518.
5. W. H. Sutton, "Investigation of Oxide-Fiber (Whisker) Reinforced Metals," General Electric Company, MSD, U.S. Army Contract DA 36-034-ORD-3768Z, Final report AMRA CR 63-01/8, June 1964.
6. C. R. Kurkjian and W. D. Kingery, "Surface Tension at Elevated Temperatures, III, Effect of Cr, In, Sn, and Ti on Liquid Nickel Surface Tension and Interfacial Energy with Al_2O_3," J. Phys. Chem. 60: 961-963 (1956).

7. W. M. Armstrong, A. C. D. Chaklader, and J. F. Clarke, "Interface Reactions Between Metals and Ceramics I. Sapphire–Nickel Alloys," J. Am. Ceram. Soc. 45: 115–118 (1962).

8. J. F. Clarke, "An Investigation of Bond Formation Between Aluminum Single Crystals and Nickel Alloys," M. S. Thesis, University of British Columbia, Canada, December, 1959.

9. E. P. Abrahamson and N. J. Grant, "β-Chromium," J. Metals 8; Trans. AIME 975–977 (1956).

10. C. J. Bechtoldt and H. C. Vacher, "Redetermination of the Chromium and Nickel Solvuses in the Chromium–Nickel System," Trans. AIME 221: 14–18 (1961).

11. A. Glasser, "The Thermochemical Properties of the Oxides, Fluorides, and Chlorides to 2500°K," Argonne National Laboratory, NNL–5740.

12. A. D. Smigelskas and E. O. Kirkendall, "Zinc Diffusion in Alpha Brass," Trans. AIME 171: 130–134 (1947).

13. M. Hansen, "Constitution of Binary Alloys, McGraw-Hill Book Company, (New York), 1958.

14. T. E. Allibone and C. Sykes, "The Alloys of Zirconium," J. Inst. Metals 39; 179 (1928).

15. W. B. Pearson, unpublished work (1951); W. B. Pearson and L. T. Thompson, "The Lattice Spacings of Nickel in Solid Solutions," Can. J. Phys. 35: 349–357 (1957).

16. D. M. Poole, and W. Hume-Rothery, "The Equilibrium Diagram of the System Nickel–Titanium," J. Inst. Metals 83: 473–480 (1955).

17. A. Taylor and R. W. Floyd, "The Constitution of Nickel-Rich Alloys of the Nickel–Chromium–Titanium System," J. Inst. Metals 80: 577–587 (1952).

18. A. Taylor and R. W. Floyd, "The Constitution of Nickel-Rich Alloys of the Nickel–Titanium–Aluminum System," J. Inst. Metals 81: 25–32 (1952).

19. A. J. Bradley and A. Taylor, "X-Ray Analysis of the Nickel–Aluminum System," Proc. Roy. Soc. (London) A159: 56–72 (1937).

20. J. E. McDonald and J. G. Eberhart, "Adhesion in Aluminum Oxide–Metal Systems," Sandia Corporation Report SC-DC-64-1488, October, 1964.

21. B. V. Tsarevskii and S. I. Popel, "Adhesion of Molten Binary Iron Alloys To Solid Aluminum Oxide," in: V.N. Eremenko (ed.), The Role of Surface Phenomena in Metallurgy, Academy of Sciences, UKrSSR, Kiev, 1961. [English Translation: Consultant's Bureau (New York), 1963, pp. 96–101.]

22. V. A. Presnov, "Physicochemical Nature of Bonds Between Dissimilar Materials," in: V. N. Eremenko (ed.), The Role of Surface Phenomena in Metallurgy, Academy of Sciences, UKrSSR Press, Kiev, 1961. [English translation: Consultant's Bureau (New York), 1963, pp. 92–95.]

23. D. Nectoux, M. C. Deribere - Desgardes, and N. Drieux, "Adhesion Phenomena," Contact of Hot Glass With Metal, Symposium Proceedings of the Union Scientifique Continentale Du Verre, Scheveningen, May 25–29, 1964, [English translation: Dr. M. Cable, pp. 19–30.]

24. J. A. Pask and R. M. Fulrath, "Fundamentals of Glass-to-Metal Bonding, VIII, Nature of Wetting and Adherence," J. Am. Ceram. Soc. 45 (12): 592–596 (1962).

25. M. Humenik, Jr., and T. J. Whalen, "Physicochemical Aspects of Cermets," in: J. R. Tinklepaugh and W. B. Crandall (eds.), Cermets, Reinhold Publishing Corp. (New York), 1960, pp. 6–49.

26. P. Benjamin and C. Weaver, "The Adhesion of Evaporated Metal Films on Glass," Proc. Roy. Soc. (London) 261A (1307):516–531 (1961).

27. R. F. Tylecote, "The Adherence of Oxide Films on Metals," J. Iron Steel Inst. (London) 195: 380–385 (1960).

28. W. H. Sutton, "Investigation of Bonding in Oxide-Fiber (Whisker) Reinforced Metals," General Electric Company, U.S. Army Contract No. DA 19–066–AMC–184(X), AMRA CR 65–01/3, March, 1965.

29. C. J. Smithells, Metals Reference Book, Vol. I, Interscience Publishers, Inc. (New York), 1955, p. 374.

Author Index

Subject Index